T0327503

INDOOR RADIO PLANNING

INDOOR RADIO PLANNING

A PRACTICAL GUIDE FOR 2G, 3G AND 4G

Third Edition

Morten Tolstrup
www.ib-planning.com, Denmark

This edition first published 2015
© 2015 John Wiley & Sons, Ltd.

Registered Office
John Wiley & Sons, Ltd, The Atrium, Southern Gate, Chichester, West Sussex, PO19 8SQ, United Kingdom

For details of our global editorial offices, for customer services and for information about how to apply for permission to reuse the copyright material in this book please see our website at www.wiley.com.

Library of Congress Cataloging-in-Publication Data applied for

ISBN: 9781118913628

Set in 10/12pt Times by SPi Global, Pondicherry, India

Printed in the UK

Contents

Foreword by Professor Simon Saunders

The compelling need for in-building wireless systems derives directly from the needs of the people who use wireless – and that means, increasingly, all of us. We spend most of our time inside buildings, whether in the office or at home, at work or at play. Typically at least two-thirds of voice traffic on cellular networks originates or terminates inside buildings, and for data services the proportion is still higher – probably in excess of 90%.

Yet for too long, most indoor service has been provided from outdoor systems requiring high transmit powers, major civil engineering works and using a relatively large amount of spectrum to serve a given traffic level. This makes great sense for providing economical initial coverage to a large number of buildings and for 'joining the dots' to enable wide area mobility. However, 'outside-in' thinking is 'inside-out', from a technical and practical viewpoint, when attempting to serve users with very high quality and coverage expectations, and for delivering high data rate services within limited spectrum. Buildings offer their own remedy to these challenges, by providing signal isolation from nearby systems and enabling the fundamental principle of cellular systems – that unlimited capacity is available from limited spectrum if the engineering is done right.

Despite these compelling benefits, in-building wireless systems have hitherto been a poor relation of the 'mainstream' macrocellular network operations. With relatively few enthusiasts and a wide range of different favoured techniques for system design and installation, the field has at times resembled a hobby rather than a professional activity. The industry desperately needs best-practice techniques to be shared amongst a wider base of individuals to serve the growing demand – there are not enough engineers for the buildings requiring service – and for these techniques to become standardised in order to drive down costs, improve reliability and drive volumes.

Given this background, I welcome the publication of this book. Morten Tolstrup is a leading practitioner in the field and an engaging and entertaining public speaker. He has written a truly practical and helpful guide to indoor radio planning, which will enable a much wider audience to convert their skills from the old world of two-dimensional networks, comprising macro cells alone, to the new world of three-dimensional hierarchical networks comprising macro, micro, pico and femto cells delivering services to unlimited numbers of users. Following the simple guidelines provided, built on years of real-world experience, will help to avoid some very expensive mistakes.

Most of all, I hope that this book will help to professionalize the industry and encourage sharing of best-practice to the ultimate benefit of end-customers for compelling wireless broadband services.

Professor Simon R. Saunders
Independent Wireless Technologist & Visiting Professor,
University of Surrey
www.simonsaunders.com

Preface to the Third Edition

7 years!

I struggle to grasp that it is about 7 years ago I wrote the original manuscript for the first edition. Seven years is about one generation in mobile technology, and although 2G is still around after about 22 years of operation, it is evident that the advent of 4G, and the high demands on data speed and spectrum efficiency, is bringing 2G to an end. Even 3G is now mostly deployed to support voice.

I am still amazed by the data speeds we see in 4G indoor sites and the adoption of all the new data services mainly driven by smart devices and cloud services.

Within a few years, the wireless mobile data transfer will surpass the data transfers on the wired network in many countries. This is a huge challenge for all of us who are designing the wireless end of the network, to ensure we can support the growing need and expectations of being online everywhere. At the end of the day, the wireless network needed to support this wireless data capacity will require a lot of wires to work… So you could say that the wireless services are dependent on a well-designed backhaul network and a well-designed air interface.

Certified DAS Planning Training

When I was writing the original book 7 years ago, I was not even sure that there was a need for a practical guide. I was clearly wrong! Since the first edition of the book, it has now become a certified training course. Together with the help of my good friends, Dan Bixby, Thompson Lewis and Jack Culbertson, the book has been turned into a 3-day training course on the basics of indoor radio planning.

Along with my colleagues, I have had the pleasure of teaching hundreds of students across the world. I love participating in these events because I always walk away with more knowledge and ideas picked up from you. And that is the whole idea – to share the knowledge with you and to get your feedback, experience and ideas on how to keep evolving and improving the book and the training material. Thanks for this.

xx

Preface

If you are interested in knowing more about the indoor radio planning certification training, you can find a link on my website: www.ib-planning.com

More on 4G, Small Cells, Applications and RF Basics

With this latest edition I have also had the opportunity to add more details and practical experience with 4G, as we have gained more practical inputs from real deployments and experience. In the last edition, 4G was still in the pre-launch phase across most of the world. More detail has also been added to 'small cells' and this has become a chapter on its own. More information on the basics of RF, a large application section, and practical installation and project implementation experience have also been added.

Useful Tool?

I still hope the book will prove a useful tool for both the novice and the experienced radio planner, and I hope I have added a few more 'blades' to the 'Swiss Army knife of RF'. Remember that if you want a copy of the Visio stencil I used for all the figures and diagrams, you can contact me via www.ib-planning.com. It is all about sharing.

Thanks!

In addition to all my friends who provided input and help with first and second editions and who I have already mentioned in the prefaces to the previous editions, I would like to thank my friend Scott Pereira for his input to Chapter 9.7.

Once again, I must thank my wife, Karin, who lives with the fact that I always swear I will not take on a new edition of the book, but who knows I will do it in the end anyway and gives me the space to complete the task. And once again, Karin has drafted the design concept for the cover of this edition.

I hope to see you out there…

Morten Tolstrup

"Good judgment comes from experience, and experience is often a result of bad judgment"

Preface to the Second Edition

This is Still Not a Book for Scientists!

This book is intended for RF planners, to serve as a practical tool in their daily work designing indoor radio distribution systems.

Based on feedback from readers of the first edition it was clear to me that I needed to add more material and in depth description of the basics of indoor systems based on using repeaters; this has grown into a new Section 4.7.

There was also a strong demand to add more detail and dedicate a full chapter to radio planning in tunnels, for both rail and road tunnels; and redundancy principles in the design focus for solving the challenge of handover zones. An entire Chapter 11 is now dedicated to tunnel radio planning.

Also, although one could argue that this actually belongs in a book about indoor radio planning, I have added the relatively new and exciting option of designing and implementing outdoor DAS. The fact is that this outdoor DAS is implemented primarily to provide indoor coverage – so yes I do think that it is important to include it in this edition, in Chapter 12.

Obviously LTE was the hot topic as I was writing the manuscript; I have added the basics on LTE, MIMO and how to implement LTE inside buildings. Naturally I cannot include all of the deep insight into LTE – for that please refer to [7] and [8]. At this point only a few deployments of LTE indoors have been carried out and these by vendors – understandably I do not want to disclose all of their secrets and results. However, I have tried to my best to share what I know at this point – I am sure that these are merely the early days of a long and exiting journey with LTE.

So with additional material amounting to more than 25% in the second edition, I do hope that you find this new edition even better than the first.

The Practical Approach

Once again, my focus in this second edition is the practical approach to how to plan and implement indoor wireless solutions, and to share some of the hard learned lessons, lessons learned by me or my good friends in the industry.

As I stated in the first edition 'I am not an expert'. I am both surprised and honored by the feedback for the first edition – thanks for that. But I need to make it clear: I am still not an expert.

I do however, have the great pleasure of working with, and meeting a lot of very experienced and knowledgeable people in the industry – you are the experts. What I learn and have published is based on projects and knowledge sharing with you good people. Thanks for that; this makes me a student, I am learning every day and I enjoy it.

So please keep sharing. . . that's how we all get better.

Keep the Originals!

Please also remember to keep the originals, so purchase your copy of a good book rather than a pirate download of a PDF copy, that's the only way to assure new books for all of us!

I hope you find this book to be a useful tool in your daily work. That was my intention.

Morten Tolstrup
January, 2011

Preface to the First Edition

This is Not a Book for Scientists

This book is intended for the RF planners, to serve as a practical tool in their daily work designing indoor radio distribution systems. It is not a complete book about all the deep aspects and corners of GSM, DCS, UMTS and HSPA networks, or all the core network systems. It is dedicated to the last 10–70 m of the network, the indoor air interface between the mobile user and the indoor mobile network.

I have spent the past 20 years working on various parts of the exciting business of cellular communication. During this time I have mostly focused on the planning of the radio interface between the network and the mobile user, with a dedicated focus on indoor radio planning. I have always tried to approach that small part of the systems that involved me, the radio interface, from a practical angle. I have struggled with most of the books available on these subjects mainly due to a theoretical level far beyond my needs. My hope with this book is to present a level of theory that is usable and accessible for a radio planner with basic radio experience.

I also need to emphasize that no matter the radio platform or standard, GSM, UMTS, HSPA or 4G, as long as the interface between the mobile and the network uses radio communication, it will always be a matter of a link calculation with a given signal-to-noise ratio for a given service requirement. After all, it is 'just' radio planning.

The Practical Approach

I am not an expert in cellular, GSM, UMTS or HSPA systems, far from it – but I have gained a lot of experience with RF design, especially with regards to indoor radio planning. An old mountaineering saying is that 'good judgment comes from experience, but experience is often a result of bad judgment'. I have made my share of mistakes along the way, and I will help you avoid making the same mistakes when designing and implementing indoor solutions.

It has been my goal to include what I believe are the most important considerations and design guidelines to enable the RF planner to design and implement a high-performing indoor distributed antenna system.

It was not my intention to provide a deep hardcore mathematical background on RF planning, but to present the most basic calculations of the various parameters that we need to consider when designing a distributed antenna system.

I hope you can use the result – this book. It has been hard but also great fun to write it and to revisit all the background stuff, projects and measurement results that are the basis for this book. I hope you find it to be a useful tool in your daily work. That was my intention.

Morten Tolstrup

Acknowledgments

Second Edition

Once again I have called in favors from the best people in the business to help me verify and check the additional chapters in this second edition.

In particular I want to thank my friends and colleagues for many years for their input and help on the repeater and tunnel planning section; Mr Henrik Fredskild – Denmark, and Mr Stephen Page – Australia. Your input and guidance has added enormous value to the new chapters in this second edition.

I also want to thank: Mr Lance Uyehara – USA, Mr Jaime Espinoza – Chile, Mrs Marianne Riise Holst – Denmark/Norway, Lars Petersen – Spain/Denmark for their input and corrections.

Also, thanks to so many of you who have given me so much positive feedback, via mail, via groups on the WEB and at conferences and meetings, etc.

When I was writing the first edition I was not sure that there was a need for a book covering the practical approach – now I know.

Thanks once again to the team from Wiley: Mark Hammond, Susan Barclay and their team, for their support in the production of this book.

Last, but not least, thank you Karin, my dear wife, once again for your patience – even though I did promise you not to take on yet another book. Thanks for your support on this project, and for once again letting me spend so many late nights, early mornings and weekends on this book.

Without your support and patience Karin, this book would not have been possible.

If you want to contact me, or should you find any errors or have suggestions for more topics for any future editions, let me know.

You can contact me via: www.ib-planning.com.

Remember that the best form of optimization you can provide for your network is to plan it correctly from day one. This also applies to indoor radio solutions.

First Edition

This book would not have been possible if it was not for my many colleagues and friends who I have spend the better part of the last 20 years with. These friends and colleagues I know mainly from my many indoor projects around the world and from other mobile operators when working on mutual indoor projects. Many hours have I spent with you guys during the design phase, site visits, project implementations and measurements, from the fanciest indoor projects to deep below ground on a tunnel project conducting verification measurements.

I want to thank Simon Saunders for contributing the foreword. Simon is one of the people in this industry I respect the most, for his dedicated work and contribution to so many fields in the industry of telecommunication.

I also thank my friends who have helped me by reviewing the book: Bernd Margotte, Lars Petersen, Kevin Moxsom, Stein Erik Paulsen and Mario Bouchard. In particular, I want to thank Robin Young for his help and inspiration on the section about noise and link budgets. Peter Walther is also acknowledged for his input on the HSPA section, and the link budget example for HSDPA.

Thanks also to the team from Wiley, Mark Hammond, Sarah Hinton and their team, for the support and production of this book.

Last, but not least, thank you Karin my dear wife, for your support on this project and for letting me spend so many late nights, early mornings and weekends on this book. Without you this project would not have been possible. Thanks also for your design concept of the front cover, and the photo for the cover.

Even though I have spent many hours on this project, checking and double-checking everything, there might be an error or two. Let me know, and I will make sure to correct it in any future editions.

You can contact me via: www.ib-planning.com

Morten Tolstrup
Dronninglund, Denmark

1

Introduction

I often think that we have now finally come full circle in the world of radio transmission. We are back to where it all started: after all, the first transmission via radio waves by Marconi in 1895 was digital, using Morse code.

These days we are heading for a fully digitalized form of radio distribution, often using Internet Protocol (IP). Most radio services – broadcast, voice transmission for mobiles and television transmission – are being digitalized and transmitted via radio waves.

Radio waves – what a discovery that truly has changed our world! The effect of electromagnetism was discovered by H. C. Ørsted in 1820. Samuel E. Morse invented his digital system, the 'Morse code', in 1840. Through copper wires the world got connected via the telegraph line, and cross-continental communication was now accessible. Marconi merged both inventions and created the basis of our modern wireless communication systems, performing the first radio transmission over an incredible distance of 1.5 kilometers in 1895. Now we live in a world totally dependent on spin-offs of these basic discoveries.

Marconi struggled to transmit radio signals over a relative short distance: a few kilometers was a major achievement in the early days. Later, radio waves were used to reach several hundred thousand of kilometers into deep space, communicating with and controlling deep space probes and even vehicles on Mars.

Would it not be fair if we could bring back Ørsted, Morse and Marconi, and honor them by showing what we can do today, using the same principles: electromagnetism, digital transmission and radio waves? I am sure that they and the many other scientists who have formed the basis of our modern communication society would be proud. No one today could even consider a world without easy wireless communication; our modern lifestyle is highly dependent on those small devices – mobile telephones.

Things in telecommunications industry are progressing fast. These days we are not happy with anything less than several Mbps over the radio interface, mobile TV, internet, email and mobile media.

Indoor Radio Planning: A Practical Guide for 2G, 3G and 4G, Third Edition. Morten Tolstrup.
© 2015 John Wiley & Sons, Ltd. Published 2015 by John Wiley & Sons, Ltd.

Back in the early 1980s I was working on NMT (1G) systems. We used analog modems and were able to achieve up to about 300 baud over the mobile phone network. That was truly amazing at the time. People could send a fax from their car and, if they could carry the 18 kg mobile cellular phone battery included, they could have a portable phone with up to 30–60 min of talk time. The cost of these types of cellular phones was equivalent to that of a small family car in the early days, so the market was limited to very few professional users. Over a few years the price dropped to about an average month's salary, and mobile phones were getting smaller and smaller. Some were even 'pocket size' – if your pocket was big and able to support a weight of about 1 kg, that was.

At some point I was told about a new futuristic mobile telephone system in the making called 2G. The plan was to convert the voice to data, and the network could support 9600 baud (9.6 kbps), 32 times more that we could do on 1G. This was an amazingly high data speed – higher than we could get over fixed telephone lines at the time. I remember being highly skeptical. Who would ever need such high data rates for mobile use and for what? Mobile TV? Absolutely mad! Man, was I wrong!

These days we are exceeding 80 Mbps, more than 8300 times faster than we could perform via NMT(1G) in 1980. In reality, we are now able to handle higher mobile data speed to one user than the total data transmission capacity of the whole mobile network in Denmark could handle then for all the users in the network!

The need for data is endless. Data rates via mobile will increase and increase, and actually the radio link is getting shorter and shorter. In order to perform these high data rates, we need a better and better radio link. The radio spectrum is getting more and more loaded, and we are using higher and higher radiofrequencies and more and more complex and quality-sensitive modulation schemes; thus the requirement for the quality of the radio link is getting more and more strict.

It is worthwhile noting that high data rates are not enough on their own. It is also a matter of services; if mobile users are not motivated by an attractive service, even the highest data rate is pointless.

The need for high data rates is motivated by user demand for mobile email, internet and multimedia services. Most 3G mobile phones are able to support video calling, but it is rarely used. This shows that, even though it is impressive from a technical viewpoint that it is possible at all, the technology has no point if the service is not attractive to mobile users. It is a fact that the most successful mobile data service to date is also the slowest data service in operation over the mobile network, transmitted via a very slow data channel: SMS (Short Message Service). SMS is still popular data service but has been surpassed by applications with IM like facebook when it comes to data services for most mobile networks. Who would have thought that mobile users of all ages from 8 to 98 would be addicted to social media, visio and audio streeming and services.

When I was introduced to SMS, I thought it might be a good service to announce voice mails etc. to mobiles, but when the first mobiles arrived that were able to transmit SMS, my thought was 'why?' Wrong again! It clearly shows that it is not only a matter of data speed but also the value of the applications and services offered to the user. The iPhone completely changed the mobile industry and network load, when it was first launched, making applications and services easy to use.

I am happy to note that one thing stays the same: the radio planning of the mobile networks. The air interfaces and especially the modulation schemes are getting more and more complex, but in reality there is no difference when seen from a basic radio planning perspective. The challenges of planning a high-performance 4G link is the same basic challenge that Marconi faced performing his first radio transmission. It is still a matter of getting a sufficient margin between the signal and the noise, fulfilling the specific requirement for the wanted service, from Morse via long waves to 80 Gbps via 4G MIMO. It is still radio planning and a matter of signal-to-noise ratio and quality of the link.

In the old days it was all about getting the radio link transmitted over longer and longer distances. These days, however, the radio link between the network and the mobile user is getting shorter and shorter due to the stricter demands on the quality of the radio link in order to perform the high data rates. Marconi struggled to get his radio transmission to reach a mile. These days we are struggling to get a service range from an indoor antenna in a mobile network to service users at 20–40 m distance with high-speed data and good quality voice service.

We are now moving towards an IP-based Cloud world even on the radio interface, and voice-over-IP. We are now using IP connection to base stations and all other elements in the network. The network elements are also moving closer to the mobile users in order to cater for the requirements for quality of voice and data.

We are now on the brink of a whole new era in the world of telecommunications, an era where the mobile communication network will be an integrated part of any building. The telecommunications industry is just about to start integrating small base stations, 'femto cells', in many residential areas in many countries around the world. People expect mobile coverage and impeccable wireless data service everywhere.

When electricity was invented and became popular, existing buildings had to be post-installed with wires and light fixtures to support the modern technology of electrical apparatus and lighting. Later it was realized that electricity probably was so popular that it was worthwhile pre-installing all the wiring and most of the appliances in buildings from the construction phase. I do believe that, within a few years from now, it will be the same with wireless telecommunications. Wireless services in buildings are one of the basic services that we just expect to work from day one, in our home, in tunnels and surely in corporate and public buildings.

The future is wireless.

2

Overview of Cellular Systems

This book is concentrated around the topic of indoor radio planning from a practical perspective, and it is not the within the scope of this book to cover the full and deep details of the 2G (GSM), 3G (UMTS) and 4G (LTE) systems and structures. This book will only present the most important aspects of the network structure, architecture and system components, in order to provide basic knowledge and information that is needed as a basis for design and implementation of indoor coverage and capacity solutions. For more details on cellular systems in general refer to [2].

2.1 Mobile Telephony

2.1.1 Cellular Systems

The concept of cellular coverage was initially developed by AT&T/Bell Laboratories. Prior to that, the mobile telephony systems were manual systems used only for mobile voice telephony. Typically implemented with high masts that covered large areas, and with limited capacity per mast, they were only able to service few users at the same time – in some cases even only one call per mast! These systems also lacked the ability to hand over calls between masts, so mobility was limited to the specific coverage area from the servicing antenna, although in reality the coverage area was so large that only rarely would you move between coverage areas. Remember that, at that point, there were no portable mobile telephones, only vehicle-installed terminals with roof-top antennas. Over time the use of mobile telephony became increasingly popular and the idea was born that the network needed to be divided into more and smaller cells, accommodating more capacity for more users, implementing full mobility for the traffic and enabling the system to hand over traffic between these small cells.

From this initial concept several cellular systems were developed over time and in different regions of the world. The first of these cellular systems was analog voice transmission, and

Indoor Radio Planning: A Practical Guide for 2G, 3G and 4G, Third Edition. Morten Tolstrup.
© 2015 John Wiley & Sons, Ltd. Published 2015 by John Wiley & Sons, Ltd.

some 'data transmission' modulated into the voice channel for signaling the occasionally handover or power control command.

Some of the most used standards were/are AMPS, D-AMPS, TACS, PCS, CDMA, NMT, GSM (2G), DCS (2G), UMTS (3G, WCDMA) and LTE (4G).

AMPS

AMPS (Advanced Mobile Phone System) is the North American standard and operates in the 800 MHz band. The AMPS system was also implemented outside North America in Asia, Russia and South America. This is an analog system using FM transmission in the 824–849 and 869–894 MHz bands. It has 30 kHz radio channel spacing and a total of 832 radio channels with one user per radio channel.

D-AMPS

D-AMPS (Digital Advanced Mobile Phone System) evolved from AMPS in order to accommodate the increasingly popular AMPS network with fast-growing traffic and capacity constraints. The D-AMPS system used TDMA and thus spectrum efficiency could be improved, and more calls could be serviced in the same spectrum with the same number of base stations.

TACS

TACS (Total Access Cellular System) was also derived from the AMPS technology. The TACS system was implemented in the 800–900 MHz band. First implemented in the UK, the system spread to other countries in Europe, China, Singapore, Hong Kong and the Middle East and Japan.

PCS

PCS (Personal Communications System) is a general term for several types of systems developed from the first cellular systems.

CDMA

CDMA (Code Division Multiple Access) was the first digital standard implemented in the USA. CDMA uses a spread spectrum in the 824–849 and 869–894 MHz bands. There is a channel spacing of 1.23 MHz, and a total of 10 radio channels with 118 users per channel.

NMT

NMT (Nordic Mobile Telephony) was the standard developed by the Scandinavian countries, Denmark, Norway and Sweden, in 1981. Initially NMT was launched on 450 MHz, giving good penetration into the large forests of Sweden and Norway, and later also deployed in the

900 MHz band (the band that today is used for GSM). Being one of the first fully automatic cellular systems in the world (it also had international roaming), the NMT standard spread to other countries in Europe, Asia and Australia.

GSM/2G

GSM (Global System for Mobile communication), or 2G, was launched in the early 1990s, and was one of the first truly digital systems for mobile telephony. It was specified by ETSI and originally intended to be used only in the European countries. However GSM proved to be a very attractive technology for mobile communications and, since the launce in Europe, GSM has evolved to more or less a global standard.

DCS/2G

Originally GSM was specified as a 900 MHz system, and since then the same radio structure and signaling system have been used for DCS1800/2G-1800 (Digital Cellular Telecommunication System). The GSM basic has also been applied to various spectra around 800–900 and 1800–1900 MHz across the world, the only difference being the frequencies.

UMTS/3G (WCDMA)

After the big global success with the second generation (2G) GSM and the increased need for spectrum efficiency and data transmission, it was evident that there was a need for a third-generation mobile system. UMTS was selected as the first 3G system for many reasons, mainly because it is a very efficient way to utilize the radio resources – the RF spectrum. WCDMA has a very good rejection of narrowband interference, is robust against frequency selective fading and offers good multipath resistance due to the use of rake receivers. The handovers in WCDMA are imperceptible due to the use of soft handover, where the mobile is serviced by more cells at the same time, offering macro-diversity.

However there are challenges when all cells in the network are using the same frequency. UMTS is all about noise and power control. Strict power control is a necessity to make sure that transmitted signals are kept to a level that insures they all reach the base station at the same power level. You need to minimize the inter-cell interference since all cells are operating on the same frequency; this is a challenge.

Even though soft handovers insure that the mobile can communicate with two or more cells operating on the same frequency, one must remember that the same call will take up resources on all the cells the mobile is in soft handover with. The handover zones need to be minimized to well-defined small areas, or the soft HO can cannibalize the capacity in the network.

UMTS has now become the global standard and has been accepted throughout the world. Several upgrades that can accommodate higher data speed HSDPA (High Speed Downlink Packet Access) and HSUPA (High Speed Uplink Data Access) can service the users with data speeds in excess of 10 Mbps.

There are several current considerations about converting the current GSM900 spectrum into UMTS900, giving a much higher spectrum efficiency, and better indoor RF penetration.

2.1.2   Radio Transmission in General

Several challenges need to be addressed when using radio transmission to provide a stable link between the network and the mobile station. These radio challenges are focused around the nature of the propagation of radio waves and especially challenges of penetrating the radio service into buildings where most users are located these days.

The challenges are mainly radio fading, noise control, interference and signal quality. These challenges will be addressed throughout this book, with guidelines on how to design a high-performing indoor radio service.

2.1.3   The Cellular Concept

After the initial success with the first mobile system, it was evident that more capacity needed to be added to future mobile telephony systems. In order to implement more capacity to accommodate more users in the increasingly more popular mobile telephony systems, new principles needed to be applied. The new concept was to divide the radio access network into overlapping 'cells', and to introduce a handover functionality that could insure full mobility throughout the network, turning several masts into one coherent service for the users.

Dividing the network into cells has several advantages and challenges. The advantages are:

- *Frequency reuse* – by planning the radio network with relative low masts with limited coverage area, compared with the first mobile systems, you could design a radio network where the cells will not interfere with each other. Then it is possible to deploy the same radio channel in several cells throughout the network, and at the same time increase the spectrum and radio network efficiency thanks to frequency reuse.
- *Capacity growth* – the cellular network could start with only a few cells, and as the need for better coverage and more capacity grew, these large cells could be split into smaller cells, increasing the radio network capacity even more with tighter reuse of the frequencies (as shown in Figure 2.1).

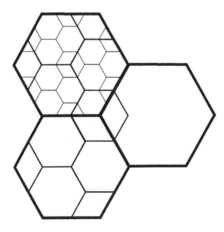

Figure 2.1   The cell structure of a cellular radio network. Cells will be split into smaller cells as the network evolves, and the capacity need grows

- *Mobility* – it is paramount for cellular networks that handovers are possible between the cells, so the users can roam through the network with ongoing connections and no dropped calls. With the advent of the first cellular systems, mobile users could now move around the network, utilizing all the cells as one big service area.

The challenges are:

- *Network structure* – when deploying the cellular structure one needs to design a theoretical hexagonal roll-out, by deploying omni, three-sector or six-sector base stations, and make sure the cells only cover the intended area. It is important that the cells only cover the intended area, and that there are no 'spill' of radio coverage to the coverage area of other cells in the network.
- *Mobility* – to offer full mobility in a cellular network you need to introduce handover among the cells, in order for the call to proceed uninterrupted when a mobile moves from one cell service area to a new one. The handover function creates a need to evaluate the radio quality of a potential new, better cell, compared with the current cell servicing the call. Complex measurements and evaluation procedures of adjacent cells are introduced in order to control the handover decision and mobility management.
- *Power control* – to service the mobile user with coherent radio coverage throughout the network area is a challenge. On the uplink from the mobile to the base station, one of the challenges is for the mobile to reach the base station. However, it is also a challenge that sometimes the mobiles can be close to the serving base station, and at the same time, the same base station using the same radio channel might also serve a mobile at 20 km distance. This can cause a problem known as 'the near–far problem', i.e. the mobile close to the base station will generate a high-level signal in the base station receiver and will simply overpower the low signal from the distant mobile. To overcome this issue, power control of both the downlink (the signal transmitted from the base station to the mobile) and the uplink (the signal transmitted by the mobile to the base station) was introduced. This not only avoids overpowering the receiving end, but also limits the overall interference in the network due to the lower average transmitted power level on both uplink and downlink. It also increases radio quality and spectrum efficiency by tighter frequency reuse among the cells. The challenge with power control is the need to evaluate both the signal strength and the quality of the radio signal received in both ends of the link, in order to adjust the transmitted power to an appropriate level, and first and foremost to make sure that the link is not overpowered.

2.1.4   Digital Cellular Systems

From only performing pure voice connections on the mobile networks around the world, users slowly started to request the possibility of sending and receiving data signals via the mobile connection. The first applications mobile data were the use of analog modems to transmit and receive fax signals over a purely analog mobile connection. Soon it became clear that there was a need for real data transmission via the mobile network.

In order to accommodate the need for mobile data transmission, better voice quality and spectrum efficiency digital technology were applied to the cellular systems. Many competing

2G (second generation) systems were developed, but on a global basis the most successful has been the GSM system.

By using digital transmission several advantages were achieved:

- Better spectrum efficiency.
- Lower infrastructure cost on the network side.
- Reducing fraud, by encryption of data and services.
- Lower terminal cost.
- Reduced transmission of power from the mobile, making it possible to have long-lasting batteries for hand-held terminals.

This book mainly focuses on indoor radio planning for 2G (GSM/DCS), 3G (UMTS) and 4G (LTE) systems. A more thorough introduction to the principles of 2G, 3G and 4G is appropriate and is to be found in the following sections.

2.2   Introduction to GSM (2G)

To perform indoor GSM radio planning, the radio planner needs some basic information about the network, signaling, etc. This section is not intended as a complete GSM training session, but rather to highlight the general parameters and network functions in GSM networks. For more details on the GSM please refer to [1].

2.2.1   GSM (2G)

2G was launched in the early 1990s, as one of the first truly digital systems for mobile telephony. It was specified by ETSI and originally intended to be used only in Europe. However, GSM proved to be a very attractive technology for mobile communications, and since the launch in Europe, GSM has evolved to be more or less the first truly global standard for mobile communication. Even though GSM is relatively old (most mobile network generations last about 7–10 years), it is still being rolled out all over the world, and in particular, the focus on high indoor usage of mobile telephony has motivated a need for dedicated indoor coverage solutions.

DCS (2G)

GSM was originally specified as a 900 MHz system. Since then the same radio structure and signaling have been used for 2G-1800. The GSM/2G basic has also been applied to various spectra around 800–900 and 1800–1900 MHz across the world, the only difference being the frequencies.

Security Features

All GSM communication is data; essentially voice is converted from analog to data and back again. Compared with the analog mobile telephony, the GSM system is much more secure and is impossible to intercept. GSM applies digital encryption, authentication of call, checking and validation of calls in order to prevent unauthorized use of the network. For the first time in mobile telephony systems, the identification of the user in the system was not tied to the mobile equipment with an internal telephony number (often an EPROM you had to replace), but to the user by the use of a SIM card (Subscriber Identity Module). The SIM is identified using an IMSI (International Mobile Subscriber Identity) – yes, International – for the first time the mobile user could roam in other countries or networks on a global basis and use their mobiles and supplementary services as if they were in the own home network.

2.2.2   2G/GSM Radio Features

Compared with the 1G analog networks at the time, such as NMT and AMPS, new and sophisticated radio network features were introduced in GSM/2G. One new feature was DTx (discontinuous transmission), which enabled the base station/mobile to only transmit when there was voice activity. This prolongs the battery life of the mobile station, but also minimizes both uplink and downlink interference due to the lower average power level. DTx detects the voice activity on the line, and during the period of no voice activity the radio transmission stops. In order to prevent the users from hanging up when there is complete silence on the link, believing the line to be dead, the system generates 'comfort noise' to emulate the natural background noise of the individual call.

The GSM system can also use frequency hopping, accommodating channel selective fading by shifting between radio frequencies from a predefined list 217 times per second. This hopping is controlled by an individual hopping sequence for each cell.

GSM Data Service

Analog networks at the time were only able to transmit data via analog modems, and only up to a maximum data speed of about 300 baud (0.3 kbps). The possibility of higher data rates on mobile networks with the advent of GSM phase one, enabling data transfer up to 9.6 kbps, was stunning. Today this seems really slow, but at the time most of us were not sure how we should ever be able to utilize this mobile 'broadband' to its full potential – remember, this was before Windows 1.0 was even launched! You were lucky if you could do 300 baud over your fixed telephone line. Over time new more sophisticated packet data transmission modulations have been applied to GSM, GPRS and EDGE, now exceeding 200–300 kbps per data call.

During the specification of the signaling of GSM, it became clear that there was a slight overhead in the data resources, and that this overhead could be used to transmit short portions of limited data – this eventually resulted in a new data service being introduced. The data service that was originally specified and launched as a paging service was SMS. SMS was originally planned to be used by the network to announce voice mails, cell information and other text broadcast services to individuals or all mobiles within an area of the network.

The first GSM mobiles were not even able to transmit SMS; no one was expecting mobile users to struggle with using the small 10-digit keyboard to key in a text message. The concept was that you could dial up a specific support number with a manned message center that could key in a text direct into the network, and charge accordingly for this service. Well, we all know today that, even though SMS has many limitations, the fact remains that this is still the most successful mobile data service and a huge revenue generator for the mobile operators.

Supplementary Services in GSM

With GSM several new services were introduced, such as call forwarding, call waiting, conference calls and voice mails. All this today seems only natural, but it was a major leap for mobile communications. It added new value for mobile users and the network operators, making mobile communication a tool you could finally use as the primary communication platform for professional business use, and a real alternative to fixed line telephone networks, especially if you had good indoor coverage inside your building.

The First Global System, from Europe to the Rest of the World

These new features introduced with GSM – SMS, data service, supplementary services etc. – were a direct result of a good standardizing work by an ETSI/GSM Mou (Memorandum of understanding). It enabled the GSM standard to spread rapidly across the rest of the world, becoming the first truly global standard – even though some countries never implemented the standard.

The GSM Radio Access Structure

The most important part of the GSM network for the indoor radio planner is the GSM radio structure and air interface. In the following the main parameters and techniques are introduced.

Idle Mode

A mobile is considered in idle mode when it is not engaged in traffic (transmitting). The mobile will camp on the strongest cell, monitor system information and detect and decode the paging channel in order to respond to an incoming call.

Dedicated Mode

Mobiles engaged in a call/transmitting data to the network are considered to be in dedicated mode. In dedicated mode a bi-directional connection is established between the mobile station and the network. The mobile will also measure neighboring cells and report this measurement information to the network for handover and mobility management purposes.

FDD + TDD

GSM uses separate frequency bands for the uplink and downlink connection (as shown in Figure 2.2). The two bands are separated by 45 MHz on GSM900, and by 95 MHz on GSM1800. GSM has the mobile transmit band in the lower frequency segment.

By the use of FDD (Frequency Division Duplex), the spectrum is divided into radio channels, each radio channel having 200 kHz bandwidth. Each of these 200 kHz radio channels is then divided in time using TDD (Time Division Duplex) into eight channels (time slots) to be used as logical and traffic channels for the users.

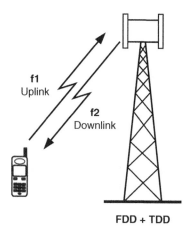

FDD + TDD

Figure 2.2 GSM uses both frequency division (FDD) in a radio channel of 200 kHz and time division (TDD) in TDMA frames of eight time slots

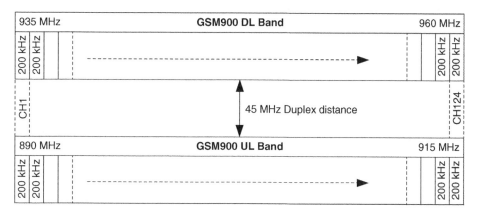

Figure 2.3 The standard GSM900 frequency spectrum. GSM uses 124 200 kHz radio channels

GSM is most commonly assigned to use on the 900 and 1800 MHz frequency bands (as shown in Figure 2.3). However depending on local spectrum assignment in different regions, other band segments are used, such as EGSM or GSM1900.

The GSM Spectrum

The GSM radio spectrum is divided into 200 kHz radio channels. The standard spectrum on GSM900 is 890–915 MHz for the uplink and 935–960 MHz for the downlink spectrum (CH1–CH124), a total of 2×25MHz paired spectra, separated by 45 MHz duplex distance. The DCS spectrum (as shown in Figure 2.4) uses 1710–1785 MHz for the uplink and 1805–1880 MHz for the downlink spectrum (CH512–CH885), a total of 2×75MHz paired spectra separated by 95 MHz duplex distance.

In some regions other spectrum allocations for GSM around 900 and 1800 MHz have been implemented, such as EGSM or DCS1900.

TDMA

In addition to the frequency separation, GSM uses time separation, TDMA (Time Divided Multiple Access); TDMA allows multiple users to occupy the same radio channel. The users will be offset in time, but still use the same 200 kHz radio channel. The radio spectrum (as shown in Figure 2.3) in GSM (900) is separated into 124 radio channels, each of these radio channels then separated into eight time-divided channels called time slots (TSL) (as shown in Figure 2.5).

Each TDMA frame time slot consist of 156.25 bits (as shown in Figure 2.6, 33.9 kbs per TSL or 270.8 kbps per frame), of which 114 (2×57) are coded data, including forward error correction. All transmitted information is transferred in blocks of 456 bits, divided into four time slot periods (456 = 4×2×57). The maximum net bit rate is per time slot is 13 kbps, excluding the error correction.

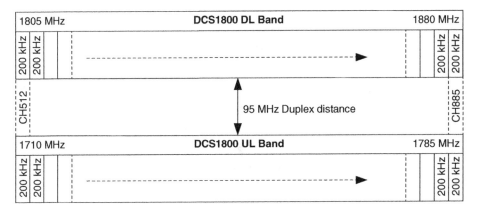

Figure 2.4 The standard DCS1800 frequency spectrum, DCS, uses 374 200 kHz radio channels

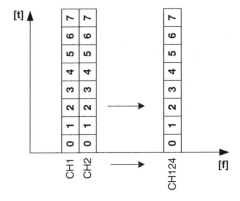

Figure 2.5 Each of the 200 kHz radio channels is time divided into eight channels, using individual time slots, TSL0–7

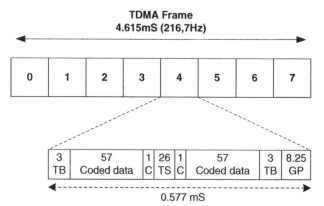

TB: Tail Bit, C: Control Bit; TS: Training Sequence, GP: Guard Period

Figure 2.6 The TDMA frame in GSM uses eight time slots, 0–7

The GSM Cell Structure

As we know, the GSM network consists of cells (as shown in Figure 2.7), the cells operate on different frequencies and each is identified by the CI (Cell Identity defined in the parameter settings in the network) and the BSIC (Base Station Identity Code). Depending on the geography and the design of the network, the frequencies can be reused by several cells, provided that the cells do not 'spill' coverage into other cells' coverage areas.

Capacity upgrade for a cell is done by deploying more radio channels in the cell; an upgrade of a cell with one extra radio channel will give eight extra logical channels that can be used for signaling or traffic. In urban areas it is normal to have six to 12 radio channels per cell. In a macro network the typical configuration is to design with three sector sites (as shown in Figure 2.7), where each mast is serving three individual cells.

Most GSM networks were initially deployed as GSM900, but in many countries DCS1800 has been implemented at a later stage, as an extra 'layer' on top of the GSM900 network to add capacity to the network to fulfill the fast growing need for more traffic channels and spectrum.

In most cases (not all), the DCS1800 network has a shorter service range due to the higher free space loss caused by the higher frequency (6 dB). If DCS 1800 is added onto the existing GSM900 network, using the initial mast separation designed for GSM900 might cause the 1800 cells not to overlap adjacent DCS1800 cells; therefore all handovers of traffic between cells have to be done via the GSM900 cells. This puts significant signaling load on GSM900 and increases the risk of dropped calls during a more complex handover. Therefore it is preferred to decrease the inter-site distance between the masts, then the network has coherent and homogeneous coverage on DCS1800. In practice DCS1800 has mostly been implemented to add capacity in urban environments, where the inter-site distance is short even on GSM900 and will provide a good foundation for DCS1800 deployment. In a dual-layer GSM network (900+1800) 'Common BCCH' (broadcast control channel) is often used to combine all signaling into one layer and free up more TSLs in the cell for more user capacity.

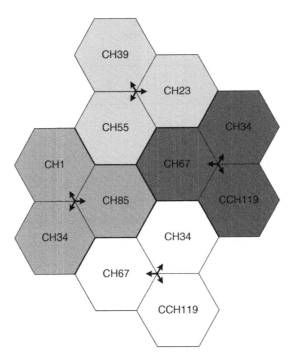

Figure 2.7   GSM cell structure with three macro base stations, each serving three cells

2.2.3   *Mobility Management in GSM*

Managing the service quality and mobility of the users in a cellular network like GSM is important. The network must insure the best possible service quality and hand over the mobiles between cells when the users roam through the network. Complex measurements for radio strength and quality have been applied to cater for the mobility management.

Radio Measurements

In dedicated mode (during a call) the base station measures the signal level and radio quality/ BER (bit error rate) of the received signal from the mobile. The mobile will also measure the signal level and the quality of the serving cell, as well as the signal levels of the neighboring cells defined by the network in the 'neighbor list'. The mobile reports these measurements (measurement reports) to the network for quality evaluation, power control and handover evaluation.

Rx-Quality

The radio quality (bit error rate) of both the uplink and downlink connection is divided into eight levels and defined as 'Rx-Qual' 0–7, 0 being the best quality. Rx-Qual 7 will start a network timer that will terminate the call if the quality does not improve within a preset time. If the Rx-Qual rises above 4 it will degrade the voice quality of a speech call to an extent that is noticeable to the user. An Rx-Qual better than 3 must be striven for when planning indoor solutions; preferably, an Rx-Qual of 0 should be insured throughout the building.

The main cause of degradation of the Rx-Qual is interference from cells using the same radio frequency, or by too low a reception level of the servicing signal.

Rx-Level

The radio signal strength is measured as Rx-Level and is reported in a range from 0 to 64. The Rx-Level is the signal level over -110 dBm. Rx-Level 30 is: $-110 + 30\,dB = -80\,dBm$.

The mobile will also measure the radio quality and signal strength of the serving cell, as well as the other adjacent cells (specified by the neighbor list received from the base station controller, BSC). The mobile measures and decodes the BSIC (base station identity code) and reports the strongest cells back to the BSC for handover evaluation/execution.

Radio Signal Quality Control

Based on the radio measurements evaluation performed by the BSC and the mobile, the network (the BSC) will control and adjust the various radio parameters, power level, timing and frequency hopping, etc., in order to insure the preset margin for radio signal level and radio quality is maintained.

A neighbor list is defined in the network and broadcast to the mobile in service under each cell, with the frequencies and details of the adjacent cells to the serving cell. This list of frequencies (the BCCH of adjacent cells) is used by the mobile to measure the adjacent cells. These measurements are used for the handover evaluation. Normally this is a 'two-way' list, enabling the mobile to handover in both directions, but it is possible to define 'one-way' neighbors, removing the possibility of the mobile handing back to the previous serving cell. This use of 'one-way' neighbors can be useful for optimizing indoor systems, especially in high-rise buildings, in order to make sure that the mobiles inside a building that camp unintentionally on an outdoor cell can hand in to the indoor system, but not hand back to the outdoor network; see Section 5.5 for more details.

Power Control

The network will adjust the transmit power from the mobile and the base station; the downlink power control for the base station is optional. The primary function of the power control is insure that the radio link can fulfill the preset quality parameters, and most importantly to make sure that mobiles close to the base station will not overpower the uplink and desensitize the base station, resulting in degraded service or dropped calls.

Adjustable target levels for minimum and maximum Rx-Level are set in the network, individually for each cell. These Rx-Levels are the reference for the power control function, and will adjust the transmit power to fulfill the requirements and keep the Rx-Levels within the preset window. Separate triggers for Rx-Qual can cause the power to be adjusted up, even if the Rx-Level is already fulfilled in order to overcome interference problems.

The power control in GSM is done in steps of 2 dB, and the power steps can be adjusted in intervals of 60 ms.

GSM Handovers

Based on the measurement reports sent from the mobile to the network, with information on the radio signal level of serving and adjacent cells, if the preset levels for triggering the handover are fulfilled, the network might evaluate that the mobile could be served by a new cell with better radio signal level and command the mobile to hand over to a new serving cell. The handover requires the establishment of a new serving traffic channel on the new cell and, if the handover is successful, the release of the old serving traffic channel. Normally the handover is not noticeable for the user, but there will be 'bit stealing' from the voice channel to provide fast signaling for the handover. This 'bit stealing' can be detected as small gaps in the voice channel, and if indoor users are located in an area where there is constantly handover between two cells ('ping pong handover'), the voice quality on the link will be perceived as degraded by the user, even if the Rx-Qual remains 0.

The handover trigger levels and values can be adjusted in the network individually for each handover relation between the cells. Various averaging windows, offsets, can also be set. For planning and optimizing indoor systems it is very important that the radio planner knows the basics of the parameter set for the specific type of base station network implemented. This will enable the radio planner to optimize the implemented indoor solutions to maximum performance, and compatibility with the adjacent macro network.

The handover in GSM can be triggered by:

- Signal level uplink or downlink, to maintain quality.
- Signal quality uplink or downlink, to maintain quality.
- Distance, to limit the cell range.
- Speed of mobiles.
- Traffic, to distribute load among cells.
- Maintenance – you can empty a cell by network command.

Intracell Handover
If the Rx-Qual is bad but no adjacent cells with higher received signal level are detected and reported by the mobile, the network may initiate an intracell handover, assigning a new TSL or a new radio channel on the same cell. This can improve the quality of the call if interference hits the used time slot or frequency.

Level-triggered Handover
Preset margins for Rx-Level can be set to trigger a handover level. Standard settings will be so that when an adjacent cell measured in the neighbor list becomes 3–6 dB more powerful, the mobile is handed over to the new cell. However, handover margins can actually be set to negative levels, triggering the handover even if the adjacent cell is less powerful.

Quality-triggered Handover
To insure that the Rx-Qual of the radio link is maintained, preset trigger levels for Rx-Qual can be set to 'overwrite' the level handover, initiating a handover if the Rx-Qual degrades below the trigger level. The typical trigger level for the Rx-Quality handover is Rx-Qual> 3. If the quality degrades beyond this level, the network will initiate a handover. It does not matter if the Rx-Level margin is fulfilled or not.

Distance-triggered Handover
The maximum cell size can be set, and handovers triggered when exceeding this distance. This can be useful for indoor cells, making sure that the indoor cell will not service unintended traffic outside the building, but remember to take into account the delay of the DAS, this will offset the cell size.

Speed-triggered Handover
It is also be possible for the network to detect if the mobile is moving fast or slow through the serving cell. This can be utilized by the network by limiting certain cells to serve only slow-moving (hand-held) mobiles. This can be very useful when optimizing indoor GSM systems, in order to avoid vehicular traffic camping on the indoor cell if the indoor cell leaks out signal to a nearby road.

Traffic-triggered Handover
It is possible to predefine a certain traffic level for a cell to start handing over calls to adjacent cells, and thereby offload the cell. This can be very useful for cells that have periodic hotspot traffic, and might save the expense of deploying extra capacity for servicing only the occasionally peak load. However this is only possible if the adjacent cells are able to service the traffic at the expected quality level.

Maintenance Handover
It is also possible to order the network to empty a specific cell. Similar to the traffic-triggered handover, the cell will push all calls that can be handled by adjacent cells out and wait for the remaining calls to terminate. During this mode the cell will not set up any new calls, and once the cell is empty, the cell can be switched off for service and maintenance. By emptying an operational cell using this method, there will only be limited impact on the users in terms of perceived network quality. This method is to be preferred when doing maintenance in the network rather than just switching off the cell and dropping all the ongoing calls.

The Complexity of the GSM Handovers

In GSM there are basically three handovers seen from the network side (as shown in Figure 2.8), involving several networks depending on the handover type.

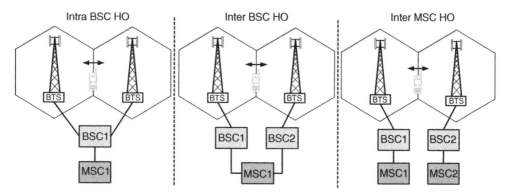

Figure 2.8 Three types of network involvement of the GSM handover: intra-BSC, inter-BSC and inter-MSC handovers (HO)

Intra-BSC Handovers

Intra-BSC handovers are between cells serviced by the same BSC. This is the least complex handover seen from the network side. Also, there are possibilities for using functions that make the handover smoother and more likely to succeed – pre-sync (chained cells), etc. These functions can be very useful in tricky handover situations, optimizing the handover zone in tunnel systems, for example.

Inter-BSC Handovers

Inter-BSC handovers are done between cells serviced by different BSCs. The handover function is basically the same as for intra-BSC handovers, but with limited functions like pre-sync, etc.

Inter-MSC Handover

Inter-MSC handovers are controlled by the MSC, and the signaling is more complex due to the implication on several interfaces and network elements. The number of inter-MSC handovers should be limited in high traffic areas (indoors).

Location Management in GSM

The GSM radio network is divided into regions (location areas). The most important motivation for dividing the radio network into several location areas is to limit the paging load. The mobile will monitor and update the network whenever it detects that it is being serviced by a new location area. This is done by the mobile by monitoring the location area code (LAC) broadcast from all the cells in the system information. In the VLR (visitor location register) in the network, the information about the LAC area the mobile was last updated in is stored. Using several location areas, the network can limit the signalling of paging messages by only having to broadcast the paging messages to the mobiles currently serviced by a specific LAC. If the concept of location areas was not implemented, the network would have had to page all mobiles in all cells each time there was an incoming call. This would have severely loaded the paging signalling resources and increased paging response time beyond an acceptable level.

Timing Advance

All the active traffic in a GSM cell is synchronized to the timing and clock of the base station, in order to insure that the time slots are sent and received at the correct time. The TDMA frame is 4.615 ms long, and each TSL is 577 μs. Radio waves travel with the speed of light, about 300 000 km/s. Analyzing the example in Figure 2.9 with two mobiles in traffic, one is close to the base station (MS1) and the other (MS2) is 20 km distant. Owing to the propagation delay caused by the difference in distance to the serving base station, the mobiles will be offset in time by about 66 μs (one-way). Both mobiles are locked to the same serving base station's synchronization 'clock'. Actually the distant mobile will in this case be offset by $2 \times 66\,\mu s$ as the signal has to travel from the base station, to the mobile and back so the synchronization of the distant mobile is offset/delayed in time by about 132ms. This delay corresponds to about

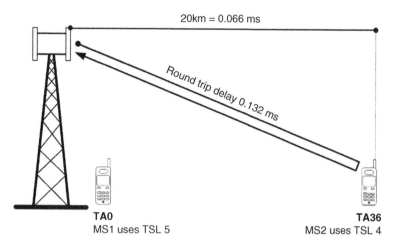

Figure 2.9 Mobiles in traffic on the cell have to compensate for the delay to the base station using timing advance

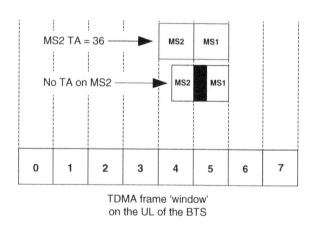

TDMA frame 'window'
on the UL of the BTS

Figure 2.10 If the GSM network did not used timing advance, traffic from mobiles at different distances from the base station would drift into adjacent TSL, due to the propagation delay

0.3 TSL, and will cause the transmitted TSL (TSL4) from the distant mobile to overlap with the next TSL (TSL5 in this example), used by the mobile closest to the base station (as shown in Figures 2.9 and 2.10).

To overcome this problem, one could consider including more guard time between the time slots, but this would come at the expense of lower spectrum efficiency. In order to preserve spectrum efficiency, the problem was solved using the concept of timing advance (TA). In practice the problem is solved in GSM by having the base station continuously measure the time offset between its own burst schedule and the timing of the received bursts from the mobiles. When the base station detects that the TSL from the mobile has started to drift close to the next or previous TSL, the base station will command a TA step-up or -down. The network simply commands the mobile to send the TSL sooner, and the offset will depend on the distance between the mobile and the serving mast. This timing advance enables the GSM system to compensate for the delay over the air interface.

The use of timing advance insures high spectrum efficiency (low guard times between time slots) and at the same time solves the issue of the timing difference between distant mobiles and mobiles near the base station.

The solution to this problem shown in Figures 2.9 and 2.10 is for the network to command the mobile MS2 to use TA36, causing the mobile to transmit early. In this example, it insures that the transmitted TSL4 from mobile station 2 will not overlap with TSL5 from mobile station 1.

The GSM system is able to handle timing offset up to 233 μs, corresponding to a maximum distance from the base station of 35 km. It has divided this timing offset into 64 timing advance steps, TA0–TA63, with a resolution of 550 m per TA step. However, since the initial GSM phase one specification, several manufacturers are now able to support 'extended cells', thereby supporting traffic up to 70 km and in some cases even further.

TA and Indoor Systems

When designing indoor-distributed antenna systems, you need to keep in mind that all the antenna distribution equipment, passive and active, will delay the signal timing, and offset the TA. Even in a passive cable, the propagation speed and velocity are about 80% of the free space speed of light of 300000 km/h. Active systems, amplifiers, optical cables etc. might offset the TA by five to eight steps, depending on the equipment, cable distances, etc.

When setting the maximum cell service range, it is very important to remember to compensate for this timing offset, introduced by the indoor antenna system. One must not be tempted to set the maximum TA to one or two per default for indoor systems, as the indoor cell might not pick up any traffic at all. The correct TA setting for the maximum service distance can easily be evaluated on-site by the user of a special test mobile station, where the TA step can be decoded in the display of the mobile, and the maximum cell service range adjusted accordingly.

2.2.4 GSM Signaling

This section will present an overview of the channel types, and basic signaling and mobility management in GSM. Not in a detailed level, but just to introduce the principles.

Logical Channels

Each TDMA frame consists of eight TSL, TSL0–7, which can be used for several types of channels. Traffic and control channels sometimes even combine channels. Several traffic time slots can be assigned to the same user for higher data rates.

Traffic Channels (TCH)

The TCH is the channel that contains all the user data for speech or data. The TCH can be a full-rate channel, or an enhanced full-rate channel with up to 13 kbs user data or good voice quality. It is also possible to divide one 13 kbps TCH into two 6.5 kbps channels, thus doubling the number of channels in the same TDMA frame. One TDMA frame can carry eight full-rate TCHs or voice calls, or 16 calls of 6.5 kbps channels; for voice this is referred to as 'half-rate' and will to some extent degrade the voice quality, although in normal circumstances it is barely noticeable for the voice user.

By the use of half-rate, the GSM network can double the voice capacity, using the same number of radio channels. The half-rate codex can be permanently enabled in the cell, or most commonly the half-rate service will automatic be activated when the traffic on the cell rises close to full load. The mechanics of the half-rate is, for example: if a cell has six ongoing calls, the network starts to convert these calls into half-rate traffic rearranging the users from the 13 kbps full rate channel to two new 6.5 kbp half-rate channels. Now, instead of taking up six TSLs, they will only use three TSLs, thus freeing capacity on the cell for new traffic.

This function for automatic allocation of half-rate channels is referred to as 'dynamic half-rate' and the trigger level for the capacity load needed to start the conversion can be adjusted in the network.

When using half-rate traffic channels; one radio channel can carry up to 16 voice calls. This is a very cost-effective tool for providing high peak load capacity, for big sport arenas, exhibition venues, etc. However it is not recommended to use permanent half-rate for high profile indoor solutions, where users expect perfect voice quality.

Control Channels

In order to handle all the signaling for the mobility management, traffic control, etc., several control channels are needed. Remember these are logical channels and might not be active constantly (most are not), but will be combined into using the same TSL as other control channels.

CCCH (Common Control Channels) and DCCH (Dedicated Control Channels)

One set of CCCH (BCCH + PCH + AGCH for downlink and RACH for uplink) are mapped either alone or together with four SDDCH logical channels into one physical TDMA time slot. The dedicated control channels, SDDCH, SACCH and FACCH, are not mapped together with other control channels. These will have a dedicated TSL or 'steal' some traffic TSL for a period of time when needed.

BCCH

The broadcast control channel (BCCH) is a downlink-only channel, transmitted from the base station to the mobile station. The BCCH broadcasts a lot of different information for mobiles: system information, cell identity, location area, radio channel allocations, etc. The BCCH is assigned using TSL0 of the first radio channel in the cell.

PCH

The paging channel (PCH) is a downlink-only channel, broadcasting all the paging signals to all mobiles currently registered in the location area (LAC). When detecting a page for the actual mobile (IMSI), the mobile will respond via the RACH. Mobiles in the same LAC will be divided into paging groups. The paging group is based on the IMSI (telephone number), so the mobile knows when to listen for specific paging signals belonging to its specific paging group. This enables the mobile to only use resources for paging reception in the short time when its own paging group is active. The mobile can then 'sleep' during the period when the other paging groups are broadcast, thus saving power and extending battery life. This is referred to as a DRX, discontinuous reception.

RACH

The random access channel (RACH) is used when a mobile needs to access the network. It transmits on the RACH as an uplink-only channel. The mobile will perform random access when attaching to the network, when it is switched on, when responding to paging, when setting up a call and when location updating from one LAC to another.

AGCH

The access grant channel (AGCH) is a downlink-only channel that is used by the base station to transmit the access grant signal to a mobile trying to access the network on the RACH.

SDDCH

The stand-alone dedicated control channel (SDDCH) is used for authentication, roaming, encryption activation and general call control. The SDCCHs are grouped into four (SDDCH/4) or eight (SDCCH/8), together with one set of CCCHs and mapped into one time slot.

SACCH

The slow associated control channel (SACCH) is linked with a specific TCH or SDCCH for radio signal control and measurements. One block of SACCH (456 bits) contains one measurement report, sent every 480 ms, reported to the BSC via the base station. The SACCH will use a part of the TCH or SDCCH for the SACCH.

FACCH

The fast associated control channel (FACCH) is for immediate signaling associated with handover commands for the specific call. The FACCH will be transmitted using the TCH, by means of 'bit-stealing' of the TCH (the small gaps in the voice stream that are noticeable during a GSM handover).

GSM Signaling Procedures

Without going into specific details, we will take a quick look at what type of signaling procedures are used in GSM in order to handle mobility and the overall function of the mobile network, such as:

- Location management.
- Paging.
- Accessing the network.
- Authentication and encryption.
- Radio signal quality control.
- Radio measurements.
- Handover.

Location Updating and Roaming

The network is divided into sub-areas called location areas, identified by a specific location area code (LAC). When a mobile detects a new LAC broadcast by the base station, it will access the network and perform a location update, so the network (the VLR) is aware which location area is currently servicing the mobile, and to which location area to send any paging signal for the specific mobile.

Paging

The mobile is not permanently connected to the network. In idle mode the mobile will monitor the PCH in the serving cell. In the case of an incoming call the mobile will respond with a paging response, after connecting to the network via the RACH.

Network Access

Every time the mobile needs to connect to the network for a location update, paging response, call set-up, SMS, data traffic, etc., the mobile has to request a dedicated channel via the RACH.

Authentication and Encryption

When performing a call set-up, the network will verify the identity of the mobile and provide a specific encryption key for the traffic. The network will encrypt the specific call, using a specific key known only to the mobile (SIM) and the network.

2.2.5   GSM Network Architecture

This section presents a short introduction to the elements of the GSM network, in a simplified example (as shown in Figure 2.11). This is not a detailed description of all network elements, but merely an introduction to the general principle of how the GSM network elements are interconnected and their primary function.

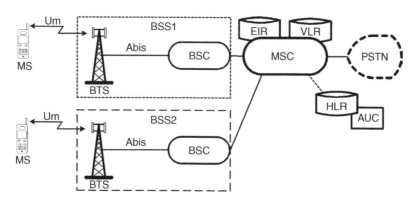

Figure 2.11   The network elements of the GSM network (simplified)

MS

The mobile station (MS) is the user terminal, telephone, data card, etc. The MS is connected to the network (BTS) via the radio interface, the Um interface.

Um, the Radio Interface

The Um is the GSM air interface, the radio channel structure of 200 kHz channels and the TDMA multiplexing presented on the previous pages, combining FDD and TDD. It normally uses the 900 or 1800–1900 MHz spectrum.

Base Transceiver Station

The base transceiver station (BTS) contains all the radio transmitters and receivers that service a specific area (cell). In macronetworks a BTS will typically service three sectors (cells). Each cell can have several radio channels on air in the same cell, serviced by individual transceivers (transmitter/receiver, TRX, TRU, etc.) A standard outdoor base station will use one transmit antenna and two receive antennas (receiver diversity) for each cell. The extra receiver antenna improves the uplink signal from the MS. Mostly, one antenna is used for transmit/receive and one for receiver diversity.

Base Station Controller

The BSC controls large groups of BTSs over a large area. The BSC will perform the power control, etc., of the base stations within its service area. The BSC will also be in charge of handovers and handover parameters, making sure to establish the new connection in the new serving cell, and to close the previously used connection. One BSC and all the BTSs connected are often referred to as a BSS (base sub system). The BSC interfaces to the BTS using the A-bis interface.

Mobile Switching Center

The mobile switching center (MSC) services a number of BSS areas, via the A-interface from the MSC to the connected BSCs. The MSC is in control of all the calls within the BSSs connected, and is responsible for all call switching in the whole area, even two calls under the same cell in the same BTS. All calls are sent back and switched centrally by the MSC. The MSC is also responsible for the supplementary services, call forwarding and charging. To support the MSC with the switching and services, it is connected to supporting network elements; VLR, HLR, EIR and AUC are connected.

Home Location Register

The home location register (HLR) is a large database that contains information about the service profile and current location (VLR address) of the subscribers (SIMs) belonging to the network registered on the particular HLR.

Visitor Location Register

The visitor location register (VLR) is, like the HLR, a database; it contains information about all the mobiles (SIMs) currently camped on the network serviced by the MSC. In the VLR we will find a copy of all the HLR information with regards to the service profile of the user, and information about the current location (location area LAC) of the mobile.

Authentication Register

The authentication register (AUC) contains the individual subscriber identification key (this unique key is also on the SIM) and supports the HLR and VLR with this information, which is used for authentication and call encryption.

Equipment Identity Register

The equipment identity register (EIR) is a database that stores information about mobile station hardware in use. If a mobile is stolen the specific hardware identity (IMEI, International Mobile Equipment Identity) can be blacklisted in the EIR, baring calls to or from the mobile.

Public Service Telephone Network

The public service telephone network (PSTN) is the outside network. It could be the fixed network *a* in this example, or another MSC or mobile network.

2.3   Universal Mobile Telecommunication System/3G

This chapter is a short introduction to the universal mobile telecommunication system (UMTS/3G). For more details please refer to References [3] and [4].

UMTS (WCDMA) was specified and selected for 3G system for many reasons, one of the main ones being that it is a very efficient way to utilize the radio resources – the radio spectrum. WCDMA has a very high rejection of narrowband interference, and is very robust against frequency-selective fading. It offers good multipath resistance due to the use of rake receivers. The handovers in WCDMA are smooth and imperceptible due to the use of soft handovers. During handover the mobile is serviced by more cells at the same time, offering macro diversity gain.

However, there are challenges to UMTS radio planning, when all cells in the network are using the same frequency. UMTS radio planning is all about noise and power control. Very strict power control is necessary to make sure that all transmitted signals are kept to a target level that insures that all mobiles reach the base station at the exact same power level. Very good radio planning discipline must be applied and cells must only cover the intended area, and not spill into unintended areas. The reason for this is that you need to minimize the inter-cell interference since all cells are operating on the same frequency; this is one of the biggest radio planning challenges when designing UMTS radio systems.

Even though the concept of soft handovers insures that the mobile can communicate with two or more cells operating on the same frequency, you must remember that the same call will take up resources on all the cells the mobile is in soft handover with, including the backhaul resources of other network elements. Therefore you must try to minimize the handover zones to well-defined small areas, or the soft HO can cannibalize the capacity in the network. This is a particular concern when the two or three cells that are supporting the soft handover are serviced by separate base stations, and different RNCs (radio network controllers) – this will cause added load to the backhaul transmission interfaces.

2G and 3G/UMTS

UMTS is often added on to existing 2G networks, so in the UMTS specification it has been insured that handovers (hard handover) can be done between 2G and 3G and vice versa.

2.3.1   The Most Important 3G/UMTS Radio Design Parameters

UMTS is a very complex mobile system, even if you 'only' need to focus on the radio planning aspects. It can be a challenge to understand the all the deep aspects regarding UMTS radio planning. Many parameters and concerns are important but the most important parameter the RF designer must remember when designing UMTS indoor solutions is that:

- 3G/UMTS is power-limited on the downlink (DL).
- 3G/UMTS is noise-limited on the uplink (UL).

The RF designer should always strive to design a solution that can secure the most power resources per user, and the least noise load on the UL of the serving cell and other cells in the network.

Remember that all cells in UMTS are on the same frequency; therefore the coverage and noise from one cell will affect the performance on other cells in the network. Do *not* apply '2G/GSM planning' tactics to 3G/UMTS planning, especially not to indoor UMTS radio planning. When planning GSM you can to some extent just 'blast' power from the outside network into a building to provide coverage. In GSM you will also often use few, high-power (exceeding +20 dBm) indoor antennas to dominate the building. What saves you on GSM is the frequency separation; you can assign separate radio channels to the different cells. This is not possible on UMTS. You might be tempted to use a separate UMTS carrier for indoor solutions and by that separate the indoor from the outdoor network, but that is 'GSM thinking' and is a very expensive decision; after all, the networks have typical only two or three UMTS carriers as the total resources for current and future services. The only solution is good indoor radio planning, controlling the noise and the power use to service the users – this is the main topic of the rest of this book.

2.3.2   The 3G/UMTS Radio Features

Compared with 2G/GSM, the 3G/UMTS radio interface is totally different, and can take some time and effort for the hard-core GSM radio planner to understand. Actually it is not complicated at all; if you just focus on the important main parameters it is easy to understand and plan indoor UMTS radio systems. There are in principle two different types of WCDMA – WCDMA-TDD and WCDMA-FDD (as shown in Figure 2.12).

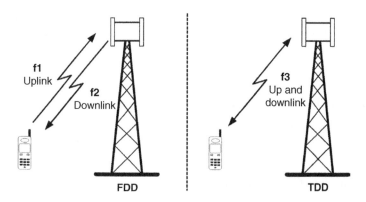

Figure 2.12   The two principle types of WCDMA frequency allocation: FDD and TDD

UMTS/3G TDD

WCDMA-TDD uses the same frequency for the UL and the DL, alternating the direction of the transmission over time. The equipment needs to switch between transmission and reception, so there needs to be a certain guard time between transmission and reception to avoid interference; after all, they are using the same frequency for both links, transmission and reception.

FDD requires a paired set of bands, one for UL and one for DL. TDD can use the same frequency for UL and DL, separated in time. The WCDMA-TDD system can be used in parts of the world where it is not possible to allocate a paired set of bands.

UMTS/3G FDD

The most used (currently only) standard for UMTS is WCDMA-FDD, which transmits and receives at the same time (constantly). Different frequency bands for the UL and DL are allocated as paired bands. The WCDMA-FDD requires a paired set of bands, equal bandwidths separated with the 95 MHz duplex distance throughout the band.

Frequency Allocation

In most regions around the world the band 1920–1980 MHz is assigned for the UL of WCDMA-FDD and the band 2110–2170 MHz is assigned for DL (as shown in Figure 2.13). The frequency allocation for WCDMA-TDD will vary depending on the region, and where it is possible to fit the TDD frequency band assignment into the spectrum.

3G/UMTS900

Commonly, most operators are assigned two or three WCDMA-FDD '2.1 GHz' carriers per license. Currently there are considerations on re-using the GSM900 spectrum, by converting the 'old' 35 MHz GSM900 band to UMTS, raising the efficiency of the spectrum for better data performance. The 35 MHz GSM spectrum will have the possibility to support seven new UMTS WCDMA channels. Furthermore the 900 MHz band will penetrate buildings much better compared with 2100 MHz, giving better indoor UMTS coverage.

The 3G/UMTS/WCDMA RF Channel

Let us have a look at some of the main features of the UMTS radio channel, and how UMTS differs from GSM.

Spread Spectrum, Interference Rejection
UMTS uses WCDMA; this is a spread spectrum signal. The narrow band information from the individual user is modulated, and spread throughout the spectrum. This distributes the energy of the user data over a wider bandwidth (as shown in Figure 2.14). Thus the signal becomes less sensitive to selective interference from narrow-band interference. Typical narrow-band interference will be inter-modulation products from narrowband services at lower frequencies, such as 2G.

Fading Resistance
A common problem in radio communication is frequency-selective fading. This occurs when reflections in the environment turn the RF channel into a multipath fading channel. The same

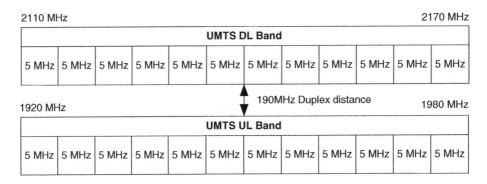

Figure 2.13   UMTS UL and DL frequency bands for the 12 FDD channels

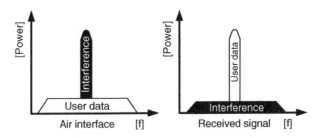

Figure 2.14   Wideband signals are less sensitive to narrowband interference

signal will have different signal paths, reflections and diffractions, thereby offsetting the phase and timing of the signal arrival at the receiving end. Most of the fading is frequency-selective (as shown in Figure 2.15), but the WCDMA structure solves the problem to some extent. Even if a small portion of the WCDMA fades, most of the energy is maintained and the communication is preserved intact.

The 3G/WCDMA RF Carrier

The bandwidth of the UMTS carrier (as shown in Figure 2.16) is about 5 MHz (4.75 MHz), and is divided into 3.84 Mcps. The chips are the raw information rate on the channel, or the 'carrier' if you like. Each user is assigned a specific power according to the service requirement and path loss to that particular user. The more user traffic there is, the more power will be transmitted and the higher the amplitude of the UMTS carrier will be (as shown in Figure 2.17).

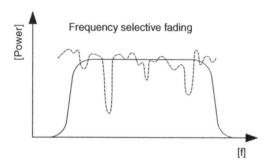

Figure 2.15 The fading channel (dotted line) affects only a small portion of the WCDMA carrier, so the frequency-selective fading is limited

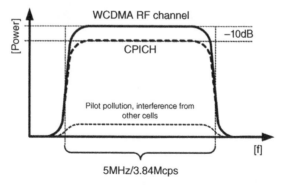

Figure 2.16 The WCDMA air channel, 5 MHz-wide modulated with 3.84 Mchips; a large portion of the power is assigned to the important CPICH channel

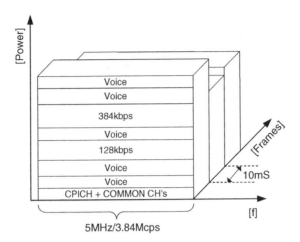

Figure 2.17   The spread WCDMA signal is transmitted in frames of 10 ms, enabling service on demand every 10 ms. One user in a voice session can get a higher data rate assigned in the next frame

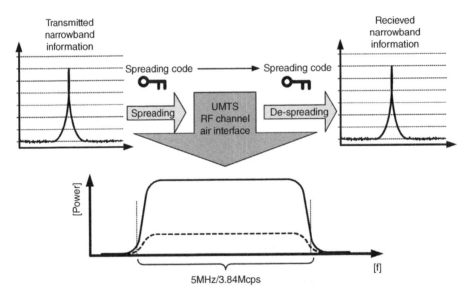

Figure 2.18   The narrowband signal is spread using a dedicated spreading code, modulated into and transmitted over the 3G/WCDMA carrier, to be despread and restored in the receiving end using the same dedicated code

Coding and Chips

Each mobile in service is assigned an individual spreading code (as shown in Figure 2.18 and Figure 2.22). All of the users transmit simultaneously using the same WCDMA frequency. The receiver is able to decode each individual user by applying the same specific spreading code assigned to each user.

The fundamental difference from 2G/GSM is that the 3G/UMTS system uses the entire 5 MHz spectrum constantly, separating the users only in the code domain, by assigning each mobile a unique code sequence. This unique code is then used to encode the transmitted data; the receiver knows the specific code to use when decoding the signal. This spreading/despreading technique enables the base station to detect the user signal from below the noise.

The coded signal is orthogonal to other users in the cell. The principle is that the code is constructed in such way that one coded signal will not 'spill' any energy to another coded user – if the orthogonality is maintained over the radio channel. Each individual user signal can only be retrieved by applying the same specific code, thereby limiting the interference between users. Essentially all transmitted UMTS information is data; each bit of this data is multiplied by a sequence of code bits, referred to as chips. The number of chips multiplied to each user bit is dependent on the service bit rate of the service assigned to each user. The principle is that the transmitted data is multiplied by the 'raw' channel code rate of a much higher frequency in UMTS 3.84 Mcps (mega-chips per second).

When multiplying the user data rate by a higher-frequency coded bit sequence, the spectrum is spread over the 3.84 Mcps carrier, and becomes a spread spectrum signal. UMTS uses a chip rate of 3.84 Mcps. This is the 3.84 Mcps that takes up the 5 MHz WCDMA radio, but the 'carrier' is 3.84 Mcps.

Frequency Reuse on 3G/UMTS

The cells in the UMTS system are operating on the same frequency (as shown in Figure 2.19), using a frequency reuse of 1. UMTS cells are separated only by the use of different primary scrambling codes; this is the 'master key' for the codes in that particular cell.

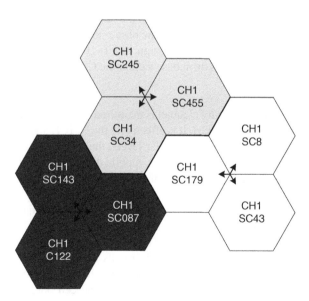

Figure 2.19 Example of three 3G/UMTS macro sites with three sectors each; all cells are using the same RF channel, but different scrambling codes

E_b/N_o

The signal quality of the user data on UMTS is defined as the E_b/N_o. The E_b/N_o is defined as the ratio of energy per bit (E_b) to the spectral noise density (N_o). E_b/N_o is the measure of signal-to-noise ratio for a digital communication system. It is measured at the input to the receiver and is used as the basic measure of how strong the signal is over the noise.

$$\frac{E_b}{N_o} = \frac{\text{Bit energy}}{\text{Noise density}} = \frac{\text{Signal bandwidth}}{\text{Bit rate}} * \frac{\text{Signal power}}{\text{Noise power}}$$

The E_b/N_o plays a major role in UMTS radio planning, as the reference point of the link budget calculation. The E_b/N_o defines the maximum data rate possible with a given noise. Different data rates have different E_b/N_o requirements: the higher the data speed, the stricter the E_b/N_o requirements.

The E_b/N_o design level for the various data services are up to the operator, and are used as the basis of the link budget calculation. A typical E_b/N_o planning level for voice will be 7 dB, and about 4 dB for a 384 kps service.

Common Pilot Channel

The common pilot channel (CPICH) does not contain any signaling. It is a pure DL channel that serves two purposes only: to broadcast the identity the cell and to aid mobiles with cell evaluation and selection. The CPICH is coded with the primary scrambling code of the cell; it has a fixed channelization code with a spreading factor of 256. The mobiles in the network will measure the CPICH power from the different cells it is able to detect. It will access the cell with the most powerful measured CPICH. The mobile will also measure and evaluate the CPICH levels from other adjacent cells (defined in the neighbor list) for handover evaluation.

A major portion of the total DL power from the base station is assigned to broadcasting the pilot channel. Typically the base station will assign 10% (−10 dB, see Figure 2.16) of the total DL power for the CPICH. For a +43 dBm base station the CPICH is transmitted at 33 dBm.

E_c/I_o

The quality/signal strength of the pilot channel is measured as E_c/I_o, the energy per chip/interference density measured on the CPICH. It is effectively the CPICH signal strength. When the mobile detects two or more CPICH with similar levels, the mobile will enter soft handover. This is essential in order to secure the link. If the mobile did not enter soft handover, the link would break down due to interference between the two cells transmitting the same channel and received at the same power level.

The mobile continuously measures the E_c/I_o of the serving cell and adjacent cells (defined in the neighbor list/monitor set). The mobile compares the quality (E_c/I_o) of the serving CPICH against the quality of other measured CPICHs. The mobile uses trigger levels and thresholds to add or remove cells from the active set, the cell or cells the mobile is engaged with during traffic. (For more details, refer to the description of the UMTS handover algorithm, Section 2.3.4.)

The CPICH defines the cell size, and the service area and cell border can be adjusted by adjusting the CPICH power of the cell. Thus it is possible to some degree to distribute traffic load to other cells by tuning cell size.

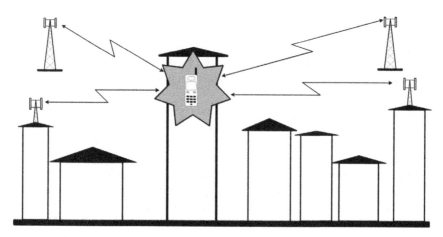

Figure 2.20   Pilot (CPICH) pollution is a big concern, when a mobile receives pilot power from distant base stations without being in handover with these base stations

A sufficient quality for the CPICH (not scrambled) will be a E_c/I_0 of −10 to −15 (energy per chip/interference) in the macrolayer. For indoor cells lower CPICH levels of −6 to −8 dB can be considered due to the high isolation of the macrocells (if the indoor system is planned correctly).

Pilot Pollution

When a mobile receives CPICH signals at similar levels from more cells it is supposed to enter soft handover – if it does not the quality will be degraded and the call might be dropped. However there can be cases where a distant cell is not defined in the neighbor list, and thus the mobile is not able to enter soft handover. This will cause interference of the serving cells' CPICH; this is referred to as pilot pollution (as shown in Figure 2.20). This is one of the main concerns when designing UMTS indoor solutions. The problem is often a big concern in high-rise buildings, predominantly in the topmost section of the building, where mobiles are able to detect outdoor cells. The problem is essentially a lack of dominance of one serving cell (see Section 3.5.3).

Geometry Factor

The geometry factor is the power received by the mobile divided by the sum of noise received by the mobile. The sum of noise is thermal noise plus interference (traffic) radiated by other base stations.

Geometry factor = received signal /(thermal noise + interference)

The closer the mobile is to the servicing base station, the higher the geometry factor will be, and the better the signal quality will be.

Orthogonality

Owing to the unique way the codes are constructed, they are orthogonal to each other. The orthogonal structure of the codes insures that all the users can be active on the same channel at the same time, without degrading each other's service; that is, if the orthogonality is perfect, or 1 (100%).

However, due to multipath propagation on the radio channel, especially in cities and urban environments, the delay spreads of the phases and time of the same transmitted signal in the receiving end will cause the codes to 'spill over', degrading the orthogonality and causing the signals to interfere with each other.

The multipath environment will degrade the orthogonality, depending on how predominant the reflections/diffractions of the signal are. The degraded orthogonality will degrade the effectiveness of the RF channel, degrading the maximum throughput on the WCDMA carrier.

In environments with low orthogonality, the data throughput of the WCDMA carrier is degraded. In the example in Figure 2.21 we can see that users inside a building serviced by a nearby macro base station typically have an orthogonality of less than 0.55. This is due to the fact that most of the signal will arrive at the mobile as reflections, due to the absence of a line-of-sight to the base station.

The degraded orthogonality causes degradation efficiency of the codes, causing the codes to 'spill over', and this 'code interference' will degrade the total throughput of the UMTS carrier. In the example from Figure 2.21, the capacity of the carrier is degraded to less than 1.5 Mbps. However, the same base station connected to an indoor distributed antenna system would have a typical orthogonality of 0.85 and would be able to provide about 2.6 Mpbs, and so is a considerable better use of the same resources. This is due to the fact that the indoor user, serviced by indoor antennas, will mostly be in line-of-sight to the antennas, with only limited reflections. By insuring a high orthogonality and efficient use of the resources, an indoor distributed antenna system can be used. This improves the throughput and efficiency of the UMTS channel, improving the business case.

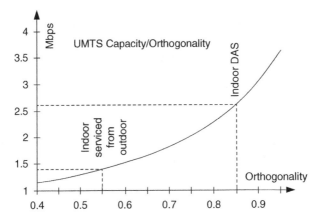

Figure 2.21   The maximum data throughput on the WCDMA carrier is dependent on the orthogonality of the RF channel. Reflections and multipath signals will degrade the orthogonality, and degrade the efficiency of the RF channel

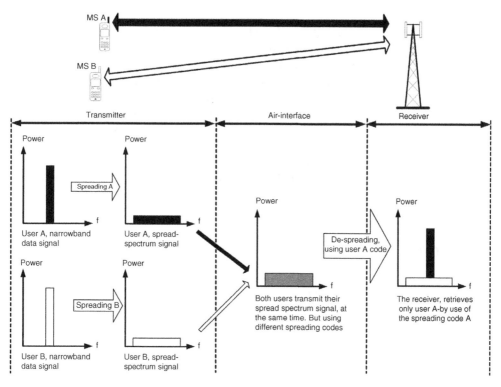

Figure 2.22 Two mobiles serviced by the same cell, using different codes, separated by individual assigned spreading codes

Table 2.1 UMTS data rate vs processing gain

User rate	Processing gain
12.2 kbps	25 dB
64 kbps	18 dB
128 kbps	15 dB
384 kbps	10 dB

Processing Gain

When using the technique with spreading and despreading the signal over a wideband carrier, a gain is obtained. This is referred to as the processing gain. The processing gain is dependent on the relation between the carrier chip rate, and the user data bit (chip) rate, as shown in Table 2.1.

$$\text{Processing gain} = \text{chip rate}/\text{user data rate}$$
$$\text{Processing gain} = 3.84\text{M}/\text{user rate [linear]}$$
$$\text{Processing gain} = 10\log(3.84\text{M}/\text{user rate}) \text{ [dB]}$$

Using these formulas you can calculate the processing gain of the specific UMTS data service

Examples

The processing gain for high-speed data, 384 kbps, is $10\log\ (3.84\text{Mcps}\ /\ 384\text{kb}) = 10\text{dB}$.

The processing gain for speech, 12.2 kbps, is $10\log(3.84\text{Mcps}\ /\ 12.2\text{kbps}) = 25\text{dB}$.

The use of orthogonal spreading codes and processing gain is the main feature in UMTS/WCDMA, giving the system robustness against self-interference. It is the main reason for a frequency reuse factor of 1, due to the rejection of noise from other cells/users.

When you know the required bit power density E_b/N_o the specific service (voice 12.2kbps + 5dB, data +20dB) you are able to calculate the required signal-to-interference ratio: voice at 12.2 kbps needs an approximately 5 dB wideband signal-to-interference ratio minus the processing gain, $5\text{dB} - 25\text{dB} = -20\text{dB}$. In practice, the processing gain means that the signal for the voice call can be 20 dB lower than the interference or noise, but still be decoded.

2.3.3 3G/UMTS Noise Control

UMTS is very sensitive to excessive noise, and noise control is essential. All traffic is on the same frequency and all signals from active mobiles need to reach the base station at the same level. If one mobile reaches the uplink of the base station at a much higher level, it will interfere with all the traffic from other mobiles in service on the same cell. Every mobile uses the same frequency at the same time.

To control this issue, UMTS uses very strict noise and power control. Traffic in UMTS is essential white noise to other users, and the amount of noise from a cell is directly related to the traffic.

Admission Control

More traffic in a UMTS cell is equal to more radiated noise in the cell. This noise will impact surrounding cells with interference, and therefore it is very important to limit the load in the cells to a preset maximum noise level in order to control the increase in noise.

Loading a UMTS cell by 50% is equal to a 3 dB increase in noise; this is a typical value for the maximum allowable noise increase. The value can be set in the network individually for each cell. Indoor cells are more isolated from the macro layer, and can in principle be loaded relative highly, maybe up to 60–65%, giving a higher capacity.

In order to evaluate the noise increase due to traffic in the cell, node B will constantly measure and evaluate the overall noise power received on the UL in order to evaluate the UL noise increase in the cell. By doing so, the base station can calculate and evaluate how much headroom is left for new traffic, this in order to control the admission of new traffic in the cell (as shown in Figure 2.23).

The total noise power on the UL of node B will be a result of the sum of the noise generated by the traffic in the cell itself, noise (traffic) from other cells and the noise figure from the base station:

$$\text{Noise}_{total} = \text{own}_{traffic\ noise} + \text{other}_{cells'\ noise} + \text{node B}_{noise\ power}$$

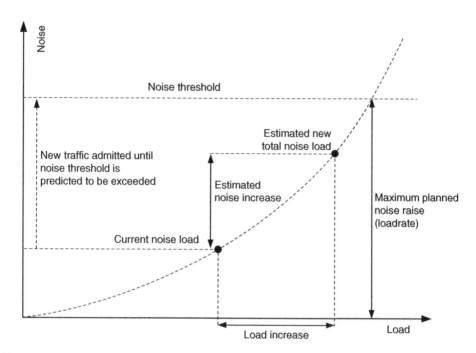

Figure 2.23   Admission control in 3G/UMTS will make sure that new admitted traffic in the cell will not cause the noise to increase above a preset level

In addition to the noise generated by the traffic in the cell, there will be a contribution from noise from users in other cells and finally the noise figure (NF) of node B and other active hardware in the system.

The base station noise figure (noise floor) is the reference for zero traffic in the cell. If the receiver only measures the noise power generated by the base station itself, the network assumes that there is no traffic in the cell/adjacent cells. The network assumes the load rate to be zero and that the cell has its full potential for servicing new traffic.

Using this noise measurement (including the hardware noise reference of the node B), the network can evaluate if the new admitted traffic will cause a noise increase that exceeds the predefined maximum UL noise increase in the cell, and the base station will keep admitting new traffic in the cell until the max noise rise is predicted or measured.

Impact of Noise Power from External Equipment

It is very important that the base station noise reference level for no traffic is updated when connecting any active equipment to the uplink of the base station. Repeaters, active distributed antenna system, etc., will generate noise power on the uplink, and these systems should be designed so that minimum noise power is injected into the UL port of the base station.

Any noise increase will cause the base station to assume this noise power to be traffic in the cell, or adjacent cells – essentially offsetting the traffic potential for the cell. It may cause the base station to not admit traffic to its full potential, and in severe cases to not admit any traffic at all. Therefore it is crucial to adjust the admission control parameters in the network, with regard to the added hardware (HW) noise power on the UL of the node B.

For most UMTS base station networks these parameters can be found in the parameter set for calibrating the system to connect to mast-mounted amplifiers or low-noise amplifiers (LNA). However the main issue is to design the distributed antenna system so that you minimize the noise load on the base station without compromising the performance of the link; this can be done using uplink attenuators; see Section 7.5 on noise control for more details.

Cell Breathing on 3G

Any traffic in UMTS is equal to the noise increase on the radio channel. We know that UMTS is very sensitive to noise increase, and therefore it is important to control the noise increase in the network. When designing UMTS systems you need to decide how much noise increase you will allow in the cell and adjacent cells. Noise from traffic in cells will also increase the noise in adjacent cells and impact the coverage and capacity of those cells.

In Figure 2.24 you can see an example of how the noise increase and power load in the cell will affect the footprint of the cell; the greater the load, the more noise, the less the coverage of the cell will be – this is referred to as cell breathing. It will be different for the uplink and the downlink, so a cell can be both downlink and uplink limited, depending on the current load profile (as shown in Figure 2.26).

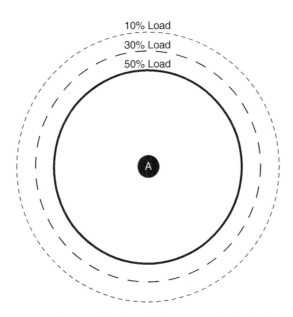

Figure 2.24 3G Cell breathing is caused by the noise increase owing to the load of the cell; more load will give more noise increase and a smaller coverage footprint

It is important that you design the system for a maximum allowable noise increase or load. If you do not, the traffic will cause the noise to spin out of control. The cell would collapse, and drop the calls on the cell border due to excessive noise increase.

Noise Increase

You are able to calculate the noise increase in the cell. Using 50% of the remaining capacity in a cell is equal to noise increase of factor 2, corresponding to 3 dB. Loading a cell from 0 to 50% load is a noise increase of 3 dB, and using another 50% of the remaining capacity, from 50 to 75%, is an additional 3 dB, a total of 6 dB, and so on.

You can calculate the noise increase as a function of the load rate:

$$\text{Noise increase} = 10\log[1/(1-\text{load factor})]$$

Typical cells will be designed to be loaded to a maximum of 50–60%, and the graph in Figure 2.25, based on the noise increase formula, clearly shows why. A 60% load rate, corresponding to a load factor of 0.6, will cause a noise increase of 4 dB. In Figure 2.25 it is very clear what happens with the noise when you exceed 80% – the noise simply rises abruptly, out of control. This excessive noise rise will cause the cell to collapse.

The capacity in UMTS is directly related to the signal-to-noise ratio (SNR; see the geometry factor, Section 2.3.2) in the cell. The capacity is proportional to the interference in the cell. Any added interference in the cell will impact the capacity.

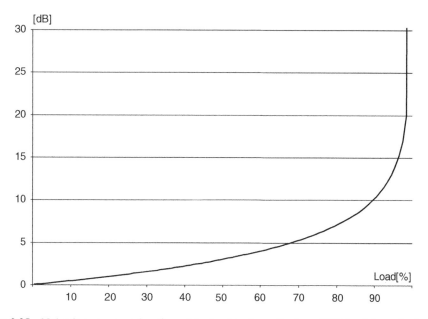

Figure 2.25   Noise increase as a function of the load in the cell; above 80% load the noise increase causes the noise to rise sharply, causing the collapse of the cell

UL and DL Load on 3G

UMTS is noise-limited on the uplink and power-limited on the downlink. In a typical cell the link tends to be noise-limited during light traffic load, and power-limited when the traffic load rises (as shown in Figure 2.26).

The service profile of traffic generated by the users in the cell will also affect the balance of the cell between being downlink- or uplink-limited. Typical data users will use higher downlink data speed, for downloads, etc.

The base station's capabilities with regards to processing power, power resources and receiver sensitivity also play a role, but the principle is that the same cell can be both uplink- and downlink-limited depending of the load profile.

2.3.4 3G/UMTS Handovers

There are several handover scenarios in UMTS. The type of handover is dependent on whether the handover is within the same node B, different node B, different UMTS/3G frequencies or even system handover to and from 2G. As we know, UMTS uses the same frequency for all the cells and the only possible way for a mobile station to hand over from one cell to another is to be connected with both cells in the area where both cells are at equal levels. Typically a handover in UMTS involves two or three cells. This handover function and algorithm is totally different compared with 2G handover zones.

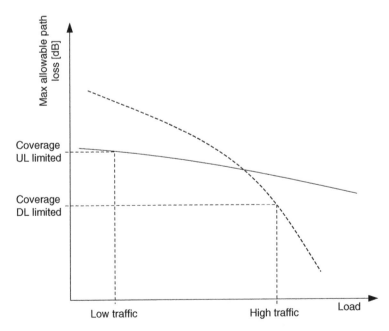

Figure 2.26   The load profile for the cell defines whether the coverage is DL or UL limited

Softer Handover

When a mobile station is within the service area of cells originating from the same node B at the same power level, the mobile station is in softer handover (as shown in Figure 2.27).

The mobile will use both RF links, and the node B will control and combine the signal stream to the RNC (see Figure 2.37 for UTRAN network elements). Two separate codes are used on the downlink, and the mobile will use rake receiving to receive the two downlink paths.

On the UL the base station will use rake receiving for the two signal paths. The mobile will provide some extra load on node B due to the double link for only one call. This extra load is limited to processing power (channel elements) and impacts only the serving base station.

Soft Handover

When a mobile is in the service area of two cells originating from different node Bs (as shown in Figure 2.28) the mobile will use one RF link to both base stations; this is macro diversity and both node Bs will send the call to the RNC (see Figure 2.37 for UTRAN network elements). In this scenario the mobile will not only load both cells with regards to power, noise and processing power (channel elements), but also double the load on the transmission link (IUR interface) back to the RNC/RNCs from both node Bs.

As in softer handover, two separate air interfaces are used; the mobile combines the downlink signals using rake receiving. The main difference is on the uplink, owing to the fact that in soft handover the RNC not the base station will have to perform the combining of the two uplink signal paths.

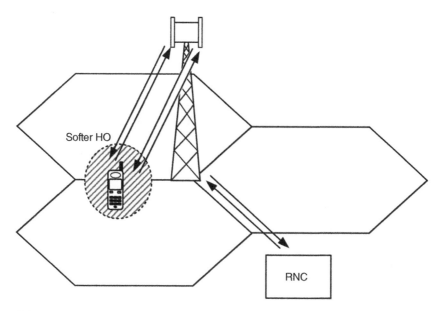

Figure 2.27 The mobile will be in softer handover when it is able to detect more than one cell from the same node B

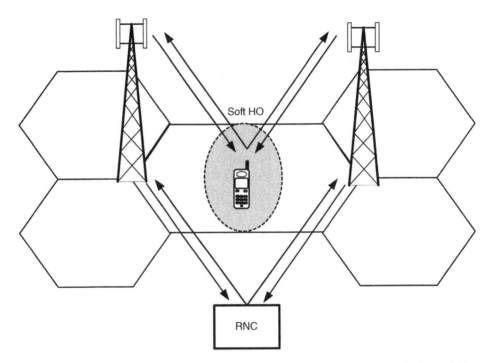

Figure 2.28 Mobiles will be in soft handover when detecting two or more cells from different node Bs

Typically the RNC will select the uplink signal path with the best frame reliability indicator, which is used for outer loop power control. For indoor radio planning for UMTS, you should aim to keep the soft handover areas to a minimum.

OBS!
Note that if a strong candidate neighbor cell is not monitored (defined in the monitor set/ neighbor list), the mobile will not be able to enter soft handover with that nonmonitored cell. If the signal from the nonmonitored cell is high, it will degrade the quality (E_c/I_0) of the serving cell, degrading the data performance, and even causing the call to be dropped.

The Soft Handover Algorithm

The handover algorithm in UMTS (as shown in Figure 2.29) evaluates the radio quality (E_c/I_0) of the serving and adjacent cells. The E_c/I_0 measurement is used by the mobile for evaluating the serving cell and handover candidate cells; the 3G system uses the following terminology for describing the handovers:

- *Active set* – this is the set of cells that the mobile is in connection with during a soft handover; it is all the cells in the active set currently engaged with traffic to/from the mobile.

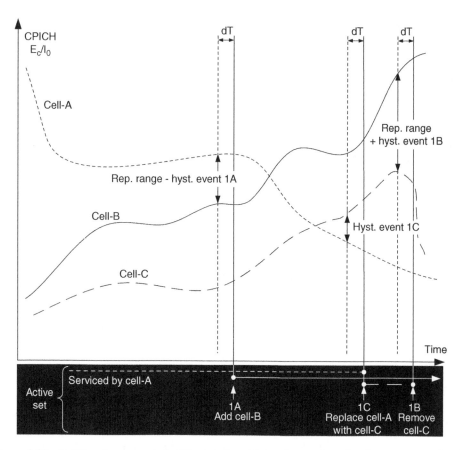

Figure 2.29 Soft handover events in 3G example with maximum size of active set assumed to be two cells

- *Monitored set/neighbor set* – these are the neighbors that the mobile must monitor (the neighbor list), but whose pilot E_c/I_0 is not strong enough to be added to the active set and engage in soft handover.

Example
This is an example of how the soft handover algorithm works (shown in Figure 2.29). Specific events are defined in 3G/UMTS to describe the status/action.

Event 1A, Radio Link Addition The monitored cell will, if the serving cell pilot E_c/I_0 > best pilot E_c/I_0 – reporting range + hysteresis event 1A, for a period of dT, and the active set is not full, be added to the active set, and engage in soft or softer handover.

Event 1B, Radio Link Removal In soft or softer handover, and if one of the serving cells pilot E_c/I_0 < best pilot E_c/I_0 – reporting range – hysteresis event 1B, for a period of dT, then this cell will be removed from the active set.

Event 1C, Combined Radio Link Addition/Removal (Replacement) If the active set is full, and no more soft handover links can be added and the best candidate pilot E_c/I_0 > worst old pilot E_c/I_0 + hysteresis event 1C, for a period of dT, then the cell with the worst E_c/I_0 in the active set will be replaced with the best candidate cell from the monitored set.

UMTS Handover Algorithm Definitions

- Reporting range: threshold for soft handover.
- Hysteresis event 1 A: the cell addition hysteresis.
- Hysteresis event 1 B: the cell removal hysteresis.
- Hysteresis event 1 C: the cell replacement hysteresis.
- dT: time delay before trigger.
- Best pilot E_c/I_0: the strongest monitored cell.
- Worst old pilot E_c/I_0: the cell with the worst E_c/I_0 in the active set.
- Best candidate pilot E_c/I_0: the cell with the best E_c/I_0 in the active set.
- Pilot E_c/I_0: the measured (averaged and filtered) pilot quality.

Hard Handovers

There are a few cases where the mobile station will need to perform a hard handover (like 2G) instead of a soft handover. When performing handover between different 3G frequencies, the mobile will do a hard handover. This could be different frequencies within the same cell, or between different cells. When the mobile station performs a system handover, i.e. a handover between GSM/DCS and UMTS, it will also be a hard handover. Hard handovers are more difficult to control and more likely to fail, and therefore these events should be limited to a minimum in order to maximize network performance and quality.

2.3.5   UMTS/3G Power Control

Compared with GSM, UMTS power control is complex and extremely strict; the motivation is the need for strict noise control. All traffic on the UL of the base station has to arrive at the exact same level, or else one high signal would overpower the traffic from all other mobiles. They all operate at the same frequency at the same time.

The power control in UMTS has three different stages: no power control, open loop power control (cell access) and closed loop power control, in dedicated (traffic) mode. Figure 2.30 shows the principle of the UMTS power control.

Idle Mode

No power control is needed when the mobile is in idle mode. The mobile will measure the received CPICH power and monitor other power control parameters (PWR_INI and PWR_ STEP info). The mobile can thereby calculate what initial power is needed when accessing the base station, in order to make sure it does not overpower the uplink of the base station and degrade the performance for other users.

Open Loop Power Control

Open loop power control (PC) is used when the mobile is in transition from idle mode to dedicated mode (setting up a call or responding to a paging signal). The mobile's initial power for the network attachment is estimated using the downlink pilot (CPICH) signal. The mobile

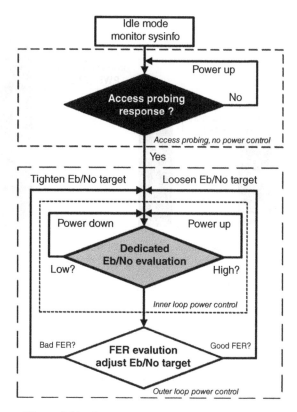

Figure 2.30   The power control algorithm in UMTS

station will send access probes to the base station when accessing the network. The mobile will monitor the system information transmitted by the base station with regards to the reference for the transmitted CPICH power. This enables the mobile station to calculate the path loss back to the base station.

Access Procedure in UMTS

MS starts with low power, to prevent UL 'blocking' the base station (BS), calculating the transmit power for the first access burst (as shown in Figure 2.31):

$$PWR\_INIT = CPICH\_Tx\_Power - CPICH\_RSCP$$

$$+ UL\_Interference + UL\_Required\_CI$$

where PWR_INI = initial access power; $CPICH\_Tx\_Power$ = transmitted CPICH power; $CPICH\_RSCP$ = received CPICH signal strength; $UL\_Interference$ = uplink interference measured by the node B; and $UL\_Required\_CI$ = uplink quality margin.

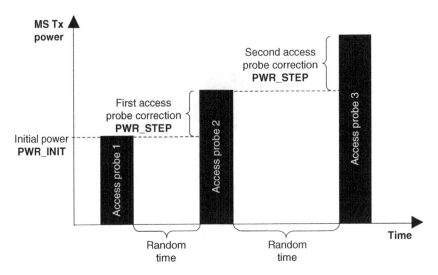

Figure 2.31   The mobile access probe in UMTS

This procedure will insure that the mobile will not use a high initial access power, increasing the noise on the UL, resulting in dropped calls for already ongoing traffic. Then the mobile station waits for response signal (access grant) from node B. If no access grant is received, the mobile assumes that the access probe did not reach the base station. Then the mobile station will increase the transmit power for the next access probe by the PWR_STEP parameter broadcasted by the cell. There will be random intervals for these access bursts, within a specified timing window, in order to prevent the access signals from several mobiles colliding repeatedly if accessing the cell at exactly the same time.

Closed Loop Power Control (Inner Loop)

Compared with GSM, where the power adjustment is done every 60 ms, the power control in UMTS is very fast. With a power control rate of about every 666 μs, the mobile transmit power is adjusted; this is a power control of 1500 Hz in closed loop power control. The closed loop power control is active when the mobile station is in traffic mode (dedicated mode). On the downlink, power control on the base station preserves marginal power for mobiles on the edge of the cell.

The Outer Loop Power Control

The purpose of the outer loop power control is to adjust the E_b/N_o target in order to fulfil the FER target (no more, no less). Owing to the multipath fading profile, this will be dependant on the speed of the mobile. Mobiles moving at faster speed will have more fading impact, and need more power resources. The FER target will be adjusted accordingly (as shown in Figure 2.32).

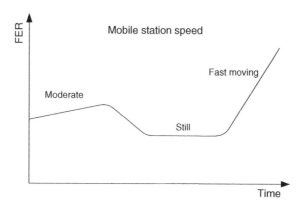

Figure 2.32   The FER target for the UMTS power control is constantly adapted to the fading channel with the speed of the mobile

UMTS Power Control Compensation for Fading

Normally a fading radio channel will affect the signal strength on the uplink. This is dangerous for UMTS because one mobile more powerful than others in the cell will affect the performance – after all, all traffic is on the same frequency at the same time. To compensate for the fading, the UMTS power control is so fast (1500 Hz) that it is able to level out the fading channel and to a large extent compensate for Rayleigh fading, at moderate speeds.

In Figure 2.33 you can see the principle: the fading radio channel (the solid line) will provide a large variance on the path loss, depending on the speed of the mobile. The power control (dotted line) will compensate for this fading, and on the uplink of the base station the topmost graph shows that the resulting radio channel is received compensated by the power control, with a typical maximum variance of ±2–3 dB.

2.3.6   UMTS and Multipath Propagation

Ideally the radio channel path should be direct, in line-of-sight (LOS), with no reflections or obstructions between the base station and the mobile. However, in the real world, especially in cities and urban environments, most RF signals reaching the receiver will have been reflected or diffracted by the clutter of the buildings. Typically only a small portion of the RF signal will be the direct signal from the base station; the main signal will derive from reflections. This is called multipath, and the environment creates multipath fading due to the phase shifts and different delays of the signals amplifying and canceling each other. This is the main source for orthogonality degradation, especially when serving indoor mobile users from the macro network.

An example of the fading environment with reflections can be seen in Figure 2.34, where T1 is the direct signal, and T2/T3 are the reflected and delayed signals.

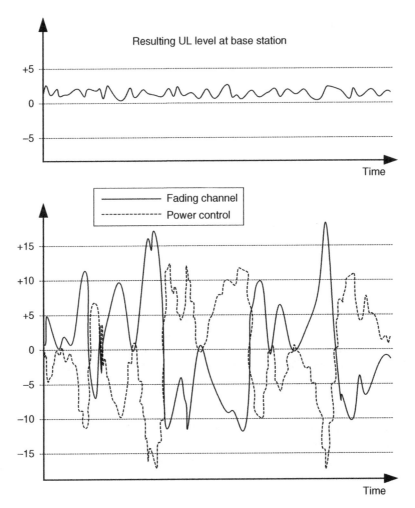

Figure 2.33   The mobile power control in UMTS is so fast and efficient that it will compensate for the fading channel, and all mobiles will reach the uplink of the base station with the same level

At the receiver T1, T2 and T3 will be offset in time and amplitude (as shown in Figure 2.35). In a normal receiver this will mean that, depending on the phase and amplitude, the signals will cancel each other out to some extent. However, the UMTS system uses a clever type of receiver, the rake receiver.

The Rake Receiver

Like the tines, or fingers, on a rake, the rake receiver has more than one point. Actually the rake receiver could be compared with three or four parallel receivers, each receiving a different offset signal from the same RF channel. The rake receiver in this example (as shown

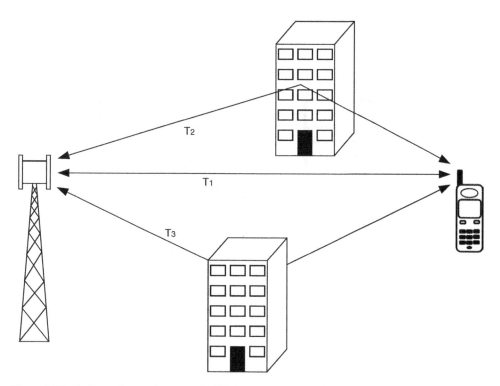

Figure 2.34 In the outdoor environment the RF channel is multipath, having more than one signal path between the mobile and the base station. The signal phase and amplitude of these signals will not be correlated

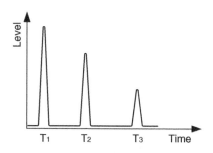

Figure 2.35 Owing to reflections from the environment, the offset signals reaching the mobile will be the different signal components of the multipath signal, with different levels and phases

in Figure 2.36) uses three fingers to receive the three differently phased signals, T1, T2 and T3 (as shown Figure 2.35), so these three signals are phase-adjusted, aligned and combined into the output signal by the rake receiver.

In order for the UMTS rake receiver to work, the delay separation between the different signals has to be more than one chip duration (0.26 µs). Delays shorter than one chip cannot be recovered by the rake receiver, and will cause inter-symbol interference.

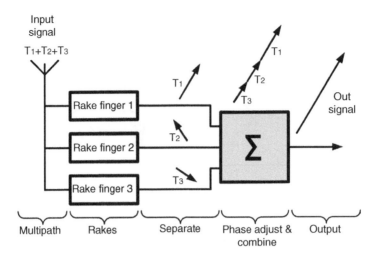

Figure 2.36   The rake receiver will be able to detect the different signals phases and amplitudes in a multipath environment. Separate phases are adjusted and the signals summed back to one signal

2.3.7   UMTS Signaling

UMTS Transport Channels

The UMTS radio interface has logical channels that are mapped to transport channels. These transport channels are then again mapped to physical channels. Logical-to-transport channel conversion happens in the medium access control (MAC) layer, which is a lower sublayer in the data link layer (layer 2).

Logical Channels

The logical channels are divided into two groups: control channels and traffic channels. The MAC uses the logical channels for data transport. The traffic channels carry the voice and data traffic.

Control Channels

- *BCCH-DL* – this channel broadcasts information to the mobiles for system control, system information about the serving cell and the monitored set (neighbor list). The BCCH is carried by the BCH or FACH.
- *PCCH-DL* – the PCCH is associated with the PICH. The channel is used for paging and notification information. The PCCH is carried by the PCH.
- *DCCH-UL/DL* – this channel carries dedicated radio resource control information, in both directions. Together with the DTCH, the DCCH can be carried by different combinations of transport channels: RACH, FACCH, CPCH, DSCA or DCH.
- *CCCH-UL/DL* – this channel's purpose is to transfer control information between the mobiles and the network. The RACH and FACH are the transport channels used by the CCCH.
- *SHCCH-UL/DL* – this channel is only used in UMTS TDD. It is a bi-directional channel, used to transport shared channel control information.

Traffic Channels

- *DTCH-UL/DL* – this is the bi-directional channel used for user traffic.
- *CTCH-DL* – this is a downlink-only channel, used to transmit dedicated user information to a single mobile or a whole group of mobiles.

Transport Channels

The user data is transmitted via the air interface using transport channels. In the physical layer these channels are mapped into physical channels. There are two types of transport channels, common and dedicated channels. Dedicated channels are for one user only; common channels are for all mobiles in the cell.

Dedicated Transport Channels

- *DCH-UL/DL, mapped to DCCH and DTCH* – this channel is used to transfer user data and control information, handover commands, measurement reports etc. to and from a specific mobile. Each mobile has a dedicated DCH in both UL and DL.

Common Transport Channels

There are six common transport channels:

- *BCH-DL* – this channel transmits identification information about the cell and network, access codes, access slots, etc. The BCH is sent with a low fixed data rate. The BCH must be decoded by all the mobiles in the cell, therefore relative high power is allocated to broadcast the BCH. The BCH is mapped into the PCCPCH (primary common control physical channel).
- *FACH-DL* – this channel transmits control information to the mobiles that are in service on the network. The FACH can also carry packet data. The FACH is mapped into the SCCPCH (secondary control physical channel).
- *PCH-DL* – this is the channel carrying the paging signals within the location area, which alert mobiles about incoming calls, SMS and data connections. The PCH is mapped into the SCCPCH.
- *RACH-UL* – this channel is used by the mobile when accessing the network with access bursts. Control information is sent by the mobile to the network on the RACH. The RACH is mapped into PRACH (physical random access channel).
- *Uplink common packet channel (CPCH)-UL* – this channel is used for fast power control, and also provides additional capacity beyond the capacity of the RACH. The CPCH is mapped into the PCPCH (physical common packet channel).
- *Downlink shared channel (DSCH)-DL* – this channel can be shared by many mobiles and is used for nontime-critical data transmission on the downlink, web browsing, etc. The DSCH is mapped to the PDSCH (physical downlink shared channel).

Physical Channels

The transport channels are mapped into physical channels.

- *CPICH* – this channel identifies the cell, used by the mobile for cell selection, and to measure the quality of serving radio link and the adjacent cells. The CPICH enables the mobile to select the best cell, and the CPICH plays a significant role in cell selection and cell reselection.
- *Synchronization Channel (SCH)* – there is both a primary and a secondary SCH. These channels are used by the mobiles to synchronize to the network.
- *Common control physical channel (CCPCH)* – there are two types of CCPCHs, the primary and the secondary:
  - *PCCPCH* continuously broadcasts the system identification (the BCH) and access control information;
  - *SCCPCH* transmits the FACH (forward access channel), and provides control information, as well as the PACH (paging channel).
- *PRACH* – this channel is used by the mobile to transmit the RACH burst when accessing the network.
- *Dedicated physical data channel DPDCH* – this channel is the user data traffic channel (DCH).
- *Dedicated physical control channel (DPCCH)* – this channel carries the control information to and from the mobile. The channel carries bidirectional pilot bits as well as the transport format combination identifier (TFCI). On the DL channel is included transmit power control and feedback information bits (FBI).
- *PDSCH* – this channel shares control information (DSCH) to all mobiles within the service area of the cell.
- *PCPCH* – this channel is for carrying packet data (CPCH).
- *Acquisition indicator channel (AICH)* – this channel is used to inform the mobile about the DCH is must use to connect to the Node B.
- *Paging indication channel (PICH)* – this channel indicates the paging group the mobile belongs to. This enables the mobiles to be able to monitor the PCH when its group is paged, and in the meantime the mobile can 'sleep' and preserve its battery.
- *CPCH status indication channel (CSICH)* – this DL channel carries information about the status of the CPCH. Can also be used for DL data service.
- *Collision detection channel assignment indication channel (CD/CA-ICH)* – this DL channel is used to indicate to the mobile if the channel assignment is active or inactive.

Power Allocation for Common Control Channels

In the network the power is allocated to the physical common control channels is defined. These power levels can be adjusted, but typical settings could be as shown in Table 2.2. Even with no traffic on the cell, 16% (−8.1dB) of the power resource is allocated for common control channels.

Measuring the UMTS Transmit Power

If you need to measure the power level at the base station antenna connector, you will need to know these settings. If the settings are like the settings above, you will measure the maximum

Table 2.2 Typical power levels for common control channels

Channel	Node B power	Activity duty cycle	BS power load
CPICH	10%	100%	−10 dB
CCPCH	5%	90%	−13 dB
SSCH	4%	10%	−14 dB
PSCH	6%	10%	−12.2 dB

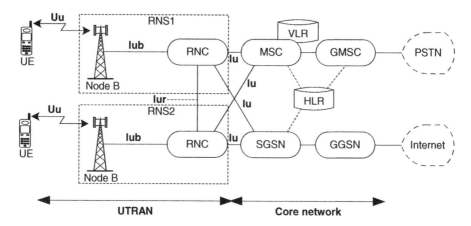

Figure 2.37 UMTS UTRAN and core network

power − 8.1 dB, so a 46 dBm base station will give a measurement of 37.9 dBm when there is no traffic on the cell. Most UMTS networks will allow you to preset a specific load in the cell, and then this reference can be used for power measurements.

2.3.8 The UMTS Network Elements

This is a short introduction to the elements of the UTRAN (UMTS terrestrial radio access network) and the key elements of the core network (as shown in Figure 2.37). This book is not intended as a detailed training book about the UTRAN and core network, but it is important that the indoor radio planner understands the general principles of the total network, especially regarding the UTRAN.

User Equipment

The mobile station is referred to as the user equipment (UE). This indicates that the UE is much more than 'just' a mobile telephone. The UE is a mobile data device (could be a mobile telephone). However, throughout this book, the UE is referred to as the mobile station, in order for the examples to be compatible with both UMTS and GSM.

Node B, Base Station

The base station in the UTRAN network is called node B. Node B consists of transceivers, processing modules that provide the 'channel elements' that service the users. The node B interfaces to the UE over the air interface, the Uu Interface. Interfacing node B to the RNC is the Iub interface.

Node B will also perform important measurements on the Uu interface, and report these measurement results (reports) to the RNC with regards to quality of the link. These are BLER, block error rate, and BER, bit error rate. They are needed in order for the RNC to be able to evaluate the QOS (quality of service), and adjust the power control targets accordingly. Also, the reception level of the mobiles is measured and reported, as well as the signal-to-noise ratio, etc.

Radio Network Controller

The radio network controller (RNC) is controlling the node Bs within its own system (RNS). For speech service, the RNC interfaces to the MSC. For packet-switched data service the RNC interfaces to the SGSN. The RNC is responsible for the load on the individual cells in the RNS, and handles admission (traffic) control, code allocation. Once a connection has been established between UE and node B, signaling to the elements higher up in the network is done by the RNC. This RNC is then referred to as the serving RNC, the SRNC.

The SRNC is responsible for handover evaluation, outer loop power control, as well as the signaling between the UE and the rest of the UTRAN. If the UE signal can be received (in soft handover) by other node Bs controlled by other RNCs, these RNCs are called DRNC, drift RNCs. The DRNC can process the UL signal from the UE and provide macro diversity, transferring the data via the Iub/Iur interfaces to the SRNC.

Radio Network Sub-system

One RNC with all the connected node Bs is defined as a radio network sub-system (RNS). A UTRAN consists of several RNSs; each RNC within the RNS is interconnected with an Iur interface.

- *Uu interface* – this is the radio interface of WCDMA, the air interface between the UE and the node B.
- *Iu interface* – this is the link between the UTRAN and the core network. The interface is standardized so a UTRAN from one manufacture will be compatible with a core network form another manufacture.
- *Iur interface* – the Iur interfaces the different RNCs; it interfaces data from soft handovers between different RNCs.
- *Iub* – between the node B and the RNC the Iub interface is used. Like the Iu interface, this is fully standardized so different RNCs will support different vendors' node Bs.

Core Network

The core network is not of interest of the radio planner on a daily basis, but a short description and overview is appropriate.

- *Mobile switching center* – the mobile switching center (MSC) controls the circuit switched connections, speech and real-time data applications for an active UE in the network. Often the VLR will be co-located with the MSC.
- *Gateway mobile switching center* – the gateway mobile switching center (GMSC) interfaces the MSC to external MSCs for circuit-switched data and voice calls.
- *Serving GPRS support node* – the serving GPRS support node (SGSN) switches the internal packet-switched data traffic.
- *Gateway GPRS support node* – the gateway GPRS support node (GGSN) switches the external packet-switched data traffic.

Visitor Location Register

The VLR is a database, with the location (location area) of all UEs attached to the network. When a UE registers in the network, the VLR will retrieve relevant data about the user (SIM) from the HLR associated with the SIM (IMSI).

Home Location Register

The HLR is a database that contains all the relevant data about the subscriber (SIM). This is the subscription profile, roaming partner and current VLR/SGSN location of the UE.

2.4   Introduction to HSPA

It is not within the scope of this book to describe all the details about HSPA. Only the issues that important for maximizing the HSPA performance inside buildings will be covered. Please refer to Reference [5] for more details about HSPA.

2.4.1   Introduction

Seamless mobility for voice and data service is of paramount importance for the modern enterprise. In particular, service for mobile telephones and data terminals is important, as most professionals leans towards 'one person, one number', also referred to as the mobile office. The service offering and quality should be seamless, no matter whether you are at home, on the road or in the office.

   Offices with a heavy concentration of high-profile voice and data users, in particular, pose specific challenges for high-performance in-building data performance using HSPA. HSPA/Super-3G is an alternative to Wi-Fi service. Mobile 3G networks are deploying HSPA on their outdoor networks, and full mobility is now an option for high-speed data. Providing sufficient in-Building coverage on HSPA poses a major challenge for the mobile network operators due to the degradation of the radio service when covering indoor users from the outdoor network. The data users with the highest demand-to-data service rate and speed are typically located indoors, so a special focus on how to solve the challenge is important.

2.4.2   Wi-Fi

Why will the users prefer to use 3G/HSPA rather than Wi-Fi? Wi-Fi is accessible in many places, it is 'free' and the data speed is claimed to be high, in excess of 50 Mbps, so why not just use Wi-Fi to provide mobile data indoors?

The Challenges for the Users, Using Wi-Fi

Lack of Mobility
There are no handovers on Wi-Fi, so users cannot roam around in the building without dropping the service. However, developments are ongoing to make handovers possible between Wi-Fi access points, and even to other systems.

Hassle to get Wi-Fi Service
It is not very user-friendly to obtain Wi-Fi service. It is often expensive and you have to get a specific user access code for every Wi-Fi provider in each building.

Wi-Fi Service is Slow
The data rate on the Wi-Fi–air interface is high, in excess of 50 Mbps. However, the actual user data rate is not limited by the air interface but by the ADSL backhaul to the Wi-Fi access point. Typical user data rates on public-access Wi-Fi services in hotels and airport are often lower than 400 kbps, far below the 54 Mbps on the typical Wi-Fi air interface.

No Guarantee of Service
Wi-Fi uses free open spectrum, so there is no guarantee that other Wi-Fi providers will not deploy Wi-Fi service in the same areas, using the same or adjacent frequencies and degrading the current Wi-Fi service. Therefore there is no quality of service guarantee using Wi-Fi.

Wi-Fi Over Indoor DAS

Why not distribute Wi-Fi over the DAS? To understand why it is not advisable to distribute Wi-Fi over the same distributed antenna system (DAS) that distributes the mobile service for GSM and UMTS, you need to understand the basic characteristics of Wi-Fi, refer to Figure 2.38.

- *Wi-Fi is distributed capacity* – the basic concept of Wi-Fi is to distribute the data capacity out to the location of the data users by the use of distributed base stations, Wi-Fi access points. By using the strategy of distribution of the Wi-Fi data capacity and deploying the Wi-Fi access points locally, combined with the limited coverage range from these access points, you will be able to support high data rates in the building. This is exactly the same

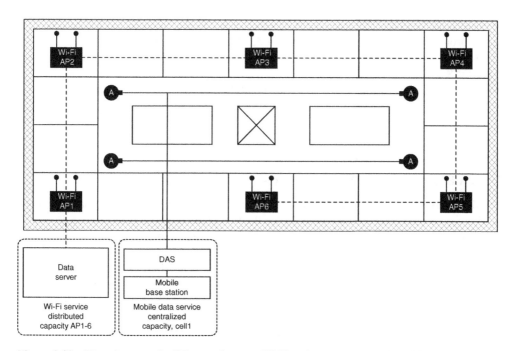

Figure 2.38 There is a generic difference between Wi-Fi and mobile data service. Wi-Fi is distributed capacity via access points. Mobile data service uses centralized capacity from the base station

principle as frequency reuse in a cellular network. If you have limited frequencies on a GSM roll-out, you do not just deploy one tall macro site in the center of a city and blast power on a limited number of frequencies. This is like Wi-Fi only having limited number of frequencies; you need to distribute the GSM base stations to raise capacity. The same principle applies to Wi-Fi: you need to distribute the access points to tender for the capacity need.

• *Hotspot data coverage* – Wi-Fi is designed to cover the hotspot data need for specific locations in the building. It is only possible to deploy three or four Wi-Fi carriers in the same area without degrading the service, owing to adjacent channel interference.

Why Not Distribute Wi-Fi Over the DAS?

It does not make sense to combine the RF signal of one to three Wi-Fi access points (APs) and distribute it via the DAS, owing to the inherent difference between GSM/UMTS/HSPA and Wi-Fi: Wi-Fi is distributed capacity via APs, and GSM/UMTS HSPA is centralized capacity, distributed via DAS.

If you distribute the Wi-Fi RF signal over the DAS you face these limitations:

• Limited number of carriers in the same cell, limited capacity.
• Not enough capacity for the data users in most buildings.
• The concept of Wi-Fi is distributed capacity, not centralized in a DAS like the capacity of mobile services such as UMTS/HSPA.

- Wi-Fi is TDD; with more than one AP in the same cell the effect is that the APs are not synchronized and the transmission will collide, degrading the performance.
- Wi-Fi APs use antenna diversity; this is important to boost the performance, which is lost in the DAS.
- The desired coverage area for the Wi-Fi and mobile signal might not correlate.
- The sector plan on the mobile system will dictate the capacity on Wi-Fi.

2.4.3   Introduction to HSDPA

HSDPA High-speed Downlink Packet Access

HSDPA is a high-speed downlink data service that can be deployed on the existing 3G network infrastructure. The operators can deploy the first phase on the existing UMTS R99 network, using the same UMTS CH (as shown in Figure 2.39) and use the power headroom not utilized by the UMTS traffic.

This provides the mobile operator with a deployment strategy that can support the need for high-speed downlink data rates, at a relative low deployment cost. Many operators will prefer to use a separate CH for HSDPA, because loading the existing UMTS R99 carrier with HSDPA will cause the network to run constantly on a higher load rate, increasing the noise on the existing UMTS CH, degrading the UMTS capacity.

Why will the operators prefer to use HSDPA?

- Typically the highest data load is on the downlink.
- HSDPA will have a trunking gain of the radio capacity.
- There is no need for distributed APs like with Wi-Fi.
- Operators can use their existing UMTS network to launch HSDPA.
- HSDPA is more spectrum-efficient compared with EDGE and UMTS.
- There is higher spectrum efficiency, more data per MHz.
- There are fewer production costs per Mb.
- There are higher data speeds, up to 14 Mbps, with indoor DAS systems.
- Mobile operators can compete with Wi-Fi using HSPA.

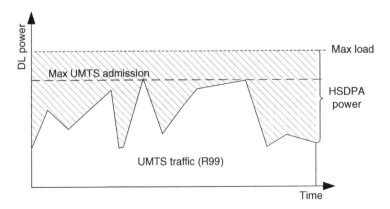

Figure 2.39   HSDPA deployed on the same RF channel as UMTS

HSDPA Key Features

HSDPA can perform high downlink data rates, provided that the radio link quality is high.

- Downlink data Rates up to 14.4 Mbps.
- No soft handover, so pico cells overlapping in the same building will produce 'self interference' (see Figure 4.38).
- Can use the same channel as UMTS by utilizing the 'power headroom' for HSDPA (as shown in Figure 2.39).
- Typical operators will launch HSDPA on a separate carrier, to minimize noise increase on the UMTS service.
- HSDPA is a downlink data service; users will use UMTS on the uplink until HSUPA is deployed.
- For the uplink data transmission from the mobile, UMTS R99 is used until HSUPA is launched in 2008 to cater for higher uplink data rates.

Indoor HSDPA Coverage

Why do you need indoor active DAS systems to have high quality HSDPA?

- The data speed is related to the SNR. A good quality radio link is needed to produce higher speeds.
- The macro layer will use excessive power to reach indoor users, causing interference and degradation of the macro layer. Only in the buildings few hundred meters from the outdoor base station will it be possible to service indoor users with HSPA coverage. In an urban environment it is not likely to exceed 360 kbps in the major portion of the area.
- Production cost will be lower per Mb for the operators, using indoor systems where the traffic is.

2.4.4   Indoor HSPA Coverage

The best performance on HSPA is achieved by deploying indoor DAS. The data speed is related to the SNR. A good radio link, with a high SNR, is needed to produce higher speeds. Using the outdoor macro layer causes excessive power to reach indoor users, causing interference and degradation of the macro layer. Only in buildings a few hundred meters from the outdoor base station will it be possible to service indoor users with acceptable HSDPA coverage. In urban environments it is not likely to exceed data speeds of 360 kbps indoors, without the use of dedicated in-building systems.

It is possible to lower the production cost per Mb for the operators by deploying indoor DAS to service HSDPA, as dedicated in-building systems help offload the macro network.

The Impact of HSUPA

HSUPA (high-speed uplink packet access) will provide the users the possibility of high-speed data rates on the uplink. HSUPA relies on the mobile's ability to reach the network and will be sensitive to bad macro network quality and to the high loss of passive in-building systems.

Therefore, an active DAS system will be essential for providing a high-quality service. The low noise of the active DAS and the high power delivered at the DAS antenna will give the highest possible data rate performance for HSUPA.

The Choice of DAS Distribution for HSPA

HSPA performance is very sensitive to the quality of the radio link, the dominance and the isolation of the indoor signal. How to insure this is explored in Chapters 3 and 4. The DAS that will perform the best is the DAS that will provide the highest downlink radiated power at the antenna, and the lowest possible noise figure. For medium to large buildings it will often be some form of active DAS that will perform the best HSPA service, owing to the attenuation of passive DAS. Be careful and design the DAS solution to maximum isolation; good indoor design strategy is very important for HSPA performance (refer to Chapter 5 for more details). Passive coax solutions do not perform at the higher data rates. Whereas using passive DAS was sometimes acceptable for 2G mobile services for voice only, the future is data, and the high losses on the coax cables will degrade the data speed.

Advantages of Using HSPA

Providing high-speed mobile data via HSPA distributed indoors via a DAS has several advantages for the mobile operator:

- *Better HSPA service* – users on 3G (HSPA) will have coherent coverage; users can roam from the outdoor network to the indoor DAS with full mobile data service.
- *Laptops are HSPA ready* – new laptops come with integrated 3G cards. Like the early days of Wi-Fi, users on HSPA will initially have to use HSPA plug-in cards in their PC, but the new generation of laptops comes with pre-installed HSPA support.
- *Less production cost per Mb* – the production cost on HSPA is significant lower than for EDGE and UMTS data service, boosting the business case for the operators.
- *Seamless billing* – the user does not have to purchase individual Wi-Fi service subscriptions in the individual buildings from individual service providers, often at a high cost. Users on HSPA can get all the data billing via their normal mobile subscription.
- *Easy to connect with HSPA* – users do not need to scan for Wi-Fi and try to connect at various Wi-Fi service providers. Operators will have automatic algorithms enabling the users just to 'click and connect'.
- *Better than Wi-Fi* – HSPA service with indoor DAS can easily outperform Wi-Fi. The DAS will distribute the HSPA service throughout the building, supporting the highest possible data rate. 3G and HSPA will service the mobile data users with higher data rates. The mobile system is not limited by the relative slow ADSL backhaul like the Wi-Fi AP. Typically HSPA will provide DL data rates exceeding 6 Mbps, when using indoor DAS.

Table 2.3 clearly shows how competitive HSDPA can be compared with Wi-Fi service. This example is a shopping mall where a mobile operator outperforms the existing Wi-Fi provider, by deploying HSPA DAS antennas in the Wi-Fi hotspots.

Table 2.3 Typical HSDPA data rate from omni antenna radiating 10 dBm in an open area of a shopping mall

HSDPA speed	Coverage radius
480 kbps	47 m
720 kbps	37 m
1.8 Mbps	33 m
3.6 Mbps	29 m
7.2 Mbps	20 m
10.7 M	15 m

2.4.5   Indoor HSPA Planning for Maximum Performance

The highest mobile data rates are typically needed only in hotspot areas of the building. Therefore you must apply a strategy to our indoor design that is based on the hotspot areas of the building. Locating DAS antennas in these areas will provide the highest possible data rate; the rest of the building can then be planned around these antenna locations. This principle is described in Section 5.4.4. Hotspot areas will typically be areas where mobile data users are able to sit down and use their PC or mobile data device.

Hotspot data areas in buildings will typically be:

Office buildings

- Conference areas.
- Meeting rooms.
- Executive floors.
- Wi-Fi zones.

Airports

- Business lounges.
- Sitting areas at gates.
- Restaurants and cafes.
- Meeting rooms.
- Staff meeting rooms.
- Wi-Fi zones.
- Administration areas.

Hotels

- Conference areas.
- Meeting rooms.
- Executive floors.
- Wi-Fi zones.
- Business rooms.
- Lobby.
- Restaurants.

Shopping malls

- Food court.
- Sitting areas.
- Cafes and restaurants.
- Staff meeting rooms.
- Administration areas.

2.4.6 HSDPA Coverage from the Macro Network

It is possible to provide sufficient levels of RF signal inside buildings to provide a strong UMTS signal and to some extent HSPA coverage, provided that the serving macro base stations are within few hundred meters of the building, and have direct line-of-sight. In reality, however, only few a buildings will fall in that category. However, apart from the signal level on UMTS, there are other important factors to consider:

UMTS Soft Handover Load

UMTS service provided from the macro layer might be acceptable seen from the user's perspective. In Figure 2.40, three macro cells are providing good indoor coverage, indicated by the three shades of gray.

The challenge for the UMTS network is the extensive soft handover area; the white area seen in Figure 2.41. This large soft handover area will increase the load on the network for

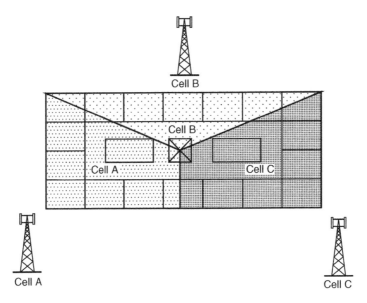

Figure 2.40 Three outdoor macro base stations near the building are providing good UMTS signal level inside the building. Each of the outdoor cells is covering about a third of the floor space

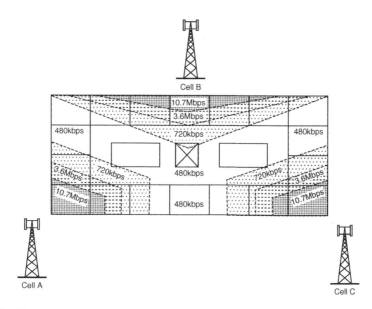

Figure 2.41 The three macro base stations give full coverage, but a larger portion of the indoor floor space is covered by more than one serving cell, the gray area

Table 2.4 HSDPA link budget example

Service	RSSI @ Cell Edge [dBm]	90% RSSI @ Cell Edge [dBm]	C/I [dB]	90% C/I confidence [dB]
DL: 120k (QP-1/4-1)	−89.7	−79.3	−14.5	−4.1
DL: 240k (QP-1/2-1)	−85.2	−74.8	−10	−0.4
DL: 360k (QP-3/4-1)	−82.2	−71.8	−7	3.4
DL: 480k (QA-1/2-1)	−80.2	−69.8	−5	5.4
DL: 720k (QA-3/4-1)	−75.7	−65.3	−0.5	9.9
DL: 3.6M (QP-1/2-15)	−74.2	−63.8	1	11.4
DL: 5.3M (QP-3/4-15)	−70.7	−60.3	4.5	14.8
DL: 7.2M (QP-1/2-15)	−69.2	−58.8	6	16.4
DL: 10.7M (QP-3/4-15)	−64.7	−54.3	10.5	20.9

UMTS traffic, for the users in the part of the building serviced by more than one cell. But for HSPA traffic the data throughput wil be degraded caused by C/I (co-channel interference), because the cells are operating on the same HSDPA frequency.

Degraded HSDPA Performance

The UMTS service will still be acceptable, but the price on the network side is the large soft handover area for about 60% of the users inside the buildings as HDPA does not use soft handover, the three serving all HSDPA cells are on the same frequency and there is a large impact on the *C/I*. In Table 2.4, you can see an example of a HSDPA link budget. In order to provide

HSDPA service at high confidence level, you need good isolation between the serving cell and the adjacent cell due to the strict *C/I* requirements.

In the example shown in Figure 2.41, only 20% of the building has a *C/I* better than 5 dB, the gray areas. Only in this area will the DL data rate be better than about 500 kbps. In the major part of the building the data rate will be lower than 480 kbps.

Solution

The only solution for a consistent and high-performing indoor HSPA service is to provide isolation between the intended serving cell and the interfering HSDPA cells from the outdoor network. In order to be able to provide HSDPA service at the higher data rate, the only solution is an indoor DAS system. The challenge is to dominate the building with the indoor coverage, and at the same time avoid leakage from the indoor DAS system. This can be an issue, especially in buildings with high signal levels from outdoor macro cells.

A viable solution is to deploy directional antennas at the edge and corners of the building and point them towards the center of the building. Using this strategy, the total indoor area is dominated by the indoor cell as long as you dominate at the border. The reason is that the indoor signal radiated by both the corner- and edge-mounted antennas and the outdoor cell will have the same signal path and attenuation from the edge of the building throughout the whole indoor area. Dominating the border of the building will give total dominance, and will at the same time minimize leakage into the macro network.

The solution is described in Section 3.5. In order to archive high HSPA data rates, good isolation between the cells is needed. The challenge with this strategy is the actual installation of the antennas in the corners of the building. Normally there will be limited access to cable ducts and installation trays along the perimeter of the building. This is even more of an issue for traditional passive DAS based on heavy and rigid coax cables. However, using an active DAS that relies on lightweight, flexible IT-type cabling to reach the antenna point, the installation of these antennas is less of a challenge.

2.4.7   Passive DAS and HSPA

Traditional passive DAS systems based on coax cables will in many case have too high an attenuation to provide sufficient HSPA data service inside buildings. Passive DAS can perform on HSPA provided that the attenuation between the base station and the indoor antenna is less than 20 dB. In practice, using passive DAS for indoor HSPA solutions is only possible for smaller buildings that can be covered by six to eight indoor antennas. For larger buildings the HSPA performance will be compromised due to the high losses of the coax cables, splitters, tappers and other passive DAS components.

HSDPA on Passive DAS

In Figure 2.42, the HSDPA service range from the indoor DAS antenna in a typical office environment is shown. The graphs represent benchmarking of the HSDPA performance of passive DAS with 20–30 dB of attenuation between the base station and the DAS antenna. This is

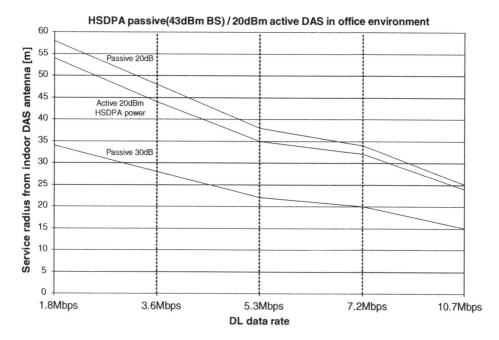

Figure 2.42   HSDPA performance passive and active DAS

compared with the HSDPA performance of the active DAS from Section 4.4, with +20 dBm output power.

The impact from the attenuation of the passive DAS is evident. Only the passive DAS with 20 dB of attenuation performs slightly better than the active DAS owing to the DL power of the +43 dBm base station. This makes passive DAS only applicable for HSDPA solutions in small buildings with less than six to eight antennas. However, the installation challenges associated with the passive DAS remains. Therefore it will be hard to draw up the correct antenna layout as shown in Section 3.5. The typical office building implemented with passive DAS and corner-mounted antennas will have losses exceeding 30–35 dB, limiting the HDSPA service range severely.

HSUPA on Passive DAS

The degrading impact on the HSUPA performance by the attenuation of the passive DAS is even more evident than the HSDPA degradation. The reason for this can be explained by basic cascaded noise theory and practice. Any loss prior to the first amplifier in the receiver will degrade the noise figure of the system, in this case the HSUPA receiver in the base station.

Figure 2.43 clearly shows the impact on the HSUPA performance, caused by the passive losses. Even a 'low loss' passive DAS with 'only' 30 dB of attenuation cannot perform at more than 700 kbps just 12 m from the indoor antenna. The active DAS will perform at about 1 Mbps up to 27 ms from the antenna, at the same location in the same environment.

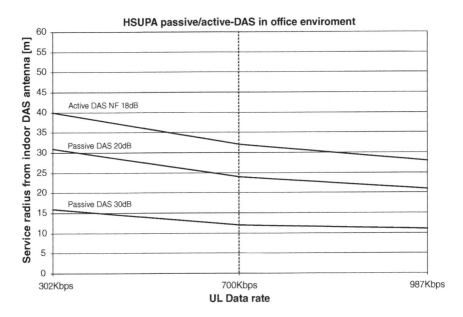

Figure 2.43   HSDPA deployed on the same RF channel as UMTS

2.4.8   Short Introduction to HSPA+

Since the introduction of HSPA in 3GPP Release 6, 3GPP has introduced Releases 7 and 8 designed to accommodate the never ending need for increased data speed.

From an indoor planning point of view the major changes have been the introduction of higher order modulation, utilizing 64QAM (descried in Section 2.5) where the air link enables an advanced antenna technology with MIMO support – using multiple antennas in the mobile terminal and in the DAS (described in Section 2.6). MIMO will have a practical impact on how we design and implement indoor DAS solutions, as described in Chapter 4.7

HSPA+ is a significant step in the evolution towards LTE (see Section 2.7).

2.4.9   Conclusion

The active DAS concept for distribution of HSPA signals has been shown to be the best for provision of high-speed data services. For both HSDPA and HSUPA, the conclusion is clear – having the DL amplifier and the UL amplifier as close as possible to the indoor DAS antenna will boost HSDPA and HSUPA performance to their maximum. The concept entails using distributed amplifiers and locating them at the antenna points, compensating for loss back to the base station. Delivering DL power at the antenna and minimizing the UL noise figure is the best solution. The calculations and real-life measurements and implementations back up the theory. Chapter 7 presents the noise calculations for this concept.

Most of the impact on our daily work as indoor radio planners will come from MIMO; we will need to plan for dual antenna installations inside buildings, at least in the hotspot areas where users are likely to use and expect the best data services. MIMO is covered in more detail in Section 2.6 and the impact on building design is addressed in Section 4.7.

2.5   Modulation

This chapter is intended to give a short introduction to the various modulation schemes used in advanced mobile communication. This chapter serves merely as an introduction, for more detailed description and in-depth mathematical descriptions please refer to [7] and [8].

The ever increasing demand on higher and higher data throughput over the air interface is a never ending challenge. As described in Section 2.6 new advanced antenna configurations have been applied to support 3G+/HSPA+ and LTE/4G in order to maximize that throughput over the air interface.

2.5.1   Shannon's Formula

Shannon established the theoretical maximum data rate over the radio channel, known as the channel capacity. By applying Shannon's RF channel calculation to a radio channel we can calculate the channel capacity (C):

$$C = BW \times Log_2(1 + (S/N))$$

C is the channel capacity
BW is the available bandwidth
S is the Signal Power
N is the Noise Power

From Shannon's formula we can see that there are two parameters limiting the capacity of the radio channel. The limiting factors are the bandwidth of the channel (BW) and the quality of the channel, the Signal to Noise ratio (S/N, SNR).

Without diving into the deep mathematical details, we can see that in order to increase the throughput of the data by increasing C, the capacity of the channel of a given bandwidth we will need to increase the Signal to Noise ratio (and by applying multiple 'parallel' channels, using MIMO as described in Section 2.6). However the high demand to S/N ratio will still challenge us, because we will face interference limited scenarios in our network to a large extent.

The scarce recourse of spectrum; the bandwidth of the radio channel has to be utilized to a maximum, and in order to increase the efficiency we can apply higher order modulation to the channel as briefly described in the following chapter.

Modulation

In order to raise the efficiency of the throughput in the radio channel, higher order modulation schemes can be applied to the channel, implying that the system will 'encode' extended information onto the channel. Thereby including additional signaling status alternatives for the information on the channel, and utilizing this for more information bits to be transmitted per modulation symbol. Let us have a look at the principle.

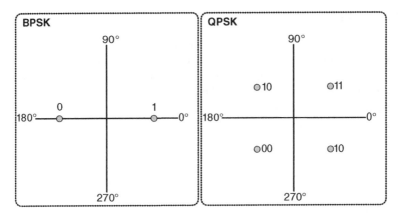

Figure 2.44 Illustrating the constellation diagram of the phases of the modulation symbols in BPSK and QPSK used in WCDMA

2.5.2 BPSK

The simplest form of digital modulation is BPSK, Binary Phase Shift Keying as illustrated in Figure 2.44. BPSK uses two phases in the channel separated by 180°.

As we can see in the 'constellation diagram' (Figure 2.44) for BPSK the information can take two different statuses in the channel '0' or '1'. This relative simple channel is very robust decoding is pretty straightforward in the receiving end, therefore this channel is pretty resistant to noise and phase shift caused by reflections and multipath (EVM), but is not very a very efficient channel in terms of throughput efficency, giving us a bit-rate of only 1bit/symbol. However, this is the 'price' for this sturdy channel; low efficiency.

2.5.3 QPSK – Quadrature Phase Shift Keying

QPSK is a more advanced form of modulation when compared with BPSK, and is the modulation type used for WCDMA (UMTS/3G). As we can see in the 'constellation diagram' (Figure 2.44) the QPSK channel in this example is encoded with four different phases/amplitudes, utilizing four points in the constellation diagram. We can see in Figure 2.44 that with four phases QPSK can encode the channel with two bits per symbol, thus doubling the capacity per symbol when compared with the BPSK modulated channel.

2.5.4 Higher Order Modulation 16-64QAM

As just described above and shown in Figure 2.44; QPSK can allow for 2 bits of information to be communicated with each one modulation-symbol interval. QAM employs both amplitude and phase in a more advanced way, so there are more states in a QAM constellation versus a PSK constellation. By extending the principle and use more phases/amplitudes to 16QAM modulation (Figure 2.45), 16 different signaling states are possible, as seen in the constellation diagram for 16QAM in Figure 2.45.

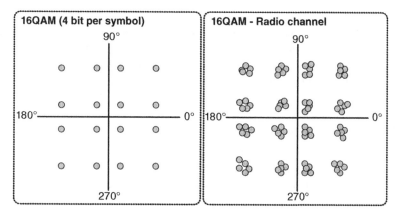

Figure 2.45   Higher order of QAM modulation, 16QAM used in HSPA and LTE, showing on the left the idea location of the constellations, but the properties of the radio channels will impact the accuracy as shown on the right (EVM)

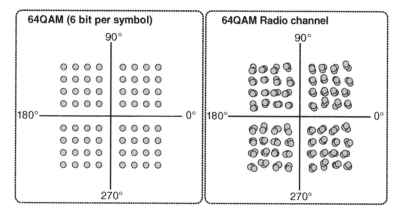

Figure 2.46   Highest order of QAM modulation currently in use, 64QAM used in 3G/HSPA and 4G/LTE, showing on the left the idea location of the constellations, but the properties of the radio channel (phaseshifts) will impact the accuracy as shown on the right. The relative short distances between the symbols used in 64QAM demands a really good RF channel for sufficient constellation accuracy, and will yield high throughout under good conditions

By applying 16QAM modulation the system will allow for 4 bits of information to be communicated for each symbol interval. This can even be extended to 64QAM (Figure 2.46), with 64 different signaling alternatives, allows for 6 bits of information to be communicated for each symbol interval. At this point in time 64QAM is the highest order modulation scheme used in HSPA+ and 4G/LTE.

These higher order modulation schemes are a very efficient use of the channel bandwidth, but it demands a really good 'clean' radio channel in order to perform 64QAM. The use of higher-order modulation offers the possibility to carry higher data speeds within a given limited radio channel bandwidth.

The price for this higher radio channel bandwidth utilization is reduced robustness to interference and noise in the radio channel. According to Sharron's formula (Section 2.5.1) 16QAM and 64QAM modulation will require a higher signal to noise ratio, $Eb/N0$ in order to perform at a maximum. This increased demand to the quality of the radio channel, is owing to the fact that the channel need to be more 'accurate' in order for the receiver to decode the many statuses in the constellation of the phases/amplitude of the signal. The radio channel will need to have a high signal to noise ratio and high phase accuracy – a good EVM.

2.5.5   EVM Error Vector Magnitude

When a QAM modulated signal like 16QAM or 64QAM is transmitted over the radio channel, there will be multipath – reflections that will distort and skew the phase/amplitude so the ideally phased/placed constellations of the information bits will be somewhat offset by the radio channel when they arrive at the receiving end, this is shown in Figure 2.45 and Figure 2.46 (right side of Figures).

It is clear that the higher order modulation; 64QAM (Figure 2.46) will need a more accurate constellation in the receiving end of the radio channel than the 16QAM signal (Figure 2.45) simply due to the shorter 'distances' between the states of the constellation, in order to the 'spill' one symbol over to the adjacent. The higher order 64QAM has a much closer proximity of the statuses in the constellation diagram, and one could easily imagine some 'spillage' between the statuses if the decoded symbols are not placed correctly in the constellation (Figure 2.46).

The higher order the modulation the better the quality of the radio channel will need to be, the quality of the radio channel will need to very good be in order for the 64QAM modulation to be effective, and if the quality of the radio channel insufficient, then the receiver would not be able to decode the constellation successfully in the receiving end. So the quality of the radio channel plays an important role in the performance of the higher order modulation schemes, and so do active equipment and nonlinear performing elements in an active DAS such as amplifiers and other signal processing elements.

The measure of the 'accuracy' of the placement of the constellation elements, compared with the ideal location in the constellation is measured in EVM; Error Vector Magnitude and the principle can be seen in Figure 2.47.

It is not within the scope of this book to describe in detail the deep mathematical principle behind measuring EVM, refer to [8] for more detail.

The Active DAS Could Compromise the Throughput

It is important to make sure that all the active DAS equipment used and implemented in your solutions for HSPA and 4G/LTE comply with the EVM specification of 3GPP in order not to compromise the data throughput of the DAS. Therefore always make sure to ask the vendor for a EVM compliance test document for all active elements. You will also need to be sure you do not overdrive these active elements, like active DAS or repeaters – this will degrade the EVM, causing a low data throughput in your DAS system. You need to assure good liniarity and transparency of the DAS.

2.5.6   Adaptive Modulation, Planning for Highest Data Speed

Obviously the higher the modulation, the more sensitive the constellation will be, due to the dense cluster of symbol locations (Figure 2.46).

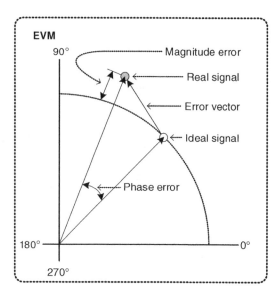

EVM

90°

Magnitude error

Real signal

Error vector

Ideal signal

Phase error

180°   0°

270°

Figure 2.47 EVM, Error vector magnitude, the difference between the ideal symbol location in the IQ constellation diagram, and the real location, that has been skewed by the radio channel. This skewing has higher impact the higher order of modulation due to the proximity of the statuses

Depending on the quality of the radio link, the 3G/4G system will adapt the modulation scheme used on the radio channel automatically in order to maximize the data throughput according to the actual radio channel quality. The quality of the radio channel is evaluated continuously and the lower the distortion of the location of the symbols is, due to reflections, etc. And the better the signal to noise ratio on the radio channel the higher the order of modulation will be used. This is called modulation.

An example of how adaptive modulation affects the coverage area of an indoor antenna can be seen in Figure 2.48 where typically, the link will perform at 64QAM close to the antenna (or antennas if using MIMO).

In the example in Figure 2.48; the area close to the antenna has good radio quality, and the modulation selected is 64QAM. At a greater distance the link will be downgraded automatically, first to 16QAM than QPSK – due to the degrading EVM of quality.

Maximizing the Throughput of the Radio Channel Will Benefit the Business Case

By applying adaptive modulation to the mobile system we maximize the capacity throughput of the spectrum and ensure that we always utilize the radio channel to a maximum. This benefits the mobile operator in terms of lower production cost per Mb and increased revenue.

This provides yet another good reason to consider the approach of 'hotspot' antenna planning when deploying antennas inside buildings, meaning that you would optimize the antenna locations to the areas where people are likely to use data services the most, like the food court in a shopping mall, business lounges and seating areas in the airports, conference venues in hotels, etc. See Section 5.4.1 for more detail on hotspot planning of antenna locations. The point is to plan the location of the antennas inside the building so that the system will perform 64QAM in these hotspot areas with the most data load, getting most

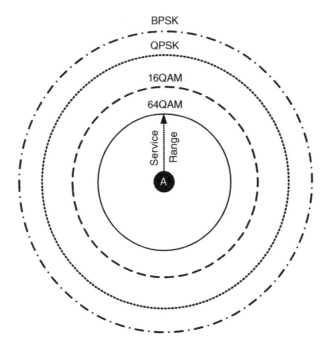

Figure 2.48 Adaptive modulation, the system will automatically adapt and use the highest possible modulation rate, thus optimizing the data speed throughout the cell. The effect makes it obvious to place antennas nearest the foreseen high load areas with many data users inside the building, maximizing the data speed for the users in the building

value and throughput for the investment, thus lowering the production cost by a more efficient use of the network when operating at the maximum data rates for the network operator.

2.6 Advanced Antenna Systems for 3G/4G

Advanced Antenna System – Introduction

Traditionally we consider a radio channel to have one input and one output and we normally consider this as one 'radio channel' – only one connection between the mobile and the network, and one in reverse (separated in Frequency FDD, or in time TDD).

For macro deployment in mobile antenna systems, we sometimes add an extra antenna on the base station uplink, traditionally implemented via a cross polarized antenna – one antenna enclosure with two input/outputs. The base station connects the combined Tx/Rx to one port and the RxDiversity to the other port. The purpose of receive diversity is to provide better resistance against fading on the UL, and the Base Station uses both signals in micro-diversity processing. Simple algorithms such as maximal ratio combining can combine the two signals optimally to improve the overall signal quality. Selection diversity is typically used for macro diversity processing such as with soft handover in CDMA networks.

This Receive diversity serves the purpose of giving better 'fading resistance' and will typically improve the fading margin on the UL by 3–5 dB, dependant on the environment (best in high multipath environments), and antenna separation.

The motivation for the improvement of the UL is that the mobile network will in many cases be UL limited (for macro deployment).

The use of diversity makes little sense in indoor DAS for several reasons, one being that typically indoor DAS has been DL limited, and the shear impact of a double antenna system was pointless (separated by 10 λ).

In order to maximize the throughput over the radio link for 3G/4G a more clever way of combining the signals from the multipath and scattering created by the local environment has been applied – yet again pushing the envelope and challenging what we thought was possible to drive over the air interface in terms of data throughput.

2.6.1 SISO/MIMO Systems

It is not within the scope of this book to provide a detailed, in-depth description and evaluation of MIMO systems, especially the mathematical background.

This is after all a practical book focusing more on the implementations out there in the real world. However, we should have a quick look on the basics with some insight as to how these advance antenna principles are being utilized and implemented in practice.

MIMO antenna systems will play an important role in DAS design, and are needed in order to pave the way for the highest possible wireless data speeds.

Inputs and Outputs

In the typical specifications, the terms 'input' and 'output' are used to describe the medium between the transmitters and receivers, this is known as the 'channel', therefore a base station with two transmitters provides two inputs to the channel, the 'MI' (Multiple Input) part, and a mobile with two receiver inputs can decodes two outputs from the channel, the 'MO' (Multiple Output) part.

This, however, is only valid if the data stream transmitted and received on both paths is individual, and is not merely a copy of the same data on both channels/streams. For MIMO, it is the channels that need to be independent and not necessarily the data streams. Diversity MIMO such as SFBC and STBC transmits redundant data in each path.

2.6.2 SISO, Single Input Single Output

The SISO radio channel (Figure 2.49) is what we may consider the basic radio channel and transmission mode for most radio based systems. There is one input and one output in the radio channel – hence 'Single Input Single Output', this SISO radio channel serves as a benchmark when evaluating performance gain in terms of throughput when using more advanced antenna systems.

Figure 2.49 Single Input Single Output (SISO), the standard radio channel as we (used to) know it

2.6.3 SIMO, Single Input Multiple Output

The SIMO configuration of the radio channel (Figure 2.50) should look fairly familiar to most radio planners, and is also known as receive diversity; the input to the channel is a single transmitter signal that feeds two receiver paths a00 and a01. Depending on the multipath fading and the correlation between the two receiver antennas gain is achieved in the form of fading resistance. The gain of this method depends greatly on the multipath environment and how efficient the receiver is in utilizing the micro-diversity enviroment. The gain using SIMO is better fading resistance thus giving better average signal to noise performance, ultimately better data throughput on the downlink only – provided that the data throughput is limited by the signal to noise ratio, as such there is no increase in the data throughput of the radio channel itself. There is an increase in the data throughput of the channel with micro-diversity as compared to SISO. However, the maximum achievable data throughput does not change.

2.6.4 MISO, Multiple Inputs Single Output

The MISO configuration of the radio channel (Figure 2.51) is a type of transmit diversity. It is used for LTE, Space Frequency Block Coding (SFBC) and is applied to the radio system in order to make the radio channel more robust and add more fading resistance. The two transmitters will send the same user data stream, but in different applied coding, thus enabling the receiver to decode the information.

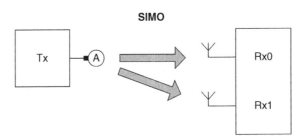

Figure 2.50 Single Input Multiple Output (SIMO) radio channel access mode, using one input and two outputs (Receive diversity)

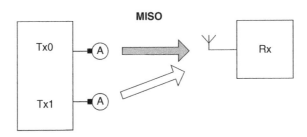

Figure 2.51 Multiple Input Single Output (MISO) radio channel access mode

2.6.5 MIMO, Multiple Inputs Multiple Outputs

The MIMO configuration shown in Figure 2.52 and Figure 2.53 uses 2 × 2 MIMO; two independent transmit signals and two separate receivers will pick up and decode the individual date streams. MIMO represents multiple individual, parallel data streams that are carried on the air interface.

The decoding of both data streams (2 × 2 MIMO) is possible thanks to scattering (short reflections) introduced by the local clutter, the clever mathematical separation of the signals using channel estimation techniques.

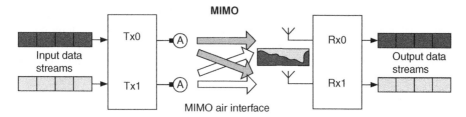

Figure 2.52 MIMO configuration, by using two transmitters and two receivers with independent data streams high data speeds on the air interface is possible

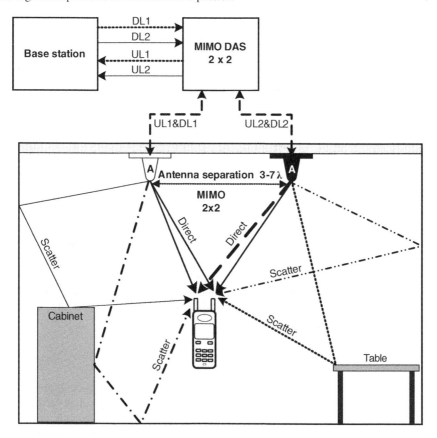

Figure 2.53 A typical Indoor MIMO environment, the scattering of signals of nearby objects inside a typical room, like tables, cabinets, walls etc. will be enough to provide the base of a MIMO radio link

The MIMO receiver applies clever matrix mathematics to separate the individual channels thanks to the scattering of the RF environment that provides the needed 'separation' of the radio channels (links).

In principle, higher orders of MIMO can be applied, so it would also be possible to consider 4 × 4 MIMO, four transmitters and four receivers on the same radio channel. This, however, raises some practical challenges when it comes to the implementation of four antennas on both the DAS and the mobile. MIMO will require an environment that has rich scattering (short reflections); the scattering will create multiple signal paths and thus increases coverage or capacity. This scattering creates orthogonal independent channels.

LTE also uses SFBC diversity processing in the 2 × 2 MIMO configurations in order to improve signal quality in scenarios where the channel will not support spatial multiplexing.

A good signal to noise ratio on all MIMO paths is required to ensure high performance, so if one of the links is degraded by bad quality, it will have a degrading effect on total performance.

Impact of MIMO on Performance

The typical MIMO for 3G/4G indoor DAS will be a 2 × 2 MIMO, implementing 'twin antennas' at all the antenna locations. In an ideal environment with perfect scattering, with perfect no correlation between the two signal paths you would be able to double the data speed. In a line of sight environment without any scattering at all; no gain in data speed would be achieved when compared with a normal SISO link. However, with the right DAS that supports true dual signal paths for the separate MIMO links (see Section 4.7) and ideal antenna configuration and under ideal RF conditions the data capacity would be doubled.

In practice however, the doubling of data capacity is never achieved uniformly distributed over the entire coverage area in the building; however, significant increase in data capacity on the air interface is possible in some areas of the building.

Indoor MIMO Better Than Outdoor MIMO

In the typical indoor environment, MIMO performance is significantly better than SISO on the RF path, when it comes to the increase in data performance. The final performance all comes down to the environment and practical installation of the antennas (see Section 4.7). MIMO indoor data performance can increase compared to SISO by more than 30% in open areas, with close to 100% data increase in high scattering/multipath areas. This makes it quite evident that MIMO plays a significant role and serves as a strong tool for mobile networks in order to achieve the required high data throughputs.

Comparing indoor MIMO performance with outdoor performance measurement campaigns indicates that the indoor data throughput using MIMO can be more than double of what can be achieved using outdoor networks.

Implementing MIMO in Practice

MIMO has been used successfully commercially by the WLAN (Wi-Fi) community for several years with the 802.11n standard. Consumer devices such as laptops, data cards and smart meters have incorporated multiple small form-factor antennas. It is the wide area wireless network (3G/4G) that has only recently decided to use MIMO in commercial networks.

In order to implement 2×2 MIMO (Figure 2.53) inside buildings, we will need to create two individual signal paths, in practice two parallel DAS antenna paths and two antennas on the mobile (see Section 4.7).

The implementation of two antennas in the mobile is outside of our control and design and must be applied in practice considering the physical limitations of mobile hardware. The recommended spatial separation varies widely in the literature and within the industry, from three to seven wavelengths, depending on the environment and assumptions. Early MIMO deployments used the more conservative six to seven wavelengths due to the relative uncertainty with the technology. However, as the industry became more comfortable with MIMO, more recent deployments used three to five wavelengths and in some cases two to three wavelengths where necessary due to space constraints. Some antenna and mobile manufacturers have come up with clever solutions for this problem, and implemented dual antennas successfully in small mobile terminals that work for MIMO, implemented by using cross polarized antennas.

The DAS however, is within our control and design, and we can implement the two parallel DAS paths required with the preferred antenna separation of $3\text{-}7\lambda$ alternativily a "dual-pol" antenna.

In practice it will be very challenging to implement MIMO for Passive DAS solutions, due to the need for two parallel sets of passive installations, cables, splitters, etc. In reality, Active DAS and femto cells would be the best choice for MIMO in real life installations; one example of the principle behind a MIMO DAS system is shown in Figure 2.53.

2.6.6 Planning for Optimum Data Speeds Using MIMO

MIMO requires a high signal to noise ratio and a high degree of multipath. Indoor environments provide a very rich scattering environment and the use of indoor DAS system provides an excellent signal to noise ratio (good isolation from other cells). Thus, indoor DAS systems are an excellent fit for indoor MIMO networks.

For years we as indoor radio planners have been struggling with interior designers and architects in order to get acceptance and permission to deploy indoor antennas. It has been a problem to get permission to install even the very visually low impact antennas, even though the same architect or interior designer will complain about the lack of indoor radio coverage or data speeds – they struggle to accept the presence of visible antennas.

As on the mobile terminal side where cross polarized antennas are utilized, similar solutions on the indoor DAS side are expected in order to limit the installation impact of MIMO when deployed inside buildings. If not it will simply not be possible to deploy MIMO in some buildings.

In order to maximize MIMO performance we would need to be sure and pay close attention to the following topics:

- **Signal quality:** As is pointed out in several sections in this book, when planning for good radio service and data speed it is a matter of providing good margin between the signal and the noise (isolation); this also applies to LTE. With a good signal to noise ratio the S/N will have to be good on both MIMO paths; this is of great importance in order to maintain 64QAM high rate modulation performance throughout the cell.
- **Indoor DAS antenna separation:** The key to good MIMO performance is to create sufficient de-correlation between the two individual MIMO paths; angular spread of the paths is of importance. The current theory and practical experience shows that an antenna

separation of 3–7λ (three to seven wavelengths on 2.6 GHz corresponds to 30 cm–80 cm), but cross polarized antennas for indoor deployment would hopefully provide a more easy practical solution for us to cope with the challenge of physical antenna separation and installation. The optimum implementation and separation of antennas will depend on the local environment inside the building, so as we gain more experience from actual implementation results from real life deployments on a larger scale, we might have different guidelines depending on the local environmental type. One can easily imagine one 'standard' for MIMO implementation in an 'open' environment, and another preference for a 'dense office' environment. Multiple antenna deployments are already used on the macro network by the use of cross polarized antennas. Antenna manufacturers are also likely to develop small for-factor dual polarized antennas for indoor deployments.

- **Uniform coverage:** As for any radio system, as described throughout this book, uniform coverage is key to high performance. So make sure that you design and implement the building with an equal distribution of coverage and not 'blast' the building with a few 'hot' antenna locations with a very high transmit power (you will still struggle with the uplink if you try!).
- **Low NF on the UL:** When implementing DAS for Indoor LTE it is important to maintain a good uplink performance – a DAS system with low loss/NF to keep mobile transmit power low; and to maintain high performance on the uplink.
- **Ease of installation:** This sounds trivial, but it is actually very important that we select a DAS deployment approach that actually can be accepted and implemented in practice. It makes no sense to select an approach that will give you 98% performance, but can only be deployed in 10% of the buildings, whereas a system that performs at 85% can be implemented in 95% of the buildings would be a better option.
- **True separation in the DAS of the two MIMO paths:** We need to ensure that we select and implement a DAS that truely separates the MIMO path throughout the system, all the way from the antennas to the inputs/outputs on the base station. If there is 'leakage' between the individual signal paths MIMO will not be truly active, with degraded performance as a result (see Section 4.7.4 for more detail).

For more detail about implementing MIMO in indoor DAS, refer to Section 4.7.

2.7 Short Introduction to 4G/LTE

In the following there will be a short introduction to the basics of LTE. It is, however, not within the scope of the book to deliver a detailed, in depth description of LTE – the following chapter will only describe some of the more basic system properties that useful to know.

For more detail about the deep inner workings of LTE refer to [7] and [8].

2.7.1 Motivation behind LTE and E-UTRAN

Due to the ever growing need for higher data speeds for mobile users, new and more efficient methods of utilizing the scarce resources of the RF spectrum is required. 3GPP is an ever evolving standard for accommodating these needs and LTE is yet another step towards higher data speeds, making sure that this new technology is compatible and can co-exist with 2G/3G.

A decade ago only a few could ever imagine that we would have a need for a mobile system that would support downlink data speeds in the range of 100 Mbps and uplink speeds up to 50 Mbps, utilizing MIMO (see Section 2.6), adaptive modulation schemes up to 64QAM (see Section 2.5) and flexible RF bandwidths from 1.4 MHz to 20 MHz channels. These days we are close to implementing this in reality, utilizing LTE.

Looking at Figure 2.54 puts it all into perspective; the ever increasing need for data speed is accommodated be a constant evolution of mobile access technology. Back in 1992 it was astonishing that you were able to achieve 9.6 kbps over a 200 KHz radio channel on GSM; since then the constant evolution of data based services imposes an ever rising data load on the mobile operator's network, at the same time the revenue per user decreases. The main challenge is that at the same time network density investment is on the increase, potentially jeopardizing the business case for the network operator. The evolution from GSM, GPRS, EDGE, WCDMA, HSPA, HSPA+ to LTE is driven by this challenge to drive the maximum data speed per bandwidth in order to accommodate the data demand, and increasing spectrum efficiency. LTE provides spectrum efficiency that is about three to four times better than HSDPA.

Via the use of wide bandwidths, advanced modulation and MIMO antenna schemes LTE is able to provide data speeds in excess of 100 Mbps on the DL and 50 Mbps on the Uplink. With regard to spectrum efficiency 4G/LTE is about three to four times better than 3G/HSDPA on the downlink and two to three times better than 3G/HSUPA on the uplink. This makes 4G/LTE a very attractive tool for network operators for better spectrum utilization.

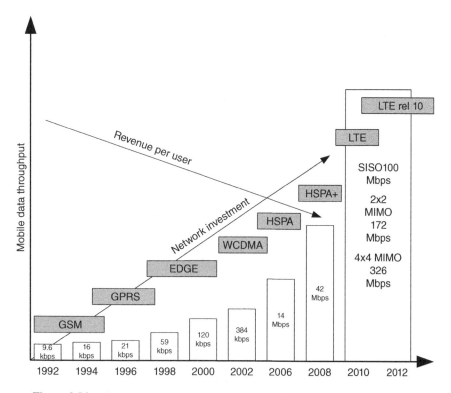

Figure 2.54   The increase of data speed over a decade, from 2G/GSM to 4G/LTE

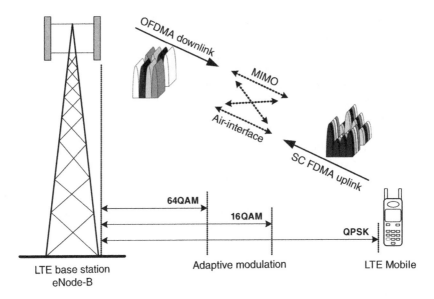

Figure 2.55   4G/LTE Utilizes advanced antenna technologies – MIMO, adaptive modulations schemes and complex modulation OFDM/SC-FDMA in order to maximize spectrum utilization and maximize performance

4G/LTE for Maximized Performance

LTE (Figure 2.55) utilizes advanced antenna technologies – MIMO as described in Section 2.6 – that employ the scattering of the local clutter in the environment (short reflections) to create multiple parallel 'links' over the air-interface that exists in the same time, frequency and space, in theory doubling the throughput of the channel (2 × 2 MIMO) provided you are in a high scattering environment. LTE also applies advanced adaptive modulation schemes – as described in Section 2.5 – in order to optimize and adapt the modulation used on both uplink and downlink to utilize constantly the highest possible modulation rate, thus maximizing data speed at any given location. These advanced features of LTE combined with HARQ (hybrid automatic repeat request), ensures very high spectrum efficiency and optimized data speeds on both uplink and downlink. All of these features combined make LTE a very strong and attractive tool for the mobile operators in order to ensure the highest possible data rate at the lowest possible cost/spectrum load. LTE utilizes two different types of modulation, one on the downlink (OFDM) and another (SC-FDMA), as illustrated in Figure 2.55.

Let us look at some of the most important features of LTE.

2.7.2   Key Features of LTE E-UTRAN

Increased Data Speed

LTE is capable of delivering really high data throughputs, adapted to the conditions of the radio link and environment. In reality 2 × 2 MIMO will probably be the option for indoor deployment when implementing MIMO.

Table 2.5   Theoretical downlink data speeds on LTE4G

Downlink peak data rate (64 QAM)			
Antenna configuration	SISO	2 × 2 MIMO	4 × 4 MIMO
Peak data rate Mbps	100	172.8	326.4
Uplink peak data rate (SISO)			
Modulation	QPSK	16 QAM	64 QAM
Peak data rate Mbps	50	57.6	86.4

Downlink peak data rate (64 QAM) Antenna configuration SISO 2 × 2 MIMO 4×4 MIMO Peak data rate Mbps 100 172.8 326.4 Uplink peak data rate (SISO) Modulation QPSK 16 QAM 64 QAM Peak data rate Mbps 50 57.6 86.4

Mobility

LTE is optimized to support a maximum data rate for pedestrian speed with a velocity of 0–15 km/h. LTE will still provide high performance data throughput for 15–350 km/h and is functional for speeds of 120–350 km/h and even speeds of 350–500 km/h to support high speed trains (depending on the frequency band, due to the 'Doppler Effect' see Section 12.5.5).

Latency

Unlike 2G and 3G data services where SMS, E-mails and internet browsing are the main drivers for data latency becomes more of an issue when you want to maintain the 1:1 data services on LTE, such as video conferencing, remote controlling, 1:1 video transmission and real time gaming, etc. So for efficient resource utilization and good quality of service it is very important to be able to provide swift shift from idle to active mode and keep data packets small. LTE provides a very low latency; idle to active < than 100 ms and on the user plane < than 5 ms.

Multimedia Broadcast Services – MBMS

In order to support point to multipoint broadcast services, mobile TV, etc. the MBMS service has been further enhanced for LTE.

Cell Size

Cell size can be up to 100 km, but typically will be a maximum of 30 km.

Spectrum Efficiency

The 4G/LTE DL spectrum utilization is about three to four times better than 3G/HSDPA and the UL about two to three times better than 3G/HSUPA.

4G/LTE RF Channel Bandwidth

Adaptable bandwidths of 5, 10, 15 and 20 MHz are supported, bandwidths smaller than 5 MHz are also supported, i.e. 1.4 and 3 MHz for FDD mode. This together with optional FDD/TDD makes LTE fit within the various spectrum allocations assigned on a global base.

Spectrum Allocation

LTE can operate in a paired spectrum for duplex operation, FDD (Frequency Division Duplex – thus it can be deployed in existing GSM or WCDMA). LTE can also be deployed in an unpaired spectrum, utilizing TDD (Time Division Duplex) This makes LTE deployments widely applicable when reusing various spectrum allocations for previous systems. One of the new bands that is now open is the old UHF band used for terrestrial broadcasting of analog TV signals – since implementation of digital TV (DTVB) parts of the 700/800 MHz spectrum are now open for LTE in some regions of the world.

Compatible with Previous Generations of Mobile Systems

The flexibility of both the RF Channel bandwidths and optional FDD or TDD operation assures compatibility with previous generations of mobile systems. This also applies to a network element base with the appropriate interfaces to 2G and 3G networks.

2.7.3   System Architecture Evolution – SAE

One of the key drivers and motivations behind LTE architecture was an overall simplification of the network structure of the supporting elements. This reduced the number of network elements in the radio access network as well as in the core network. This simplification also leads to enhanced performance, in particular reducing latency in the network. The reduced number of network elements also helps in reducing the overall network investment, thus reducing the production cost for the mobile operators – an ever increasing challenge. Obviously, interoperability and compatibility with existing 2G, 3G and 3.5G networks must be maintained.

2.7.4   EPS – Evolved Packet System

The EPS (Evolved Packet System) is divided into the radio access network and core network and will be described briefly in the following. For a detailed description refer to [6] and [7].

Evolved UMTS Radio Access Network – E-UTRAN

The Evolved UMTS Radio Access Network (E-UTRAN – see Figure 2.56) consists of a single network element; the bases station – in LTE called the Evolved Node-B (eNode-B). As the name suggests the eNode-B is more advanced than the Node-B used for UMTS.

The eNode-B services all the user plane controls and protocols so as to enable and support communication and handling of the mobiles; radio resource management, admission control, secluding encryption algorithms, cell broadcast, data compression and decompression of the user data streams. The eNodeB incorporates some of the functions of the RNC (Radio Network Controller for 3G/UMTS networks); this decentralization helps improve latency.

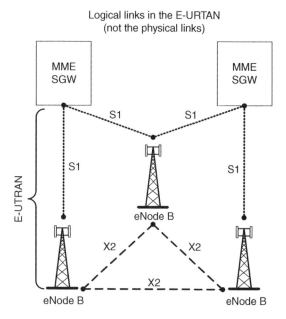

Figure 2.56   A simplified diagram illustrating the principle of the E-UTRAN (Evolved UMTS Radio Access Network) logical links, not the physical links between the elements of the radio access network in LTE

Interface Description of the E-UTRAN

The logical (not physical) connections of the E-UTRAN are shown in Figure 2.56. The S1 (S1-MME) interface connects the eNode-B to the core network; the function of the S1 is to support communication between the eNode-B and the Mobility Management Entity (MME) and the Serving Gateway (SGW) via the S1-U.

To support direct logical connection between the eNode-Bs internally the X2 interface is used and serves as a direct 'tunnel' of packet communication during handovers between the cells to minimize load in the core network and reduce latency and optimize overall performance during handover.

2.7.5   Evolved Packet Core Network – EPC

The two principal functions of the EPC (Evolved Packet Core) are handling of high speed data packages and handling of mobility management in the network. These functions are controlled and handled by the SGW (Serving Gateway) and the MME (Mobility Management Entity) – see Figure 2.57. This architecture is optimized to handle fast network control and exchange of packets – to maximize performance and maintain low latency in the network.

The EPC (Figure 2.57) consists of the following basic network elements.

Serving Gateway – SGW

The function of the SGW is as a router that forwards and routes the packages of user data. The SGW is also capable of providing quality of service management used by other network

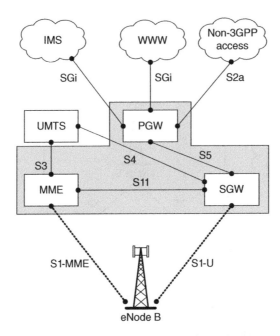

Figure 2.57 A simplified diagram illustrating the principle of the E-UTRAN (Evolved UMTS Radio Access Network) logical links, not the physical links between the elements of the radio access network in LTE

elements and tendering for some of the basic UL/DL services. During handovers the SGW acts as the local anchor point for inter eNode-B handovers. The SGW can also act as the anchor point for handovers between UMTS and LTE and other functions, such as packet buffering in idle mode and initiation of network triggered service requests.

Mobility Management Entity – MME

The Mobility Management Entity serves as the primary key signaling node in the Evolved Packet Core Network. The MME is responsible for idle mode mobile location, paging, roaming and user authentication. The MME also selects the appropriate SGW during the bearer activation/deactivation process.

Packet Data Network Gateway – PGW

The Packet Date Network Gateway serves, as the name suggests, as the interface point of data communications to external networks. The PGW provides filtering of data via deep packet inspection. In order to handle mobility management/handovers to non 3GPP networks, WiMAX, CDMA, etc the PGW serves as the anchor point.

2.7.6   LTE Reference Points/Interfaces

This is a brief description of the reference points and LTE interfaces (see Figure 2.57).

- **S1-MME**
  The S1-MME serves as the reference point for the control plane protocol between E-UTRAN and the MME.
- **S1-U**
  The S1-U Reference point is employed between E-UTRAN and SGW for the per-bearer user plane tunneling and inter Node-B path switching when in handover.
- **S11**
  The SS11 is the reference point between MME and SGW.
- **S3**
  This is the interface between SGSN and MME and enables user and bearer data transmission for inter 3GPP access communication.
- **S4**
  The S4 interface provides the user plane with mobility and control support between the other 3GPP networks and the SGW.
- **S5**
  The S5 interface provides user plane communication and tunnel management between the SGW and PGW. It is used to support mobility and if the SGW needs to connect to another.
- **S2a**
  The S2a interface provides support for communication between the LTE network and trusted non-3GPP IP access networks; WiMAX, CDMA, etc.
- **SGi**
  The SGi interface is the reference point between the LTE networks PGW and other public or private packet data networks.

2.7.7   The LTE RF Channel Bandwidth

LTE can be deployed using various RF channel configurations, as shown in Figure 2.58. Thus, an LTE operation can have a selected bandwidth of 5, 10, 15 and 20 MHz; bandwidths smaller than 5 MHz are also supported, i.e. 1.4 and 3 MHz for the FDD (Frequency Division Multiplex)

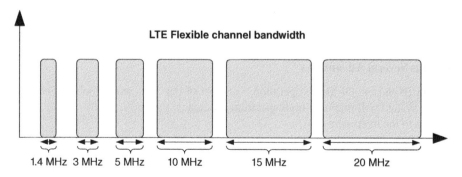

Figure 2.58   The 4G/LTE RF carrier can be deployed in different bandwidths, so the network operator can utilize and fir it into existing spectrum. LTE can be deployed as TDD of FDD

mode. This ensures that LTE can fit into most spectrum allocations in a paired spectrum for FDD (separate UL and DL bands) or be implemented in an unpaired spectrum for 1.4 and 3 MHz channels using TDD (Time Division Multiplexing, employing the same frequency for UL and DL separated in time).

This enables a high degree of flexibility, and re-using of existing bands used for the 2G and 3G systems; and even to co-exist with these technologies in adjacent spectra.

Do Not Mix TDD and FDD on the Same Indoor DAS

Please note that you would normally not be able to operate both FDD and TDD on the same DAS in bands with close proximity. The FDD part of the system will inject unwanted in/band signals to the TDD system when the TDD system is in receive mode. It is important to watch out for this, and also when deploying separate DAS in the same building, one for TDD and one for FDD. The required isolation between the systems needs to be maintained. This can also be an issue on DAS where you operate UMTS 2100 and LTE 2300 and 2600 on the same DAS/ antennas – strict filtering may be required.

2.7.8   OFDM – Orthogonal Frequency Division Multiplexing

OFDM (Orthogonal Frequency Division Multiplexing) is a multi carrier modulation that relies on multiple sub-carriers for transmission, as shown in Figure 2.59. OFDM has (in theory) perfect isolation between the sub-carriers/users of the cell, but degradation of the radio channel will affect this in reality.

It is unlike traditional FDM (Frequency Division Multiplexing) that relies on single carriers for the modulation, with a need for a specific guard band between the carriers' OFDM. OFDM breaks down the user data and transmits this data over multiple data streams; each data stream is modulated into individual sub-carriers. These orthogonal spaced sub-carriers are transmitted in parallel. The key with OFDM is to distribute the high rate data streams over multiple parallel lower bandwidth sub-carriers. The lower bandwidth sub-carrier with a cyclic prefix provides more resiliency against multipath distortion.

The orthogonal spacing of these sub-carriers provides a very efficient utilization of the RF frequency spectrum, as shown in Figure 2.59; for LTE the sub carrier spacing is 15 KHz.

OFDM Has Several Advantages

- In NLOS (Non Line Of Sight) operation, i.e. most of the cases are in mobile communication, the system will maintain high spectrum efficiency, thus utilizing the scarce spectrum resources to the maximum.
- Ideal for operating in a multipath environment, thus ideal for 95% of the typical mobile environment.
- The design of the receiver and the equalizer is much simpler as compared with other options.
- Supports adaptable RF bandwidths, thus the various channel segments used for LTE.
- High spectrum efficiency.

2.7.9   OFDMA – Orthogonal Frequency Division Multiple Access

4G/LTE uses OFDMA (Orthogonal Frequency Division Multiple Access) in the Downlink path from the base station to the user. This is a multi-carrier system that allocates radio resources to multiple users over a number of sub-carriers spaced orthogonally in the spectrum, as shown in Figure 2.59. The use of orthogonal channel spacing provides for a very efficient use of spectrum and good inter carrier isolation. The system splits the LTE carrier into many individual sub-carriers spaced 15kHz apart. Each individual sub-carrier is modulated using QPSK, 16-QAM or 64-QAM depending on the quality of the radio channel. OFDMA will assign individual bandwidth to the users as per the need for bandwidth. Unused sub-carriers are not transmitted thereby reducing power transmission, limiting interference and power consumption. OFDMA is based on OFDM (Figure 2.60 and Figure 2.61, shown with only four sub-carriers) but differs from OFDM when it comes to scheduling and resource assignment. In OFDM the entire bandwidth belongs to a single user for a period of time; in OFDMA multiple users will share the same bandwidth at each point in time.

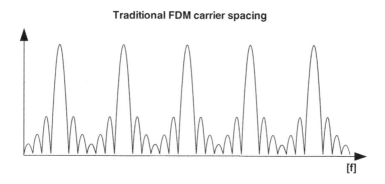

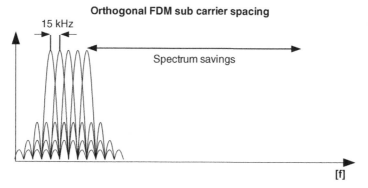

Figure 2.59   Compared to the traditional FDM channel spacing, orthogonal spaced sub carriers used in 4G/LTE (OFDM) is a much more efficient use of spectrum

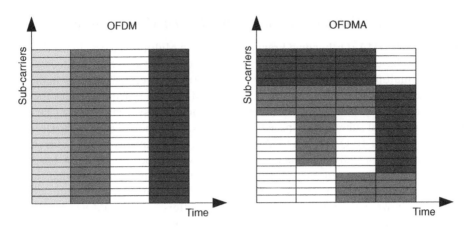

Figure 2.60   OFDM – OFDMA, each block represents one burst, each shade of gray, one user. OFDMA enables the usage of all available capacity, so users can share the available bandwidth

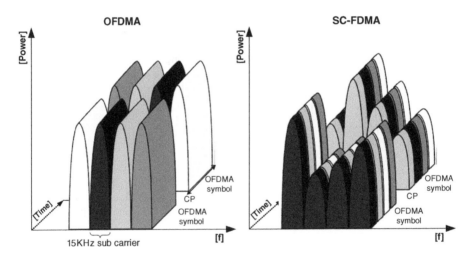

Figure 2.61   In OFDMA, each sub-carrier carries unique information, in SC-FDMA the information is spread over multiple sub-carriers

2.7.10   SC-FDMA – Single Carrier Frequency Division Multiple Access

In the Uplink path, from the mobile to the base station, 4G/LTE uses SC-FDMA (Single Carrier Frequency Division Multiple Access), as shown in Figure 2.61 (shown with only four sub-carriers). In SC-FDMA the user data is spread over multiple sub-carriers; this is unlike the downlink channel that uses OFDMA where each sub-carrier services one user. In OFDMA, used in the downlink, the users' data is also spread over multiple sub-carriers. The primary difference between OFDMA and SC-FDMA is that in SC-FDMA, the signal is coded to look like a single carrier modulated at a higher data rate. The resulting SC-FDMA signal will then have the peak-to-average ratio of the single modulated carrier versus the higher peak-to-average ratio of a multi-carrier waveform. SC-FDMA will have than lower average power when compared with OFDMA and thus will not drain the battery in the mobile; the circuitry in the mobile to drive SC-FDMA is also less complicated and less expensive to produce.

2.7.11 LTE Slot Structure

Let us have a brief look at the principles of the structure of the LTE slots, resource elements and resource blocks (0.5 ms) in Figure 2.62. The smallest frequency/time unit for downlink transmission is called a resource element. Seven of these resource elements and symbols is one slot, each resource element corresponds to one OFDM 15 kHz sub-carrier during one OFDM symbol interval, as shown in Figure 2.62. One group of 12 contiguous 15 kHz sub-carriers corresponds to 180 kHz, that is one resource block. Data for each user is allocated (scheduled) in terms or resource blocks, at a minimum of two consecutive resource blocks, called scheduling blocks.

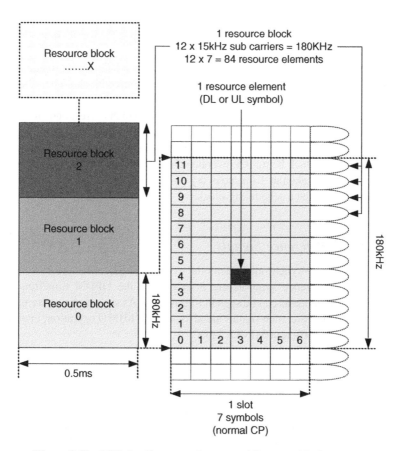

Figure 2.62 LTE slot, Resource element and Resource block structure

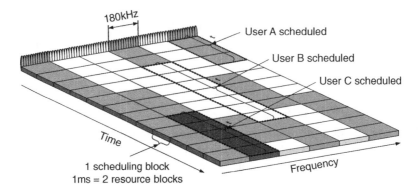

Figure 2.63 The different user's traffic is assigned certain sequences of recourse blocks, a minimum of two in so called scheduling blocks. This is a very dynamic and adaptable way of allocating the resources in the cell and maximizes the throughput

2.7.12 User Scheduling

Each of the different users is assigned (scheduled) a number of resource blocks in the time-frequency grid, see Figure 2.63 – the higher the required data speed is for the individual user the more resource blocks a user is assigned.

In Figure 2.63 we can see three different users assigned to a LTE carrier, made possible by scheduling for each user a specific series and number of resource blocks. The scheduling is a complex procedure; this is merely a brief introduction.

Each resource block has a time slot duration of 0.5 ms, which corresponds to six or seven OFDM symbols. The smallest resource unit that the scheduler can assign to a user is one scheduling block. One scheduling block consists of two consecutive resource blocks.

So, with one resource block being 0.5 ms, one scheduling block is 1 ms. Thus, scheduling of resources/users can be taken at 1 ms, so every two resource blocks, 180 kHz wide and in total 1 ms in length, is called a scheduling block.

2.7.13 Downlink Reference Signals

In order for the mobiles to synchronize, demodulate and aid detection and evaluation of the LTE channel certain reference symbols are inserted into the OFDM time/frequency grid among the resource elements, as can be seen in Figure 2.64. As shown, these reference signals have a fixed location. The reference signals are also used for MIMO operation, in order for the LTE system to detect the different MIMO 'links'

2.7.14 The 4G/LTE Channel

The LTE Channel (Figure 2.65) can be deployed in several RF channel bandwidths, as described above. The more RF spectrum that is assigned to the LTE RF Channel (1.4, 3, 5, 10, 15 or 20 MHz), the more transmitted Resource blocks can be carried, thus providing higher capacity.

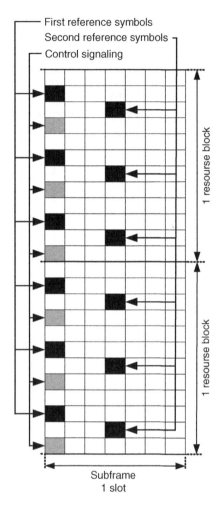

First reference symbols
Second reference symbols
Control signaling

1 resourse block

1 resourse block

Subframe
1 slot

Figure 2.64 The LTE system inserts certain reference signals in some of the resource elements in order to aid cell detection, evaluation and synchronization

Thus, the number of Active Resource blocks (RB) is linked to the RF Channel bandwidth and is one of the capacity considerations when designing LTE based systems. The maximum used bandwidth corresponds to the number of resource blocks multiplied by 180 kHz.

For more detailed information about Resource blocks, frame structure, refer to [7] and [8].

2.7.15 LTE Communication and Control Channels

Data transport and signaling in LTE are transmitted by means of a protocol stack via the air interface using three different types of channel. LTE uses Logical, Transport and Physical channels; each type of channel is defined by its own set of functions and attributes. The following is a brief introduction to these channels and their function.

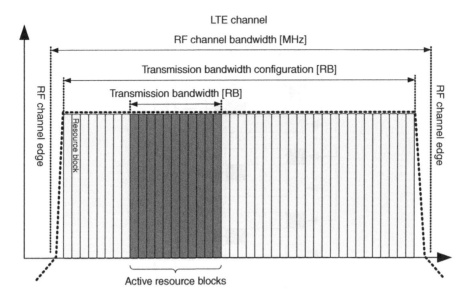

Figure 2.65 The relationship between RF channel bandwidth, transmission bandwidth configuration and transmission bandwidth

It is beyond the scope of this book to cover all the detail of the frame and channel structure, refer to [6] and [7] for more detailed information.

LTE Logical Channels

Logical channels can carry both user data and control data. As the name suggests the Logical channel is not a 1:1 physical connection between the elements in the network.

- **Control Channels**
  Control channels are only used to transmit control plane information:
  BCCH – Broadcast Control Channel
  This Downlink Channel is used to broadcast system control information shared by all users in the cell. The BCCH broadcasts network identity, configuration of the cell, access information, etc.
  CCCH – Common Control Channel
  This channel is used to transmit control information between the mobile and the network. It is used in idle mode and during call setup before radio resource management is established in dedicated mode.
  PCCH – Paging Control Channel
  This is a downlink only channel that broadcasts paging information – triggering the mobile to respond and initiate connection to the network when paged.
  DCCH – Dedicated Control Channel
  This bi-directional point to point channel transmits dedicated control information between the mobile and the network.
  MCCH – Multicast Control Channel
  This is a point to multipoint downlink channel used when transmitting MBMS (Multimedia Broadcast Services) to support point to multipoint broadcast services, mobile TV, etc.

- **Traffic Channels**

  These, as the name suggests, are the channels that carry the communication data itself between the mobile and the network. There are several types of traffic channels:

  DTCH – Dedicated Traffic Channel

  This is a dedicated point-to-point traffic channel reserved for transmitting the data for one user/mobile. The DTCH also carries out signaling for that particular call.

  MTCH – Multicast Traffic Channel

  This is a point-to-multipoint downlink channel used to carry data from the network to several mobiles, carrying MBMS (Multimedia Broadcast Services), like mobile TV, etc.

- **LTE Transport Channels**

  The transport channels are used in both Uplink and Downlink and are described briefly in the following:

  BCH – Broadcast Channel

  This downlink broadcast channel is transmitted as a common channel and has a pre-defined transport format defined by the requirements. This channel transmits the system information for the cell/network.

  DLSCH – Downlink Shared Channel

  The DLSCH carries downlink signaling and traffic; depending on the nature of the data in the channel it will support dynamic as well as semi-static resource allocation with optional support of DRX (discontinues reception) of the mobile. Error control in the channel is also supported by HARQ (Hybrid Automatic Repeat request) as well as dynamic link adaption by adapting the modulation (see Section 4.12.10).

  PCH – Paging Channel

  This channel is associated with the PCCH (Paging Control Channel) and carries the paging messages to the mobiles in idle mode. The PCH supports DRX (discontinues reception) for the mobiles in order to benefit from prolonged battery life.

  MCH – Multicast Channel

  The MCH is associated with the multicast services from the upper layers in the system.

  RACH – Random Access Channel

  The RACH is used during random access from the mobiles when establishing connection to the system. This is a shared uplink channel and merely supports the initial call set up when mobiles access the cell.

  ULSCH – Uplink Shared Channel

  The ULSCH carries dedicated and common signaling as well as dedicated traffic data on the uplink. Like its downlink counterpart, the DLSCH, it can adapt the modulation accordingly.

- **LTE Physical Channels**

  The LTE physical channels are the 'carriers' of the transport channels over the LTE radio interface. This is a brief description of the LTE physical channels:

  PBCH – Physical Broadcast Channel

  The PBCH transmits cell identity and system information repeatedly via the PBCH in cycles of 40 ms.

  PCFICH – Physical Control Format Indicator Channel

  The PCFICH provides the mobiles with information about the configuration of the number of symbols used on the OFDM used for the PDCCH.

PDCCH – Physical Downlink Control Channel

The PDCCH provides the mobile with information about the uplink and downlink resource allocation of the PCH and the DLSCH.

PHICH – Physical Hybrid ARQ Indicator

The PHICH is used to carry the Hybrid ARQ ACK/NAK as the response to the transmissions. ACKs and NAKs confirm the delivery of valid data, or request a retransmission of the erroneous blocks received. The ACKs and NACKs are a part of the HARQ procedure.

PDSCH – Physical Downlink Shared Channel

The PDSCH carries the DLSCH and the PCH.

PMCH – Physical Multicast Channel

The PMCH carries the MCH and multicast/broadcast information.

PUCCH – Physical Uplink Control Channel

The PUCCH is used to carry the uplink control information ACK/NAK as the response to the uplink transmissions. ACKs and NAKs confirm the delivery of valid data, or request a retransmission.

PUSCH – Physical Uplink Shared Channel

The PUSCH carries the ULSCH application signaling and user data.

PRACH – Physical Random Access Channel

The PRACH carries the RACH, random access preamble probes transmitted by the mobile when initiating network access.

2.7.16   Radio Resource Management in LTE

The following serves only as a brief introduction, for more detail refer to [6] and [7].

Cell Selection

Cell selection in idle mode is similar in concept to 2G/3 where the mobile camps on the cell with the strongest signal, decoding the system information broadcast via the BCH as well as the signal strength and quality of the decoded cell (see Figure 2.66). There are network configurable parameters used to determine threshold levels for initiating new searches. Once the mobile detects a valid cell that fulfils the requirements it starts accessing the cell using Random Access. In dedicated (connected) mode the LTE mobile employs the handover procedure.

LTE also supports DRX – discontinuous reception that allows the mobile to 'power down' and only listen for paging signals at assigned intervals, in order to save battery life.

The DC Carrier

The 4G radio channel can be deployed at 1.3, 3, 5, 10, 15 or 20 MHz channel bandwidth, and when you switch on a 4G mobile in a random network, the mobile will have to camp on any given 4G channel without knowing the exact bandwidth. Therefore, there is a single 'empty' sub channel constantly transmitted as a beacon signal in the center of the carrier (Figure 2.66).

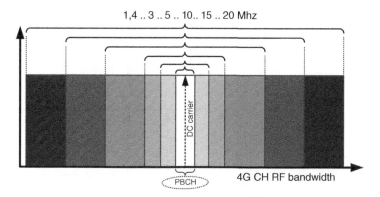

Figure 2.66   The 'DC carrier' for cell selection is placed in the center of the 4G carrier, no matter the bandwidth. There is the same basic signaling structure around the center as well

This DC carrier aids synchronization of the mobile and guides the mobile to lock on to the specific 4G cell during the initial selection of the network. All the required synchronization signals and system information are also located at the center of the carrier – utilizing the same 'location' and structure (62 sub-carriers) for any 4G radio channel bandwidth, from 1.4 to a 20 MHz 4G carrier. This makes it possible for the mobile to initially camp on a random network when switching on without any pre-knowledge of the frequency, bandwidth nor configuration of the exact 4G carrier.

The PBCH

The physical broadcast channel (PHCH) in the center of the 4G carrier (Figure 2.66) carries all the needed information for the mobile to establish the initial access to the network. The PBCH is located around the DC carrier as described in the previous section.

Random Access

The random access procedure in LTE is similar in concept to UMTS (see Figure 2.67). After decoding and selecting the best cell the mobile will – when it needs to connect to the network; in order to respond to paging, perform location update or initiate a new call – start the random access procedure; and initiate cell access utilizing preambles with low transmit power in order to probe the access to the cell, ramping up the transmit power gradually – so as not to overshoot the uplink of the base station resulting in uplink receiver blocking (see Section 4.12.10). The initial access procedure is based on a specific algorithm. The principle is shown in Figure 2.67. At a certain point the network detects the mobile and responds on the downlink using the PRACH channel and the mobile continues signaling and data transmission – using the last power stem from the preambles and the initiated uplink power control.

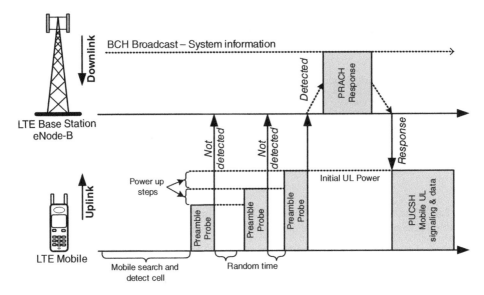

Figure 2.67 LTE uses random access procedure, using cell selection, random access preambles like those we know from other mobile systems like 3G

Another important part and result of the access procedure is to establish the timing advance required in order for the mobile to synchronize its transmission timing according to the time difference (distance) to the serving cell.

LTE Handover

LTE uses hard handovers in both downlink and uplink; the handover evaluation is based on measurements made by the mobile of the reference signals, level (RSRP Reference Signal Received Power) and quality (RSRQ Reference Signal Received Quality) of the cell. The core network oversees all handovers.

The LTE handover is evaluated and controlled by the eNode-B and MME. The handover evaluation is based on measurements done and reported by the mobile. The LTE handover is designed to be 'lossless' in terms of data package, the data packets are forwarded from the serving eNode-B to the handover target eNode-B. The LTE system also supports measurement reporting, evaluation and handover to and from lower order systems, 3G (UMTS) and 2G (GSM).

Intra- and Inter-RAT Handovers

In order to support heterogeneous networks (see Section 13.2), 4G handover must support handover functionality between different radio access technologies (RATs). In the example in Figure 2.68, we can see the principle of inter and intra-RAT handover. Intra-RAT handovers are between the same RATs, and inter-RAT handovers are between different RATs. Therefore the type of measurement reporting the network can ask the mobile to support will also support measurements across different RATs.

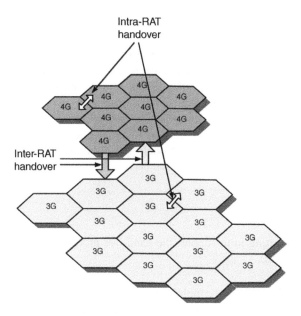

Figure 2.68 Both intra and inter-RAT handovers must be supported in order to assure handover between 4G and 3G (and 2G) to support heterogeneous networks

The 4G Handover Algorithm and Procedure

The 4G handover is, as described in the previous section, based on evaluation of radio channel measurement reports and conditions. The measurements are sent via the radio resource control (RRC) protocol. The handover algorithm will then evaluate the handover decision, and evaluate the success of the actual handover.

The handover procedure is based on the following actions:

1. *Handover preparation*
   The handover preparation phase includes:
   - Radio measurements of RSRP and RSRQ.
   - Weighting and filtering of these radio level and quality measurements.
   - Measurement reporting to the network. Evaluation of handover according to the algorithm.
   During the handover preparation phase, the user data flows between the mobile and the network. In the handover preparation phase, measurement control etc. are also maintained by the serving cell, which defines the mobile measurement reports and parameters. The actual handover decision is then made by the serving base station (eNodeB) which requests a handover to the target cell and performs admission control. The handover request is then acknowledged by the target eNodeB.
2. *Handover decision and execution*
   The actual handover execution phase will be triggered when the source eNodeB sends a handover command to the mobile as a result of the measurement evaluation and triggers.

Throughout this phase, the data is forwarded from the source to the target eNodeB, which buffers the packets. The mobile will synchronize to the target cell and perform a random access to the target cell to obtain uplink resources, timing advance and other needed parameters. The mobile will send a 'handover confirm' message to the target eNodeB, and thereafter the target eNodeB will start sending the forwarded data to the mobile.

3. *Handover execution / completion*

In the final stage of the handover, the target eNodeB updates the MME that the user plane path has changed. The SGW is then notified to update the user plane path. The user data now starts flowing to the new target eNodeB. All radio and control plane resources are released in the source eNodeB.

Triggering of the Handover

The measurement reporting in to support the handover procedure can be performed by the mobile in different ways:

1. Serving cell and a list of cells the mobile are asked to scan.
2. Serving cell and cells in the frequency set (detected by the mobile).

As in 3G, the handovers / selection in 4G are defined by 'event types', and the mobile triggers these events when measuring serving and neighboring cells accordingly. Different hysteresis and timing triggers are defined to avoid 'ping-pong' handovers.

The mobile can be configured to perform a series of measurement reports after a specific trigger event (regular reporting event). The mobile can also be configured to report periodic measurements immediately (not triggered by a specific event)

Events in 4G Handovers
A1 – Serving cell is better than the absolute threshold RSRP level.
A2 – Serving cell falls below the threshold RSRP level.
A3 – Neighbor cell has higher RSRP level (including a defined offset).
A4 – Neighbor cell RSRP is higher than defined threshold RSRP level.
A5 – Serving cell falls below the threshold, and the neighbor cell is higher than the RSRP threshold level

Inter-RAT Handovers (4G/3G)
B1 – Neighbor cell is better than the threshold.
B2 – Serving cell falls below threshold and neighbor cell is above the threshold

Details of the Signaling Flow in the 4G Handover Procedure

In Figure 2.69 there is a simplified example of handovers between two 4G cells, from Cell A (4G) to Cell B (4G) performing an intra-RAT handover in an 'A3 Event', and an inter-RAT handover 'B1 Event' from Cell B (4G) to Cell C(3G).

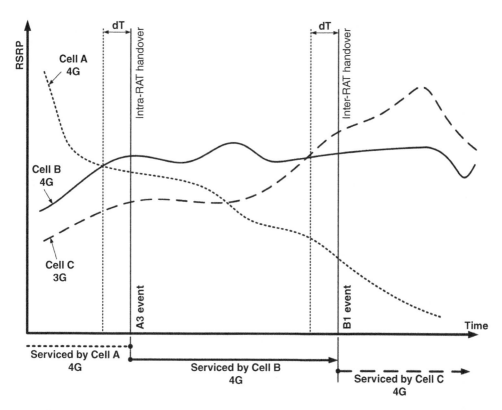

Figure 2.69 Simplified handover example of intra-RAT 4G handover with 'A3 event' and inter-RAT handover with 'B1 event' from 4G to 3G

Let us have a brief look at the handover procedure and events in the network, with Figure 2.70 as a reference, performing an intra-RAT handover between two 4G cells.

1. *Measurement configuration is sent to mobile*
   Network informs mobile on how to perform measurements.
2. *Mobile sends measurements*
   Based on the measurement configuration, the mobile sends the measurement reports back to the network.
3. *Handover decision*
   At a certain point, once the measurements received from the mobile by the serving eNodeB fulfill the conditions to hand over the call to the candidate eNodeB, the serving eNodeB will decide to initiate handover.
4. *Handover request*
   Once the measurements trigger the conditions for handovers, the serving eNodeB will request permission to perform handover to the candidate eNodeB.
5. *Admission control*
   After receiving the HO request, the candidate eNodeB will start admission control to ensure the appropriate radio resources.

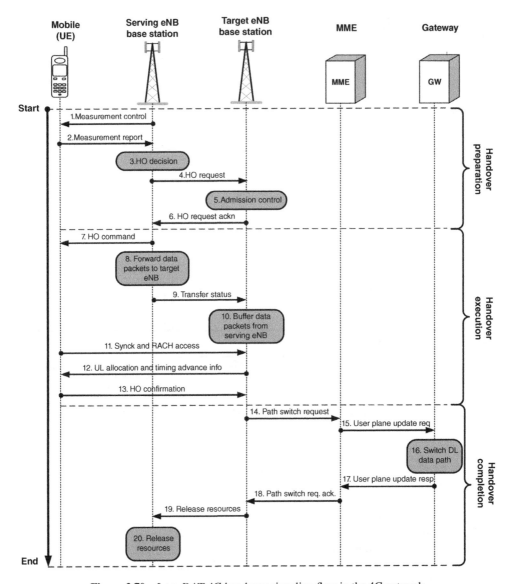

Figure 2.70   Intra-RAT 4G handover signaling flow in the 4G network

6. *Handover request acknowledge*
   Once resources to service the new call have been established, the target candidate eNodeB
   sends an HO request acknowledge to the source eNodeB.
7. *Handover command*
   The source eNodeB assigns the handover command to the mobile and starts to forward
   data to the target eNodeB.
8. *Forward packets*
   The serving eNodeB starts forwarding the traffic data to the target candidate eNodebB.

9. *Status Information*
   Source eNodeB transfer status information to eNodeB.
10. *Start buffering*
    The target eNodeB now starts buffering the traffic data from the servicing eBodeB, to ensure a smooth transition to the new eNodeB with no loss of data.
11. *Synchronization*
    Mobile synchronizes to the target eNodeB and performs RACH to access the cell.
12. *Timing Advance and UL path*
    Target eNodeB establishes UL path and timing advance adjustment information.
13. *Handover Confirmation*
    The mobile sends handover confirmation to the target eNodeB.
14. *Path switch request*
    The target eNodeB sends a path switch message to the MME, updating the network of the change of serving cell.
15. *User plane update request*
    The MME sends a user plane update request to the serving gateway.
16. *Switch downlink*
    The serving gateway switches the data path for the downlink to the target side.
17. *User plane update response*
    The serving gateway sends the user plane update to the MME.
18. *Path switch request acknowledge*
    The MME confirms the path switch.
19. *Release resource*
    The target eNodeB confirms success of the handover to the source eNodeB and commands the source eNodeB to release resources.
20. *Release of resources*
    After receiving the release resource command, the source eNodeB releases radio and processing resources to the mobile, which now has full service of the new cell.

Power Control in LTE

The LTE downlink channel has no power control, the eNode-B utilizes adaptive modulation-coding schemes to adjust the throughput downlink transmit power level per resource block, with the goal of reducing the inter-cell interference with neighboring cells.

The LTE uplink channel uses a relatively slow power control compared to the 1500 Hz power control used by UMTS. The main goal of uplink power control is to reduce the mobile power consumption, to reduce interference leakage to neighboring cells and to avoid the near-far problem when users are closed to the eNode-B antennas/DAS antennas. LTE power control is rather complex and ties in with the used channel bandwidth. Power evaluation is based on the PUCCH (Physical Uplink Control Channel) and uplink reference signals.

Timing Advance in LTE

As shown for GSM, the frame nature of LTE must use Timing Advance in order for the mobile to synchronize to eNode-B for uplink transmissions so as to prevent overlap in time. LTE can provide timing advance commands to support a 100 km cell range.

4G Design Levels

In 2G we designed for the Rx-Level, and in 3G we designed for the common pilot channel (CPICH) level. In 4G, RSRP is the reference and a design target level of –95 dBm to –85 dBm seems to be standard among most mobile operators, and –80dBm in high-demand areas, with even –75 to –70 dBm RSRP in VIP areas. Like any other RF system, the actual design target level will depend on the actual interference level in the building, and you have to adjust the design level to overcome this interference (see Section 3.5.6 for more details):

- Basement and low-interference areas: –100 dBm to –95 dBm RSRP
- Office and high-use environment, with limited interference: –90 dBm RSRP
- High rises, high-use critical areas with some interference : –85 dBm RSRP
- High-interference areas, even higher than –85 dBm RSRP design level

Receiver Blocking on 4G
To avoid saturation of the receivers in 4G, it is highly recommended not to exceed –25 dBm RSRP.

Isolation Requirements

Like other radio systems, we need to maintain a good 'clean' serving cell signal, with good isolation to any other cells on same frequency. In 4G we have a frequency reuse of 1 (all cells using the same frequency). For well-isolated, high-capacity cells in a DAS, it is recommended to be about 10–15 dB more dominant than other cells in most of the coverage area. Obviously there will be 'transition' areas where you need to have overlap. Make sure to plan these hand-over areas in places where there is limited capacity load.

RSRP, Reference Symbol Transmit Power

The RSRP is used to measure the RF level of the cell for mobility management. The RS is also used to identify the individual MIMO paths. The RSRP is the linear averaged RS signal level over the six reference symbols (RSs) in each resource block (RB; see Figures 2.64 and 2.71).

In order to calculate the link budget, we first need to be able to calculate the transmitted power of the RS. The RS transmit power relates to the bandwidth of the 4G channel (the number of sub-carriers). This is also evident in Figure 2.71, showing a 5 MHz 4G channel with 25 physical resource blocks (PRBs), each with 12 sub-carriers, giving a total of 25 × 12 = 300 sub-carriers.

We can calculate the power per sub-carrier in relation to the total power of the transmitter.

Example
Let us calculate an example of a 15 MHz 4G channel and a composite power of the transmitter of 43 dBm.

In Table 2.6 we can see that a 15 MHz channel has 75 RBs; each RB has 12 sub-carriers, so there is a total of 75 × 12 = 900 sub-carriers.

Therefore each of the 900 sub-carriers gets assigned 1/900 of the total power.

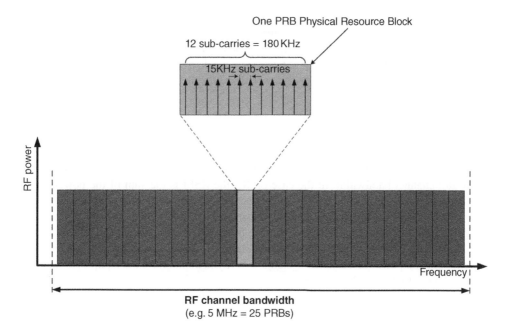

Figure 2.71   Physical resource blocks (PRBs) vs RF channel bandwidth

Table 2.6   RF bandwidth/resource blocks showing RS transmit power

4G channel bandwidth (MHz)	No. of resource blocks	No. of sub-carriers	Total power / RS power (dB)
1.4	6	72	18.6
3	15	180	22.6
5	25	300	24.8
10	50	600	27.8
15	75	900	29.5
20	100	1200	30.8

We will now be able to calculate the power assignment to each sub-carrier in relation to the total power:

$$\text{Power drop per sub-carrier of 900 sub-carriers} = 10 \times \log(900) = 29.5 \text{dB}$$

Now we can calculate the transmitted power per sub-carrier:

$$\text{TxPWR RSRP} = \text{total power} - (\text{power drop per sub} - \text{carrier})$$
$$\text{TxPWR RSRP} = 43 \text{ dBm} - 29.5 \text{ dBm}$$
$$\text{TxPWR RSRP} = 13.5 \text{ dBm}$$

In Table 2.6 we can see the same calculation results, of each 4G channel bandwidth from 1.4 to 20 MHz. Like any other shared resource, the PPC Power per Carrier drops the more carriers the same amplifier will have to support. Therefore we need to back off from the full composite power according to the Table 2.6 to assure we do overload the power amplifier, and drives it into a nonlinear performance area and degrades the quality of the transmitted 4G service.

4G RSSI Signal Power

The RSSI is the total composite received strength of a radio channel, a 'raw' RF measurement over the full channel bandwidth. In the case of 4G, the RSSI is the sum of the power of all active sub-carriers. Therefore the RSSI is dependent on the bandwidth of the carrier; the wider the bandwidth, the more sub-carriers will be transmitted.

When designing 4G systems we take reference in the RSRP. We can calculate the expected RSSI of the full 4G channel when we know the RSRP level. This is very useful when measuring signal strength and power levels in 4G systems, e.g. if you do not have the equipment that can decode and measure the RSRP directly.

Let's try to calculate the RSSI when we know the RSRP we are designing for, using the same approach as the previous example and using the values in Table 2.6 from a 5 MHz 4G carrier

Example
We can see in Table 2.6 that a 5 MHz 4G carrier has 25 resource blocks; each 12 sub-carriers a total of $25 \times 12 = 300$ sub carriers. This gives a power offset of $10 \times \log(300) = 24.77$ dB. The means that the RSRP is 24.77dB lower than the RSSI.

If we were designing for –85 dBm RSRP then the expected RSSI will be 24.77dB higher:

$$RSSI = RSRP + offset$$
$$RSSI = -85\,dBm + 24.77\,dB = -60.23\,dBm\ \left(in\ full\ 5\,MHz\ bandwidth\right)$$

Hotspots in 4G

As for any DAS, planning the antennas close to the hotspots ensures the most efficient utilization of the network resources, thus increasing the capacity of the cell. It also ensures that most mobiles will transmit less power to reach the DAS. This improves the interference scenario of neighboring cells both inside the building and possibly in the surrounding macro network – increasing the overall network performance and capacity in the area gives better data performance, so it is highly recommended to have MIMO implemented in these antenna locations at least, if not the whole DAS. Less transmit power from the mobiles also means longer battery life, less power consumption – a more 'green' approach.

In many applications, 4G will support the highest data rates, while voice and other 'slow' services might be supported via 3G and not 4G. Often the 4G mobile operator will compete against the Wi-Fi services in the building, and possibly use them as an offload. Therefore we

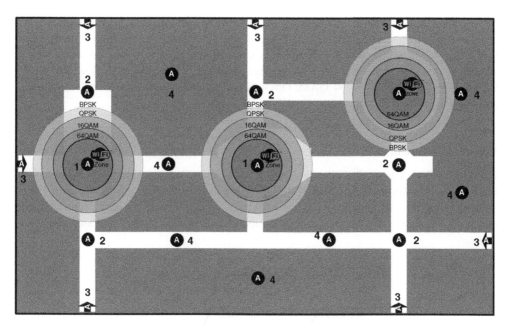

Figure 2.72 4G hotspots, and MIMO/modulation schemes

Table 2.7 Theoretical data throughput on 4G channel bandwidth MIMO

	Downlink	Uplink
Bandwidth	**2 × 2 MIMO**	**1 × 2 SIMO**
5 MHz	37 Mbps	18 Mbps
10 MHz	73 Mbps	36 Mbps
20 MHz	150 Mbps	75 Mbps

need to pay close attention to the exact location of the hotspots in the building. This is where the Wi-Fi access points will be located (if the Wi-Fi designer has done a good job). In Figure 2.72 we can see the principle of 'hotspot 4G planning' – we place the 4G DAS antennas in the same spots as the Wi-Fi access points, ensuring we can match the performance. We could leave the design at that, and cover the rest of the space with 3G. Most probably there will already be an existing 3G DAS, and we would want only to 'fill in' with 4G in the hotspot areas. For more details about hotspot planning and strategy, refer to Section 5.4.1. It will be advisable to make the most of these 4G hotspot locations in terms of reporting, evaluation and handover to and from lower-order systems, 3G and 2G.

MIMO Throughputs on 4G Indoors

We have covered the basics of MIMO in Chapter 2.6.5, but how does MIMO perform in real-life installations indoors, compared with the theoretical performance in Table 2.7?

I had to opportunity to see several deployments and to evaluate actual performance. Most of these deployments were implemented using X-pol MIMO antennas due to practical limitations. And it is clear that MIMO really increases the throughput regardless of whether we are very close to the antennas or we are on the edge of coverage, and also regardless of whether we are running 16 QAM or 64 QAM. Obviously the highest signal levels and modulation scheme will give the best performance and data throughput.

Frequency Reuse in 4G

In Figure 2.73 we can see that all cells in the 4G network will operate on the same frequency. Each cell will be assigned a portion of the spectrum for the specific capacity needed for that coverage area. The assignment of the bandwidth for each cell can be static, using fractional allocation, or adaptable, pending the needed capacity in a dynamic way using soft frequency reuse.

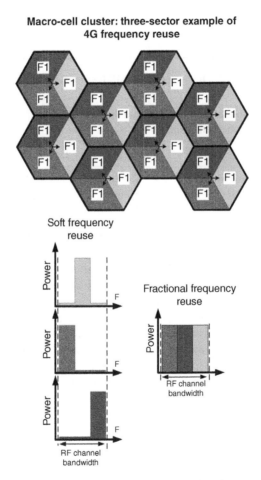

Figure 2.73   4G frequency re-use in 4G can be fractional reuse or soft reuse

Fractional Frequency Reuse

A portion of the spectrum will permanently be assigned to each cell – this will limit the maximum throughput to each user to the pre-assigned bandwidth of each cell. The pre-assigned bandwidth does not need to be symmetrical, so each cell could have different assignment of capacity according to the load in the particular coverage area.

Soft Frequency Reuse

With soft frequency reuse, the bandwidth assignment for each cell is dynamic, so the bandwidth will be adjusted according to the load profile of the specific area. Each cell will then be able to utilize the full bandwidth and maximize the throughput. Transmit power assignment will be adjusted to individual PRBs to minimize inter-cell interference. This will limit the maximum throughput to each individual user, especially limiting the throughput for users near or on the cell edge, and users close to the cell will have high peak throughput.

4G Knowledge

Although this chapter merely scratches the surface of 4G/LTE, I hope that some of the key fundamentals have helped you to understand some of the basics. It is not within the scope of this book to cover all details of the 4G system, and I recommend that you seek more detailed knowledge on the Internet, via training and books etc

3

Indoor Radio Planning

There are numerous challenges, both from a business and a technical perspective, when designing and implementing indoor coverage solutions. Indoor radio planners carry a major responsibility for the overall business case and performance of the network. In many countries 80% of users are inside buildings, and providing high-performance indoor coverage, especially on higher data rates, is a challenge. It is much more than a technical challenge; the business case must also be evaluated, as well as future-proofing of the solutions implemented, among other considerations. As an indoor radio planner, it is important not to focus only on the technical challenge ahead, but to look a few steps ahead.

3.1 Why is In-building Coverage Important?

There are many reasons for the mobile operator, both technical and commercial, for providing sufficient in-building coverage. The technical motivations are typical; lack of coverage, improvement of service quality, need for more capacity, need for higher data rates and to offload the existing macro network. In 3G & 4G networks, the need to offload the existing macro network is an especially important parameter. The need for higher-speed data rates inside buildings also plays an important factor. It is evident that you will need dedicated in-building (IB) solutions to provide high-speed data service on 3G/4G and especially when deploying 4G high-speed data services.

This book will focus mainly on the technical part of the evaluation, and design of IB solutions. However, even the most hardcore technical RF design engineer must realize that the main driver for any mobile network operator must be to increase the revenue factor. The purpose is to maximize the revenue of the network and to lower the production cost of the traffic. The cost of producing a call minute (CM) is a crucial factor for the mobile operator, and so is the production cost per Mb of data transmitted in the network.

Indoor Radio Planning: A Practical Guide for 2G, 3G and 4G, Third Edition. Morten Tolstrup.
© 2015 John Wiley & Sons, Ltd. Published 2015 by John Wiley & Sons, Ltd.

3.1.1 Commercial and Technical Evaluation

First and foremost, the mobile operator must do a business evaluation, before even considering investing in any in-building coverage solution. The operator must use standard tools and metrics to evaluate the business case, in order to be able to calculate the revenue of each individual user of the different user segments. This will enable the operator to compare the business case on all the individual indoor coverage projects, in order to prioritize the projects.

This evaluation must be based on a standardized evaluation flow (see Section 5.1.1), using standard forms templates and metrics in order to secure a valid comparable business case for each indoor coverage project.

3.1.2 The Main Part of the Mobile Traffic is Indoors

Depending on what part of the world you analyze, it is a fact that the bulk of the traffic originates inside buildings. Therefore, special attention to the in-building coverage is needed, in order to fulfill the user's expectations and need for service. This is especially the case in urban environments, and the focus from the mobile users is on higher and higher data rates.

3.1.3 Some 70–80% of Mobile Traffic is Inside Buildings

In most cities it is very interesting to note that typically a few important buildings (hotspots) will produce the major part of the traffic. In some cities more than 50% of the traffic originates from about 10% of the buildings. These buildings are referred to as the hotspots. These buildings will typically be shopping malls, airports and large corporate office buildings.

3.1.4 Indoor Solutions Can Make a Great Business Case

Especially for 3G, the power load per user (PLPU) is an important factor owing to the fact that downlink power on the base station is directly related to the capacity. The higher the PLPU, the higher the capacity drain from the base station per mobile user will be. This results in relative high production costs for indoor traffic on 3G when trying to service the users inside buildings, from the outdoor Macro network.

Not only will the coverage, quality and data speed be better on 3G with dedicated indoor coverage solutions, but the PLPU will be much lower due to the fact that with an indoor system the base station will not to have to overcome the high penetration loss of the building (20–50 dB). In addition, when servicing indoor 3G users from the macro base station, the signal will rely mostly on reflections in order to service the users, degrading the orthogonality.

Implementing indoor coverage solutions is a very efficient use of the capacity (DL power) of the base station, and of the data channel. You can reduce the production cost per call minute or Mb, using IB coverage solutions and also reduce the overall noise increase in the network. *The production cost per call minute or Mb on 3G can be cut by 50–70% using IB coverage solutions.*

3.1.5   Business Evaluation

Even the most technical dedicated engineer must appreciate that the main reason for providing IB solutions is to make a positive business case. The designer of the IB solution carries a major responsibility, on the one hand, for a well-designed and high-performing technical solution, but also a solution that needs to be future-proof, and will make the investment worthwhile. This is a fine balance between investment and technical parameters. Engineers are often tempted to overdesign the solutions 'just to be sure' – but the cost of doing this is high.

3.1.6   Coverage Levels/Cost Level

Selecting the correct coverage design level for the indoor design is crucial, for the performance of the indoor system, the data throughput performance and the leakage from the building. We will take a closer look at these more technical parameters later on in this book, but there is more to it than 'only' the technical part.

The radio planner must also realize that the design levels come at a cost, and this has a direct impact on the business case. The higher the coverage level, the higher the cost of the system, obviously due to the need for more antennas and equipment for the indoor system, more equipment, more installation work and maintenance costs.

3.1.7   Evaluate the Value of the Proposed Solution

Before considering any indoor coverage solution, you must carefully evaluate the value of the proposed solution. You will need to answer these questions:

1. Will the investment make a positive business case?
2. When will the investment begin to pay back?
3. Is the selected solution optimum for future needs:
   - Higher data rates.
   - New services.
   - More operators.
   - More capacity?
4. Can the selected solution keep up with the future changes in the building:
   - Reconstruction.
   - Extension?
5. Will the solution offload the macro layer, and free needed capacity? This must be part of the business case for the indoor solution; it is added value if you free up power or capacity on the outdoor network that can service other users.

6. Are there strategic reasons for providing the IB Coverage Solution:
   - Competitive edge over other operators.
   - Increased traffic in other parts of the network.
   - International roamer value (airports, harbors, ships, ferries, hotels, convention centers)?
7. Can dedicated corporate buildings be covered, in order to secure the business for the whole account:
   - Better coverage.
   - Better quality.
   - Better capacity.
   - Higher data rates.
   - More loyalty from the users?

3.2 Indoor Coverage from the Macro Layer

Why not just use the macro coverage to provide the needed indoor coverage? When designing a cellular network, especially in the first phase of rollout, many radio designers initially try to cover as many buildings from the macro layer as possible. This is despite knowing that most of the traffic originates from inside buildings.

To some extent, and in certain areas, this strategy makes sense. In many cases you are able to provide reasonably good overall indoor coverage from the macro base stations, but it is a fine balance and a compromise.

In a typical suburban environment you need to rely on a very tight macro grid with an inter-site distance of no more than 1–2 km, depending on the services that are offered. In urban environments the inter-site distance can be down to 300–500 m to provide the deep indoor penetration needed for 2G and 3G. In many cases even this tight a site grid is not sufficient to provide the higher-data-rate EDGE coverage on 2G, and is not sufficient for providing higher data rates on 3G (64–384 kps). On 3G especially, 4G can be a major concern, when covering from the outside network into the buildings.

As you can see from traffic data from many real-life examples like the one shown in Figure 3.1, even a tight macro grid will in many cases be insufficient to service the indoor users, and certainly not the data users on the higher 4G data rates in particular, but also even voice users on 2G and 3G.

3.2.1 More Revenue with Indoor Solutions

This is an example of the traffic production (Figure 3.1), after implementing solid indoor coverage in a shopping area. This shopping area consists of two major shopping malls, as well as many small outside shops. Prior to implementing the indoor coverage in the two shopping malls, the whole area was covered by a tight grid of standard three-sector 2G macro sites. The macro sites are separated by about 450 m, providing an outdoor coverage level measured on-street of minimum −65 dBm (2G-900) in most of the area.

However, even this high a coverage level from the macro sites was not enough to service all the potential traffic in the area, and thus not able to cater for the actual service need. This is clearly documented by the results shown in Figure 3.1.

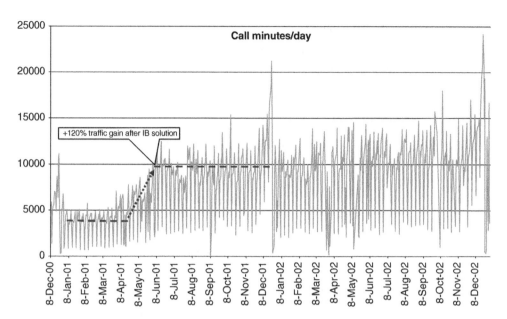

Figure 3.1 The traffic in the area covered by macro sites is more than doubled when implementing indoor coverage

Traffic Boost

This example in Figure 3.1 clearly shows that, in the traffic statistics for the total traffic production for the whole shopping area, there is an instant 'jump' in the traffic production when the indoor coverage system in the two shopping malls is set into service – the traffic rises 120% instantly.

The total traffic in the area exceeded the existing traffic on the macro (prior to the IB coverage being implemented) plus the new traffic on the IB solutions. Not only did the indoor system pick up more traffic, it also boosted the traffic in the neighboring macro-serviced area, by carrying more traffic into this area due to the increased service level, and reduced the number of dropped calls to less than 0.5% in the area after the implementation.

The growth rate of the traffic in the area with 100% IB coverage also showed an annual gain of traffic of about 30%. This is more than double the growth rate in a similar reference area.

Even High Coverage Level from the Macro was Not Enough

Even with an existing macro coverage level of –65 dBm there had been users that were not covered, that you could now capture with the indoor solutions. This clearly shows that IB coverage solutions are important revenue generators for the operators. This is, of course, provided that the indoor coverage solutions are implemented in the right area, covering the right buildings, and are designed correctly.

3.2.2 The Problem Reaching Indoor Mobile Users

Why is it a problem for macro sites to cover inside buildings? Let us explore some of the challenges for the macro coverage penetrating deep inside buildings and providing sufficient coverage and service level where the users are located.

Urban and Suburban Environments Rely on Reflections

In urban and suburban environments macro coverage will typical reach the users by reflection and diffraction – multipath propagation (see Figure 3.2). The delay profile will typical be 1–2 μs. Only a minor part of the traffic is serviced by the direct signal in line-of-sight to the base station antenna, having only the free space loss plus penetration loss of the building (as shown in Figure 3.2).

In most or all cases the resulting signal at the receiver will be a result of the multipath radio channel – a 'mix' of different signals with different delays, amplitudes and phases. The result is a multipath fading signal (as shown in Figure 3.3), with a fading pattern that mainly depends

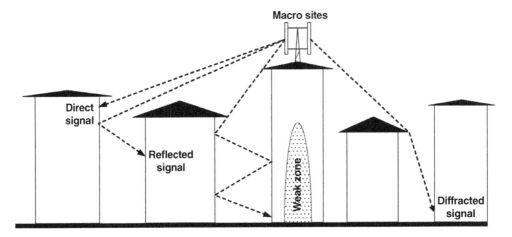

Figure 3.2 Macro sites rely mainly on reflections in urban and suburban environments to provide indoor service

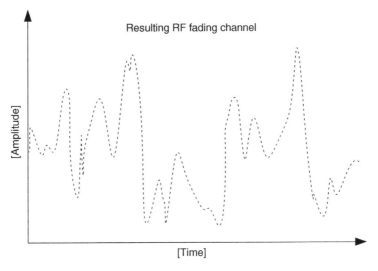

Figure 3.3 Typical multipath radio fading channel

on the environment and the speed of the mobile. Surprisingly, very often a building with a macro site on the roof can have coverage problems at the core of the same building, especially on the lower floors of the building and near the core.

This peculiar problem occurs due to the fact that coverage inside the building relies on reflections from adjacent buildings. On the topmost floors the coverage might be perfect, but the further down, problems starts to occur. Starting with lack of coverage in the staircase and in the elevators, in many cases the inner core of the building might have performance problems, especially with data service.

If a building with the rooftop macro has no, or only low, adjacent buildings, this problem can be a major issue, due to the lack of buildings to reflect the signal back into the servicing building. However the problem can be easily solved by deploying a small indoor solution, filling in the black spot. See Section 4.6.3 for details.

The Mobile Will Handle Multipath Signals

If the reflections are shorter than about 16 µs, the equalizer in the 2G mobile will to some extent cancel the reflection (if longer, it will impact the quality as co-channel interference). The rake receiver on 3G will, if the reflections are offset more than one chip duration (0.26 ms), recover and phase-align the different signal paths and 'reconstruct' the signal to some extent, by the use of the rake receiver and maximum ratio combining; see Section 2.3.6 for more details.

3.3   The Indoor 3G/HSPA Challenge

When providing radio coverage for mobile users inside buildings, you are facing several radio planning challenges. It is mainly these challenges that motivate the need for indoor coverage solutions. Radio planners with indoor 2G planning experience need to be very careful not to apply all the radio planning strategy gained on 2G when designing indoor 3G/HSPA solutions. If you are not careful, you will make some expensive mistakes and compromise high-speed data performance, as well as the business case for these indoor solutions.

3.3.1   3G Orthogonality Degradation

The 3G RF channel efficiency is sensitive to degradation of the 'RF environment', typically caused by multipath reflections. Without going into mathematical details, the efficiency of the 3G RF-channel is expressed using the term 'orthogonality'. The higher the orthogonality of the radio channel is, the higher the efficiency of the radio link is.

The Data Efficiency on the 3G Channel

The same 5 MHz (3.84 Mcps) 3G RF channel can in some cases carry high data rates, in excess of 2 Mbps, provided that the users are in line-of-sight to the serving cell, and only if a minor portion of the signal is reflected energy. This is the typical indoor scenario, with a

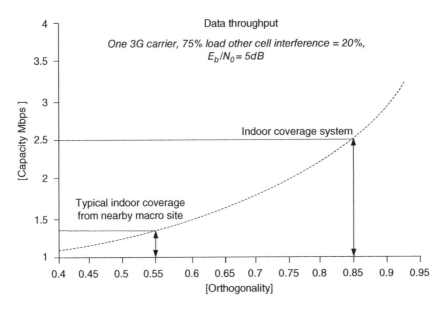

Figure 3.4 Orthogonality affects the efficiency of the 3G RF channel

dedicated indoor coverage solution. In that case the orthogonality can be as high as 0.85–0.90, so the channel is very efficient in carrying high-speed data efficiently (as shown in Figure 3.4).

However when covering 3G indoor user from the macro layer in an urban or suburban environment, the RF channel relies on many reflections, diffractions and phase shifts – a multipath channel. This will degrade the efficiency of the radio channel, and under these circumstances the orthogonality can be as low as 0.55.

Degraded Orthogonality, Higher Costs for the Operator

The degradation of the orthogonality when servicing indoor users from the macro layer is a major concern for the mobile operators. This concern is due to the degraded efficiency of the channel, resulting in a lower data throughput. In practice a particular 3G base station will be able to serve line-of-sight users with high data throughput, whereas indoor users served by the same base station will only be serviced at lower data rates, due to the degraded orthogonality.

The orthogonality is directly related to the production cost per Mb, and thus directly related to the business case of the operator. In addition to the less efficient radio channel, there are other negative effects, including higher power load per user due to the high penetration loss into the buildings. We will elaborate more on these effects later in this chapter.

Degraded Indoor HSDPA/HSUPA Service from the Macro Layer

The modulation on HSPA service is very sensitive to interference and degradation of the radio channel. HSDPA and HSUPA need a high-performing RF link, in order to support the highest possible data rates. In reality this means that HSDPA/HSUPA will only be served in the buildings in direct line-of-sight to the serving macro, and only in the part of the building facing the nearby macro site. To provide coherent and high-performing indoor HSPA coverage, dedicated indoor coverage solutions are needed.

The Macro Layer Will Take a Major Impact from Indoor Traffic

As just described, the orthogonality is degraded when servicing indoor users from the outdoor macro base stations. This is mainly due to the reflections and diffractions and phase shifts of the signals from the clutter in the area, buildings, etc.

More than Orthogonality Degradation

In addition to degraded orthogonality, there are several additional degrading factors to take into account, in order to evaluate the impact on the macro layer from users inside buildings (as shown in Figure 3.5).

The link loss increases with the distance from the base station, the free space loss and the additional penetration loss into the building. The penetration loss depends on the wall type, thickness and material. Typical penetration losses of the outer wall of a building can vary from 15 to 50 dB or more, depending on the type and the frequency.

In addition to the penetration and free space loss from the base station, the clutter inside the building will also add to the attenuation of the RF link. This is especially true for the users

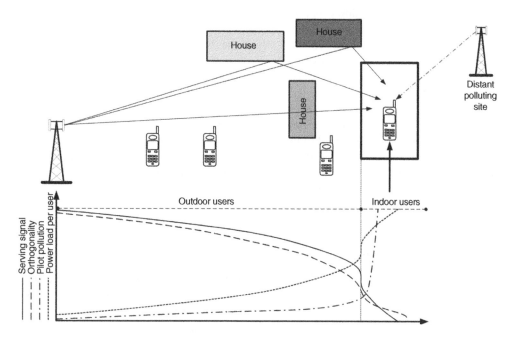

Figure 3.5   The degradation of the 3G channel, and power load when servicing indoor users from the macro layer

at the core deep inside the building and those on the side of the building opposite the servicing macro base station. These indoor users will have a relative high link loss, and demand a high power resource on the base station.

In order to maintain the downlink, the base station has to assign more power to these high-loss users inside the building. These mobiles will also need to power up their transmit power, in order to maintain the uplink.

3.3.2   Power Load per User

The end result is a high power load per user for users with high link losses inside buildings. It is a fact that the capacity resources of the 3G base station are directly related to the power resource. The higher the PLPU is, the higher the power drain off the base station and the higher the capacity drain off the base station will be per indoor user. This high power drain per indoor user has a big impact on the remaining power (capacity) pool of the base station, needed to service other users. This directly relates to the business case.

3.3.3   Interference Control in the Building

To add even more complexity to the matter, the higher up a building the users are, the more likely it is that they will receive nonintended distant base stations (pilot pollution). This is especially a concern in high-rise buildings in the topmost floors that rise over the clutter of other buildings in the area (see Section 3.5.3). This is a particular problem for those users that are close to windows, opposite to the side of the building of the serving base station. These users will be able to detect pilot signals from distant base stations. This pilot pollution will degrade the quality; the E_b/N_o of the 3G signal. For 2G the Rx-Qual (bit error rate) will be degraded, and this might result in dropped calls even though the signal level is relatively high. 3G macro cells close to the building must always be in the neighbor list (monitored set) as soft handover will be active whenever the pilot signals from outside the building are high, in order to prevent pilot pollution and dropped calls. However, soft handover will also pose a potential problem; when the mobile inside the building enters soft handover, more links are used to maintain the same call, so all the degraded orthogonality and power load will in reality hit even more outdoor base stations, draining even more resources from the macro network. HSDPA/4G planning mandates special attention to this problem, as all cells are on the same frequency and interference from other cells will degrade the HSDPA/4G data rate.

3.3.4   The Soft Handover Load

As we know, all serving cells in 3G are on the same frequency, only separated by codes. Therefore 3G has to use soft handover (SHO), as shown in Figure 3.6; this is the only way to shift the calls smoothly when the user roams from one cell to another. Soft handover is a main feature of 3G, securing the traffic transition between the cells, but it comes at a cost. During soft handover the mobile takes up resources on all the cells engaged in soft handover. Typically, with two to three cells, one mobile in soft handover will load the network with a factor of 2–3. However, one must distinguish between softer handover, which occurs within the cells on the

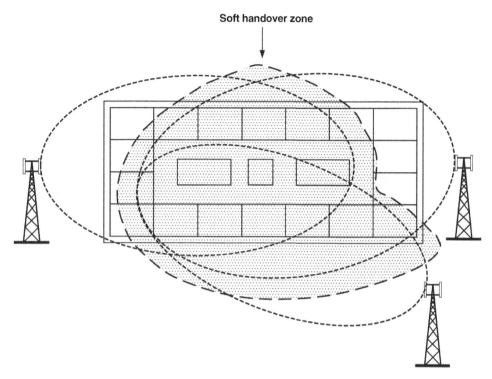

Figure 3.6   Three 3G cells from the macro layer provide excellent indoor service, but most of the building is in soft handover

same site, and soft handover between cells on different sites. The latter takes more resources due to the need for double backhaul to the RNC/RNCs, see Section 2.3.4 for more details.

3.3.5   3G/HSPA Indoor Coverage Conclusion

From the topics just covered, it is clear that it is very costly to cover indoor 3G users from macro base stations because of the impact of power drain of the macro layer, the degraded RF channel due to low orthogonality, the pilot pollution in high-rise buildings and the load increase due to soft handover (lack of dominance) when more macro cells are servicing users inside a building.

Even Perfect Indoor Coverage from 3G Macro Sites can be a Problem

In the example in Figure 3.6 there are three surrounding macro sites providing good, deep indoor coverage. These sites are close to the building, and the signal level throughout the building is perfect.

Therefore all users inside the building are serviced with high RF quality and high-speed data service. However, the users are covered by three different macro cells with a considerable overlap. This is a big concern; the major part of the traffic inside this building will permanently be in soft handover.

In this example one mobile in the soft handover zone will be in simultaneous communication with all three cells. Therefore the same transmission will load three cells, and also put a factor 3 load on the backhaul network. This has a big impact on the outside network, due to the capacity load, increasing the production cost of the traffic for mobiles in the SHO area.

The probability of soft handover decreases when the margin between the serving cell and the other cells in the monitored set/neighbor list increases. As a guideline, one needs a margin of about 10–15 dB to avoid soft handovers – that gives room for fading.

Traffic inside buildings comes at a high cost when serviced from the macro base stations. This might be a minor problem if this is a building with only few users, then the impact on the surrounding macro network will be marginal. In the case of hotspot buildings with high traffic density, like a shopping mall, a big corporate building or an airport, the impact on the macro network will be severe. It is important to dedicate an indoor coverage solution for these types of buildings, to secure a dedicated dominant indoor signal, with enough capacity to accommodate the traffic.

Cover the Hotspot Buildings from the Inside, from Day One

The effects that occur when you try to cover indoor users from the outside are really important to keep in mind; the fact is that only a few hotspot buildings in a city can overload the macro network. The users inside the building might have perfect service, but the fact is that the macro base stations might not have any resources left to maintain service to other users outside these hotspots buildings.

This might not be evident in a 3G macro network with only a minor load of traffic, like in the roll-out phase of the 3G network. However it might become serious when the traffic increases in the network. Then it might be too late, and if you wait to address the 3G indoor coverage issue until then, you will compromise the user perception of the 3G service and the network quality. Users might turn to other operators and the business case for the network will have been compromised. Yes, the radio planner has a direct impact in the business case for the network; it is a big responsibility, one that is not to be underestimated.

3.4 Common 3G/4G Rollout Mistakes

It is very tempting to place a nearby macro site close to one of these hotspot buildings, especially when coming from a 2G radio planning background. Often the result is that the side of the building facing the macro site will have really high signal level. However, eventually there might still be some areas on the far side and in the basement that still have an insufficient coverage level.

3.4.1 The Macro Mistake

The classic example is a shopping mall; these are usually important hotspot buildings with many users, and therefore on the top of the list when prioritizing the roll-out plan. A macro site just across the street from the shopping mall could be a perfect solution, or is it? Typically the operator will realize that sufficient indoor coverage is still lacking in large areas inside the building, especially for HSPA data service.

The capacity needed inside the building might also exceed the resources on the macro site, mainly due to the high power load per user inside the building. Then the operator realizes that the next step is an indoor 3G/4G solution.

3.4.2   Do Not Apply 2G Strategies

Trying to cover a shopping mall from the outdoor macro mast is tempting, but this is the 2G experience that kicks in; this is how the 2G network has been and is still being planned. However, this mindset is expensive to use for 3G/4G radio planning, because after implementing the indoor 3G system in the building, the operator will now realize that about 60–80% of the traffic inside the building is in handover-zones, due to the high signal coming into the building from the nearby macro site, causing a lack of dominance. This will also severely degrade the data performance in the building due to high interference between the cells. This problem can easily be solved on 2G, by using different frequencies, but 3G/4G uses the same frequency on all cells. You might be tempted to assign a dedicated channel for the indoor 3G solution, but once again this is 2G planning, because by doing so you have used a huge part of your capacity. Typically you have only two or three 3G channels in total: one assigned for 3G, one for HSPA and if you have a third channel assigned to the indoor solution, there is no way to utilize this channel for future capacity needs.

The best solution will then be redirecting or removing the sector of the macro that covers the shopping mall. However, often this sector will also service other smaller buildings and areas that cannot be serviced without this sector. Often the only valid solution is to remove the sector and deploy a smaller site in the new problem area outside the shopping mall.

3.4.3   The Correct Way to Plan 3G/4G Indoor Coverage

The correct way to do 3G/4G indoor and macro planning, is to plan the network 'from the inside out' not from the outside in. This is especially important in the high-capacity areas, that is, areas with hotspot buildings. Operators doing 3G/4G roll-out need to realize that a portion of the roll-out cost should be reserved for indoor coverage in the most important high-traffic buildings. Deploying indoor systems in these hotspot buildings will save a lot of cost, grief and hassle in the long run.

This will for sure be the most economical strategy, providing better data service with higher data rates and higher quality in the network.

3G/4G networks should be planned, from the inside out.

Better Business Case with Indoor 3G/4G Solutions

It is a fact that the production cost when servicing the users inside the hotspot buildings from the macro base stations is very high, even if the service inside the hotspot buildings seems to be perfect. This is due to the less efficient RF link from the macro base station, mainly due to the high power drain/interference raise per user that uses up the capacity of the 3G/4G base station.

Depending on the scenario, a cost reduction of up to 65% can be achieved by covering the 3G/4G users from inside the building with indoor coverage systems, rather than using nearby macro base stations to provide solid indoor coverage.

It is not realistic to cover every building from the inside, but the hotspot buildings at least must be considered from the first roll-out plan. These buildings will have a big impact on the macro layer. If you cater for them, the macro layer can use their power resources to service all the other buildings and areas, and the revenue in the network will be boosted.

Most operators do a roll-out plan for their macro network three to five years in advance; the most important buildings should be a part of that plan, and this will give the best performance and the best business case.

3.5   The Basics of Indoor RF Planning

No matter what the radio service, whether it is 2G, 3G, 4G or other technologies, there are some basic design guidelines one must apply in order to design a high-performing indoor coverage solution.

3.5.1   Isolation is the Key

If you must select one parameter and one parameter only that truly defines the most important success parameter when designing an IB solution, it must clearly be the 'isolation'. Isolation is defined as the difference between the IB signal and the outdoor network, and vice versa.

Users in office buildings are typically close to the windows; therefore the dominance of the indoor system must be maintained throughout the building, even right next to the windows.

3.5.2   Tinted Windows Will Help Isolation

Modern energy-efficient windows with a layer of thin metallic coating will attenuate the macro signal servicing the building. This type of window will attenuate the RF signal from the outdoor network, and create a need for a dedicated indoor solution.

The positive side effect of this 'problem' is that, once the indoor system is installed, these metallic-coated windows will actually help the design, giving us good isolation. These types of windows with a thin layer of metallic coating will typically attenuate the signal by 20–40 dB, depending on the radio frequency, and the incident angle of the radio signal.

Recently new types of windows are being used, or film applied to existing windows in high-profile office buildings. These buildings are being fitted with 'Wi-Fi-proof glass'. This prevents hackers with laptops outside the building camping on the Wi-Fi service from within the building. This type of window or film attenuates the radio signal even more, by up to 50–70 dB.

This will actually help in producing a good indoor radio design: these windows, together with the aluminum-coated facade that is typical for many modern corporate buildings, are relatively easy to plan, with regards to isolation (until someone opens a window…). With 40–70 dB of isolation, even a nearby macro site will be shielded efficiently, and you can rely on omni antenna distribution inside the building, as shown in the example in Figure 3.7,

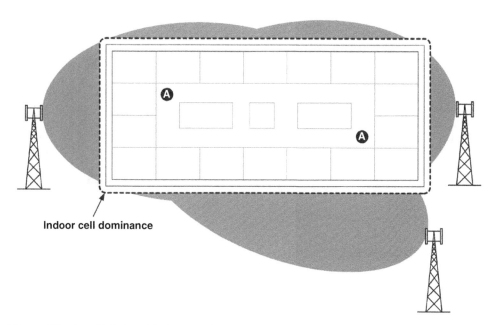

Figure 3.7 A well-designed indoor solution will be dominant throughout the building, but not leak access signal to the surrounding network

without leakage from the building to the outside network. This makes the 'handover zone' easy to design really close to the building, so that the indoor system does not service outside users or leak interference outside the building.

3.5.3 The 'High-rise Problem'

In some high buildings, typically older buildings with normal windows (no metallic coating), you can experience very high interference levels from the outside macro network; even strong signals from distant macro bases will reach indoor users at surprisingly high signal levels. This is mainly due to the low (or no) attenuation from the windows, providing only limited or no isolation between macro base stations and the area inside the building along the windows. Despite these physical factors, you need to insure isolation from the high-level outdoor signal and dominance of the indoor coverage system throughout the building; you need to carefully consider the strategy for designing and implementing the indoor coverage solution.

The traditional approach would be to deploy omni antennas in the walkways near the core of the building (like in Figure 3.7), but this can cause an unwanted side effect. These 'central' omni antennas must radiate high RF levels, in order to overcome the high signal from the nearby macro base stations, and make the indoor cell also dominant along the windows of the building.

The unwanted side effect of this strategy is that the high power from the indoor system will leak high levels of signal from the indoor cell into the macro network, increasing the noise, and degrading the quality and capacity in the outside macro network.

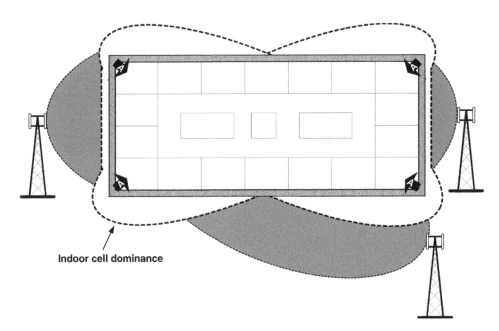

Figure 3.8 The topmost floors in a high-rise building pose specific challenges to isolation due to powerful macro signals. The key is dominate at the border of the building perimeter, in this example using directional antennas mounted in the corners pointing towards the center of the building

Solution to the 'High-rise Problem'

The solution is to make the indoor cell the dominating cell throughout the building. The indoor cell needs to dominate the total area from along the windows and all the way into the center of the building, without leaking signal out to the outdoor network. You can achieve this by deploying directional antennas along the border of the building, and direct them to the center of the building (Figure 3.8).

This design will insure that the indoor signal is stronger than the outdoor signal, even with very high signals from the nearby macro base station. This is due to the proximity and line-of-sight of the user to the nearby serving indoor antenna. Thus the indoor cell is dominant at the border of the building, and when the antennas are pointing inwards to the center of the building, the indoor cell will be dominant throughout the indoor area. This due to the fact that both the outdoor signal and the indoor signal are penetrating the same indoor signal path, so the isolation between the outdoor and indoor signal will 'track' into the center of the building.

Plan for Perfect Isolation

Exactly how efficient the 'corner antenna' strategy is, is shown in the simulation in Figure 3.9. The graph shows how efficient the solution can be when you use the approach of directional antennas pointing inward to the center of the building.

This is a simulation, with the indoor directional antennas mounted suspended in free space, having only the front-to-back isolation of the antenna, 8 dB. In a real-life installation, the

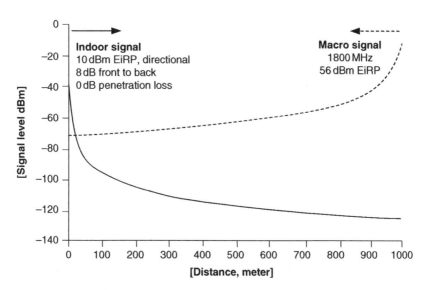

Figure 3.9   The graph shows how efficiently the signal is contained inside the building, with minimum leakage to the outside macro network

antennas will be mounted on a corner pylon or on the outer wall, above the window, so the front-to-back isolation ratio will be 5–20 dB better. In either case it is clear that this is a very efficient way to provide indoor dominance, even in buildings with little penetration loss, and even in buildings with high signals from the outdoor network.

Installation is a Challenge

Placing antennas in the corners pointing towards the center of the building as shown in Figure 3.8 might be a perfect radio planning solution to the 'high-rise problem', but one practical challenge remains: installing the antennas at this location can be an issue. In particular, the installation of cables reaching these 'perimeter' antennas can be a problem, especially when using passive distributed antenna systems that rely on rigid passive cables. It can be difficult to find an appropriate cable route, and end up with an expensive installation, but using active distributed antenna systems, relying on thin cable infrastructure, is a possible solution to the problem (see Section 4.4.2).

What Level of Isolation is Needed?

You know now that the main parameter when doing indoor radio planning is isolation of the macro layer, but how much isolation do you need in reality?

2G
The 2G system will be less restricted in terms of the isolation issue, due to the fact that different frequencies are assigned for the indoor and outdoor cells. However, isolation is still important, especially when planning indoor solutions in high-rise buildings in high-capacity macro

networks. This is the typical example in cities; in these areas you have to deal with the fact that the indoor frequencies might be in use even by nearby macro sites. When planning 2G voice, the co-channel isolation must be more than 11 dB in order to secure the quality – well, this is the theory, but in reality it depends very much on the actual fading environment, so it is advisable to add a margin of 6–10 dB. When planning for 2G data, EDGE MCS-6, you will need a co-channel isolation of more than 17 dB plus a margin! You must plan for a very good signal quality inside the building; refer to Chapter 8 for more details on EDGE planning levels.

Also for 2G, dominance is a concern in order to ease the optimization, and to avoid 'ping-pong handovers' where users toggle between the outdoor and the indoor cell. Even with perfect Rx-Qual the user perceived quality during the 'ping pong' handover is degraded. This is due to the 'bit stealing' of the traffic channel in order to perform fast signaling for the 2G handover. This often results in bursts of 'clicks' on the audio on the call, degrading the voice quality.

3G/4G

As you know, on 3G all users are using the same RF channel, and in order to maintain the link when two or more serving cells are at the same signal level, the mobile enters soft handover, loading more than one link and taking up resources on several network elements and interfaces. If there is no dominant indoor coverage system, then typically two or three base stations will service a user inside an urban building. Even implementing indoor solutions you need to make sure that the mobiles inside the building will only be served by the indoor base station.

You must make sure that this base station is dominant throughout the building; the less dominance, the higher probability there is of soft handover. As a general guideline, you should make the indoor cell 10–15 dB more powerful inside the building than any outside macro signal. However this is a fine balance; you must design the indoor solution in order to make sure that the indoor system does not leak too much signal outside the building, thus 'pushing' the soft handover zone outside the building.

Luckily modern building design will help us, with aluminum-coated exterior walls, metallic-coated windows, etc. In some cases buildings are being installed with 'WLAN-proof windows' in order to protect the IT system within the building from hackers outside, trying to lock in on the Wi-Fi signal. Attenuation of more than 50 dB is in many cases a fact.

However, first of all you need to apply really good radio planning when designing indoor 3G/4G coverage systems. Use corner-mounted antennas to overcome the high-rise problem and generally distribute more antennas inside the building, radiating lower power, to achieve a good uniform signal level. You must definitely avoid 'hot' antennas, i.e. few antennas radiating very high power (In access of 25–30 dBm) to cover large indoor areas. These 'hot' antennas will often cause more problems than solutions, mainly due to leakage from the building into the nearby macro network, raising the noise load on nearby base stations and degrading the capacity.

3.5.4   Radio Service Quality

The performance of any radio link, 2G, 3G and 4G included, is not related to the absolute signal level but rather to the quality of the signal. This is described as the signal-to-noise ratio (SNR), and the bigger the ratio between the desired signal – the signal from the serving indoor cell – and the noise – the unwanted signals from the macro layer – the better performance the link will have, and the higher is the data rates that can be carried on the radio link.

The good radio designer, doing indoor coverage solutions, will make an effort to make sure the design is maximized to have the best possible isolation from the outside network. Leaving some areas of the building in service from the macro sites is tempting, in the areas where the macro coverage is strong. However, this temptation should be avoided if possible; the result on 3G will be large areas of soft handover, large areas with degraded 4G performance.

3.5.5   Indoor RF Design Levels

Frequently operators use only one design level for indoor coverage and one level only. However, as you know, the quality and data throughput of the radio signal is dependent on the SNR, not the absolute signal level. How does this impact the radio planning design for indoor systems?

As you know, isolation plays a major role in the performance of the indoor system. Therefore the actual design level for the signal from the indoor system must be adapted to the particular building, the existing macro coverage inside the building and the isolation of the building. Using more than one design level in the indoor design procedure will also help assure the business case. It is very expensive to provide too high a coverage level in areas of the building where it is not needed.

3.5.6   The Zone Planning Concept

The advice is not to focus on the absolute signal level, and not to use only one planning level. The planning level must be adapted to the project at hand. In practice you might need to use two or three different planning levels in the same building, depending on the 'baseline' of the existing coverage from the surrounding macro sites. An example to describe the need for adaptation of the design level is evident if you look at the typical high-rise building in Figure 3.10.

Divide the Building into Different Planning Zones (areas)

For a high-rise building like the example in Figure 3.10 you can typically divide the building into three different planning zones, each zone-area having special considerations and design requirements. The following design levels are guidelines, ultimately RF surveys will dertermin the best planning level.

Zone A: Coverage Limited Area

The isolation to the outside network in this part of the building is really good, which is no surprise this part of the building is sub-ground level. Typical isolation is better than 70–80 dB and the design level for the indoor cell can be relative low, because you do not need to account for macro interference affecting the performance.

The coverage level for the indoor cell and the noise floor of the indoor cell itself are the main driving factors. One part of Zone A that needs careful consideration is the entrance and

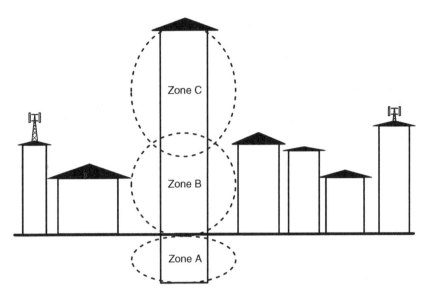

Figure 3.10 Owing to different levels of interference inside the building, it is wise to adjust design levels accordingly, dividing the building into different zones, each zone having individual design levels; this can save cost and maintain RF performance

exit of the building, if Zone A contains an underground parking area. You need to insure sufficient handover level, and allow time for the users entering and exiting the building.

Typical design levels for zone-A are:

- –85 dBm for 2G RxLev.
- –90 dBm CPICH level for 3G.
- –87 dBm RSRP level for 4G.

Zone B: Coverage and Interference Limited Area

This midsection of the building is often served by the nearby macro cells. The building is to some extent isolated from the interference coming from distant macro sites, by the neighboring buildings. When planning indoor coverage for zone B, you need to overcome relatively low interference coming from distant macro sites, and to insure that the indoor cell will be the dominant server, overpowering the coverage from the macro cells that currently cover this area. Typically a medium signal level is needed, to provide sufficient dominance/isolation in order to avoid 'ping-pong' handovers (2G) and to limit the soft handover zones (3G).

You must be careful with leakage from zone B, and make sure that the indoor DAS system does not service outdoor users near the building, pedestrians or nearby cars. Typical design levels for zone-B are:

- –70 dBm for 2G RxLev.
- –80 dBm CPICH level for 3G.
- –75 dBm RSRP level for 4G.

Zone C: Interference Limited Area

The topmost part of the high-rise building is typically above the local clutter of neighboring buildings. Users along the perimeter of the building at the windows will have clear line-of sight to many distant macro sites. These distant base stations will cause co-channel interference and all the usual high-rise problems. In order to overcome the raised noise floor and pilot pollution due to the unwanted signals from all the distant macro sites, you need a high planning level.

Interference and pilot pollution can come from even very distant macro sites, up to 10–20 km away, and therefore these cells will be impossible to frequency/code plan inside the building. The interference has a side effect; with no indoor system the mobiles will typical run on a relative high uplink power, to overcome uplink interference on the base station, especially when camping on distant, unintended servers. This will generate noise increase on the macro network and on the indoor cell, degrading the capacity and data service on 3G/HSPA, and degrading quality on the macro base station. The high transmit power from the mobile will also expose the indoor user to higher levels of radiation.

Typically the users will experience high signal level in the display of the mobile, but suffer degraded service, bad voice quality, reduced data service, dropped calls or even totally lack of service, due to the high interference level. Typical design levels for zone C are:

- −60 dBm for 2G RxLev.
- −70 dBm CPICH level for 3G.
- −70 to −65 dBm RSRP level for 4G.

Be Careful with Leakage from Zone C

However in extreme cases you will need to plan for even higher signal levels from the indoor system. Always be careful when radio planning in zone C and be sure not to leak out excessive power from the building, due to the high indoor design levels. Use corner-mounted antennas facing towards the core of the building, and use many low to medium power antennas uniformly distributed throughout zone C. Do not be tempted to crank up the radiated power in excess of 15–20 dBm in zone C, this might cause excessive leakage into the macro network, degrading performance and capacity.

Beyond Zone C

For very high buildings, there is actually a zone D. In this zone the interference from the outdoor network starts to decrease. However, you still need to be careful about leakage from the building from Zone D.

3.6 RF Metrics Basics

Although most of the audience for this book understands the basics of RF, I had several requests to add a few basic RF facts, focusing on gain, dB, dBm, Watts, delay, some of the main parameters and merits when designing a DAS. These are basic RF metrics you should know like the back of your own hand when you are doing radio planning. And it can be good for all of us to refresh these basics from time to time.

3.6.1 Gain

Gain describes the relative difference between the input signal power and the output signal power of a device. Normally we consider gain as a positive value per default, always expecting more signal power out, than we feed the device – however you could also have negative gain, normally we call this for attenuation.

If we look at the amplifier example in Figure 3.11 we have a device into which we feed a signal power of 1 milliWatt (mW or 1/1000 of a Watt [W], corresponding to 0.001 W). We can see that the amplifier gives 1 W of output power, i.e. 1000 times more than the signal power we feed into it, a factor of 1000.

Gain can also be negative, i.e. we get less signal power out of the device than we feed into it. We normally use the term 'attenuation' for negative gain. An example of a device with negative gain is the attenuator in Figure 3.12. We feed the attenuator an input signal of 1 W (1000 mW), and get 1 mW (0.001 W) output power, so the output signal power is 1000 less than the signal power feed it.

3.6.2 Gain Factor

The gain factor is the output power in relation to the output power of a device. If we continue the example from above in Figure 3.11 We feed the amplifier 1mW, the output power is 1W (1000mW), the amplifier gain the signal power 1000times. So we have a gain factor of 1000 when we compare the output with the input power. This is the so-called gain factor (F):

$$\text{Gain}(F) = P_{out}/P_{in} = 1000/1 = 1000 \,(\text{Figure } 3.11)$$

The gain factor is also used to describe negative gain (attenuation). In the attenuator example in Figure 3.12, we can see that when we input 1 W signal power to the device (attenuator), the output power is merely 1 mW (0.001 W). In other words, the signal power drops by 1000 times.

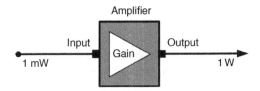

Figure 3.11   Amplifier and gain

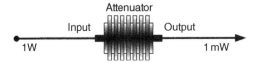

Figure 3.12   Attenuator and loss

As for the amplifier in Figure 3.11 we can calculate the gain factor for the attenuator in Figure 3.12:

$$\text{Gain}(F) = P_{out} / P_{in} = 1/1000 = 0.001 \left(\text{Figure } 3.12\right)$$

In Section 7.2 we can calculate the cascaded noise factor of a complex RF system, and in the Friis formula we need to express the gain and noise in factors.

3.6.3   Decibel (dB)

When designing RF systems it is convenient to express the relation between different signal powers as a logarithmic ratio rather than in factors (F) in order to avoid handling impractical long numbers and decimal figures. Using dB we can easily handle very large and small numbers. By comparison, the decibel (dB) is a dimensionless unit, used for quantifying the ratio between two RF power level values, such as signal-to-noise ratio (S/N), input and output power of a device, component. Hence, dB cannot express absolute values of a power level.

The decibel(dB) number is 10 times the logarithm(base 10) of the ratio between the two signal powers:

$$\text{dB}\left(\text{decibel}\right) = 10 \times \log_{10}\left(F\left[\text{factor}\right]\right)$$

In Table 3.1 we can easily see how convenient it is to use dB compared with factors, and as a radio planning engineer you will need to know the basis of dB extremely well.

Using our basic knowledge from earlier, we can now calculate the gain in dB of the amplifier in Figure 3.11, now that we have established the gain factor (F):

$$\text{Gain}\left(\text{dB}\right) = 10 \times \log_{10}\left(\text{gain factor}\right)$$

A gain (F) of 1000 is therefore equal to:

$$\text{Gain}\left(\text{dB}\right) = 10 \times \log_{10}\left(1000\right) = 30\,\text{dB}$$

We can also calculate the negative gain (attenuation) in dB of the attenuator Figure 3.12. A gain (F) of 0.001 is therefore equal to:

$$\text{Gain}\left(\text{dB}\right) = 10 \times \log_{10}\left(0.001\right) = -30\,\text{dB}$$

As described in the earlier example (Figure 3.12) with the 30 dB attenuator, gain can also be negative, i.e. a loss. Another example is given in Figure 3.13, which shows a 1:4 power splitter.

Table 3.1   Showing the dB vs. ratio comparison of two signal powers

dB	Factor
100 dB	10 000 000 000
90 dB	1 000 000 000
80 dB	100 000 000
70 dB	10 000 000
60 dB	1 000 000
50 dB	100 000
40 dB	10 000
30 dB	1 000
20 dB	100
10 dB	10
6 dB	3.981 (4)
3 dB	1.995 (2)
1 dB	1.259
0 dB	1
−1 dB	0.794
−3 dB	0.501
−6 dB	0.251
−10 dB	0.1
−20 dB	0.01
−30 dB	0.001
−40 dB	0.000 1
−50 dB	0.000 01
−60 dB	0.000 001
−70 dB	0.000 000 1
−80 dB	0.000 000 01
−90 dB	0.000 000 001
−100 dB	0.000 000 000 1

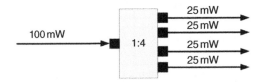

Figure 3.13   Four-way splitter

Example

When we feed the splitter 100 mW, this power is split equally four ways and on the output of each of the four ports we have 100 mW/4 = 25mW (Actually a typical splitter will also have some insertion loss, say 0.2dB or so. We will disregard the 0.2 dB insertion loss in this example, but do include all the insertion and connector losses in large passive DAS or your design could be several dB off.)

$$\text{Gain}(F) \text{ is } P_{out} / P_{in} = 25\,\text{mW} / 100\,\text{mW} = 0.25$$

$$\text{Gain}(dB) \text{ is } 10 \times \log_{10}(\text{gain factor}) = -6\,dB$$

3.6.4   dBm

dBm is another fundamental standard measure that you, as a radio planner, need to understand completely. Unlike dB, dBm is an absolute power value. dBm is related to dBWatt (dBW). This is the RF power in relation to the absolute RF power level of 1 Watt (in a 50 Ω load) and is therefore an absolute unit, to be used when measuring absolute power levels.

dBm (dBmilliWatt) relates to an RF signal power of 1 milliWatt (1 mW), so 0 dBm is 1 mW in a in 50 Ω load. As dBm relates to 1 mW and dBW to 1 W, the relation between dBm and dBW can be established:

$$dB \, ratio = 10 \times \log_{10}(1/0.001) = 30 \, dB$$

In other words, 1 W is 30 dB more than 1 mW, and 0 dBm (1 mW) equals –30 dBW, and 1 W equals 30 dBm.

If we consider the power levels in Table 3.2, it is evident that it is convenient to use dBm to describe RF power levels owing to its ability to express both very large and very power values in a short form. In Table 3.2, we can see that by using dBm, we can describe the high transmit power of a base station, e.g. 43 dBm (20 W), and the low receiving power of the signal at the mobile of –90 dBm, rather than describing the values in Watts. With 20 W transmitted by the base station and 0.000 000 001 mW received at the mobile, it should be obvious that 43 dBm and –90 dBm are much easier to handle and there is less risk of making mistakes.

Table 3.2   The relationship between RF power levels described in dBm and Watt

dBm	Level
46 dBm	40 W
43 dBm	20 W
40 dBm	10 W
30 dBm	1 W
20 dBm	100 mW
10 dBm	10 mW
0 dBm	1 mW
–10 dBm	0.1 mW
–20 dBm	0.01 mW
–30 dBm	0.001 mW
–40 dBm	0.0001 mW
–50 dBm	0.000 01 mW
–60 dBm	0.000 001 mW
–70 dBm	0.000 0001 mW
–80 dBm	0.000 000 01 mW
–90 dBm	0.000 0000 01 mW
–100 dBm	0.000 000 0001 mW
–110 dBm	0.000 000 000 01 mW
–120 dBm	0.000 000 000 001 mW

3.6.5 Equivalent Isotropic Radiated Power (EiRP)

Equivalent isotropic radiated power is used to express the radiated power from an antenna, taking into account the passive losses and the directivity of the antenna, in reference to an omnidirectional antenna with 0 dBi of gain (see Chapter 5.3.7 for basic antenna theory) The EiRP is the power transmitted in the main lope of the antenna, the direction of the antenna radiating maximum power:

$$\text{EiRP} = (\text{RF power from the RF source}) - (\text{loss in the system}) + (\text{directivity of the antenna})$$

Example

If we have a DAS remote unit with 10 dBm of RF transmit power, transmitted via a 2.14 dBiomni antenna fed by a coax cable with 1 dB of loss, we can calculate the transmitted EiRP as follows

$$\text{EiRP} = 10\,\text{dBm} - 1\,\text{dB} + 2.14\,\text{dBi} = 11.14\,\text{dBm}$$

The EiRP is used as the base for estimating the coverage range from the antenna, as we do in Chapter 8 when we construct the link budget for the system.

3.6.6 Delays in the DAS

Normally timing delays in a DAS is not a big concern if you design with passive DAS in a relatively small solution. Delays in the DAS will cause the base station to conclude that the mobile is more distant from the DAS antenna than it actually is. This is because all DAS systems, passive or active, will have propagation delays caused by the fact that velocity over the media (cables) and digital signal processing is always slower than free space propagation over the radio waves.

   If we consider a passive DAS such as that shown in Figure 3.14 (top), we have a base station connected to a passive coax cable and then a single antenna. The mobile is exactly under the DAS antenna, but the base station will actually conclude that the mobile is at an offset distance from the antenna owing to the delay of the cable.

Delays of a Passive DAS

Let us try to calculate the offset between the actual location of the mobile (just under the antenna) and where the base station thinks the mobile is located. In the example (Figure 3.14, top) the delay (velocity) of the RF signal on the passive cable is 88% of the normal radio propagation (C = speed of light = 300 000 km/s). The velocity of the signal on the cable is $C \times 0.88$.

   If we have a theoretical passive cable of 200 m with no delay (such as over the air), the base station will conclude that the mobile is at a distance of 200 m. However, we need to consider that the passive cable has a velocity of 300 000 × 0.88 = 264 000 km/s, or a delay factor of 300 000/264 000= 1.1363.

   Therefore the base station will conclude that the mobile is at a distance of 200 m × 1.1363 = 227 m (the 'ghost' mobile in Figure 3.14) and not at '0 m', just under the antenna.

   Normally the delay of a typical passive DAS is not a big concern, due to the relatively short cables in a typical small passive DAS deployment. However, we might have to consider the

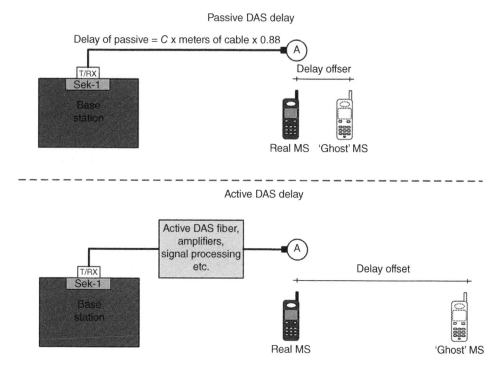

Figure 3.14 DAS delay and 'ghost' mobile, real location vs. delays, espicially when digital functions are used in the DAS

delay when we define 'max cell size' and 'timing window' in the settings for an indoor DAS cell, for lager systems like tunnels etc. You sometimes want to use these 'timing window' parameters in the cell settings to make sure that an indoor DAS cell does not pick up any unintended traffic outside the buildings. Therefore it is easy to be tempted to conclude that you set the maximum cell size to say 100 m to avoid unintended users from outside, but actually the DAS already offsets the cell size by 227 m. Hence the cell will not pick up any traffic, as even the mobiles just under the DAS antennas will be outside the 100 m 'timing window' as a result of the propagation delay of the passive coax cable.

Delays of an Active DAS

Delays of the DAS have an impact on the cell size when we are designing active DAS solutions with longer fiber distances and complex digital processing devices, such as media converters, DSP-amplifiers etc.

The actual delay is based on the signal processing in the active digital part of the DAS and also the velocity of both the fiber and the passive cable. Considering the example in Figure 3.13 (bottom) we can make a short example of an active DAS. Consider the following data:

- Active Digital DAS delay from datasheet = 8μS.
- Velocity over the fiber = 0.681 (factor 1.468).
- 400 m of fiber to the DAS remote.
- 30 m of coax between the DAS remote.

We can now calculate the distance offset for each element and then put it all together:

- Active DAS delay from datasheet = 8μS. We know that light travels at 300 000 km/s, so for 8μS the distance travelled is 300 000 × 8 μS = 300 000 × 0.000 008 = 2.4 km distance offset.
- Velocity over the fiber = 0.681 (delay factor 1.468) so 400 m of fiber will correspond to 400 m × 1.468 = 587 m offset.
- 30 m of coax between the DAS remote with a velocity of 0.88 – this is a delay factor of 1.136, corresponding to 30 × 1.136 = 34 m.

The total distance offset of the active DAS in the example above due to the delay factors of each element will therefore be:

$$\text{Total distance offset} = (\text{offset due to the active DAS}) + (\text{fiber offset}) + \text{coax offset}$$

$$\text{Total distance offset} = 2400\,\text{m} + 587\,\text{m} + 34\,\text{m} = 3021\,\text{m}$$

The total delay of the active DAS system, including the passive fiber and coax, will therefore offset the 'location' of the mobile in Figure 3.14 (bottom), although the mobile is physically located directly below the DAS antenna, and one could assume it to be at 0 m distance. The reality is that the base station considers the mobile to be at the 'ghost' location at a distance of about 3021 m due to the offset.

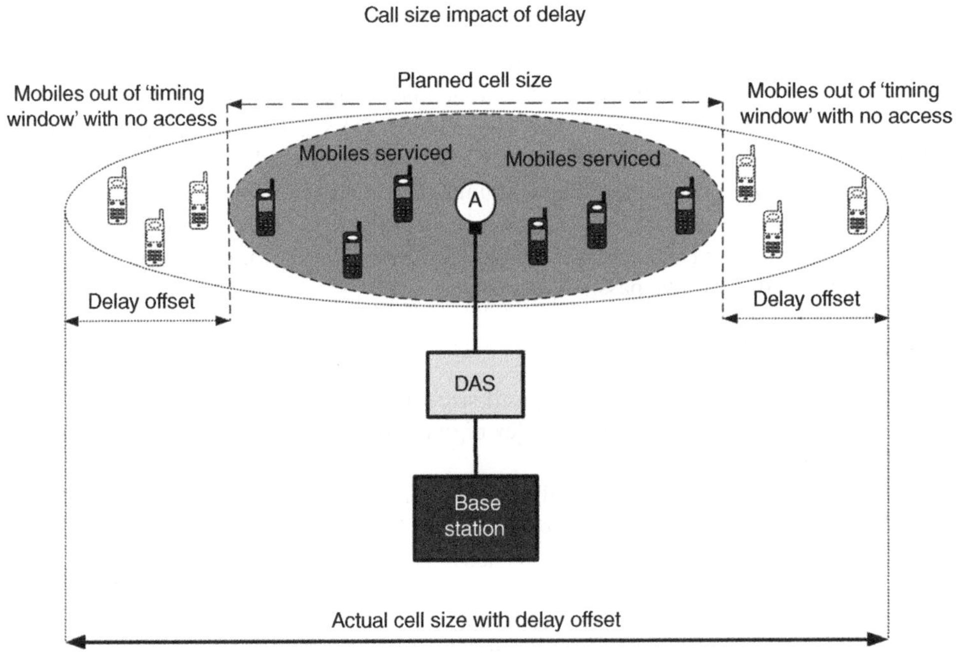

Figure 3.15   Cell size vs. delays in the DAS

3.6.7   Offset of the Cell Size

As described in Section 3.6.6, the delay of a DAS, especially an active DAS, will increase the cell size and thereby offset the location of the mobiles. This effect can be seen in Figure 3.15. The result of this offset could very well be that when you put a new cell into service in the network and you define it as an indoor solution, having a preset default 'timing window'/'cell size' in the network of, say, 1000 m maximum in the network for indoor cells (as we don't want them to be available to service users outside the building), beyond a certain distance from the DAS antenna some mobiles will not be able to set up new calls, and often all mobiles will not set up any calls whatsoever, because they fall outside the 'timing window' of the cell due to the delay of the complete DAS. To solve the issue and still make sure you are not servicing any mobiles at a distance from the building, you need to do the offset calculation as in Section 3.6.4 and use that as the 'zero' distance, and then add your desired cell size on top, thus making this the setting in the network for the maximum planned cell size.

The really big concern and pitfall in relation to DAS and delay of cell size is with active DAS and long fiber distances and passive cable sections (see Section 11.11.1 for a tunnel DAS example).

4

Distributed Antenna Systems

The lesson learned from Section 3.5 is clear; you need to distribute a uniform dominant signal inside the building, from the indoor cell, using indoor antennas in order to provide sufficient coverage and dominance. In order to do so you must split the signal from the indoor base station to several antennas throughout the inside of the building.

Ideally these antenna points should operate roughly at the same power level, and have the same loss/noise figure on the uplink to the serving base station. The motivation for the uniformly distributed coverage level for all antennas in the building is the fact that all the antennas will operate on the same cell, controlled by the same parameter setting. In practice passive DAS will often not provide a uniform design to all antennas; you might have one antenna with 10 dB loss from the base station, and in the same cell an antenna with 45 dB loss back to the base station, and the actual parameter setting for handover control etc. on the base station might not be able to cater for both scenarios. Therefore uniform performance throughout the distributed antenna system is a key parameter in order to optimize the performance of the indoor coverage system.

4.1 What Type of Distributed Antenna System is Best?

There are many different approaches to how you can design an indoor coverage system with uniformly distributed coverage level; passive distribution, active distribution, hybrid solutions, repeaters or even distributed Pico cells (Small Cells) in the building. Each of these approaches have their pros and cons, all depending on the project at hand. One design approach could be perfect for one project, but a very bad choice for the next project – it all depends on the building, and the design requirements for the current project, and the future needs in the building.

Seen purely from a radio planning perceptive you should ideally select the system that can give the most downlink power at the antenna points and the least noise load and loss on the

---

Indoor Radio Planning: A Practical Guide for 2G, 3G and 4G, Third Edition. Morten Tolstrup.

uplink of the base station, and at the same time provide uniform coverage and good isolation to the macro network. On top of the radio planning requirement, other parameters like instal- lation time and costs, surveillance and upgradeability play a significant role. In practice the service requirements and the link budget (see Section 8.1.3) will dictate how much loss and noise you can afford and still accommodate the service level inside the building you are designing for.

4.1.1   Passive or Active DAS

Traditionally passive distributed antenna systems have been used extensively for 2G in the past many years. Therefore naturally many radio planners will see this as the first choice when designing indoor coverage for 3G/4G systems. However, it is a fact that, for 3G and especially for 4G active distributed antenna systems will often give the best radio link performance and higher data rates. The main degrading effect from the passive systems is the high losses, degrading the power level at the antenna points and increasing the base station noise figure on the higher frequencies used for 3G/4G. 3G and 4G can perform really high-speed data trans- mission, but only if the radio link quality is sufficient, and passive systems will to a large extent compromise the performance; see Section 7.2 for more details on NF increase due to passive loss.

Another big concern with passive distributed antenna systems is the lack of supervision. If a cable is disconnected the base station will not generate any voltage standing wave radio (VSWR) alarm, due to the high return loss through the passive distributed antenna system. Distributed indoor antenna systems are implemented in the most important buildings, serving the most important users, generating revenue in our network. Surely you would prefer to have surveillance of any problems in the DAS system.

On the other hand, passive systems are relatively easy to design; and components and cables are rigid and solid, if installed correctly. Passive distribution systems can be installed in really harsh environments, damp and dusty production facilities, tunnels, etc., places where active components will easily fail if not shielded from the harsh environment. Passive distribution systems can be designed so they perform at high data rates, even for indoor 4G solutions – but only for relative small buildings, projects where you can design the passive distributions system with a low loss, short distances of coax.

4.1.2   Learn to Use all the Indoor Tools

It is important that the radio designer knows the basics of all the various types of indoor cov- erage distribution solutions. In many projects the best solution will be a combination of the various types of distribution hardware. Good indoor radio planning is all about having a well- equipped toolbox; if you have more tools in your toolbox, it is easier to do the optimum design for the indoor solution. Having only a hammer might solve many problems, but only having a hammer in you toolkit will limit your possibilities. If you only know about passive distribu- tion, learn about the possibilities and limitations of active distribution, repeaters and Pico cells. This will help you design high-performing indoor coverage distribution systems that are future-proof and can make a solid business case. After all that's why you are here – to generate revenue in the network.

4.1.3   Combine the Tools

Indoor radio planning is not about using one approach only. Learn the pros and cons of all the various ways of designing indoor coverage, and then you will know what is the best approach for the design at hand. Often the best approach will be a combination of the different solution types.

4.2   Passive Components

Before you start exploring the design of passive distributed antenna systems, you need to have a good understanding of the function and usage of the most common type of passive components used when designing indoor passive distributed antenna systems.

4.2.1   General

Inside buildings you must fulfill the internal guidelines and codex that apply for the specific building. In general you will be required to use fire-retardant CFC-free cables and components.

Be very aware of how to minimize any PIM (passive intermodulation) problems. Also be sure that the components used fulfill the required specification, especially when designing high-power passive distributed antenna systems. The effect of combining many high-power carriers on the same passive distributed antenna system using high-power base stations will produce a high-power density in the splitters and components close to the base stations. Use only quality passive components that can meet the PIM and power requirements, 150 dBc or better specified components (see Section 5.7.4).

4.2.2   Coax Cable

Obviously coax cable is widely used in all types of distributed antenna systems, especially in passive systems. Therefore it is important to get the basis right with regards to cable types, and losses. Table 4.1 shows the typical losses for the commonly used types of passive coaxial cables. For accurate data refer to the specific datasheet for the specific cable from the supplier you use. Often there will be a distance marker printed on the cable every 50 cm or 1 m, making it easy to check the installed cable distances.

Calculating the Distance Loss of the Passive Cable

It is very easy to calculate the total loss of the passive coaxial cable at a given frequency.

Example
Calculating the total longitudinal loss of 67 m of $\frac{1}{2}$ inch coax on 1800 MHz, according to Table 4.1:

$$\text{total loss} = \text{distance}\,(\text{m}) \times \text{attenuation per meter}$$
$$\text{total loss} = 67\text{m} \times 0.1\text{dB} / \text{m} = 6.7\text{dB}$$

Reduce the Project Cost When Selecting Cable

The main expense implementing passive indoor systems is not the cable cost, but rather the price for installing the cable. Installing heavy rigid passive cable can be a major challenge in a building. In particular, the heavier types of cable from size $\frac{7}{8}$ inch and up are a major challenge. These heavy cables literally take whole teams of installers; after all the cable is heavy, and a challenge to install without dividing the cable into shorter sections.

Carefully consider the price of installing the cable against the performance. You might be alright with 2 dB extra cable loss, if you can save 50% of the installation costs by just selecting a cable size thinner. On the other hand, do be sure that the distribution system will be able to accommodate the higher frequencies and data speeds on 3G and HSPA.

4.2.3 Splitters

Splitters and power dividers are the most commonly used passive components in distributed antenna systems, dividing the signal to or from more antennas. Splitters (as shown in Figure 4.1) are used for splitting one coax line into two or more lines, and vice versa. When splitting the signal, the power is divided among the ports. If splitting to two ports, only half-power minus the insertion loss, typically about 0.1 dB, is available at the two ports. It is very

Table 4.1 Typical attenuation of coaxial cable

Cable type	Frequency/typical loss per 100 m (dB)		
	900 MHz	1800 MHz	2100 MHz
$\frac{1}{4}$ inch	13	19	20
$\frac{1}{2}$ inch	7	10	11
$\frac{7}{8}$ inch	4	6	6.5
$1\frac{1}{4}$ inch	3	4.4	4.6
$1\frac{5}{8}$ inch	2.4	3.7	3.8

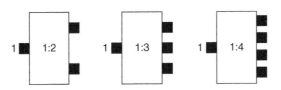

Figure 4.1   Coax power splitters/dividers

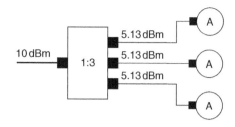

Figure 4.2   Power distribution of a typical 1:3 splitter

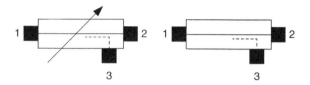

Figure 4.3   Taps, adjustable and fixed

important to terminate all ports on the splitter; do not leave one port open. If it is unused, terminate it with a dummy load.

Example
You can calculate the loss through the splitter:

$$\text{splitter loss} = 10 \log\left(\text{no. of ports}\right) + \text{insertion loss}$$

For a 1:3 splitter (as shown in Figure 4.2), the attenuation will be:

$$10 \log(3) + 0.1\,\text{dB} = 4.87\text{dB}$$

In this example, when we feed a 1:3 splitter 10 dBm power on port 1, the output power on ports 2–4 will be $10 - 4.87 = 5.13$ dBm

4.2.4   Taps/Uneven Splitters

Tap splitters (as shown in Figure 4.3) are used like splitters, used to divide the signal/power from one into two lines. The difference from the standard 1:2 splitter is that the power is not equally divided among the ports.

This is very useful for designs where you install one heavy main cable through the building, and then 'tap' small portions of the power to antennas along the main cable. By doing so, you reduce the need to install many parallel heavy cables, and still keep the loss low.

This is an application that is commonly used in high-rise buildings, where you install a heavy 'vertical' cable and tap off power to the individual floors (as shown in Figure 4.14). By adjusting the coupling loss on the different tappers by selecting the appropriate value, you can actually balance out the loss to all the floors in the high-rise building, providing the required uniform coverage level.

Table 4.2 Typical taps and their coupling losses

Type	Loss port 1–2	Loss port 1–3
1/7 Tap	1 dB	7 dB
0.5/10	0.5 dB	10.5 dB
0.1/15	0.1 dB	15.1 dB
Variable	0.1–1.2 dB	6–15 dB

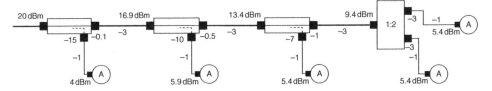

Figure 4.4 Typical configurations of tappers on a distributed antenna system to keep a uniform coverage level for all antennas over a large distance

Figure 4.5 RF attenuator

Taps come in various types, the principle being that there is a low-loss port (1–2) and then a higher-loss port (3), where you 'tap' power of to a local antenna/cluster of antennas. In a high-rise building, you install one vertical heavy $\frac{7}{8}$ or $1\frac{1}{4}$ inch cable in the vertical cable riser, and then use tappers on each floor to feed a splitter that divides out to two to four antennas fed with $\frac{1}{2}$ inch coax. Standard tappers are available with the values shown in Table 4.2.

Example of Use
In this example (as shown in Figure 4.4), we can see that, even over long distances (200 m at 2G-1800), using a $\frac{7}{8}$ inch main cable, and different types of tappers and a splitter, we can keep a relative constant attenuation of all of the antennas within a variation of 1.5 dB, even though the longitudinal loss of the main cable varies up to 12 dB. It is evident that it is possible to balance out the loss efficiently when using tappers.

4.2.5 Attenuators

Attenuators (as shown in Figure 4.5), attenuate the signal with the value of the attenuator. For example a 10 dB attenuator will attenuate the signal by 10 dB (port 2 = port 1 – attenuation). Attenuators are used to bring higher power signals down to a desired range of operation, typical to avoid overdriving an amplifier, or to limit the impact of noise power from an active distributed antenna system (see Section 10.2).

Typical standard attenuator values are 1, 2, 3, 6 10, 12, 18, 20, 30 and 40 dB. When you combine them, you can get the desired value; variable attenuators are also available, but typical only for low power signals. Note that, when attenuating high power signals for many carriers, typically for multioperator applications you should use a special type of attenuator, a 'cable absorber', to avoid PIM problems.

4.2.6   Dummy Loads or Terminators

Terminators (as shown in Figure 4.6), are used as matching loads on the transmission lines, often on one port of a circulator, or any 'open' or unused ports on other components. In applications that are sensitive for PIM, the better option is to use a cable absorber (−160 dBc)

4.2.7   Circulators

The circulator splitter (as shown in Figure 4.7) is a nonreciprocal component with low insertion loss in the forward direction (ports 1–2, 2–3 and 3–1) and high insertion loss in the reverse direction (ports 2–1, 3–2 and 1–3).

The insertion loss in the forward direction is typically less than 0.5 dB and in the reverse direction better than 23 dB. You can get 'double stage' isolators with reverse isolation better than 40 dB if needed.

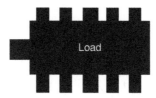

Figure 4.6   Standard 50 Ω dummy load or terminator

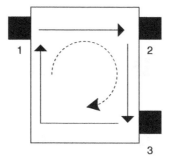

Figure 4.7   RF circulator

Examples of Use

The circulator can be used to protect the port of a transmitter (as shown in Figure 4.8) against reverse power from reflections caused by a disconnected antenna or cable in the antenna system.

A common application for circulators in mobile systems is to use the circulator to separate the transmit and receive directions from a combined Tx/Rx port (as shown in Figure 4.9). This is mostly used for relatively low power applications due to PIM issues in the circulator. For high-power applications it is recommended to use a cavity duplex filter to separate the two signals.

Circulators can also be used to isolate transmitters in a combined network for a multioperator system (as shown in Figure 5.28).

4.2.8 A 3 dB Coupler (90° Hybrid)

The 3 dB coupler shown in Figure 4.10 are mostly used for combining signals from two signal sources. At the same time the coupler will split the two combined signals into two output ports. This can be very useful when designing passive distributed antenna systems.

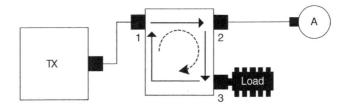

Figure 4.8   Circulator used for protecting a transmitter against reflected power

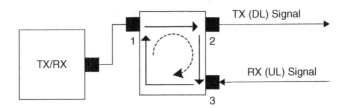

Figure 4.9   Circulator used as a duplexer, separating Rx and Tx from a combined Rx/Tx line

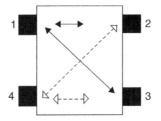

Figure 4.10   A 3 dB coupler

The 3 dB coupler has four ports; two sets of these are isolated from each other (ports 2 and 3/1 and 4). If power is fed to port 1, this power is distributed to ports 2 and 3 (−3 dB). Port 4 will be powerless provided that ports 2 and 3 are ideally matched. Normally a terminator will be connected to port 4.

Example of Use

If you need to combine two transmitters or two transceivers (TRXs/TRUs), you can use a 3 dB coupler (as shown in Figure 4.11). However, if you need to combine the two transmitters and at the same time distribute the power to a passive distributed antenna system with several antennas, you should connect one part of the DAS to port 2 and the other to port 3 (as shown in Figure 4.12). Thus you will increase the power on the DAS by a factor of 2 (3 dB). This method is to be preferred, and will increase the signal level by 3 dB in the building, rather than burning the 3 dB in the dummy load on port 3 (as shown in Figure 4.11) – this will only generate heat!

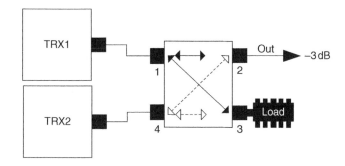

Figure 4.11   A 3 dB coupler used as a two-port combiner

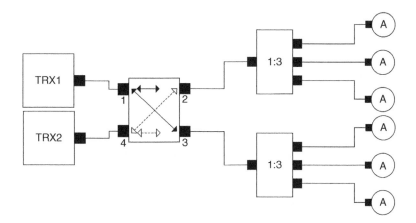

Figure 4.12   Combining two TRX and splitting out to a distributed antenna system

4.2.9   Power Load on Passive Components

One parameter that is very important to keep in mind when designing with passive components is not exceed the maximum power rating that the passive component can handle. This is a particular problem for high capacity or multioperator passive DAS solutions, where you combine many carriers and base stations at high power levels into the same passive distribution system.

Calculating the Total Power from the Base Stations

How do we calculate the total power?

Example, Calculating Composite Power on a Passive DAS

We have a multioperator system in an airport; there are three 2G operators connected at the same point, and each operator has six TRX. The base stations output 46 dBm into the distributed antenna system. In total we have $3 \times 6 = 18$ TRX

Worst case is that all carriers are loaded 100%; therefore each 2G radio transceiver transmits a full 46 dBm constantly on all time slots. We need to sum all the power, but first we have to convert from dBm to Watt:

$$P(\text{mW}) = 10^{\frac{\text{dBm}}{10}} = 40\,000\,(\text{mW}) + = 40 \text{ W}$$

We have 18 carriers of 40 W each; the total composite power is:

$$\text{Total power} = 18 \times 40 \text{ W} = 720 \text{ W}$$

Then back to $\text{dBm} = 10 - 1\log\left(720\,000[\text{mW}]\right) = 58.6\,\text{dBm}$.

Therefore we need to insure that the passive components can handle 720 [Watt]/58.6 dBm constantly in order to make sure the system is stable.

The PIM Power

PIM is covered in Section 5.7.4. However, let us calculate the level of the PIM signals. If we take a passive component, with a PIM specification of -120 dBc, the maximum PIM product will be 120 dB below the highest carrier power. For example, continued from the previous example, we can expect the worst case PIM product to have a signal level of $46\,\text{dBm} - 120\,\text{dB} = -74\,\text{dBm}$.

This is a major concern; -74 dBm in unwanted signal, especially if the inter-modulation product falls in the UL band of one of the systems, it will become a very big problem that will degrade the performance of that system/channel.

Exceeding the maximum the maximum power rating on a passive component will make it even worse. Be sure to use passive components with very good specifications when designing high-power, high-capacity solutions, and be absolutely sure to keep within the power specifications of all the components.

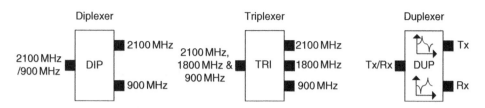

Figure 4.13 The typical filters used to separate frequency bands: diplexer and triplexer. Also, the duplexer used to separate uplink and downlink

4.2.10   Filters

When designing indoor solutions there are basically two types of filters that you will encounter, the duplexer and the diplexer or triplexer, as shown in Figure 4.13.

Duplexer

The duplexer is used to separate a combined TX/RX signal into separate TX and RX lines. Bear in mind the isolation between the bands as well as the insertion loss and the PIM specifications.

Diplexer/Triplexer

The diplexer will separate or combine whole bands from or with each other, for example, input combined 2100 and 1800 MHz and output separate 2100 and 1800 MHz bands. Bear in mind the isolation between the bands as well as the insertion loss and the PIM specifications.

A three band version that can separate or combine 900, 1800 and 2100 MHz is also available, called a triplexer. Some manufacturers even do combined components that contain both a diplexer or triplexer and a duplexer.

4.3   The Passive DAS

Now that we know the function of all the passive components, we are able to make a design of a passive distributed antenna system. Passive DAS systems are the most used approach when providing indoor solutions, especially to small buildings.

4.3.1   Planning the Passive DAS

The passive DAS is relatively easy to plan; the main thing you need to do is to calculate the maximum loss to each antenna in the system, and do the link budget accordingly for the particular areas that each antenna covers. You will need to adapt the design of the passive DAS to the limitations of the building with regards to restrictions to where and how the heavy coax can be installed. Often the RF planner will make a draft design based on floor plans before the initial site survey, and afterwards adapt this design to meet the installation requirements of the building. In fact, the role of the RF planner is often limited to installation planning, not RF planning, when designing passive DAS.

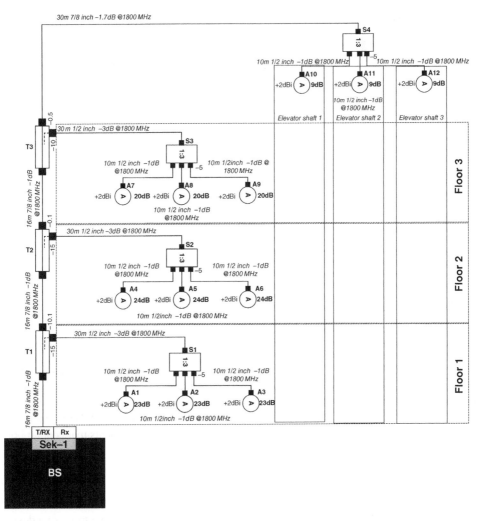

Figure 4.14   Typical passive DAS diagram with the basic information and data

It is very important that you know all the cable distances and types so that you can calculate the loss from the base station to each individual antenna. Therefore you must do a detailed site survey of the building, making sure that there are cable routes to all of the planned antennas. When doing passive DAS design, you will often be limited and restricted as to where you can install the rigid passive cables. Frequently, the limitations of installation possibilities will dictate the actual passive DAS design, and because of this the final passive solution will often be a compromise between radio performance and the reality of the installation restrictions. You need the exact loss of each coax section in the system, in order to verify the link budget (see Chapter 8), and place the antennas.

A typical passive DAS design can be seen in Figure 4.14, showing a small office building. This design relies on a main vertical $\frac{7}{8}$ inch cable, using tappers on each floor that tap off power to the horizontal via $\frac{1}{2}$ inch coax cables to 1:3 splitters on each floor.

The advantage is that the heavy $\frac{7}{8}$ inch cable can be installed in the vertical cable raiser where there is easy access and the installation of the rigid coax will be relative simple. On the office floors it is more of a challenge to installing coaxial cable as no cable trays are available. By using 'thin' $\frac{1}{2}$ inch coax for these horizontal cable runs, we can simply strap the cable onto the frame of the suspended ceiling with cable ties, making the installation relatively fast and inexpensive.

The coverage in this example is to some extent uniform; there is acceptable balance between the antennas serving the three office floors. It was decided to radiate more power into the top of the three elevator shafts, in order to penetrate the lift-car, and provide sufficient coverage inside the lift for voice service.

When doing the diagram (Figure 4.14), all information must be documented, including losses of components and cables, type numbers, component numbers and total loss to each antenna. This design documentation must contain all relevant information, and must be available on-site in case of trouble-shooting.

Trouble-shooting is an issue with passive systems. You will need to use a power meter connected to selected points throughout the passive DAS in order to disclose any faults, in case of a fault on the system. You could use the 'one meter test' from Section 5.2.9 for the worst faults, but you need to connect a power meter to the DAS to be absolutely sure of the power level.

This is the main downside of using passive DAS: trouble-shooting is painstaking. Also, even realizing that there is a fault in the system is pretty much a question of customer complaints from the building; even severe faults on the passive DAS will give no alarm at the base station.

4.3.2 Main Points About Passive DAS

There are many arguments for and against the use of passive DAS. Remember that passive DAS is just one of the tools in the indoor radio planning toolkit, sometimes passive DAS will be the best choice, sometimes not. The clever indoor radio planner will know when to use this approach, and when not to.

The advantages of passive DAS are:

- It is straightforward but time-consuming to design.
- Components from different manufacturers are compatible.
- It can be installed in harsh environment.

The disadvantages of passive DAS are:

- There is no surveillance of errors in the system – the base station will not give VSWR alarm even with errors close to the base station due to high return loss.
- It is not flexible for upgrades (split into more sectors, higher frequencies).
- High losses will degrade data performance.
- It is hard to balance out the link budget for all antennas, and to get a uniform coverage level.
- It requires a high-power base station and dedicated equipment room for site support equipment, power supply, etc.

The fact remains that passive DAS is the most implemented type of DAS on a global basis. However the need for 3G and 4G service and even higher speed data service in the future will affect the preference for selecting DAS types.

The attenuation of the passive DAS is the main issue in this context: frequencies used for future mobile services will most likely get higher and higher, and the modulation schemes applied for high-speed data services are very sensitive to the impact of the passive cable loss. This will degrade the downlink power at the antenna, and on the uplink the high noise figure of the system caused by the passive losses will limit the uplink data speeds. Surely passive systems will continue to be used in the future, but only for small buildings with a few antennas, and the losses must be kept to a minimum.

4.3.3 Applications for Passive DAS

Passive DAS is the most widely used distribution system for indoor coverage systems for mobile service. Passive DAS can be used for very small buildings with a low-power base station and a few antennas, all the way up in size to large airports, campuses, etc.

The main challenge in using passive DAS is the installation of the rigid cables, which have a high impact on the installation cost, and might limit the possibilities as to where antennas can be installed. This could be an issue in solving the high-rise problem covered in Section 3.5.3. The building will more or less dictate the design, because of the installation challenge.

Degraded data service can be an issue if the attenuation of the system gets too high, especially on 3G/4G. This problem can be solved by dividing the passive DAS into small sections or sectors, each serviced by a local base station. However, this is costly and often ineffective use of the capacity resources; refer to Section 6.1.9 regarding trunking gains. The extra backhaul costs, interface loads on the core network and software licenses to the equipment supplier also add to the cost of distributing the base stations.

High user RF exposure with passive DAS. Mobiles inside the building will radiate on relatively high power, due to the fact that the mobile has to overcome the passive losses by transmitting at a higher power level (see Section 4.13). Thus the mobile will expose the users to higher levels of electromagnetic radiation.

Maintenance and trouble-shooting are challenging in passive DAS systems. Be sure to use a certified installer, and do not underestimate the importance of proper installation code and discipline when installing coax solutions and connectors.

'Passive systems never fail' – this is not true. Most likely, you just do not know that there is a fault! Trust me, when an installer is on top of a ladder at four o'clock in the morning, mounting the 65th coax connector that night, he might be a little sloppy. Even small installation faults can result in severe problems, inter-modulation issues, etc. Do not underestimate the composite power in a passive system when combining many carriers and services into the same system at high power.

4.4 Active DAS

The principle function of an active distributed antenna system is that, like a passive distributed antenna system, it distributes the signal to a number of indoor antennas. However, there are some big differences. The active distributed antenna system normally relies on thin cabling, optical fibers and IT type cables, making the installation work very easy compared with the rigid cables used for passive systems.

The active distributed antenna system consists of several components, the exact configuration depending on the specific manufacturer. All active distributed antenna systems will to some extent be able to compensate for the distance and attenuation of the cables.

4.4.1 Easy to Plan

The ability to compensate for the losses of the cables interconnecting the units in an active distributed antenna system makes the system very easy and fast to plan, and easy to implement in the building.

Whereas, when designing a passive distributed antenna system, you need to know the exact cable route and distance for each cable in order to calculate the loss and link budget, when designing active distributed antenna systems it does not matter if the antenna is located 20 m from the base station or even 5 km. The performance will be the same for all antennas in the system; the active DAS system is transparent. This 'transparency' is obtained automatically because the active system will compensate for all cable losses by the use of internal calibrating signals and amplifiers. This is typically done automatically when you connect the units and commission the system. Therefore the radio planner will not need to perform a detailed site survey. It does not matter where the cables are installed, and the system will calibrate any imbalance of the cables. Nor does the radio planner need to do link budget calculations for all the antennas in the building; all antennas will have the same noise figure and the same down-link power, giving truly uniform coverage throughout the building. These active DAS systems are very fast and easy to plan, implement and optimize.

It is a fact that modern buildings are very dynamic in terms of their usage. Having a distributed antenna system that can easily be upgraded and adapted to the need of the building is important. It is important for the users of the building, the building owner and the mobile operator. The active DAS can accommodate that concern, being easy and flexible to adapt and to upgrade. There is no need to rework the whole design and installation if there are changes and additions in the system; there is always the same antenna power, whatever the number of or distance to the antennas.

4.4.2 Pure Active DAS for Large Buildings

Ideally in an active DAS there will be no passive components that are not compensated by the system. Therefore the active DAS is able to monitor the end-to-end performance of the total DAS and give alarms in case of malfunction or disconnection of cables and antennas. These active DAS systems can support one band–one operator, or large multioperator solutions.

No Need for High Power

The philosophy behind the purely active DAS architecture is to have the last DL amplifier and the first UL amplifier as close to the antenna as possible. Co-located with the antenna is the remote unit (RU), avoiding any unnecessary degrading losses of passive coax cables.

When using this philosophy, having the RU located close to the antenna, there is no need to use excessive downlink transmit power from the base station to compensate for losses in passive coax cables; therefore the system can be based on low to medium transmission power from the RU, because all the RU downlink power will be delivered to the antenna with no losses.

Better Data Performance on the Uplink

Purely active DAS has big advantages for the uplink data performance. Having the first uplink in the RU, with no losses back to the base station, will boost the data performance. This is very important for the performance of high-speed data, the higher EDGE coding schemes on 2G, high-speed data on 3G and in particular, 4G performance.

The main difference between the passive and active DAS on the uplink performance is that, even though the active DAS will have a certain noise figure, it will be far lower compared with the high system noise figure on high-loss passive DAS systems. The effective NF performance is basic radio design; refer to Chapter 7 for more details on how the loss and noise figure affect the system performance. Some full active DAS claims wide band support, we need to consider the risk on the UL when unwanted (not serviced by the DAS) signals, "hits" with high level on the UL.

Medium to Large Solutions

By the use of transmission via low-loss optical fibers, a typical active DAS can reach distances of more than 5 km. The cable between EU and RU up to 250 m makes these types of solutions applicable in medium to large buildings, typically large office buildings, shopping malls, hospitals, campus environments and tunnels.

Save on Installation and Project Cost

The active DAS will typically only require about +10 dBm input power from the base station; there is no need for a large, high-power base station installation, with heavy power supply, air-conditioning, etc. A mini-base station can be used to feed the system, and the system components are so small that an equipment room can be avoided; everything can simply be installed in a shaft.

Less power consumption due to the need for less power from the base station with no ventilation saves on operational costs, and makes the system more green and more CO_2 friendly. The fact that the installation work uses the thin cabling infrastructure of an active DAS can also cut project cost and implementation time.

Time to deployment is also short compared with the traditional passive design. You are able to react faster to the users' need for indoor coverage, and thus revenue will be generated faster, and the users will be more loyal.

The Components of the Active DAS

In order to understand how you can use active DAS for indoor coverage planning, you need to understand the elements of the active DAS. Some active DAS systems use pure analog signals; other systems convert the RF to digital and might also apply IP transmission internally.

The names of the units, numbers of ports, distances and cable types will be slightly different, but the principle is the same (Figure 4.15). Typically these types of DAS systems will be able to support both 2G and 3G/4G, so you will need only one DAS system in the building to cater for all mobile services and operators.

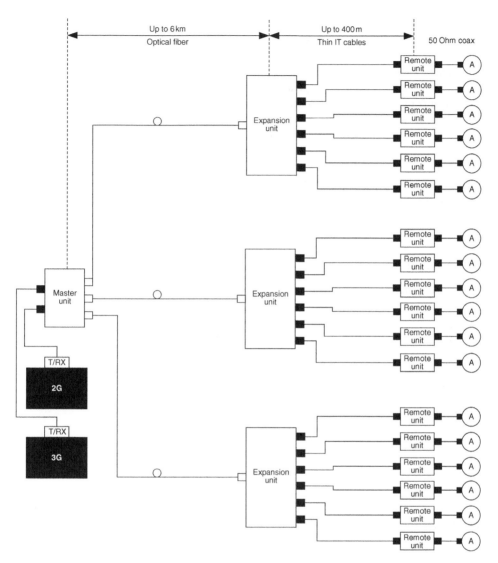

Figure 4.15 Example of a pure active dual band DAS for large buildings; up to 6 km from the base station and antennas with no loss

Main Unit

The main unit (MU) connects to the low-power base station or repeater; the MU distributes the signals to the rest of the system via expansion units (EU). The MU will typically be connected to the EU by optical fibers. The MU is the 'brain' of the system and also generates and controls internal calibration signals in the system together with internal amplifiers, and converters adjust gains and levels to the different ports in order to compensate for the variance of internal cable loss between all the units.

The MU will also monitor the performance of the DAS system, communicating data to all units in the DAS. In the event of a malfunction or a warning it is able to send an alarm signal to the base station that enables the operator to exactly pinpoint the root of the problem and resolve the problem fast.

Detail about specific alarms is typically good; the system will normally pinpoint the exact cable, antenna or component that is the root of the problem. Thus the downtime of the DAS can be limited, and the performance of the system re-established quickly.

It is possible to see the status of the whole system and the individual units at the MU, using LEDs, an internal LCD display or via a connected PC – it all depends on the manufacture of the system.

It is also possible to access the MU remotely via modem or via the internet using IP, perform status investigation, reconfiguration and retrieve alarms in order to ease trouble-shooting and support.

Expansion Unit

The EUs are typically distributed throughout the building or campus and are placed in central cable raisers or IT X-connect rooms. The EU is connected to the MU using optical fibers, typically separate fibers for the UL and the DL. The EU converts the optical signal from the MU to an electrical signal and distributes this to the RU.

Ideally the EU will also feed the DC power supply to the RUs via the existing signal cable in order to avoid the need for local power supply at each antenna point (RU). In many cases LEDs will provide a status of the local EU and subsystem (the RUs).

Optical Fiber Installations

Some systems can use both single mode fiber (SMF) and multimode fiber (MMF), and some systems only SMF. It is important to consider this when planning to reuse already installed fiber in the building, since old fiber installations are typically only MMF. Installation of fibers and fiber connectors takes both education and discipline, so be sure that your installer has been certified for this work, and always refer to the installation guidelines from the manufacturer of the DAS.

Remote Unit

The RU is installed close to the antenna, to keep the passive losses to a minimum and improve the radio link performance. The RU converts the signal from the EU back to normal DL radio signals and the radio signal from the mobiles on the UL is converted and transmitted back to the EU. The RU is located close to the antenna, typically only connected to a short RF jumper. This will insure the best RF performance and the possibility of active DAS to detect when the antenna is disconnected from the system.

The RU should be DC fed with power from the EU in order to avoid expensive local power supply at each antenna point. In addition, the RU should be designed with no fans or other noisy internal parts, in order to enable the system to be installed in quiet office environments.

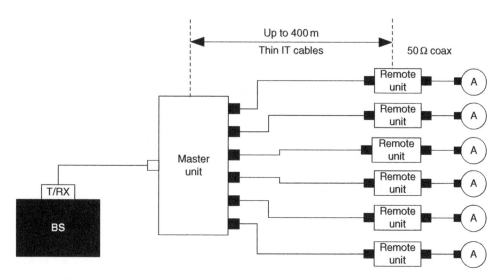

Figure 4.16 Example of a pure active DAS for small buildings; up to 400 m distance between the base station and the antennas with no loss

The RU is connected to the EU with a thin coax, or CAT5 cables or similar 'thin cabling', making it very easy and quick to install compared with the rigid passive coax cables used for passive DAS.

4.4.3 Pure Active DAS for Small to Medium-size Buildings

Even though that active DAS is normally considered applicable for large buildings, small buildings can be designed with pure active DAS (Figure 4.16), thereby utilizing the benefits of having a high-performance DAS, with light cabling infrastructure, a small low-power base station and full surveillance of the DAS.

Main Unit

The system consists of one MU connected directly to RUs with thin coax, CAT5 or other 'IT-type' cables; there is no use of optical fibers. This small system has all the same functions and advantages with regards to auto calibration, uniform performance and improved radio link as the large systems. Even the detailed alarming is the same; full monitoring of the system, including the antennas, is possible.

For medium-sized buildings, you can feed more of these systems in parallel to obtain more antenna points. This makes this medium system a very cost-effective and high-performing system.

Cost-effective Installation

Both of these pure active systems are very cost-effective and easy to install. Often the IT team of the specific building can carry out the installation for the operator, thus sharing the cost load with the operator. In this way the building owner or user pays for the installation and the operator pays for the equipment and infrastructure. This will boost the business case, and create a more loyal customer, who is now a part of the project. This also gives advantages for system performance, as the end user can provide more detailed help with inputs as to where in the building it is most important to plan and implement perfect coverage, and also in understanding why certain areas have a lower signal. However, it is important to note that the units in active DAS cannot always be installed in tough environments, as moisture, vibrations and dust might cause damage. Therefore it is very important to check the manufacturer's guidelines and installation instructions. Many systems can be installed in harsh environments, if shielded inside an appropriate IP-certified box.

Applications for the Pure Active DAS

The large version of the pure active DAS as shown in Figure 4.15 is a very versatile tool for large buildings or campuses. The low-impact installation of the pure active DAS, using and reusing thin 'IT-type' cabling, makes this solution ideal in corporate buildings, hotels and hospitals. The installation is fast and easy, making it possible to react to requests for indoor coverage swiftly. The concept of having the RU close to the antenna boosts the data performance in these high-profile buildings, where 3G/HSPA service is a must.

Radiation from mobiles and DAS antenna systems are a concern for the users of the building. The pure active DAS has the RUs installed close to the antenna; this boosts the uplink data performance. In addition there is a side effect: because the system calibrates the cable losses, there is no attenuation of the signal from the antenna to the base station. Hence the mobile can operate using very low transmit power, because it does not have to compensate for any passive cable loss back to the base station in order to reach the uplink target level used for power control. This makes this approach ideal for installations in hospitals, for example. See Section 4.13 for more details on electromagnetic radiation.

The small version of the pure active DAS shown in Figure 4.16 has all the advantages of the large system, but it does not rely on fiber installation. This takes some of the complexity out of the installation process. The system is ideal and cost-effective for small to medium-size buildings. Given that the system can reach more than 200 m from the MU to the RU, if installed in the center of a high-rise tower it could cover a 300 m-high building without any need for fiber (reserving some distance for horizontal cables). The systems are fully supervised all the way to the antenna; there is full visibility of the performance of the DAS.

Reliability is a concern, due to the number of units distributed throughout the building. You must select a supplier that can document good reliability and mean time between failures (MTBF) statistics since access to so many active elements in the building can be a concern, when servicing the DAS. You also need to make sure that the location of all the installed units is documented, so they can be accessed for maintenance.

You must be careful when installing these types of systems in moist, damp, dusty environments and shield them accordingly.

4.4.4   Active Fiber DAS

The increasing need for more and more bandwidth over the DAS to support multiple radio services, 2G, 2G, 3G, 4G, Wi-Fi, Tetra, PMR etc. has motivated a need for the indoor fiber DAS system to support a wider bandwidth to accommodate all the radio service, as shown in Figure 4.17.

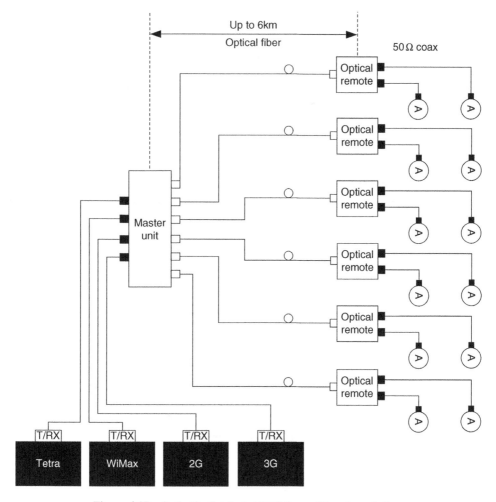

Figure 4.17   Optically distributed DAS for multiservice solutions

Although it is debatable if it is feasible to combine many radio services with quite different design requirement into the same DAS, it is a fact that it is an installation advantage to have only one system and one set of antennas. The compromise will often be the performance on one or more of the services, and a costly DAS.

If all services require 'full' indoor coverage, the consequence is that you will have to design according to the lowest dominator. For example, if the DAS design turns out to show a coverage radius for 2G = 24m, 3G = 21m, but 4G = 10m, you have to place the antennas according to the 4G requirement, thus overdesigning the 2G and 3G system. In fact you often have to adjust the gain of the various services accordingly, to prevent signal leak from the building.

Another concern is the sector plan. Be sure that the DAS allows you to granulate all the different radio systems into independent sector plans, it is very likely you dont need the same sector lay-out for all the services.

The Components of the Fiber DAS

Main Unit
The MU interfaces the fiber DAS to the different base stations. The MU distributes the signals directly to the optical remote units by the use of optical fibers. In many cases this fiber is a composite cable, containing both the fiber and the copper cable for power supply to the remote unit. Alternatively the power is fed locally at the optical remote unit.

The MU is the 'brain' of the system and also generates and controls internal calibration signals in the system together with internal amplifiers, and converters adjust gains and levels to the different ports in order to compensate for the variance of internal optical cable loss between the MU and remote units.

The MU will also monitor the performance of the DAS system. In the event of a malfunction or a warning, it must be able to send an alarm signal to the base station, which enables the operators to exactly pinpoint the root of the problem and resolve the problem fast. Details about specific alarms is typically good; the system will normally pinpoint the exact cable, antenna or component that is the root of the problem, thus the downtime of the DAS can be limited, and the performance of the system quickly re-established.

It is possible to see the status of the whole system and the individual units at the MU using LEDs, an internal LCD display or via a connected PC – it all depends of the manufacture of the system. It is also possible to access the MU remotely via a modem or the internet using IP, and perform status investigation or reconfiguration.

Optical Remote Unit
The ORU is installed throughout the building; it converts the optical signal from the MU back to normal RF and the RF UL signal from the mobiles is converted and transmitted to the MU. The ORU will typically need to operate on medium to high-output RF power and will often have two or more antenna connections, for the DAS antennas.

Daisy-chained Systems

A variant of the optical DAS system can be seen in Figure 4.27, where the optically remote units can be daisy-chained, thereby avoiding the star configuration of Figure 4.17, where all optical remote units need an individual fiber or set of fibers back to MU. This makes the optical system very applicable in tunnels, street DAS and high-rise buildings.

The downside with the daisy-chained system is that, if the fiber is cut, or the optical by-pass in one optical remote fails, then the rest of the DAS on the chain will be out of service.

Applications of the Optically Active DAS

This system is ideal for systems where there is a need for multiple services, other than 'just' 2G and 3G/4G. However, combining all these radio services into the same DAS is a challenge when it comes to inter-modulation and composite power resources.

Power supply to the ORU can also be a challenge; when using the composite cable that has both the copper cable for the DC power and the fiber cable, you need to be careful about galvanic isolation between the buildings and grounding. The concept of having the ORU close to the antenna boosts the data performance in these high-profile buildings, where 3G/4G service is a must.

Radiation from mobiles and DAS antenna systems are a concern for the users of the building. The optically active DAS has the ORU installed close to the antenna; this will boost the uplink data performance. In addition there is a side effect; because the system calibrates the cable losses, there is no attenuation of the signal from the antenna to the base station. Hence the mobile can operate using very low transmission power, because it does not have to compensate for any passive cable loss back to the base station in order to reach the uplink target level used for power control. This makes this approach ideal for installations in hospitals, for example. See Section 4.13 for more details on electromagnetic radiation.

The systems are fully supervised all the way to the antenna; there is full visibility of the performance of the DAS. Reliability is a concern, due to the number of distributed unit throughout the building. It is essential to select a supplier that can document good reliability and MTBF statistics. You also need to make sure that the location of all installed units is documented, so that they can be accessed for maintenance.

Care must be taken when installing these types of systems in moist, damp or dusty environments and they must be shielded accordingly. This is a particular challenge in tunnel installations, where the daisy-chained version of this type of DAS is often used.

4.5   Hybrid Active DAS Solutions

It is important to distinguish between the pure active DAS that we covered in the previous section and hybrid DAS solutions. As the name suggests, a 'hybrid' DAS is a mix of an active DAS and a passive DAS.

The passive part of the hybrid DAS will, as for the pure passive DAS, limit the installation possibilities, impact data performance on 3G/HSPA and to some extent dictate the design, due to installation limitations.

4.5.1   Data Performance on the Uplink

The basic of RF design is to limit any loss prior to the first amplifier in the receive chain. The fact is that the passive portion of the hybrid solution between the antenna and the hybrid remote unit (HRU) will degrade the UL data performance. This is explained in more detail in Chapter 7. The impact of the passive portion of the hybrid DAS becomes a concern for 3G and HSUPA performance.

4.5.2   DL Antenna Power

Even though the typical hybrid DAS produces medium to high power levels out of the HRU, the power at the antenna points will typically be significantly lower. The reason for this is that the power is attenuated by the passive DAS between the HRU and the antenna, but if you only service a few antennas and keep the losses low, you can obtain relatively high radiated power from the antennas.

4.5.3 Antenna Supervision

The small passive DAS after the HRU will give a high return loss back to the HRU; therefore it is often a problem for the HRU to be able to detect any VSWR problems due to the attenuation of the reflection from a disconnected antenna. This, in practice, makes the antenna supervision nonexistent.

4.5.4 Installation Challenges

In order for the HRU to provide high output power levels, the HRU power consumption is quite high. Therefore you will typically need to connect a local power supply to the HRU; alternatively you can use a special hybrid cable from the MU to the HRU that contains both the fiber and copper wires feeding the DC power to the HRU. The need to use a local power supply for each HRU will add cost and complexity to the system, and more points of failure. Most likely, you will not be allowed to use any local power group, but are requested to install a new power group for all the HRUs in the DAS. Therefore, using the hybrid DAS type with composite fiber cables that also accommodate copper wires for power supply of the HRU is often to be preferred; however, there might be a concern in campus installations where you need to make sure that there is galvanic isolation between the buildings. Therefore copper cables might not be allowed to be installed between the buildings, due to grounding issues. Disregarding this might cause severe damage to the DAS and the buildings, and in the worst case start a fire.

Owing to the high power consumption, the HRU normally has active cooling, and often a fan to help with ventilation. This might limit the installation possibilities because of the acoustic noise, which is often restricted to the vertical cable raiser.

4.5.5 The Elements of the Hybrid Active DAS

Refer to Figure 4.18.

Main Unit

The MU connects to the low-power base station or repeater, and distributes the signals to the HRUs in the system. Typically the MU is connected to the HRU using optical fibers. The MU is the controlling element, the 'brain' of the system and also generates and controls internal calibration signals in the system, and then adjust gain and calibration levels to the different ports in order to compensate for the internal cable loss between the MU and HRU. However the system cannot include the passive DAS after the HRU; this part still relies on manual calculation and calibration.

The MU will also monitor the performance of the active part of the DAS system, and in the event of a malfunction is able to send an alarm signal to the base station so the operator can resolve the problem quickly. The system will not be able to detect any problems on the passive DAS, as there will normally be no antenna surveillance. It is possible to see the status of the active part of the system at the MU, on LEDs, an internal LCD display or via a connected PC.

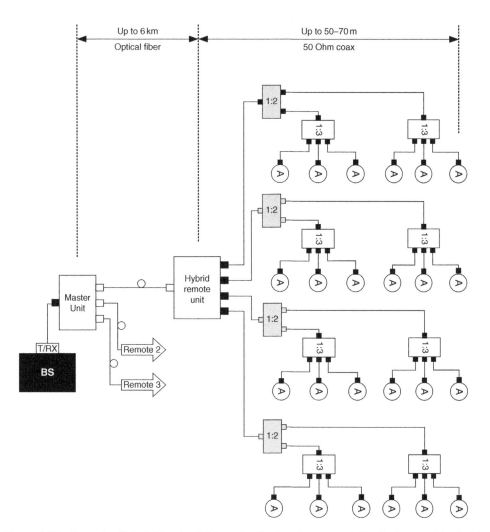

Figure 4.18 Example of a hybrid active DAS, a mix of active elements and distribution, combined with a passive DAS

It will often be possible to access the MU remotely via modem or the internet using IP, perform status investigation or reconfiguration, and retrieve alarms in order to ease troubleshooting and remote support.

Hybrid Remote Unit

The HRU is installed throughout the building. It converts the optical signal from the MU back to normal RF, and the RF UL signal from the mobiles is converted and transmitted to the MU. The HRU will typically need to operate on medium- to high-output RF power, in order to compensate for the losses in the passive DAS that feed the signals to the distributed antennas connected to the HRU.

Applications for Hybrid DAS

Hybrid DAS solutions are ideal solutions where you need high output power at the remote unit. This could be in tunnels; for 'T-Feed systems' see Section 4.8. It could also be sports arenas and multioperator systems where the composite RF power of the remote unit must be shared by many channels.

However, you must be careful with the high output power, and make sure that the uplink can track the coverage area of the downlink, or else the DAS will be out of balance. You must also be careful not to use the high power to only power one 'hot' antenna in the building; this could cause electromagnetic radiation (EMR) concerns (see Section 4.13) and cause interference in the surrounding network.

The relative high cost of the hybrid DAS system normally makes it applicable only for only large structures with high traffic, and high revenue basis.

4.6   Other Hybrid DAS Solutions

Often it can be effective to combine different type of DAS designs in a project, using passive DAS in one part of the building close to the base station, and active DAS in other more distant areas. This will often be a cost efficient option.

It is also possible to combine DAS solutions with macro sectors; for example an outdoor macro site might also be connected to an indoor DAS in the same building, for example where the outdoor sites are located on the roof.

The combination of different concepts will often enable the radio planner to design the DAS as economically as possible, and at the same time maximize performance. Once again it is all about using more tools in the toolbox; do not always rely on only one type.

4.6.1   In-line BDA Solution

It is possible to add an in-line bidirectional amplifier (BDA) to a passive DAS in order to boost the performance of both the uplink and downlink on distant parts of the DAS, as shown in Figure 4.19. However it is a fact that all passive attenuation prior to the BDA, between the BDA and antenna, will seriously impact the noise figure on the system, limiting the uplink performance, which is especially a concern for 3G/4G indoor designs. Preferably the BDA should be installed as close as possible to the antenna. Refer to Section 7.2 for more details on how to optimize the BDA design.

As the name suggests, the BDA is a two-way amplifier with two lines, two amplifier systems, one for the DL one for the UL, and a filter system at both input and output.

In some applications the DC power to the BDA can be fed via the coax. This is very useful for tunnel solutions (see Section 4.9.2) and will save costs and complexity for power distribution. In that case it is important to use passive components that are able to handle the DC power that are feed over the coax cable.

Owing to the remote location of the repeater, the only way to get an alarm back to the operations center is to use a RF modem located at the BDA location. Obviously this concept only works if the RF modem does not have to rely on the coverage signal from the BDA itself!

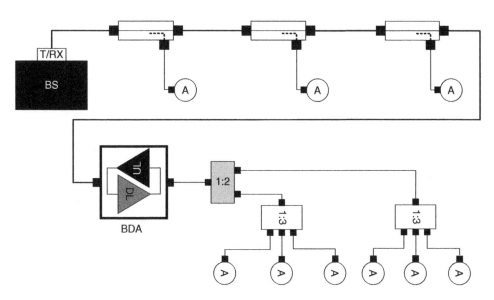

Figure 4.19 Example of a hybrid DAS, a passive system with a BDA added

It is possible to cascade BDAs, i.e. have more than one BDA in the DAS system. This could be done in a daisy-chain structure or even in parallel. However, it is important to be careful about noise control for this type of application. The noise increase in cascaded BDA systems can cause major problems with uplink degradation and have a serious impact on the performance of the base station and the network if you are not careful in the design phase. Refer to Chapter 7 with regards to noise calculation, and how to optimize the BDA design.

4.6.2   Combining Passive and Active Indoor DAS

Often the most ideal solution for an indoor project would be to combine the best parts of passive and the best part of active DAS design (Figure 4.19). Passive DAS is cost-effective in basements, easy to install in the open cable trays available in basements and parking areas, and with low distances and only servicing a few antennas, the RF performance can be good. Active DAS is often more expensive, but it has the edge on performance at longer distances, and is easier to install in the more challenging parts of the building.

In the typical indoor project you will find that the equipment room you are assigned for the base station is located in the basement. Therefore it would be natural to cover the areas near the base station, parking areas, basement, etc., using a passive DAS with few antennas. This is 'zone A' (see Section 3.5.6), so you can get by with a relatively low coverage level.

In the areas far from the equipment room, however, the office floors in the topmost part of the building, the high loss on a passive DAS might degrade the performance. Also, the installation challenge with rigid heavy passive cables might be an issue, and too expensive. Then the natural choice would be to use passive DAS close to the base station, but active DAS in the more challenging areas. Then you can use the best of both applications, and avoid the downsides of any of them – now that is good use of the 'radio planning toolbox'.

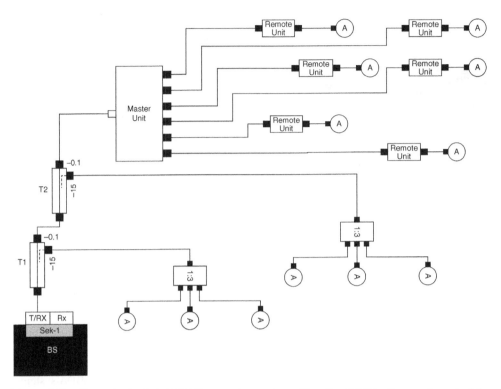

Figure 4.20 Example of a hybrid DAS, a passive system with active DAS in the remote part of the building

Mind the Noise Power

There is one issue you need to be careful about when combining active DAS with a passive DAS, as shown in Figure 4.20 that is the noise power that the active DAS will inject into the passive system (base station). If you have a scenario where the uplink in the passive part of the DAS is the limiting factor, then the noise rise caused by the active DAS could degrade the uplink coverage in the passive part of the DAS.

However the problem can be solved by carefully choosing the correct value of UL attenuator on the main hub of the active system. Refer to Section 7.5 for more detail on how to control the noise power.

4.6.3 Combining Indoor and Outdoor Coverage

Often you will find that a building where a macro base station is located on the rooftop has surprisingly poor indoor coverage near the central core, especially on the lower floors. This is mainly due to the fact that the radio power is beamed away from the building by the high-gain antennas; the coverage inside the building has to rely on reflections from the adjacent buildings and structures. This is a problem especially in high-rise buildings that are not

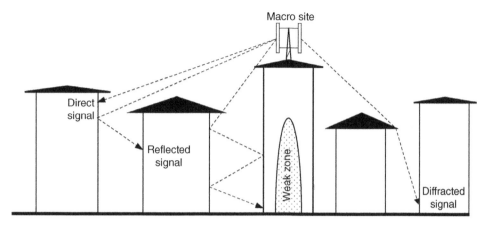

Figure 4.21 RF coverage in the building with the rooftop macro is low due to the need for reflections to provide indoor coverage

surrounded by high adjacent structures that can reflect the RF signal back into the building (as shown in Figure 4.21). In these cases it might make sense to use the rooftop macro base station as a donor for an indoor DAS. This saves the costs of an indoor base station, backhaul, etc. The capacity is trunked between the indoor and outdoor areas. If the traffic profiles between the two areas are offset, it can be a very efficient use of resources; see Section 6.1.9 for load sharing of the traffic profiles.

Minimize the Impact on the Donor Macro Sector

When you split a rooftop macro cell into also serving an indoor DAS there are concerns that need to be addressed. One option is to split the power to one of the outdoor sectors to a passive DAS. However, this approach costs power and coverage area of the outdoor cell, especially if you need to feed a large indoor DAS.

The solution could be to tap off a fraction of the power (0.1 dB) to the macro cell to an active DAS (as shown in Figure 4.22). The advantage of this approach is that the outdoor coverage for the donor cell is maintained. The active DAS typically needs only +5 dBm input power, so only very little power needs to be tapped off from the outdoor sector.

Mind the Noise Power from the DAS

There is one issue you need to be careful about, which is the noise power that the active DAS will inject in the UL of the macro donor sector. This can desensitize the receiver in the base station, limiting the uplink performance and impacting the UL coverage area, and on 3G the noise load will offset admission control. The problem can be solved by installing an attenuator on the uplink port of the active DAS and very carefully choosing the correct value of the attenuator. Refer to Section 7.5 for more detail on how to control the noise power and to Chapter 10 on how to optimize the performance.

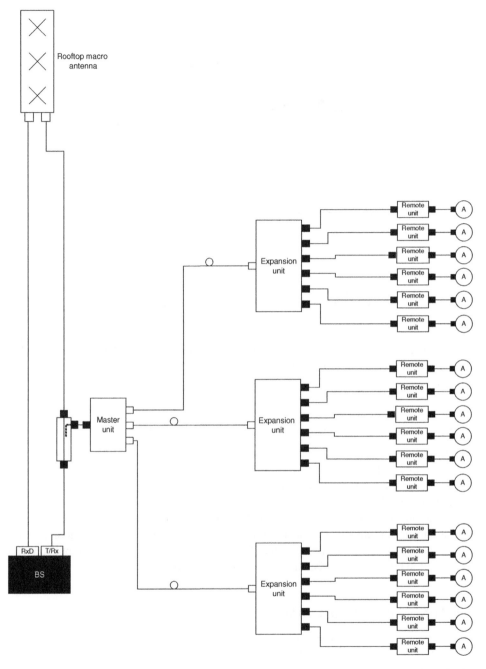

Figure 4.22 Tapping off a fraction (0.1 dB) of the power to the outdoor sector is enough to feed an active indoor DAS, improving the utilization of the base station

Applications for Tapping of a Macro Base Station

Only a fraction of the power from the macro base station is needed, which will be more than enough to feed a large active indoor system. This makes this solution shown in Figure 4.22 applicable to many cases, like shopping malls and sports arenas.

However it is a fine balance; you must remember that all the downlink power and capacity used inside the building will also be radiated from the rooftop macro sector, causing interference load and noise increase in the macro sector. By configuring the settings correctly and carefully planning the indoor system to provide high levels of indoor coverage, it should be possible to minimize this effect, and make sure that the power control on the base station will power down the signals to the indoor users to a minimum.

4.7 Indoor DAS for MIMO Applications

In Section 2.6 we covered the basics of advanced antenna systems, utilizing MIMO so that we could optimize the data performance in the building. At this point in time it seems unrealistic from a practical viewpoint to implement more than 2×2 MIMO inside buildings, due to installation restrictions. As mentioned in Section 2.6 more advanced indoor antennas, utilizing cross polarized antennas, might ease the challenge of implementing MIMO in real life.

The key to good MIMO performance is a total separation of the MIMO links throughout the DAS, from the two antennas all the way to the base station, so it is not possible to use the same passive infrastructure at any point in the indoor DAS – that would destroy the isolation between the MIMO paths in the system.

4.7.1 Calculating the Ideal MIMO Antenna Distance Separation for Indoor DAS

Recent studies and calculations have provided the 'ideal' antenna separation of 3–7λ; this also applies when implementing indoor MIMO DAS solutions in order to benefit from a maximum de-correlation between the MIMO paths created by the scattering of the indoor environment (Figure 4.23).

Using the guideline of a MIMO antenna separation of 3–7λ we can calculate the ideal separations.

We can calculate the wavelength (Lambda = λ) of a given RF frequency:

$$\text{Wavelength } \lambda \left[\text{meters}\right] = 300 \,/\, \text{frequency} \left[\text{MHz}\right]$$

Using this formula we can calculate the ideal DAS antenna separation for MIMO deployment inside a building, for the typical 4G/HSPA+ frequencies.

Looking at the ideal antenna separation for MIMO in Table 4.3 we can easily conclude that practical implementations of the physical antenna separation for indoor DAS becomes a real issue all the more so on the lower frequency bands due to the longer wavelengths at lower frequencies.

As we can see in Table 4.3 MIMO separation of five wavelengths at 700 MHz would take an antenna separation of 2.14 meters.

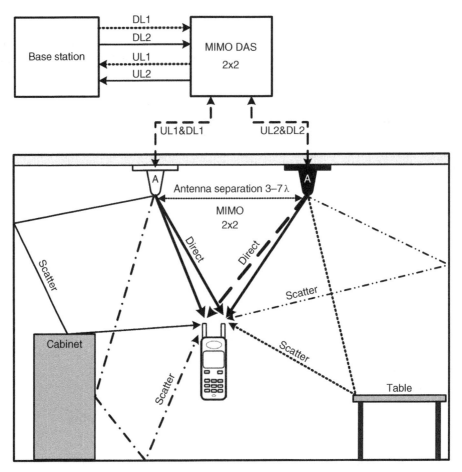

Figure 4.23 The typical MIMO deployment; we must watch out for antenna separation, and at the same time make sure both antennas have good SNR and the delay spread of the DAS is not too big relative to the two signal paths on the DAS. The ideal antenna separation distance is dependant on the frequency/wavelength and also the type of local environment

Table 4.3 MIMO antenna distance

Frequency [MHz]	MIMO distance @ 3 λ [m]	MIMO distance @ 5 λ [m]	MIMO distance @ 7 λ [m]
460	1.96	3.26	4.57
700	1.29	2.14	3.00
850	1.06	1.76	2.47
950	0.95	1.58	2.21
1850	0.49	0.81	1.14
2150	0.42	0.70	0.98
2350	0.38	0.64	0.89
2600	0.35	0.58	0.81
3500	0.26	0.43	0.60

4.7.2 Make Both MIMO Antennas 'Visible' for the Users

We want to make sure that we have sufficient de-correlation between the two antenna paths of 2 × 2 MIMO systems; we must ensure that we do not introduce too big a difference in the path loss in order not to impact upon power control and dynamic range. We must make sure that we do not place the antennas in the same cluster too far apart, which would introduce other potential issues, especially the the 'near-far' effect, where a mobile is close to one antenna.

In practice this will mean that we must make sure that both antennas are 'visible', for example when implementing MIMO inside buildings, like hallways – and make sure that both antennas will beam down the corridors – as illustrated in the example in Figure 4.25.

It is very tempting to implement the antenna clusters as illustrated in Figure 4.24, thus utilizing MIMO between the two antenna clusters A & B themselves. However, we need to be careful; there is great potential for creating a 'near-far' problem when the MIMO antennas are too far apart. This will impact upon power control and might degrade the performance of the cell and minimize MIMO performance. The potential problem is evident in Figure 4.24; the mobile is near antenna cluster A – but is only in line of sight to the white antenna, the other part of the MIMO link is provided by antenna cluster B. This creates several concerns; the link loss to antenna cluster A is much less than the link loss to cluster B due to the distance and the increased number of walls for antenna cluster B. Another concern is the potential for inter symbol interference (HSPA+) due to the highly likely delay difference in the DAS to the antennas in cluster A relative to cluster B. It is unlikely that you have the same cable distance to the two clusters in the DAS, thus skewing the timing/phase between the two clusters of antennas, resulting in a degrading performance – potentially limiting the areas where the users can be serviced by 64QAM. This results in less throughput per area in the building, thus working against the whole purpose of implementing MIMO in the first place; to increase the data speed per area of the building.

A better way of implementing the MIMO clusters of antennas inside the building can be seen in Figure 4.25; here both antennas in both clusters are placed to utilize the corridors, for

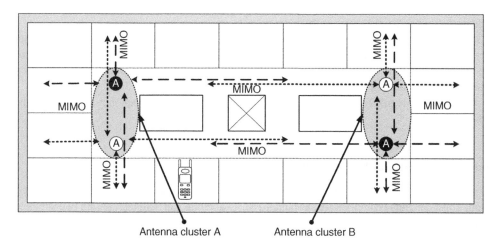

Figure 4.24 Example of layout of two clusters of MIMO antennas; at first glance it seems like a good idea, however we must watch out for potential 'near-far' issues

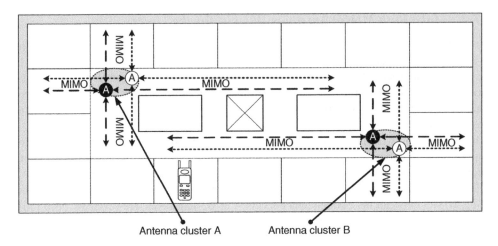

Figure 4.25 Example of layout of two clusters of MIMO antennas; at first glance it seems like a good idea, however we must watch out for potential 'near-far' issues

both antennas in both directions. This will result in a much more uniform RF level performance between the two MIMO paths, at the same time taking advantage of the scattering inside the building. This also results in less inter symbol interference, maximizing the areas where high order modulation 16/64QAM can be utilized, boosting performance inside the building.

It is highly recommended that one keep this in mind when planning the antenna clusters for MIMO in hotspot areas inside the building – where data use is high, and where you must strive for the best possible performance. For more detail about hotspot antenna planning inside buildings, refer to Section 5.4.1.

Other MIMO Concerns

We also must make sure that we have a good signal to noise ratio on both MIMO paths, if not then performance would suffer on both paths.

In real life the ideal antenna separation would also depend on the delay spread and scatter of the local environment; it might well be that we will gain more experience in the future, so we could recommend one separation distance for an open indoor environment, another for a more dense environment with more walls, etc.

Also please note that for multiband solutions it will be the lowest frequency that dictates the antenna distance.

Indoor DAS Antennas Designed for MIMO

Most manufacturers of indoor DAS antennas also produce antennas designed for MIMO operation. These antenna types look like a standard SISO antenna, but actually contain two separate antenna elements inside the same single enclosure (radome). The DAS feeds the two antennas via separate antenna connectors. The major advantage is that these MIMO antennas only appear as one single antenna installation, unlike the examples in Figures 4.24 and 4.25. This is

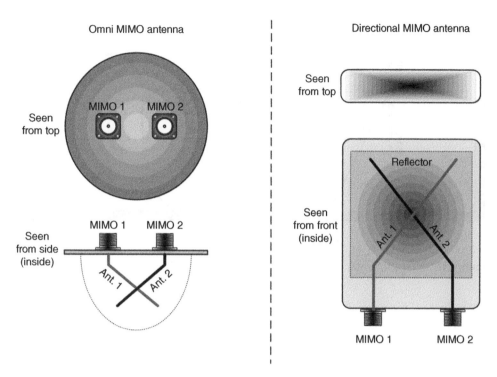

Figure 4.26   X-pol antenna in one Radome – Omni and directional

important when obtaining permission for the physical installation of the DAS. So in real-life installations, it helps with implementing the DAS antennas needed for MIMO support, as you halve the number of antennas needed in the building, with considerably lower visible impact in comparison with using two separate antennas at each of the antenna locations to support MIMO.

However, one might be concerned about the de-correlation performance and whether the MIMO performance will be effective enough. The concern has to do with the lack of physical separation of the antenna elements, as the two antennas reside inside the same radome, which could compromise the required inter-antenna distance according to Table 4.3. However, this inter-antenna distance is based on two antennas with the same polarization. The MIMO antennas shown in Figure 4.25 are certainly close to each other, but they use different polarizations with 90° relative offset (+/– 45°). This polarization separation will actually ensure sufficient MIMO separation of the scattered signals in the building and create sufficient de-correlation between the two individual MIMO paths. The deployments I have see with this approach so far have also been performing well on MIMO. We usually call these type of antenna 'X-pol' or 'cross-polarized' antennas.

Uniform MIMO performance

One of the big performance advantages of these X-pol antennas is the fact that they will typically have uniform MIMO performance. Therefore, an X-pol MIMO omni antenna will have 360° MIMO performance in the service area, relative to the antenna location (see Figure 4.27),

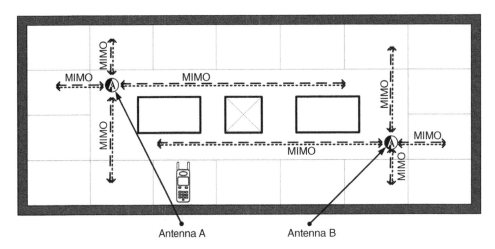

Figure 4.27   Floor plan (office) with MIMO and X-pol, same MIMO performace in 360° service area

unlike the two distance antennas used separately, as in Figures 4.24 and 4.25, which will typically have lower MIMO performance in the direction in which the two antennas 'line up' with the direction of the mobile. However, this also depends on the local environment: in a vcluttered environment, there will most likely be sufficient reflections to compensate, but this could be more of a challenge in a relatively open area. Figure 4.27 shows two MIMO omni antennas deployed in a typical office floor. Having only a single X-pol MIMO antenna to deploy will make it more likely that you will have an installation location for the antenna with clear 'visibility' for both MIMO paths down the hallways, and, in addition, the X-pol MIMO antenna will have 360° MIMO performance.

MIMO upgrades of SISO systems

There are occasions when it would be convenient if you could upgrade existing SISO DAS to MIMO without the need to double the DAS infrastructure, install more antennas and pull more cables throughout the structure. Ideally, you would select every other antenna for MIMO1 and the alternate antennas for MIMO2 etc., thus interleaving the existing SISO antennas to create a MIMO DAS in an existing installation. The success of this approach depends on several factors:

- It must be ensured that the footprints of the interleaved antennas overlap in a sufficient area – in the areas with no overlap there will only be SISO performance.
- The C/I of both antennas in a MIMO system must be sufficient; the performance will be dictated by the C/I of the worst antenna.

If we consider the example in Figure 4.28, an original SISO design for an office area, we could use one antenna for MIMO 1 and the other for MIMO 2 – the MIMO area will be limited to where the two antennas overlap with sufficient signal level. In this example, most of the

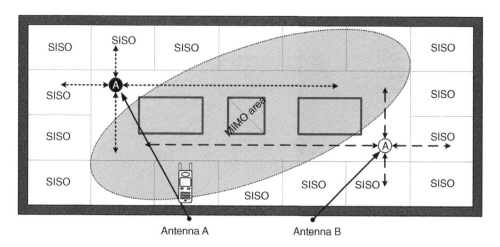

Figure 4.28   Floor plan (office) with MIMO "overlap" of two SISO antennas

MIMO serviced area falls in the hallways and service areas, not in the offices where we expect most of the data load, and where the benefit of MIMO is needed. So the upgrade principle might not be very attractive in this type of application.

But there might be other building types where this approach is more applicable (such as the example in Figure 4.29), typically with large open spaces that can ensure a good overlap of the two MIMO antennas. In the areas with insufficient coverage from one of the antennas, there will only be SISO (Figure 4.29).

We just concluded, based on the office floor application in Figure 4.28, that we would have a limited MIMO performance area, due to the limited overlapping area of the two MIMO antennas. However, if we use the same approach and upgrade an existing SISO DAS by 'interleaving' the existing antennas in open areas, where it is likely to have greater overlap, we might have a more extended service area for MIMO.

Figure 4.29 shows an example where the upgrade of an existing SISO system might perform well. This is a conference auditorium with four existing SISO antennas, labeled 1–4.

We have now upgraded this DAS to use two antennas (antennas 1 and 3) for one MIMO branch, and the other two (antennas 2 and 4)for the other MIMO branch. As shown in Figure 4.29, most of the area actually performs as MIMO (gray area) where there is full service overlap – only in the more distant corners are antennas 1 and 2 not able to reach and overlap with antennas 3 and 4 (white area) and only antennas 3 and 4 cover SISO service. The large MIMO area (gray) is mostly due to the large overlap, thanks to the open aspect of the venue. A similar approach can be seen in Section 14.2.3 where we look at a large, open, hall structure, like a warehouse, in more detail. This approach could also be applied in parking areas, although it is less likely that you will need MIMO to support high-speed data services in parking areas.

It is always recommended to do a real MIMO design if you start from scratch and are designing and installing a new DAS, but sometimes, in open areas of the buildings, you might get by with an upgrade from SISO to MIMO as described. In areas such as exhibition halls, conference areas, and sports arenas, a relatively simple upgrade like this might suffice (for further detail, see Section 14.2.3).

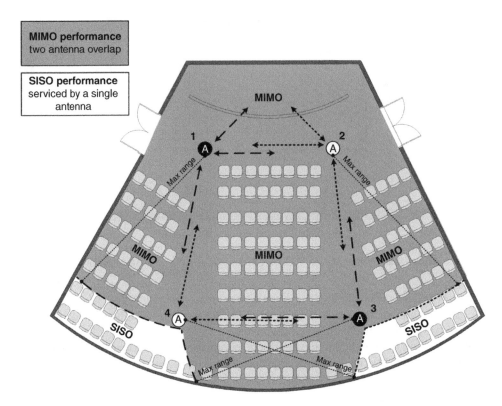

Figure 4.29   Floor plan (large conference hall) with MIMO "overlap"

Although you do not add more antennas with this approach, you will obviously need to pull extra cables to support the two fully separated branches – one cabling line/active DAS for MIMO A and one for MIMO B. If the two branches interconnect at any point in the installation, by accident, the MIMO performance will lost for that cell. So be careful with the documentation and installation discipline and supervision. I sometimes recommend that the installer mark all the different components in two different colours, to avoid mixing up the two individual MIMO paths.

4.7.3   Passive DAS and MIMO

The whole purpose of introducing MIMO inside buildings is to maximize data performance on both the downlink and the uplink. As we have learned in Chapter 7; the more loss we have between the base station and the DAS antenna the worse the Noise Figure will be and would limit data performance on the uplink. So, even considering Passive DAS for MIMO inside buildings (Figure 4.30) contradicts the goal itself. It is a fact that you will need total isolation of the different paths in the DAS for successful MIMO operation, in reality you would need to install two parallel Passive DAS installations, increasing cost and implementation time. On top of this, it is a challenge to ensure that you use the same cable distance on both links in the Passive DAS to avoid inter symbol interference that will degrade both the uplink and downlink (HSPA+).

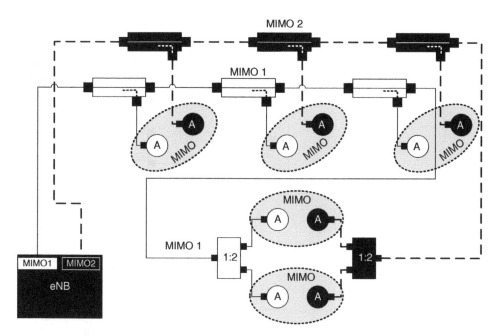

Figure 4.30 An example of a Passive MIMO DAS; in real life Passive DAS would rarely be considered for supporting MIMO, especially in large indoor DAS deployments

You would still be able to operate 4G on a passive DAS with single antennas (SISO mode, see chapter 2.6.1), provided the link budget is confirmed – but obviously not MIMO unless you install a parallel DAS.

4.7.4   Pure Active DAS for MIMO

As described in Section 4.4.3 pure Active DAS has many advantages: ease of installation, 100% visibility on the O&M state of the system – most of all that the remote unit contains the last downlink amplifier and the first uplink amplifier in the system, thus maximizing the downlink coverage and providing a good Noise Figure on the uplink – boosting data speed in the cell.

Pure active DAS, as is shown in Figure 4.31, is a perfect choice for HSPA+/4G MIMO DAS deployments for several reasons.

First of all, we have all the advantages as described in Section 4.4.3 connected with MIMO performance.

The MIMO DAS illustrated in Figure 4.31 supports two independent MIMO paths for both the downlink and the downlink (2×2 MIMO). The eNode-B connects with 2×2 MIMO to the Master Unit, from the Master Unit throughout the system – via the Expansion Unit to the Remote Unit the system relies on one set of optical fiber, cable and units. Within the system, the 2×2 MIMO paths are kept 100% separate with perfect isolation. As illustrated in Figure 4.31, only the interface between the Master Unit and the eNode-B, and the interface to the two MIMO antennas are two independent cable systems. Within the system the two MIMO paths are kept separate by mixing down the two paths to separate intermediate frequencies for

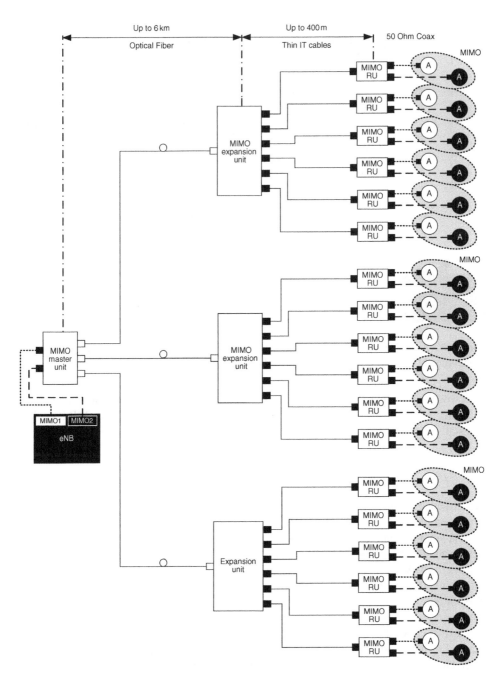

Figure 4.31   The typical Pure Active DAS for MIMO - the two MIMO paths are kept separate when transported in the Active DAS, by means of separate intermediate frequencies for Analogue DAS, or separate data streams for digital active DAS

analogue active DAS, for digital distributed DAS in separate data streams. It is recommended that one keeps the system 'symmetrical' throughout – so the same cable distances for the jumpers and interface between the Master Unit and the base station (eNode-B) and the same type and lengths of jumpers from the remote units to the antennas.

4.7.5   Hybrid DAS and MIMO

In Section 4.5 we covered the basic functions of the Hybrid DAS, i.e. a distributed antenna system that is a mix of an active DAS (with high power Remote Units) feeding a passive DAS. We can also use a Hybrid DAS to support MIMO, as illustrated in Figure 4.32.

Being a Hybrid DAS it has all the disadvantages of the Passive DAS and all the advantages of the Active DAS; the two MIMO paths (2 × 2) are fed to the Master Unit – separated and transmitted over optical fiber to the high power Hybrid Remote Unit. After the Remote Unit we connect two parallel Passive DAS that distribute the signals inside the building. As described earlier it is important to employ a symmetrical design and implementation of the passive system in order not to introduce any time skewing of the two paths.

4.7.6   Upgrading Existing DAS to MIMO

In practice we will encounter a frequent need for upgrading existing DAS installed in buildings. Keeping in mind that we need to make sure that we have a symmetrical deign of the two MIMO paths, the question is whether this is possible at all; to reuse the existing DAS as one path in the MIMO system and install a new DAS in parallel. In practice, it would be worthwhile to consider a complete new DAS design and new implementation for the following reasons:

- The DAS has been installed for some time, likely designed for previous generations of mobile systems, thus the link budget and antenna locations do not match the requirements and link budget for (3G)HSPA+/4G.
- The (Passive) DAS was designed for lower frequencies, so it will have high losses degrading both the downlink power and the uplink Noise Figure, compromising 4G performance on the higher frequencies.
- The lack of documentation might make impossible to install a 'mirror' DAS, a truely 1:1 copy that will fulfill the requirements of good symmetric 2 × 2 MIMO operation.
- The antenna locations do not support dual installation of antennas.
- The existing DAS might be supporting FDD, and you want the new DAS to support TDD – so the constant downlink signal from the FDD system might block the 4G-TDD system when in receive mode.
- The existing antenna locations might not provide sufficient isolation to perform on the higher modulation schemes needed for 4G. For more details of optimizing the isolation refer to Section 3.5.3.

Given these arguments and other good reasons, it is normally worthwhile considering wiping the slate clean and simply designing and installing a completely new system – often a pure Active DAS to support 4G/HSPA+(3G) MIMO systems. This is especially so when considering relatively old DAS implementations designed originally for 2G.

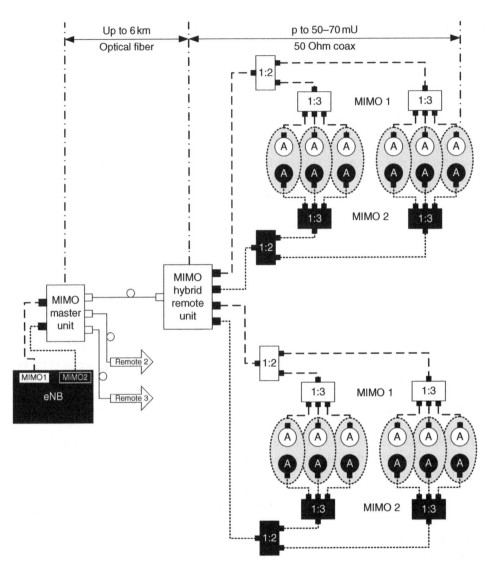

Figure 4.32 The typical Hybrid DAS for MIMO operation; a mix of an active DAS connected to a Passive 'twin' DAS after the remote units

4.8   Using Repeaters for Indoor DAS Coverage

A dedicated base station is often used to find indoor DAS solutions. This is a relatively easy and straightforward approach; you simply dedicate a base station with specific capacity and resources to the indoor DAS. Selecting this type of an approach using a dedicated base station is an easy choice for DAS solutions that demands medium to high capacity. The decision is made by evaluating the potential new traffic when implementing the solution (see Chapter 6) and evaluating the business case based on this evaluation, comparing the investment of base station and DAS plus the supporting transmission, power supply, installation, etc. It is a fact that all Indoor DAS solutions should be implemented based on a business case evaluation (see Section 5.1). However, sometimes it turns out that the traffic in the indoor area that requires coverage improvement does not justify the deployment of a new base station with all the cost associated with this, as well as the backhaul/transmission and upgrade on network elements that supports this new base station.

Where you need only to improve coverage and not add significant capacity inside the building then feeding the DAS with an off air repeater could often be a viable option.

Repeaters typically serve the purpose of providing sufficient signal level inside a building, providing dominance inside a building that lacks single server dominance and in general to improve the data throughput of the service inside buildings.

A repeater needs no connection to the network in terms of transmission interface, but has to rely on an available air signal in the area from an existing base station, pick up this service and re-radiate this existing cell inside the building via a DAS.

Purpose of the Repeater

You could say that the basic purpose of the repeater is to circumvent some of the path loss between the serving cell and the mobile inside the building. So if the repeater installation can compensate for some of the link loss we can improve the link budget and get a higher quality of service inside the building, perfect!

Repeaters are a strong tool, you can solve many problems, but if you are not careful, repeaters can create many more problems than you actually solve – because even if you can see an instant improvement of the RF service signal level inside the building being serviced by the repeater, the impact on the donor cell and other cells could still be negative and affect performance.

Repeaters Should be a Strategic Decision

Quite often repeaters are deployed to provide a remedy for one or more problems, such as VIP location, an indoor DAS where the mobile needs to implement a swift solution to solve the problem. But before implementing that first solution, one should consider the general approach carefully, the different repeater types that could be used and, moreover, remember that the repeater will be another network element, that should be monitored.

There have been far too many cases of networks where a few repeaters has been deployed in 'panic' to solve a pending problem, soon to be followed by yet more similar deployments, only to realize that one to two years down the road, you might have ten to thirty repeaters deployed. Along the way you might have learned some lessons. And now, you realize that you forgot to consider the general impact of repeaters on your network, the impact on the performance of your existing macro net, how to optimize the setting, how to select the appropriate repeater – and one of the most important parameters, remote monitoring of these new network elements

Monitoring Repeaters

Repeaters, like any other active network elements, should be integrated in the mobile operator's network performance monitoring system (the NOC/OMC). Potentially, a faulty repeater could degrade the service quality in a widespread area of your network if it is not configured correctly, or if it becomes faulty, etc. Therefore you must make sure that you can monitor performance, the status of the remote repeater, and if needed switch the repeater system completely off remotely until these problems are solved. It is not recommended to implement any repeaters in your network that you cannot monitor or control and monitor 24/7.

Also, changes in the macro Donor cell might trigger a need for changes of the repeater settings and configuration, such as change of frequency, etc. Most repeaters are remotely controlled via an internal wireless modem and rely on a standalone platform in the network operations centre for control and monitoring of all repeaters implemented. So when selecting repeaters, it is important to always consider the impact of the full implementation over several years, as well as the crucial monitoring system and platform.

4.8.1   Basic Repeater Terms

Let us start with the basic terms in a typical repeater's application (see Figure 4.32).

The Repeater

The Repeater itself (Figure 4.33) is a Bi Directional Amplifier (BDA) that will amplify the Uplink and Downlink of a specific part of the spectrum, simply passing the signal from the outside network to the indoor system, and in reverse for the Uplink amplifying the signal from the mobile inside the building. It is highly recommended that one selects a repeater where the gain on the UL and DL can be manually configured independently. It is also recommended that one selects a repeater where one can select the exact channels/bandwidth of spectrum one requires to support via the DAS inside the building. A more detailed description of the different types of repeaters can be found in Section 4.8.2.

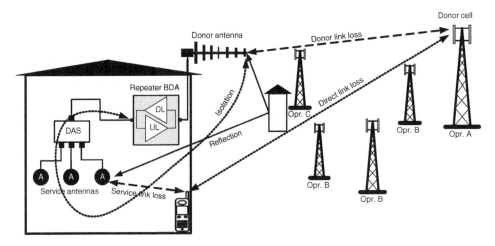

Figure 4.33   Principle terms of a typical repeater application designed to provide indoor coverage and Capacity

The Donor Antenna

The Donor antenna (Figure 4.33) is also sometimes referred to as the 'Pick-up Antenna' this antenna is typically a highly directive antenna that is directed towards the existing macro cell – the Donor cell that is selected to provide signal to the indoor system via the repeater. Since we are normally relying of line of sight service from the Donor cell we must make sure and use the same polarization for the donor pick-up antenna as used by the Donor cell. This can be a challenge if the Donor cell uses 'air combining' – utilizing a cross polarized antenna, one polarization fed by TRX one, the other polarization fed by TRX two. Then you might pick up a good signal on one of the TRX, but when traffic is shifted to the other; the polarization loss of signal level will impact upon performance.

The exact installation location of the Donor antenna in the building's structure is crucial for the performance of the repeater-DAS solution. Not only will we have to ensure that we can reach the desired service level, but we must also place the antenna in such a way that we maximize the isolation between the service antennas inside the building and the Donor antenna. Moving the location of the Donor Antenna itself in the building is often a part of the optimization process, and you have to take advantage of structures in the building and surrounding buildings in order to maximize the Donor Signal, minimize unwanted signals and maximize isolation by carefully selecting the optimum type of donor antenna as well as the ideal location on the building for the installation. You should strive to have line of sight to the Donor cell from the donor/pick-up antenna.

The Donor Link Loss

The Donor Link Loss (Figure 4.33) is obviously an important part of the link budget calculation and you typically will measure the signal level from the Donor cell in order to calculate the Donor Link Loss for your link budget.

Be careful when you measure the signal level at the location of the donor antenna in order for you to estimate initially the required repeater gain and directivity (gain) of the donor antenna. If you use a standard portable mobile phone with test software and apply the directivity (gain) of the donor antenna and use that as a basis for your link calculation you are likely to make a mistake. A mobile with a simple antenna with low directivity will pick up a multi path signal and combine this with a signal level due to the internal equalizer/rake reciever in the receiver. However, when using a highly directive (high gain) antenna as a donor antenna you have virtually no multi path, and a lower donor signal level.

Bear in mind that the link loss will not be stable, but be prone to fading and reflections, etc. This dependstypicaly on the local environment, and the link will normally be more unstable over longer distances. It is highly recommended that one rely only on line of sight Donor cells, in order to minimize fading and to have a more stable system.

The Donor Cell

The Donor cell (Figure 4.33) is the existing macro cell that is selected to provide coverage and capacity to the indoor DAS via the repeater. This Donor cell will then share its existing coverage area to include the indoor area that is added to the service area via the repeater and indoor DAS.

Keep in mind that you bring the cell inside the building via the repeater and the DAS, so if you select a cell that normally does not service this geographical area, you will need to adjust neighbor lists, etc accordingly. The Donor cell might also suffer from UL noise loading from the repeater and DAS – depending on the NF of the repeater-DAS, the UL gain you set in the repeater and the link loss. The impact of this noise load in terms of noise power increase is covered later in this chapter (Section 4.8.3) and in general in Chapter 7. It is important to select a suitable donor cell with a sufficient signal level, but also with a sufficient quality. If the signal to noise ratio on the Donor Cell is bad, no matter what the pickup level, the repeater will never be able to improve quality. It is always recommended that one selects the Donor with the best S/N quality, rather than on the highest signal level and one should strive to achieve a quality of donor signal of a minimum 20 dB signal to noise ratio or better.

Donor Dominance

It is very important to realize that the pickup antenna (Figure 4.33) is a crucial part of the repeater solution, and potentially the problem! The donor antenna is the Achilles heel of the solution, not only should it be of sufficient gain to pick up the appropriate donor cell with the desired quality, but also directive enough to ensure that you do not pick up several potential donor cells in the same signal range. If this happens you create a large soft handover zone inside the building (3G) or C/I problems that result in a degraded service. Therefore it is recommended to make sure that one has at least 10 dB single cell dominance, preferably more than 15 dB of the desired serving cell.

Indoor DAS

The indoor DAS (Figure 4.33) could be any type of DAS system as described in Chapter 4: Passive, Active or a combination (Hybrid). The indoor DAS will then service the mobiles inside the building by re-radiating the selected Donor Cell picked up by the donor antenna, amplified by the repeater on the downlink, and vice versa on the uplink. We will have to calculate a separate link budget for the indoor service range – to estimate the maximum allowable link loss on from the service antennas to the mobile, the Service Link Loss – more detail on Link Budget calculations can be found in Chapter 8. One should strive to have a minimum of 20 dB signal to noise ratio on the Donor cell that the repeater picks up.

Gain

The repeater will have separate amplifiers for the Uplink and Downlink with separate gain settings – more detail on this is to be found later in this chapter. Be careful not to offset the balance between UL and DL gain too much, this can affect power control in the network.

Isolation

Isolation is a very important issue when designing repeater solutions – isolation is defined as the loss (in dB) between the output of the repeater and the input of the repeater.

This includes all gain and losses of donor and service antennas, cable losses, DAS gain (active DAS)/loss (Passive DAS), etc.

Service antennas are a crucial parameter when designing repeater solutions. One could consider this as the Achilles heel of repeater solutions. In order for the repeater to radiate a desired power via the DAS inside the building we must make sure that the isolation will be better than specified by the repeater manufacturer. Typical requirements would be a minimum of isolation of 10 to 15dB plus the gain setting. The margin is required in order to prevent the amplifiers in the repeater from oscillating, via positive feedback – by the repeater amplifying its own signal.

In practice, you must ensure that the isolation will remain unaffected by changes in the local environment near the donor or service antenna.

There are instances where a repeater solution will be affected by the opening of windows in a building. Due to the metallic coating on the window, there was excellent isolation – until the window was opened, then the repeater turned down the gain automatically and the users lost service inside the building.

You can check the isolation by applying a test transmitter/Sitemaster to the DAS (at the repeater output connector) and measuring the power you get at the input at the repeater. However, leakage between the test transmitter and donor antenna is an issue. In practice, one might consider deploying a small test transmitter at the expected 'worst case' service antenna inside the building, and measuring the power one picks up at the repeater input. One must use the directivity of the donor antenna and physical objects in the building such as elevator towers, ventilation shafts, etc, to help maximize isolation; this can sometimes be a considerable challenge.

Some repeater types come with intelligent isolation enhancing features that will limit the need for physical isolation considerably.

Isolation concerns for repeater solutions that have to rely on high gain can be quite challenging, and even reflections from distant objects; buildings, hills. etc more than several kilometers away can be a problem.

Direct Link Loss

It is very important to be aware of any areas inside the building where the mobile might be able to 'see' the direct signal from the Donor cell as well as the Donor cell via the repeater system. This will cause a multipath signal and one must ensure that the delay between the two signal paths can be handled by the receiver and the equalizer. For 3G/WCDMA systems the rake receiver will use one finger for each of the signals; this can be observed on the test mobile, on 2G it can be identified by frequent, rapid changes in TA (Timing Advance) steps – due to the delay offset by the repeater in its coverage area – adding to the TA used by the mobile.

Service Link Loss

As when you design a standard indoor DAS you will have to do link budget calculations to estimate the UL and DL service range for the required wireless service on the DAS. The DL is pretty much the same calculation as that for a normal DAS; you have a specific output power from the DAS antenna, then you can calculate the expected service range. On the UL you will have to include the NF and gain performance in the DAS, the repeater and the link back to the base station.

Linearity

For DAS systems inside a building that are to provide a high speed data service using 64QAM such as 3G/HSPA and 4G it is very important to make sure and select a high quality repeater that supports this modulation accuracy, if not then the service inside the building will be limited in terms of throughput, not due to lack of signal level but to lack of phase quality to support 64QAM.

Output Power

Output power is obviously an important parameter when designing an indoor repeater solution that feeds a DAS, especially if you are feeding a passive DAS (Section 4.3) where the repeater needs to drive all the losses in the passive cables, splitters, etc. It is less important if the repeater feeds an Active DAS (Section 4.4) where the DAS itself has integrated amplifiers. One has to keep in mind that even though the output power of the repeater is specified to a certain level, the actual output power, in practice, is a result of the input signal strength, the gain and the limitation on the isolation.

Example: Calculating the Output Power from a Repeater

Let us look at a simplified example of a 3G/WCDMA Repeater application and estimate the output power we can get in practice (Figure 4.33).

Repeater Data

Output Power @ 1 CH : 35 dBm (CPICH 25 dBm)
Gain 60–90 dB in 1 dB Steps
Isolation is measured to 86 dB
Measured CPICH from the Donor at the repeater input (including donor antenna gain, etc) : −75 dBm.

This repeater drives a passive DAS with 17 dB of loss to the Service antennas and we want to achieve a CPICH power of 5 dBm from the Service antennas.

We can then calculate the desired CPICH power we need from the repeater (assuming CPICH is −10 dB)

$$\text{Desired CPICH Power} = \text{DAS Loss} - \text{CPICH@ServiceAntennas}$$

$$-55.5 \text{ Desired CPICH Power} = 17\text{dB} - 5\text{dBm} = 12\text{dBm}$$

Surely this should be possible, considering that the CPICH Power from the repeater is specified to 25 dBm, but we must take the isolation and the gain into account.

In order to achieve 12 dBm we can calculate the needed gain, knowing the input power to the Repeater.

$$\text{Needed Gain} = \text{Input Signal} - \text{Output Power}$$

$$\text{Needed Gain} = 12\text{dBm} - (-75\text{dBm}) = 87\text{dB}.$$

The repeater has up to 90 dB of Gain so this should be possible? Actually no, it is not possible!

Considering that we have measured the isolation to 86 dB, and the repeater manufacturer requires that the gain must be set at 10 dB lower than the isolation we will be limited to a gain setting of 76 dB (86 − 10).

We can now calculate what the maximum CPICH power will be when considering the isolation:

$$\text{CPICH Output} = \text{CPICH Input} + (\text{Isolation} - 10)$$

$$\text{CPICH Output} = -75\text{dBm} + (86\text{dB} - 10\text{dB}) = 1\text{dBm}.$$

Thus we can conclude that the CPICH power in this case will be limited by lack of isolation and not the power capabilities of the repeater. In this example we needed +5 dBm CPICH power from our DAS antennas, but can achieve a maximum of +1 dBm.

So in order to make the system comply with the design specifications we must optimize the isolation or add a few more antennas to the indoor DAS.

4.8.2   Repeater Types

Repeaters come in various types: high power, low power, mini repeaters, band selective, channel selective, 'Automatic' configurable versions, etc. Let us have a look at the main features and concerns with some of the more common repeater types.

It is a very basic and important question: what type of repeater should you select for your application?

Obviously, this will depend on the precise applications, so let us have a look at some of the common concerns, pros and cons of the various types of repeater.

Wideband Repeaters

Wideband/broadband repeater is a generic term used to describe a repeater that will cover a broad section of spectrum, typically the whole section of a specific band. So, a broad band 3G 2100 repeater will typically cover the entire 60 MHz of the 2100 MHz 3G spectrum. If we consider the repeater deployment scenario in Figure 4.33 where we are designing a repeater solution to provide coverage inside a building for Operator A; we should be careful and consider not choosing a Wideband repeater. The reason is that Operator A is the most distant base station, and the nearby base stations from Operator B and Operator C that are relatively close to the building where we are deploying the repeater solution.

In this example the potential issue is quite evident when considering the wideband support illustrated in Figure 4.33. The risk is that the donor antenna that points towards the Donor cell from Operator A (Figure 4.33) also points directly at several unwanted cells from Operator B and Operator C and selecting a wideband repeater for this application could result in a rise in noise and receiver blocking due to the high signal levels from the unwanted signals, that fall within the supported spectrum of the wideband repeater. The risk is potentially also present on the UL, the noise generated by the UL amplifier could cause a rise in noise in alien base stations very close to the donor antenna. This will also impact upon base stations not belonging to the operator who deploys the solution.

So, we can conclude that a wideband repeater is less suitable for a single operator solution. However, sometimes we have to provide coverage for several operators feeding the same indoor DAS; for this application it is tempting to use a wideband repeater to cover all operators. Nevertheless, it will be recommended to use individual channel selective repeaters to feed the system – in this case gain settings, etc. can be set individually and adapted for each of the donor links.

Band Selective Repeaters

As the name indicates, a band selective repeater will support a sub-section of a specific spectrum. Normally, a band selective repeater can be configured to support any section of a given spectrum; one will have to configure the 'band start' and 'band stop' frequency and the repeater will then support the spectrum within these settings for band support limits. Typically a mobile operator will configure a band selective repeater to support the specific band that specific operator uses, this will ensure that the mobile operator is more or less independent of frequency changes in the donor base station, frequency hopping, etc. When using this type of repeater we must still be careful, as the example in Figure 4.34 shows; in this example we have configured a band selective repeater to support Operator C, and as we can see in Figure 4.34

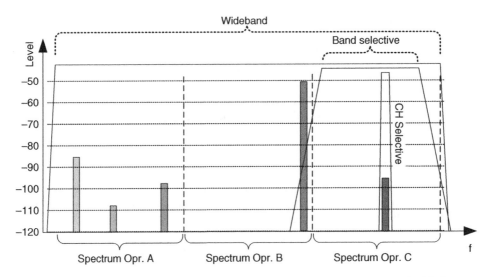

Figure 4.34 The different repeater options in terms of band support, as compared with the set up in Figure 4.33

there is a relatively low donor signal from Operator C of about 95 dBm. Near the building where the repeater supporting Operator C is being deployed, Operator B has a strong base station nearby that is using a frequency just on the lower edge of Operator C's band. The band selective repeater has some 'roll off' in terms of gain below the selected 'band start' frequency. In this case it will mean that the strong signal from Operator B will also be amplified and feed to the DAS solution, or even worse it might skew the uplink gain, add noise or even block the receiver of the repeater. The solution to this problem would be to move the 'band start' frequency up so the strong signal from Operator B will not pass through. But this will also mean that the gain on the lower part of Operator C's spectrum will be low, and this could be a problem if the donor base station changes frequency. For 3G systems where operators typically have a coherent band of one to three channel bands selective repeaters could be an ideal solution, bearing in mind the potential ACIR problem with adjacent frequencies. Some 2G operators have segments of sub-bands interleaved between the segments of other mobile operators, in this cases band selective repeaters can be a challenge to deploy.

Channel Selective Repeaters

In many cases the Channel Selective Repeater will be an ideal source to feed an indoor DAS. As illustrated in Figure 4.34, one can select the exact donor channel one wants to pick up for your service. This means that one can be virtually unaffected by the other signals illustrated. This makes the system relatively easy to configure and optimize; however one must remember that if the Donor cell changes frequency the repeater will have to be reconfigured, if not the indoor DAS that is feed by the repeater will be out of service. In some networks, such as 2G, one must also pay attention and ensure that the repeater is not fed by a donor that uses frequency hopping on channels not supported by the repeater. For 3G systems one will typically be sure to support both the first carrier (used currently for R99 3G/voice) as well as the second channel used normally for HSPA. For 2G systems one must make sure that the repeater will support all channels used by the Donor cell, both the channel for the BCCH as well as the frequency used by all the THC channels in the cell. Some more advanced GSM7DCS repeaters can actually decode System information and track and follow the associated TCHs when there is a change of frequency in the Donor cell.

'Lunchbox' Repeaters

There are many small repeaters on the market, typically of lunch box size, with a DC power supply, integrated service antenna and a connector to connect to an external donor antenna. These repeaters apply different levels of 'automatic configuration' in terms of gain; some have some sort of 'intelligence' designed to limit support to specific operators, etc.

In general, these repeaters should only be considered for 'single room coverage', and given their automatic settings are not ideal to feed a DAS since you have no control of any setting or configuration. One is not a big fan of these types of solution, but acknowedges that they are fast and 'easy' to deploy – however, remember that any repeater will have an impact on the Donor cell in terms of noise load on the UL, and one would prefer to be in control of that impact, and not leave it to a low cost 'automatic' device. But then again, new versions and types are constantly being introduced so the future might bring some good solutions even for this type of repeaters. In recent times, the market has seen a new type that utilizes an internal RF link (using the Wi-Fi open spectrum) to link up with a 'donor unit', located near the window with one or several 'service' units that provide fill in coverage at the core of the building.

Frequency Shifting Repeaters

Primarily to avoid the problem with lack of isolation between the donor and the service antenna a special type of repeater can be used; 'the Frequency Shifting Repeater'. As the name indicates, this repeater uses different input and output frequencies. Actually, this repeater solution is a set of units, one unit located at the Donor cell; this unit is hard wired to the donor site, converts the donor cell band to another frequency channel on another band (a band that is licensed to the operator) and transmits this converted band over the air-link, to the remote site (located at the building/are we want to cover). The remote repeater unit then picks up the converted channel, re-converts to the same frequency/channel of the Donor cell and transmits this to the indoor DAS. This solves the isolation challenge due to the fact that the input frequency band/channel and the output frequency band/channel operate on different frequencies. This makes it possible to apply high gain repeaters, with high output power, in applications where assuring sufficient isolation is a challenge. Obviously, the mobile operator must have the right to operate on both bands, the 'service channel' i.e. the channel used by Donor cell and thus the remote repeater system, as well as the band used to transmit/link the Donor cell and the remote repeater system.

4.8.3   Repeater Considerations in General

Repeaters are a strong tool, with relatively low investment and are quick to deploy, but like any other good tool it is important to know how, when and where to use them – and when not to.

Generically, a repeater (Figure 4.33) is merely an amplifier, a bi-directional amplifier that will extend a coverage footprint from a Donor cell into an area which the Donor cell cannot reach by its own, in our case inside a building. Repeaters come in various types, as described in this chapter, but there are some main performance considerations that are important to remember when designing with repeaters.

Not 'Install and Forget'

Repeater based solutions are not 'install and forget' solutions, one needs to monitor and track the performance of the repeater and the DAS carefully. Not only can a malfunction in a repeater solution cause problems for the users being serviced by the repeater inside the building, but a repeater could also potentially degrade the performance of the Donor cell if any errors occur or if the repeater solution is not installed or tuned correctly. Even if you have good coverage inside the building, the repeater could still degrade the performance of the Donor cell. So once a repeater is deployed it is recommended that one monitors the performance of the Donor cell closely over some weeks in order to evaluate the impact of the repeater.

It is highly recommended only to user repeaters where one has operation and maintenance access and alarm monitoring, and if needed, shut it down remotely – this functionality is typically implemented via an integrated wireless modem, and controlled by the mobile operator from the operation and maintenance centre. It is a well-known problem for many mobile operators that they can lose track of their repeater installations and performance. Some repeaters might actually be installed without their knowledge and a flaw in the repeater or its configuration can have a significant impact on the Donor cell's performance.

Noise Issues and Uplink Capacity

A repeater is a set of amplifiers, one amplifier for the uplink, one for the downlink. We know that these internal amplifiers both come with a specific gain and a noise figure. This noise figure will generate noise power that will be broadcast in the service area (inside the building) and noise power transmitted back to the donor cell on the outdoor network. The UL noise power generated by the repeater is typically the main concern and must be evaluated carefully when configuring the UL gain settings in the system, especially when the repeater is connected to an Active DAS where you also need to adjust gain settings (for more detail on noise and noise power calculations refer to Chapter 7), Potentially, the increased noise load in the UL in the Donor cell could cause the UL foot print of the donor cell to shrink losing coverage area on the edge of the cell, an effect as illustrated in Figure 4.35 and Figure 4.36.

For WCDMA/3G solutions we know from Chapters 2 and 10 how important it is to limit the impact of noise on the UL due to the fact that 3G is a noise sensitive system, and ultimately increases noise on the UL thus decreasing UL capacity.

Minimizing Noise Rise in the Donor Cell

As illustrated in Figure 4.35, a repeater will generate UL noise that will impact upon the uplink coverage area of the Donor cell. The idle load in the 3G Donor cell could be affected so badly by the noise generated and injected by the repeater that it will offset Admission Control and limit UL capacity (see Chapter 10 for more detail). In practice, you sometimes

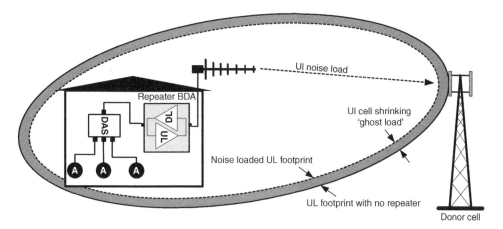

Figure 4.35   Noise impact on the UL of the Donor cell, the noise power generated on the UL of the Donor Cell will cause UL cell shrinkage

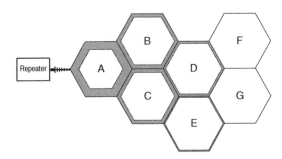

Figure 4.36 Noise impact of a faulty repeater on the Donor cell and throughout the macro network, if a repeater generates too much UL noise power, a wide spread area of the macro network can be impacted on the UL

have to make your repeater fed DAS uplink limited so as to avoid loading the Donor and other cells with the noise power from the repeater (see Figure 4.36). For indoor DAS where a repeater is feeding an Active DAS – the noise power out of the active DAS will be amplified by the repeater – this will add its own NF to the system and if this is not carefully evaluated when setting the gain in the DAS and on the repeater, the impact of this noise power can cause degradation of UL capacity in a wide area of the network, as illustrated in Figure 4.36.

It might be that a channel selective repeater is used to pick up Cell D (Figure 4.36), but, as illustrated, the UL noise power of the repeater will cause UL shrinkage on Cell A, B, C and E.

It is important to remember that the repeater, depending on type, will cause noise power on adjacent channels and spectrum, which may affect other operator's cells.

As a rule of the thumb, one should aim for the repeater system to be designed to keep the noise floor radiated back to the BS receiver at least 6 dB below the BTS receiver's (multi-coupler's) noise floor.

In general, we must remember that the purpose of the repeater is to overcome the link loss between the Donor and the repeater. The rise in noise is the noise power of the repeater system (NF + UL + gain of donor antenna and cable system), therefore the noise power generated at the input of the Donor cell would be any excessive gain in margin over the actual link loss.

Careful consideration is needed if you consider daisy chaining several repeaters, i.e. one repeater repeating the signal from another repeater, no matter if the repeaters are coupled via the air, or hardwired through cables: you need to be very careful with the potential cascaded noise ramp up (Chapter 7). This would be the case if an off-air repeater is feeding an Active DAS – that in itself could be considered as a repeater system. In practice, it is very important to configure the gain of the UL on the Active DAS that feeds the repeater correctly. If we set the UL gain too high on the active DAS, the noise power injected in the UL of the off-air repeater will be much too high and amplified by the repeater and ultimately injected over the air-link on the base station uplink, limiting the UL capacity/performance of the Donor cell.

DL Power Offset and Power Control

One could argue that repeaters add capacity, to the DL that is – as we have just learned we must configure the repeater system carefully so as not to cannibalize UL capacity.

On the DL, however, the added DL power will free up power from the Donor cell, if the Donor cell was providing coverage in the area where we implement the repeater.

One must be careful when configuring the DL gain and UL gain of the entire system, and strive to keep the balance between UL and DL gain due to Power Control concerns. Depending on the type of system (2G, 3G, 4G, etc) the effect is different – but offsetting the radiated power in the cell from the broadcast power that is also included in the System Information will impact upon Random Access to the cell during call set-up and power control in general, see Chapter 10 for more detail.

Other Concerns

When re-radiating a cell inside a building via a DAS feed by a repeater, Diversity and MIMO performance will be impacted (voided). This might also be a concern when evaluating the service requirements. Delays on the RF link must also be considered. One might pick up a distant Donor cell and add delay to the signals due to the repeater and DAS – this will skew the timing of the cell, affect the service range and One might have to widen the search window of the Donor cell. For more detail on delay refer to Chapter 10, and Section 11.12 for an example of how to evaluate the impact of delay caused by repeaters.

Obviously we must also be careful if we rely on very distant donors, and make sure that neighbor cells are defined and updated according to the new service area.

4.9   Repeaters for Rail Solutions

Rail coverage of passenger trains is a priority for many mobile operators, and as the rise in business travel via rail services increases, so does the demand for a good wireless service onboard trains.

However, it is a challenge for mobile operators to provide the sufficient signal and dominance needed to maintain high data services inside trains, sometimes even to provide basic voice services. Even if base stations are deployed at frequent intervals along the train line, or if the mobile operator relies on an outdoor DAS along the rail track (for more detail on Outdoor DAS refer to Chapter 12) the problem remains to get sufficient RF penetration through the train in order to service the users inside the train carriage.

Issues with penetration losses arise mainly from the fact that the train itself is a long metal tube, and modern trains even have metallic coating on the windows (see Section 11.2 for more detail).

Often the mobile operator has only one way to solve this issue, and that is to install an on-board repeater on the train that can compensate for the loss in the train.

The next chapter will highlight some of the many issues related to this challenge, of installing a repeater and DAS onboard a train.

4.9.1   Repeater Principle on a Train

Repeaters in general are a challenge, most of the concerns have been covered in the previous chapter, but one must realize that when deploying a mobile repeater solution onboard a train it all becomes even more complex.

The principle of a typical train repeater application can be seen in Figure 4.37; a donor/pick-up antenna, typically Omni Directional, will link the repeater onboard the train to the

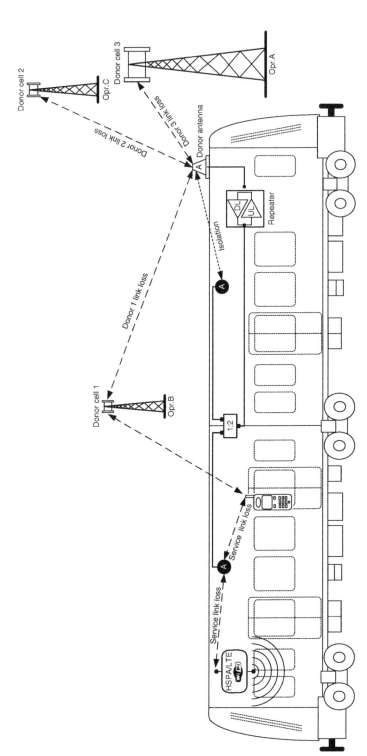

Figure 4.37 Repeater deployments on a train; small passive DAS fed by a repeater. Also a mobile Wi-Fi access point is supporting Wi-Fi users inside the train, utilizing mobile data as backhaul

outside macro network. One or more service antennas and/or radiating cable inside the train will then connect the users to the outdoor network; normally a radiating cable could be an alternative to the service antennas. Sometimes a mobile wireless Access Point could be deployed in the train carriage. This device provides a Wi-Fi service onboard for data users that do not rely on the direct data service via the mobile network. The mobile Wi-Fi Access Point backhauls the data traffic via the mobile data service provided through the repeater, thus relying on a level of high data service from the mobile network.

4.9.2   Onboard DAS Solutions

The rolling stock of a railway operator is a multiple of different sections of trains that sometimes operate as small individual trains consisting of two carriages, and sometimes several of these two-section trains connect together and form a single train of four, six, eight or more of these smaller sections.

Frequency span is also a challenge, for multiband solutions spanning from 800 MHz to 2600 MHz a 1/2 inch radiating cable could be a good alternative – even when installed in a plastic cable tray against a metallic ceiling. However, one always carry out a mock up installation in order to verify the performance in the real life enviroment.

Several Repeaters/DAS on the Same Train

To cope with this mobility challange and the need for adaptability one needs to install a repeater and DAS solution inside every two (or one) carriage. This will make sure that each repeater and DAS system will be operational no matter which configuration the rail operator will use for the train service.

However, this will also mean that sometimes you will have two to eight independent repeaters and DAS systems onboard the same train, operational in the same cell/cells at the same time. This will require careful consideration and also raise some concerns when it comes to radio planning.

4.9.3   Repeater Features for Mobile Rail Deployment

When using repeaters to feed a normal DAS inside a building we can control and measure the signal levels, gain settings, isolation, etc. We can evaluate the impact on the cells from the mobile operator, or operators that feed the signal; this is not that easy when designing repeaters for rail deployment.

In Section 4.8.2 we evaluated the different repeater types and what type of repeater is most applicable for various deployments. Typically a repeater deployment on a train will be a multi operator solution, so a wide band repeater is required.

However, given the dynamic nature of the radio conditions that will be encountered when feeding a mobile repeater solution, the evaluation of applicable repeaters for mobile rail deployment is much different.

Let us address a few of the main challenges when deploying mobile repeaters onboard a train.

Isolation Requirements

In general, a standard isolation of the gain +10 dB could work, but given the physical proximity of the donor antenna (typically omni directional) cell and the service antennas inside the train, the isolation will often be pretty low, limiting the possible gain. One must also take into consideration the fact that the isolation can vary greatly, due to reflections from objects near to the train; a prime example will be when a train enters a tunnel, then the isolation will plunge – thus limiting performance.

One has actually seen an example of an onboard rail repeater/DAS deployment presented at a conference, where everything worked perfectly – after some fine tuning of gains, etc. However, the downside was that every time the train doors opened the isolation dropped and the repeater started to feed-back (oscillate) and calls dropped, both inside the train and, even worse, also in the nearby cells due to UL interference in nearby base stations caused by the oscillating repeater!

Gain Settings and Service Levels

Considering the typical repeater deployment, as seen in Figure 4.37, it is evident that the donor signal level feeding the repeater can vary by 80 dB or more, depending on the location of the train, relative to the Donor cells. It is also very easy to appreciate that quite often one operator will have a donor base station a few meters from the train and at the same time other operators will have much more distant Donor cells feeding the same wideband repeater solution and DAS. The difference between these Donor cells could easily be more than 80 dB – and given the limitations in terms of dynamic range and the receiver blocking concerns of the repeater, some automatic functionality is required controlling the gain of the repeater and DAS.

It will never be 100% perfect, and it will always be a compromise, but it is recommended that one use a repeater where one can limit the gain to a pre-set maximum. This pre-set gain will balance out the losses of the train and provide just enough signal inside the train to maintain service. Then, on top of this pre-set gain, some sort of automatic gain control is desirable to some extent. However, one must realize that when the train is very close to one operator's donor antenna, and far from other operators' antennas, this could mean that the automatic gain control will turn the gain down so low, in order to avoid receiver blocking, that there is not sufficient signal level from the more distant donors. Some manufacturers offer special repeater hardware dedicated for rail deployment, with individual sections of repeaters, one for each of the mobile operators – with adaptive gain for each individual operator.

Timing and Delay

As we know, repeaters will add delay to the signals, like any amplifier, and will skew the timing of the cell service. Normally this will not be a major concern, but one must realize that if one is feeding the repeater with a signal that is radiated by a tunnel DAS or Outdoor DAS where there might already be concerns with the timing offset due to delays of the Donor DAS (see Section 11.11 for more detail) then the extra delay from the on-board repeater could be the final offset that takes the system over the limit.

O&M

Where repeaters and DAS systems are deployed in rail carriages, just to get access to a faulty unit for repairs is a concern, and normally this will have to be scheduled for the next time the train is in for service. But how will one actually know that there is a problem? Yes, make sure to use repeaters where one can connect and control alarms and settings, etc., via a wireless modem, and as a minimum have an 'SMS alarm system' that will provide you with an SMS with the name/number of the repeater if there is an alarm. There are devices available where you can pre program alarm labels to a range of inputs, and the device will then send an SMS accordingly if the specific input is triggered. Also, make sure that the modem will relay all alarms from the repeater back to the network in the event of a power failure – so that the system will know that one lost power to the system – and did not lose modem functionality. In practice, a short battery backup to the wireless modem should be implemented, and one alarm trigger on the loss of DC power to the repeater system.

GPS Control

Sometimes one might deploy a rail repeater system onboard a train that passes through other countries or regions where it is not desired to have the repeater operational. This could be in an urban area where the signal levels are high enough to service users inside the train without the repeater being activated, and thus also eliminating noise load on the uplink of the macro network. It could also be the case that the repeater moves into another country where you do not have permission or desire to maintain the service. In these cases you can deploy a repeater that is controlled by GPS location. You can define exactly when and where you want the repeater to be activated. Some solutions even provide an option to use GPS control of the sup-ported bands and services when passing over international borders.

Wi-Fi and Data Services

With some applications it is desired to provide an onboard Wi-Fi service for data users. Some PCs might not be able to log on directly to the 3G/4G data service, but would have to rely on Wi-Fi. Providing mobile Wi-Fi coverage is possible when using a mobile wireless WLAN Access Point, as shown in Figure 4.37. The Mobile Access Point backhauls the traffic using the EDGE/3G/4G, via the repeater and services the PC users inside the train carriage via Wi-Fi. Naturally the data speed on Wi-Fi depends on the data offerings and quality via the repeater.

4.9.4   Practical Concerns with Repeaters on Rail

Given the special nature of the environment of the application when you deploy repeaters onboard trains there are a number of practical concerns that are very important to keep in mind.

Equipment Certification

Most rail operators will demand that ALL the hardware you want to deploy inside and on a train comply with specific standards that apply to equipment that is to be used in railway installations. Obviously there is radio compatibility with other communication systems installed on or in the train, but there are many other concerns: fire safety, CFC free cabling, labeling of all equipment, cables, etc. One example is the donor antenna; one must make sure to use only an approved rail antenna that is designed to withstand that harsh environment on top of a train. One example might be that you fail to realize that an antenna installed on a train must be able to take a hit from one of the electrical power lines that feed the train with power, without endangering any passenger or equipment inside the train!

So be sure to comply with all the appropriate standards, it can be fatal – literally – not to do so.

Mechanical Issues

Do not underestimate the impact of the mechanical stress on repeater and DAS equipment due to the vibrations and temperature extremes onboard a train.

Installation space is very restricted, but do try to find a location with relative easy access, stable temperature and mount the active repeater and DAS equipment in shock/vibration absorbing brackets.

There are many examples of repeater solutions that virtually fell to pieces after a few months' operation in a train installation, due to vibration and mechanical stress.

So on top of the normal RF testing that one may apply when approving new equipment in your network; be sure to evaluate the mechanical craftsmanship of the repeater and other equipment that is intended to be deployed onboard the train. Also, pay attention to the craftsmanship of the installation work itself, the fixing of cables, antennas, etc.

Power Supply

The lack of a good quality power supply onboard a train is one of the biggest challenges, when it comes to the reliability of the solution. Make sure that all active equipment can handle the high spikes, dips and general noise that come on the power supply line to the repeater.

Conclusion on Repeaters

Hopefully this chapter on repeaters and repeater deployment has provided some insight on repeater applications and what to consider when designing these solutions.

Repeaters are a strong tool, can be cost efficient and in some cases the only remedy to solve a coverage problem onboard a train or a ship.

Repeaters can be an excellent solution to feed an indoor DAS, but it takes skill and careful configuration and planning to maximize performance, and to minimize the impact on the Donor cells.

4.10   Active DAS Data

Active DAS systems consist to a large extent of amplifiers and repeater/BDA systems; therefore it is important for the RF planner to understand the basic data of these system components. Many of these standard metrics are used to benchmark the radio performance of different manufacturers and systems. Make sure that the data you are comparing all use the same standard benchmark reference.

The amplifiers used in active DAS, repeaters and BDAs have to be very linear, in order not to distort the signal and degrade the modulation. The more complex the modulation used, the higher demands are on linearity and performance. This is very important for the higher coding schemes on 2G EDGE, 3G and especially 4G. Be careful when selecting equipment used for indoor DAS systems, since performance of the system is often directly related to the price. The most basic parameters and merits are described in this section.

4.10.1 Gain and Delay

Gain

Gain is the amplification of the system (as shown in Figure 4.34), the difference between input signal and output signal power. The power of the output signal is:

$$\text{output signal} = \text{input signal} + \text{system gain } (\text{dB}), \text{ or input} \times \text{gain factor } (\text{linear value})$$

Gain is typically stated in dB. For a system with a factor 2 power gain, for example, 1 W input (+30 dBm) will lead to 2 W output.

Power gain in dB can be calculated as:

$$\text{Gain } (\text{W}) = 10 \log (\text{gain factor}) = 3\text{dB}, \text{ thus the output power will be} +33\text{dBm}$$

For voltage, gain factors can be converted to dB using:

$$\text{Gain } (\text{voltage}) = 20 \log (\text{gain factor})$$

Gain can be negative; often it will be referred to as attenuation or loss.

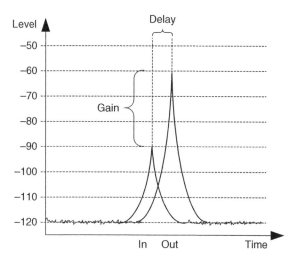

Figure 4.38 Input/output signal of an amplifier vs time

Delay

Delay is the time difference between the input and output signal (as shown in Figure 4.38). In practice this will offset the timing for mobile systems. In 2G the timing advance will be offset (increased), and the synchronization window on the 3G base station network may have to be adjusted wider to accommodate the delay introduced.

Note that you have to include both the delay of the active elements, and the delay of cables (also optical) due to the decreased velocity of the propagation on the cable, for large systems. (See Chapter 11.11 for more details.)

dBm

The signal level in RF design is described as absolute power related to 1 mW (in 50 Ω) and expressed in dBm.

$$P(\text{dBm}) = 10 \log\left(\frac{P(W)}{1\text{mW}}\right)$$

$$P(\text{mW}) = 10^{\text{dBm}/10}$$

4.10.2   Power Per Carrier

Amplifiers in active DAS systems, repeaters and BDAs are normally composite amplifiers. This means that the same amplifier amplifies all carriers throughout the bandwidth. All carriers share the same amplifier resource; the result of this is that the more carriers the amplifier must support, the less power can be used for each carrier. The sum of all the powers will remain the same, hence the name composite power.

Every time you double the number of carriers, the power decreases about 3 dB per carrier, depending on the efficiency of the amplifier. For an example, see Table 4.4.

4.10.3   Bandwidth, Ripple

Bandwidth

Normally, when defining the bandwidth of an amplifier, it is the 3 dB bandwidth (as shown in Figure 4.39) that is referred to. The 3 dB bandwidth is the band that supports the amplification with a gain decrease of maximum 3 dB.

Table 4.4   Power per carrier from an active DAS

Number of carriers	Power per carrier
1	20 dBm
2	17 dBm
4	14 dBm
8	10 dBm

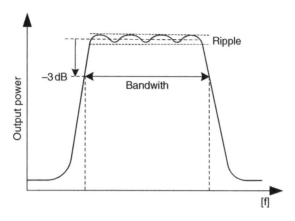

Figure 4.39 Output signal of an amplifier, over the whole operating bandwidth

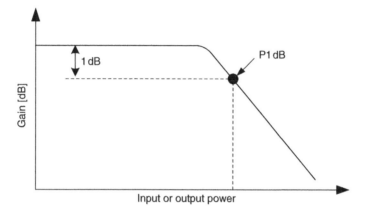

Figure 4.40 The P1dB compression point

Gain Ripple

Gain ripple describes the variance in gain over the bandwidth (as shown in Figure 4.39).

4.10.4   The 1 dB Compression Point

The 1 dB compression point (P1dB) is a measure of amplitude linearity. This figure is used for defining output power capabilities (as shown in Figure 4.40). The gain of an amplifier falls as the output of the amplifier reaches saturation; a higher compression point means higher output power. P1dB is at an input (or output) power where the gain of the amplifier is 1dB below the ideal linear gain. P1dB is a convenient point at which to specify the output power rating of an amplifier.

Example
If the output P1dB is +20dBm, the output power from this amplifier is rated at +20dBm maximum.

Avoid Intermodulation Problems

Reducing the output power below the p1dB reduces distortion. Normally manufacturers back off about 10 dB from the P1dB point: an amplifier with 20 dBm P1dB is normally used up to +10 dBm.

4.10.5   IP3 Third-order Intercept Point

IP3

IP3 is a mathematical term (as shown in Figure 4.41). It is a theoretical input point at which the fundamental (wanted) signal and the third-order distorted (unwanted) signal are equal in level to the ideal linear signal (the lines A and B).

The hypothetical input point is the input IP3 and the output power is the output IP3. IM3 'slope' (B) is three times as steep (in dB) as is the desired fundamental gain slope A.

Unlike the P1dB, the IP3 involves two input signals. The P1dB and IP3 are closely related: roughly $IP3 = P1dB + 10dB$

Testing IP3

IP3 is used as a merit of linearity or distortion. Higher IP3 means better linearity and less distortion. The third-order inter-modulation products are the result of inter-mixing the inputs by the nonlinearities in the amplifier:

$$fIM3\ 1 = 2 \times f1 - f2$$
$$fIM3\ 2 = 2 \times f2 - f1$$

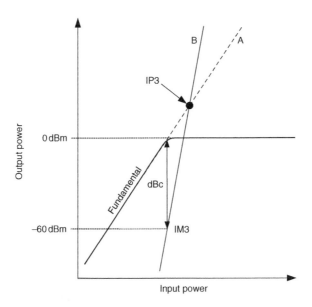

Figure 4.41   The IP3, third-order intercept point

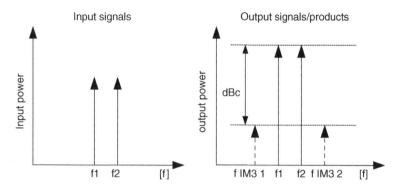

Figure 4.42 The two-tone test of IM3

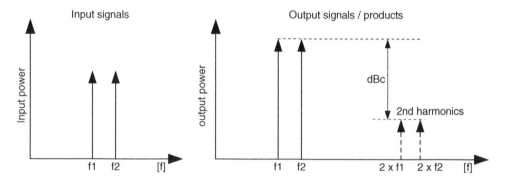

Figure 4.43 Harmonic distortion

The two-tone test (as shown in Figure 4.42) is often used to test IP3. Third-order inter-modulation products are important since their frequencies fall close to the wanted signal, making filtering of IM3 an issue.

4.10.6   Harmonic Distortion, Inter-modulation

The harmonic distortion (Figure 4.43) specifies the distortion products created at integers of the fundamental frequency; dBc means dB in relation to the carrier.

4.10.7   Spurious Emissions

Spurious emissions (as shown in Figure 4.44) are emissions, which are generated by unwanted transmitter effects such as harmonics emission or inter-modulation products.

4.10.8   Noise Figure

The noise figure is the noise factor described in dB, and is the most important figure to note on the uplink of an amplifier system. The NF will affect the DAS sensitivity on the uplink.

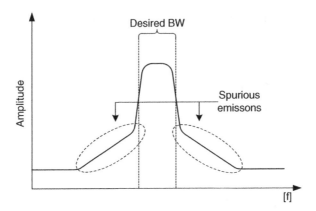

Figure 4.44   Spurious emissions from a transmitter

The noise factor (F) is defined as the input signal-to-noise ratio divided by the output signal-to-noise ratio. In other words the noise factor is the amount of noise introduced by the amplifier itself, on top of the input noise.

$$\text{noise factor } (F) = \frac{\text{SNR}_{(input)}}{\text{SNR}_{(output)}}$$

$$\text{noise figure}\,(\text{NF}) = 10\,\log\,(F)$$

The effect of noise and noise calculations will be described in more details in Chapter 7.

4.10.9   MTBF

Failures are a concern when installing distributed active elements in a building. All components, active or passive, will eventually fail; the trick for the manufacturer is to insure that the expected failure is after the expected operational time of the system (as shown in Figure 4.45). Typically there will be an expected lifetime of a mobile system of 10–15 years. It makes no sense to design systems that can last for 130 years that are too expensive. Having multiple active elements scattered around a building, can be a service access concern when implementing "full active DAS".

'Infant Mortality'

The manufacture will perform a 'burn-in' test of the equipment in order to insure that the 'infant mortality' is cleared from the shipped equipment.

Operational Period

For an active element, it is assumed that during the useful operating life period the parts have constant failure rates, and equipment failure rates follow an exponential law of distribution. MTBF of the equipment can be calculated as:

$$\text{MTBF} = 1/(\text{sum of all the part failure rates})$$

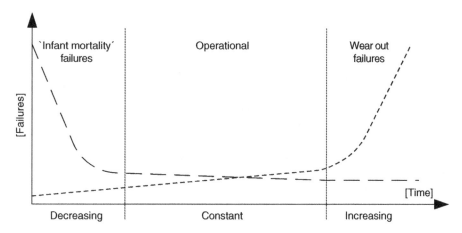

Figure 4.45   The MTBF curve – 'the bathtub curve' – of distribution of failures

Probability of Failures

The probability that the equipment will be operational for some time T without failure is given by:

$$\text{reliability} = \exp^{-T/\text{MTBF}}$$

Thus, for a product with an MTBF of 450000 h, and an operating time of interest of 7 years (61320 h):

$$\text{reliability} = \exp^{-61320/450000} = 0.873$$

There is an 87.3% probability that the system will operate for the 7 years without a failure, or that 87.3% of the units in the field will still be working after 7 years.

This is a useful guideline to estimate number of spares needed for your installed base of equipment.

4.10.10   Dynamic Range and Near-far Effect

When designing indoor DAS it is important to realize that the amplifiers in remote units, repeaters and front end of base stations have their limitations. Normally we are concerned with the sensitivity in the low end of the RF level range; however, often the higher signals pose more of a challenge when designing and implementing indoor DAS, due to the close proximity of the mobiles to the DAS antennas. Take especially care when implementing "wideband" DAS solutions. Then the UL can be "hit" by terminals not in service and thus in power control by the DAS.

Let us look at the two most important parameters; receiver dynamic range and receiver blocking.

Receiver Dynamic Range

Receiver dynamic range is the ability of a receiver to detect a low signal without degradation of the quality while a strong signal is present at the same time. This is a very important parameter to consider when evaluating the active parts of a DAS, a repeater, the base station, etc. One could say that the dynamic range of receiver is the range/span of signal levels, low to high, over which it can operate without compromising the quality of the detected signals. The low end of this range is determined by receiver sensitivity, dictated by the noise figure of the receiver and the high end is determined by its ability to withstand overload and distortion with strong signals present at the input at the receiver; this is determined by the receiver's IP3 (see Section 4.12.5).

So, in short, the dynamic range of a receiver (the uplink in the remote unit or in the input of the repeater) is essentially the range of signal levels from low to high over which it can operate without problems.

Receiver Blocking

The presence of strong RF signals could be 'in-band signals', i.e. signals within the bandwidth support of the DAS/base station, not necessarily at the operational spectrum of that particular operator, and could lead to distortion and blocking of the receive path of the DAS/base station, affecting performance of the receiver – especially for lower power signals. The ability of the active components in the DAS to cope with high signals without blocking is an important parameter to look out for, and this is one of the parameters that distinguishes a good quality active DAS/repeater from another – so it is important that a receiver stays within its linear operational range in order not to create any intermodulation distortion at high signal levels; a high IP3 is therefore important (see Section 4.12.5).

Some active DAS has an integrated limiter (ALC, Automatic Level Control) integrated in the remote unit, that will limit the signal feed internally to the amplifier in the uplink. The ALC will kick in before the receiver starts generating intermodulation – this function also applies to many repeater solutions; however, it is important to confirm that this ALC is actually effective prior to the first amplifier stage – if not, it is of little use, if the first stage of the amplifier is already generating intermodulation.

Let us have a look at a real life example in Figure 4.46, considering three different scenarios. This is a typical office building with a central Omni directional antenna on the floor serviced by mobile operator A; three mobiles are currently active on this floor; MS1, MS2 and MS3. There is also a macro cell present at the floor serviced by Operator B.

Scenario 1 – All Mobiles in the Indoor Cell, in Power Control

In this scenario (Figure 4.46 and Figure 4.47), all three mobiles; MS1, MS2 and MS3, are in traffic mode, all camped on the indoor cell serviced by mobile Operator A. Thanks to Power Control the system ensures that the receive level on the uplink of the base station is kept at a relatively constant level (Figure 4.47). In this example the target signal level is −80 dBm. This is very important in WCDMA/3G where the power from all mobiles should be kept at the

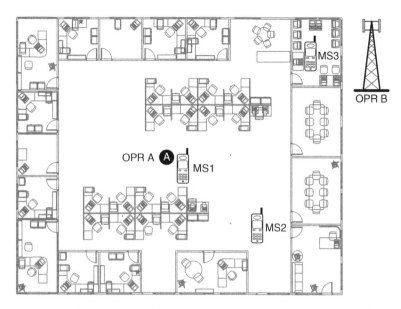

Figure 4.46 Three mobiles in a typical indoor environment with a centrally placed Omni directional antenna connected to an indoor DAS that services Operator A, and one outdoor cell also providing service from Operator B

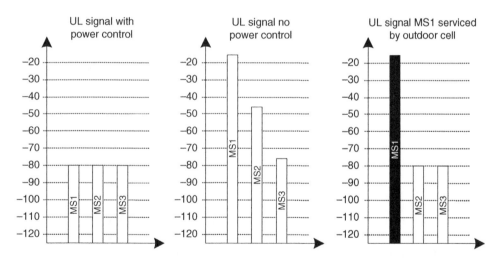

Figure 4.47 The uplink signal level in the indoor DAS described in the three scenarios

same level, due to the fact that all cells and mobiles are on the same RF channel, so one mobile could easily degrade the performance of the whole system, cause intermodulation in the receiver, possibly receiver blocking, and thus even cause dropped calls (Scenario 2). Thanks to uplink power control this is avoided. However, there is a limit to power control – even at the lowest possible mobile transmit power, a mobile that is very close to the indoor DAS antenna could cause concern and potential problems – therefore mobile systems specify a MCL

(Minimum Coupling Loss). The MCL is, as the name suggests, the lowest allowable path loss between the base station and the mobile that the system can handle without any problems even with the lowest possible transmit power from the mobile.

Scenario 2 – All Mobiles in the Indoor Cell, No Power Control

This is the same situation as Scenario 1, all three mobiles, MS1, MS2 and MS3, are in traffic mode all camped in the indoor cell serviced by mobile Operator A (Figure 4.46 and Figure 4.47). In this scenario there is no Power Control active – this will mean the all three mobiles are at maximum transmit power. The different path loss due to distance and wall attenuation (for MS3) causes a big offset in uplink receive power at the indoor base station, as illustrated in Figure 4.47. For a WCDM/3G system this is devastating – MS1 will simply block out the signal for all other mobiles in the cell; MS2 and MS3 will simply drop their calls – however, in real life this is avoided thanks to Power Control. However, all mobiles might not be in Power Control by the indoor cell (Scenario 3).

Scenario 3 – One Mobile not in Power Control

In this scenario (Figure 4.46 and Figure 4.47), MS2 and MS3 are in traffic mode all connected to the indoor cell service by mobile Operator A. These two mobiles are in power control, hence the uplink signal levels of the base station are at the desired, controlled uplink signal level. However, in this scenario mobile MS1 is camped in the outdoor cell from Operator B. This is a concern, MS1 is blasting out full transmit power in order to compensate for the increased path loss to the outdoor base station from Operator B (Figure 4.46). Even though MS3 is operating on a different frequency, the fact is that it is still in the supported frequency range of both the indoor DAS and the bandwidth of the base station receiver, and the uplink amplifier of the remote unit connected to the indoor Omni antenna. Therefore it is very likely that with the mobile blasting full power to reach the outdoor network, MS3 can cause receiver blocking or adjacent channel interference; see Section 5.9.2 for more detail about ACIR (Adjacent Channel Interference Ratio).

Try to Avoid the Potential Problems

The problems with the 'near-far' effect and receiver blocking can be avoided if one makes sure to keep safe from the MCL; in reality one should strive to be sure not to place indoor antennas where the users can get very close to the antenna. In practice this is a challenge due to physical limitations – but try at least to keep the antennas distant (a few meters) from locations where users are stationary. In extreme cases one might need to add extra loss between the indoor antenna and the active element, remote unit, repeater or base station, to avoid these problems. These problems are yet another reason as to why we should consider multi operator solutions inside buildings (see Section 5.10). The increased complexity of several technologies (2G, 3G, 4G, TDD/FDD) deployed in spectra with close RF frequency proximity, makes this challenge even harder – and giving careful attention to these potential problems is important.

4.11 Electromagnetic Radiation, EMR

EMR is a concern for mobile users all over the world. From time to time there are heated debates in the media, but indoor radio planners need to stand above these often emotional discussions and try to be neutral and objective. Above all, you need to accept and respect that this is in fact a concern for the users, however unlikely you yourself believe the danger to. You need to comply with any given EMR standards and guidelines that apply in the region you work in and make sure that the indoor DAS systems you design and implement fulfill the approved regulations in the country where the systems are implemented. Often you will have to accept that a neutral party conducts post-implementation on-site measurements in the building, and certifies that the design is within the applicable EMR specification.

Different regions around the world use different standards and regulations, and these regulations change over time. Find out exactly what standard applies within your country.

4.11.1 ICNIRP EMR Guidelines

Currently many countries use the guideline laid down by the ICNIRP (International Commission on Non-ionizing Radiation Protection), who are recognized by WHO. This guideline specifies the maximum allowed EMR exposure for the general public 24 h/7 days/52 weeks of the year (higher levels are allowed for professional users, some regions of the world uses lower values).

A measurement specification (EN50382) specifies how these measurements should be conducted and that a mobile transmitter must radiate lower than:

- 900 MHz, maximum 4.5 W/m^2.
- 1800 MHz, maximum 9 W/m^2.
- 2100 MHz, maximum 10 W/m^2.

The measurement should be averaged over 6 min.

Guidelines are different for different countries, and will be adjusted over time. Check exactly what applies in your case; some countries apply a standard that is much stricter than the ICNIRP levels.

You cannot relate these values to a specific receiver level in dBm, due to the fact that the ICNIRP levels are power density, and therefore will be the sum of all powers on air within the measured spectrum. Measuring in dBm using a test mobile receiver will only indicate the specific level for that one carrier you are measuring, not the sum of power from all carriers.

Example
This is an example of a measurement at a real-life installation of a DAS, using 2G 1800 MHz 18 dBm radiated from the omni DAS antenna, four TRX with full load:

$$50 \text{ cm distance from the antenna} : 0.630 \text{W}/\text{m}^2 \ \left(\text{average over } 6\,\text{min}\right)$$
$$200 \text{ cm distance from the antenna} : 0.0067 \text{W}/\text{m}^2 \ \left(\text{average over } 6\,\text{min}\right)$$

The measurement clearly shows that, in practice, you are well below the maximum allowed ICNIRP levels.

DAS Systems are Normally Well Below These Levels

Typical DAS systems, passive or active, are well below these levels. However, I have seen examples of high-gain outdoor sector antennas, connected directly indoors to a high-power base station, and users able get so close that they can touch the antenna! This is clearly not a correct design; this 'hot' antenna might blast away, providing a lot of coverage in the nearby indoor area, and most likely leaking out interference outside the building. In this extreme case you could exceed the limits. However, common sense must be applied both with regards to good indoor radio planning and for minimizing EMR from the indoor DAS.

When you are being questioned by the users in the building whether it will be safe to work every day underneath the installed indoor DAS antenna, you should always ask yourself the question whether you would want to sit underneath that antenna 24/7 and design accordingly.

Is it Safe?

Well that is the question; the current EMR guidelines are defined with a large margin to any level that might cause any known effect on humans. Often you will be asked if you can prove that these EMR levels are safe. This is an understandable question to ask, but science cannot prove a noneffect, only an effect, and until now no effect has ever been documented when adhering to these guidelines for the design levels.

To use an analogy, we cannot prove it is safe to drink a glass of milk each day for 50 years! Still we drink milk every day. We might be able to prove that milk was hazardous, if we could conclude an effect due to a level of toxin found in the milk, but with no toxin found, there is nothing we can do but to accept that it is most likely safe to drink milk. We do have a responcibility to always assure a respectfull dialouge on these sensitive subjects.

4.11.2   Mobiles are the Strongest Source of EMR

Most users inside buildings are concerned about the radiation from the DAS antenna, but it is a fact that the main source of human EMR when using mobile phones is not the indoor DAS antennas, but the mobile handset. This is due to the proximity of the handset to the mobile user. Even if a high-power outdoor base station may generate 700 W and the mobile only 2 W, the determining parameter is the distance to the antenna. Having the mobile close to the user's head will expose the user to more power than would an outdoor base station even as close as 50 m! Indoors the margin is even clearer as shown by the measurement just documented in the example above.

It is a fact that the main source for EMR exposure is the mobile, and the trick is to keep the mobile at the lowest possible transmit power. This is the only way to minimize the exposure of users inside the building. The power control function in the mobile network will adjust the transmit power from the mobile automatically, in order to insure that the received signal level on the uplink at the base station is within the preset level window for minimum and maximum receive levels.

This power control insures that the more attenuation there is on the radio link (buildings, walls etc.) and distance between the mobile and the base station, the higher transmit power the mobile will use to compensate for this loss. Therefore the only way to minimize the EMR exposure is to make sure to keep the attenuation on the link as low as possible. The more indoor antennas you have in the building, the lower the link attenuation is to the users and the lower transmit power indoor users will be exposed to.

4.11.3   Indoor DAS will Provide Lower EMR Levels

The effect is that, even in buildings very close to an outdoor base station, and where the mobile coverage seems perfect, the mobile will typically operate at or close to the full transmit power. The high downlink power from the base station might provide high signal levels received by the mobile, but the power control depends on the uplink level at the base station, and the mobile transmits at far lower levels than the high power transmitter at the base station.

Indoor antenna systems with low attenuation will help. By deploying an indoor DAS you can create lower path loss for the mobile to the base station, and the mobile will operate on lower transmit power.

Less Radiation with Active DAS

Installing a passive DAS inside a building will to some extent bring down the mobile transmit power, but it is a fact that the mobile still needs to overcome the attenuation of the passive DAS, thus operating on a relatively high power even on an indoor system close to the DAS antenna. The mobile transmit power obviously depends on the attenuation of the passive DAS, but with 20 dB of attenuation on the passive DAS, the mobile has to transmit 20 dB higher power compared with the active DAS from Section 4.4.2.

By deploying an active DAS inside buildings, the attenuation between the mobiles and the indoor antennas is low. The result is that the mobile will run at or close to minimum transmit power. This is due to the active DAS being a 'zero loss' system; it is an active system where all the losses in cables are compensated by small amplifiers close to the antennas. This only applies to pure active DAS; using hybrid DAS the mobile will still have to compensate for the passive losses prior to the HRU (as shown in Figure 4.18).

Note that power control is triggered by the received level, and the noise figure of the active DAS will not cause the transmit power from the mobile to be adjusted up.

Example, Lower Mobile Transmit Power with Active DAS

The low mobile transmit power, using an active DAS system, is evident in this graph (as shown in Figure 4.48). The graph shows the typical mobile transmit power inside an office environment vs the distance to the indoor DAS antenna.

This is a 2G-1800 example, where the minimum received level on the base station is set to −75 dBm, so the base station will adjust the transmit power of the mobile in order to reach this uplink level.

On the graph (as shown in Figure 4.48), the pure active system is compared with typical passive systems with 20, 30 and 40 dB of attenuation from the base station to the DAS antenna.

As shown, the transmit power from the mobile covered by a passive DAS will only use the lowest power level when it is located very close to the antennas, whereas the same mobile in the same type of environment on an active DAS will stay on the lowest possible transmit

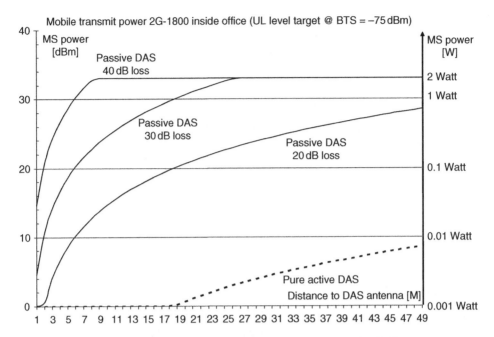

Figure 4.48 Mobile transmit power on passive and active DAS

power even up to a distance of 19 m from the indoor antenna, and stay at a low level compared with the passive DAS.

The mobile connected to the passive DAS will ramp up transmit power even close to the DAS antenna, and this is due to the passive attenuation. This is evident on the graph, even for the 'low loss' 20 dB attenuation of passive DAS.

4.12   Conclusion

It is evident that the mobile connected to the pure active system consistently maintains an output power below 0.01 W, and the mobile connected to a passive system can easily reach 1 or even 2 W (as shown in Figure 4.48). Using a traditional passive distributed antenna system will to some extent help with radiation from mobiles, especially mobiles being serviced by antennas with relatively low loss, close to the base station room. However, the fact is that, due to the losses in the passive system, the mobile has to compensate for the losses in the passive cable dB for dB, resulting in higher transmit power from the mobile and thus higher EMR exposure of the users.

Even if you often need to install an uplink attenuator between the active DAS and the base station to minimize the noise load of the base station, it is clear that the active DAS will keep radiation from the mobiles to the lowest possible power.

Both passive and active DAS will help bring down the transmit power from the mobiles, if the alternative is to rely on coverage via the outdoor macro net. All mobiles have to apply a certain radiation limit (SAR value), so even when operating on the highest power level, no mobile is dangerous.

5

Designing Indoor DAS Solutions

Before starting to design the first indoor solution, it is highly recommended to develop a well-structured and documented workflow for the task ahead. This procedure should include every aspect of the process from start to finish. This will insure a uniform workflow from design to design, and make sure that all these solutions (investments) follow the same process, helping to prioritize the projects and the investment. Please refer to Chapter 15 for a more detailed description of the technical part of the planning, instalation and deployment of the DAS project.

5.1 The Indoor Planning Procedure

Operators need a well-structured procedure to evaluate the business case and implementation process of indoor solutions. Often the need for an indoor coverage solution in a particular building is initiated by the sales and marketing department of the operator responsible for that particular area or customer. The procedure must include all aspects of the process, in order to make it visible for all working within the operator: how, why and when an indoor DAS is to be implemented; who is responsible for what part of the process; what documents are needed; and the general workflow of the process.

There are many valid methods to use when organizing the workflow; one typical structure could be like the one shown in Figure 5.1.

5.1.1 Indoor Planning Process Flow

Briefly, these are the main parameters: in- and output of the different parts of the process (as shown in Figure 5.1).

Indoor Radio Planning: A Practical Guide for 2G, 3G and 4G, Third Edition. Morten Tolstrup.
© 2015 John Wiley & Sons, Ltd. Published 2015 by John Wiley & Sons, Ltd.

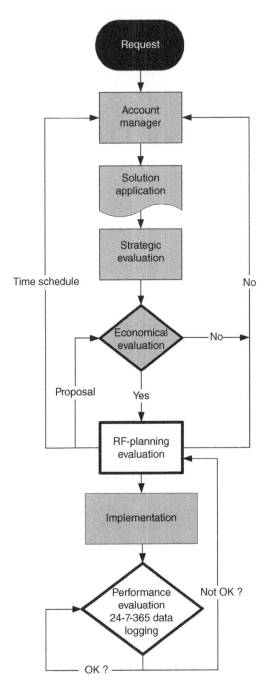

Figure 5.1   One way of structuring the indoor planning process

Input from the Sales Team/Key Account Manager

The process starts with the requirements from the potential customer, preferably at an early stage, before the customer starts to use mobiles on a wider scale. The sales team then provides an application to get approval to implement a dedicated solution. There must be a clearly defined revenue goal.

The application should consist of:

Business input

- Number of users.
- Types of users.
- Types of service requirements, data speeds, etc., needed.
- Duration of the contract.
- Expected airtime per user.
- Expected on-site traffic in the building.

Documentation of the building

- Floor plans.
- Markings on floor plan showing the coverage needed in the different grades and areas:
  ○ 100% coverage;
  ○ 90% coverage;
  ○ areas where it is 'nice to have' coverage.
- Markings on the floor plan showing different environment types:
  ○ dense areas with heavy walls;
  ○ open office areas;
  ○ storage rooms.
- Pictures of the building (often photographs can be found on the company's web page).
- Details and drawings of planned reconstructions or extensions of the buildings.
- Details for the contact person, who is responsible for approving the installation.

Floor Plans are the Base of the Design

Often it can be hard to obtain usable floor plans, but in most countries it is mandatory to have some sort of site plan in A3 format located at the central fire alarm, so that the fire department can find their way around the buildings.

These plans often provide excellent basis for the planning and can be used for:

- Design documentation.
- Installation documentation.

5.1.2 The RF Planning Part of the Process

The RF planner uses the input from sales to make a draft design. Often this draft design can be done by an experienced radio planner without doing a site visit. This of course depends on the quality and detail of the input, the size of the solution and the experience of the RF planner.

The RF planner provides the following output:

- Floor plans with suggested placement of antennas and equipment.
- Diagram of the DAS.
- Equipment list.
- Estimated implementation costs.
- Estimated project time.

These outputs from the RF planner are used as the final input to the 'Coverage Committee', which is responsible for a 'go' or 'no-go' implementation of the solution. The Coverage Committee, which is also responsible for the roll out budget for indoor solutions, has technical, sales and marketing representation.

5.1.3   The Site Survey

Prior to the site survey, the RF planner has done a draft design, using a link budget tool, RF propagation simulation and experience. To do the final design, the RF planner uses the draft design as a basis for a site survey, and adjusts the draft design according to the results of the site survey.
   The purpose of the site visit is to:

- Get the solution approved by the building owner.
- Collect information regarding equipment rooms, installation challenges cable ducts, etc.
- Take necessary photographs for the installation team, and for the RF planner.
- Take photographs of rooftop 'line-of-sight' possibilities to other sites, if microwave transmission is to be used.

   Participating in the site visit should be:

The RF planner

- The RF planner is the project manager.
- After the survey, the RF planner will provide the final design to be approved by the building owner, and used by the installer.
- The RF planner might need to take measurements of the existing coverage provided by the macro layer.
- The RF planner also might need to do RF survey measurements inside the building, to verify the draft design.

The acquisition manager

- The acquisition manager is responsible for the constriction permits, legal contract, etc.

The installer

- The installer is responsible for the implementation.
- He will provide 'as-built' documentation after implementation.

The local janitor and local installer of IT and utilities

• He knows all the details and cable ducts.
• He and the operator's installation team will be in direct contact regarding the installation, once the RF planner has provided the final design.

The building owner

• The building owner will approve the design, antenna placements, etc., as the team walks the site, together with the janitor and IT department.

Process Control

It is important that all the involved parties, all members of the Coverage Committee and the implementation team, are all working with predefined timeframes, and well-defined input/output documents and procedures in order to control the process.

5.1.4   Time Frame for Implementing Indoor DAS

For a normal indoor planning process, the typical timeframes would be:

• Sales/key account meeting with the end-user, providing inputs and documentation
  ◦ 1–2 weeks
• Draft design from RF
  ◦ 1–2 days
• Site visit
  ◦ 2–6 hours
• Final design and documentation
  ◦ 1–4 days after site visit
• Implantation start-up
  ◦ 1–2 weeks after site visit

Transmission

Based on experience, it is often the implementation time of transmission to the BS that is the show stopper. Therefore it is very important that the transmission department is advised as soon as the solution is approved by the Coverage Committee, to avoid the typical situation where an implemented solution is still awaiting transmission 5–8 weeks after implementation.

5.1.5   Post Implementation

The RF designer or installer is responsible for doing a walk test of the system once it is operational. The coverage is documented on floor plans, using post-processing software. The RF designer should also contact the end-user, to check that the coverage is as expected, and monitor the performance of the system using the statistical tools available, evaluating the live network data and checking neighbor lists, quality, capacity, etc. These results (the walk test and the performance tools) are used to fine-tune the RF parameters of the cell, enhancing the system performance.

An important check point is that the traffic produced is within the expected range, based on the input from sales, and the RF planner provides feedback on the total performance to the account manager. After 2 weeks, the responsibility for the performance of the cell is handed over to 'operations'; after this it is their responsibility to monitor the cell. The RF designer receives and approves the invoices from the installers, so he knows (and can learn) if the estimated price was within scope, and be better placed to predict project costs for future systems.

5.2   The RF Design Process

5.2.1   The Role of the RF Planner

After we have had a quick look at the general process of the total indoor implementation process, we will have a closer look on the design tasks for the RF planner; after all, this is the purpose of this book.

Draft Design and Site Survey

Based on the design inputs provided by sales, the RF planner will do a draft radio link budget and prepare a draft design prior to the site survey. Thus the radio planner can check all the planned antenna locations, and adapt the design accordingly. Based on the experience from the site survey, the radio planner will be able to adapt the draft design to the reality and restrictions in the building, in order to make the final design. During the RF survey, it is important to check the type of walls, take notes on the floor plans of the different types, etc.

Take Photographs

The RF planner should bring a digital camera, take lots of photos and mark the position of each photo on the floor plan; this will help in making the final design. It is advisable to take a photo of each antenna location, in order to document the exact antenna location.

It is a good idea to bring a laser pointer, and point at the exact planned antenna location when you take each photo. The red dot from the laser will be very clear on the photograph. In the design documentation, each photograph is then named according to the antenna number, A1, A2, etc. This helps the installer to install all the antennas in the correct location and avoid expensive mistakes.

5.2.2   RF Measurements

RF measurements are a crucial part of designing and verifying indoor coverage solutions. It is important to know the 'RF-baseline' (the existing coverage) both inside and outside the building in order to establish the correct design level and parameters to use when designing the indoor DAS.

Log and Save the Data

It is highly recommended to always use a measurement system that allows you to log the measurement data on a PC for post analysis. You should preferably use a system that can navigate and place the measurements on a matching floor plan, and indicate the measurement result by color or text.

These floor plans, with plots of signal level, quality and HO zones are also crucial documents to prove that the system implemented fulfills the agreed design criteria, and are very useful as a reference for trouble-shooting at a later stage. These measurement results should be saved in a structured data base system. This can be very valuable experience in future designs.

5.2.3   The Initial RF Measurements

The measurement routes for the initial measurements needed in order to design the radio system are shown in Figure 5.2. The first measurement that is needed is the outdoor measurement 1 in Figure 5.2, in order to determine the outdoor level and the servicing cells. It is very important to determine the outdoor signal level in order to design the HO zone between the indoor solution to the outdoor network. This important HO zone will cater for the handover of the users entering and exiting the building.

Measurement 1 in Figure 5.2 is also used to estimate the penetration loss into the building, when compared with measurements 2 and 3 in Figure 5.2, and can be useful to calculate the isolation of the building. In this way you can estimate the desired target level for the indoor system.

Measurements 2 and 3 in Figure 5.2 serve the purpose of obtaining an RF baseline of the existing coverage levels present in the building. It is very important to establish this prior to the design and implementation of the indoor solution, in order to select the correct design levels for the DAS design, according to any interference from outdoor base stations.

Measure the Isolation

Using the measurement method described above, you can calculate the isolation of the building and plan accordingly. Typically you will need to perform measurement 2 at least on the ground floor, middle floor and topmost floor. In a high-rise building it is advisable to repeat measurement 2 on every fifth floor.

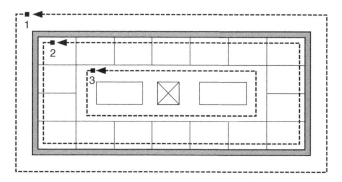

Figure 5.2   Initial RF survey measurement routes

When designing a 2G system, you should design for dominance of the indoor cell. When designing 2G, the indoor system should preferably exceed the signal level of any outdoor macro cells present in the building by 6–10 dB. This will ensure that the indoor cell is dominant and prevent the mobiles from handing over to the outside macro network. This can sometimes be a challenge to fulfill, especially with nearby macro sites adjacent to the building; there are tricks on 2G that 'lock' the traffic to the indoor cell, even if the indoor cell has a lower signal level (see Section 5.5.1), but it is preferable to solve the dominance problem with careful radio planning upfront.

For 3G designs these measurements are also very important. Based on the measurements you should try to design the indoor system to be 10–15 dB more powerful, in order to avoid extensive soft handover zones. It is important to minimize the soft handover zones in the building to avoid cannibalizing capacity for more than one cell. In particular, hotspot areas in the building with high traffic density should not have any soft handover zones.

5.2.4 Measurements of Existing Coverage Level

Measurements of existing coverage, penetration losses and verification measurements are a crucial part of the RF design. These measurements provide the RF designer with valuable information to be used for the design and optimization of the indoor solution. In addition to the initial measurements just described in Section 5.2.3, several other measurements are equally important. There are several types of measurements that needs to be done.

Channel Scans

You should always consider performing channel scans on the same floors as measurement 2, preferably close to the windows (or even open the window if possible), in each direction of the building. A channel scan can be performed by most measurement tools using a test mobile.

Typically the user defines the start and end channels, and the mobile will scan and measure all the channels in that specified range. For 3G you use a code scanner that logs all the decoded cells, scrambling codes and CPICH levels. The purpose of this channel scan is to measure potential pilot polluters, find unexpected neighbors and establish the baseline for the noise level on the radio channel.

Example Channel Scan 2G

In the example shown in Figure 5.3 the test mobile has been programmed to scan from CH01 to CH33, all the 2G channels of the specific 2G operator CH1–CH32 planning the indoor system, including the first adjacent channel of the next operator in the spectrum CH33. It is evident that the high signal levels of CH29 (−43 dBm), CH06 and CH26 are powerful. They are probably nearby macro sites across the street from the building. You cannot isolate against these powerful cells, and so you need to take them into account when defining and optimizing the neighbor list and HO zones.

The scan is also useful to select appropriate frequencies for the indoor system. In this case CH20, CH02 and CH31 look like good candidates to use inside the building. However, be

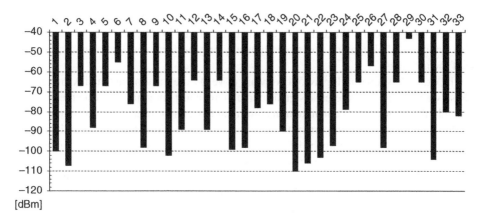

Figure 5.3   Example of channel scan measurement on 2G CH1-33

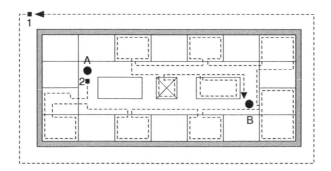

Figure 5.4   Typical RF survey route, measuring in the same area as the proposed antenna location

careful; the reason for these channels being low in level could simply be lack of traffic on these channels at that particular moment in time. You must always verify the channel you propose to use inside the building in the macro frequency planning tool. On 3G and 4G you would scan for the power levels of the codes of the available cells.

5.2.5   RF Survey Measurement

In some cases you should also verify the RF model that is used to simulate the coverage in the particular building. This is especially important for the first 10–20 projects. By then you will have gained enough experience and measurement results to adjust and trust the propagation model you have selected (see Section 8.1.5) and your 'RF instinct' for that particular environment.

Performing the RF Survey Measurement

The typical measurement set-up can be seen in Figure 5.4. In this example two proposed antenna locations are verified in the draft design, A and B. The design concept with these two antennas is to use the 'corridor effect' to a maximum; using the corridors to distribute the signal around the more solid core of the building and reach the users in the offices along the perimeter of the building. Since the two antennas are placed in the same type of environment, with a symmetrical placement, you would typically only perform verification measurement on one of the antennas.

In the example shown in Figure 5.4 we want to verify the location antenna A, and in order to do so we must place a reference signal source there, a test antenna. The test antenna should preferably the same type of antenna, transmitting at the same power as you plan to use in the final design. It is also very important to place the antenna as close as possible to the intended installation position. Preferably the RF survey transmit antenna should be placed in the ceiling in the actual position or at least as close as possible.

The location of the test antenna is very important because the performance of antennas will be highly influenced by their immediate environment. The performance can be different if it is placed on a tripod 1.5 m off the floor, compared with the real position in the ceiling, upside down. This is because the local structures and environment around the antenna will affect the directivity performance of the antenna and you want the RF survey to be as close as possible to the actual conditions of the final installation.

Preferably you should perform measurements on the adjacent floors under and above the floor with the survey antenna installed. This is to establish the floor RF separation, which is useful when estimating whether 'interleaving' of the antennas is possible (see Section 5.3.9). Normally you will not take measurements in the area very close to the antenna or the nearest rooms because:

1. You will know that the coverage in the area only 5–10 m from the antenna is good. If not, it will be evident in the measurement results further from the antenna.
2. In many cases, measuring close to the antenna will saturate the receiver of some test mobiles, and this will distort the real measurement results, e.g. one mobile might not be able to measure signals higher than −40 dBm (2G example), and therefore measurement values higher than this value will only be logged as −40 dBm, and will skew the calibration of the model.

5.2.6   Planning the Measurements

The planned RF measurement route must represent all areas planned to be serviced by the antenna location. It must be planned to measure with a good overlap to areas where other antennas are intended to provide service. One example can be seen in Figure 5.4, where the purpose of the measurement is to verify the planned location of antenna A.

It is very important to conduct measurements close to the windows on each side of the building and in particular in corner offices. These areas are prone to 'ping-pong' handovers (2G/4G) and soft handover zones (3G) and this might lead to degraded quality due to interference and pilot pollution. Therefore you must carefully select a design level based on the measurements with special focus on these areas to make sure you can overpower the macro signals, and at the same time avoid leakage of signal from the building.

It is also important to conduct a measurement outside the building to estimate the leakage from the building. However, for obvious reasons, this can normally only be done during measurement of the ground floor antenna verification. It will not be possible to measure every antenna location on each floor; typically only one or two antenna locations are selected in each type of different environment to verify the model used.

These RF survey measurements are a crucial part of the experience you want to gain as indoor radio planners. After you have conducted a few of these measurements and analyzed the results, you will soon gain trust in your model and experience of when to trust the RF model. Over time you will be able to 'see' the RF environment, and will be able to decide when it might be necessary to perform RF survey verification measurements when designing future projects and when to trust your 'RF vision'.

Save your measurement results in a database, to gain a knowledge database of penetration losses for future projects, and for fine-tuning your simulation models. This will enable you to do statistical analysis of the results in a structured manner, which is useful for reference and for sharing experience.

Use Calibrated Measurement Tools

When conducting these RF reference measurements it is very important to use a calibrated transmitter and receiver-calibrated test mobile. A test mobile is similar to a standard mobile, but it has been calibrated and enabled with special measurement software. This software enables the user to perform detailed measurements on the network. These measurements can be conducted in both idle and dedicated modes. Typically the user will be able to read the basic RF measurement information in the display of the mobile. This will typically be channel number, cell-ID, RF-quality, information about neighboring cells, signal strength and data rates. The simplest test mobiles let you read this information in their display; more sophisticated models let you save the measurements for post processing.

Log and Save the Measurements

It is important to use a test mobile that is able to save the RF measurements on a connected PC. In addition to the propriety data format that enables the user to save the results in a file for post processing within the software package that comes with the test mobile, it is important that the measurement system software also enables you to export the RF measurement results in a text format so you can import the results in standard software like MS Excel. This enables you to do various post processing analyses of the measurements (this is how the measurements in Figures 4.26 and 4.28 were documented).

Log the Measurements According to the Measured Route

It is highly recommended to use a measurement software package that enables you to import the floor plan of the building where the measurements are being performed. This enables the user to mark reference points on the floor plan, and the software then distributes the measurement samples between the markers along the route. This might even be the same software platform you use for your RF planning tool, merging it all into the same tool.

The navigation works typically by marking the next point you are heading to on the floor plan, and clicking when you arrive. Then the tool distributes the measured samples between the points. Typically this will be the corner points all along the route. Be sure to keep a constant pace from way mark to way mark, in order to secure an even distribution of the measurement samples along the route.

Some of the simpler measurement tools only allow you input 'waypoint' markers in the measurement file for the position reference. These tools will be alright for a measurement in a tunnel, where you can use typical reference points such as 100 m, 200 m, 300 m, since you know the direction is from X to Y (there being no Z in a tunnel) and, by keeping a constant speed, you can log the results accurately to the position. However, for a measurement in a complex building it is close to impossible to use only waypoints for reference, unless you break the measurement up into many files, each covering a specific room in the building.

Do Not Bias/offset the Measurement Results

Make sure that the measurement receiver for these measurements is in a neutral set-up. Do *not* carry the measurement receiver upside-down in your pocket when conducting the survey measurements. You need to be sure that the antenna is unobstructed in all directions, in order not to skew the measurements. During the post analysis of the measurement, you can always add body loss and other design margins, but the measurement must be done as neutrally as possible.

When performing RF-survey measurements be sure that you use a clean RF channel in the spectrum for the survey transmitter, so no interference from other base stations will distort the measurements. One of the worst examples i have seen in real-life, was an engineer claiming the DAS was under designed. Actually it turned out that the antenna on his test-mobile were broken of, and there was a hole in the mobile where the antenna used to be! (he still claimed to be an RF engineer though!)

5.2.7 Post Implementation Measurements

After implementation you will also need to perform a measurement in order to document the 'as-built' system. This will also help you to find any antennas in the system not performing as expected. You will need to measure all floors on all levels of the building, preferably measuring 'edge-to-edge', and do samples of all the different areas of the building.

These measurements can also help you calibrate your design tools and models, for future designs. Preferably the radio planner will do these measurements at least for the first 10–20 buildings, in order to gain experience of how the building, walls and interiors affect the signal and propagation of the signal. A copy of these measurement results should be kept as a part of the on-site documentation; this is useful for trouble-shooting on the system in the future.

The Simplest Measurements, Post Installation

The simplest measurement you can perform on an indoor coverage system is to stand below each and every antenna in the building, and average out about 20 s of samples, just using the display of the phone. Note this result on the floor plan as a reference for checking the radiated power from the antennas, using the free space loss formula to estimate if all antennas are performing as expected.

You will also need to perform measurements of more locations distributed throughout the building, especially in the areas where you expect to find the lowest signal. Typical this would be the corners of the building, the staircases and the elevators. Do not forget to measure the executive offices as it is nice to be absolutely sure everything is alright in that area – there is no point in claiming that the indoor system performs to the required 98% coverage area, if the last 2% is in the CEO's office! See Section 5.3.10 for more details.

5.2.8   Free Space Loss

Free space loss is a physical constant. This simple RF formula is valid up to about 50 m distance from the antenna when in line-of-sight inside a building. The free space loss does not take into account any additional clutter loss or reflections, hence the name. The free space loss formula is:

$$\text{free space loss (dB)} = 32.44 + 20(\log F) + 20(\log D)$$

where F=frequency (MHz) and D=distance (km). The free space loss for some standard frequencies will be as shown in Table 5.1.

As we can see in Table 5.1, the loss on 950 MHz is 32 dB at 1 m, and each time we double the distance or frequency we add 6 dB more in free space loss (as shown in Figure 5.5). We can also see that the path loss on 1850 MHz is 6 dB more than on 950 MHz. The difference in frequency between 2G-1800 (1850 MHz) and 3G (2150 MHz) is only about 1 dB due to the relatively small difference in frequency.

5.2.9   The One Meter Test

After implementation of the solution, one can be in doubt whether an antenna is performing as expected or not. Often you want to connect a power meter or spectrum analyzer to the antenna system to check if the antenna is fed the correct power level. However this is often

Table 5.1   Examples on free space losses (rounded to the nearest dB)

Free space loss	1 m	2 m	4 m	8 m	16 m
800MHz	31dB	37dB	43dB	49dB	55dB
950 MHz	32 dB	38 dB	44 dB	50 dB	56 dB
1850 MHz	38 dB	44 dB	50 dB	56 dB	62 dB
2150 MHz	39 dB	45 dB	51 dB	57 dB	63 dB
2600MHz	41dB	47dB	53dB	59dB	65dB

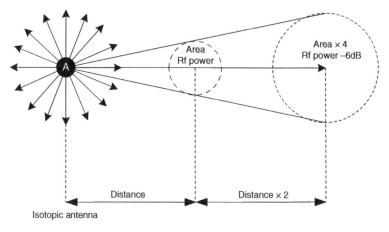

Figure 5.5   It is a physical constant that each time you double the distance, the free space loss is increased by 6dB

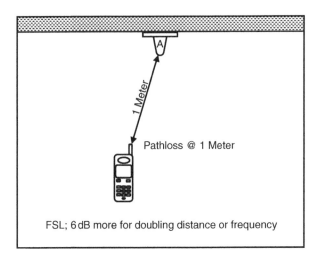

Figure 5.6   The '1 m test' is a fast way to estimate the antenna power

challenging due to the installation, with the antenna connector hidden above the ceiling and difficult to access.

Another issue is that this will still not check if the antenna itself is working or not. Often you will not consider the antenna to be a likely point of failure, Experience, however, documents that this is a common issue in indoor antennas systems. A very practical way to test if indoor antennas are performing correctly is to use the '1 m test', based on the free space loss (see Figure 5.6).

Performing the One Meter Test

To test if the antenna is radiating the expected power is simple; all you need to do is to apply the free space loss. After all, this is a physical reality, and a good guideline for a fast, efficient test (Figure 5.6).

Example 1800 MHz
The test mobile measures −35 dBm at 1 m distance

$$\text{The radiated power from the antenna} = -35\,\text{dBm} + 38\text{dB} = +3\,\text{dBm}$$

Note that some mobiles 'saturate' at higher signal levels. You will often need to 'average' some samples. The radiation pattern of the antenna also impacts the level, so be sure to measure in the expected 'main beam'.

The One Meter Test is Not 100% Accurate

This method is not to be considered 100% correct, but it is very useful for a quick verification, and one can easily estimate if the power is off by more than 6 dB. For an accurate measurement, however, it is recommended to connect a power meter or spectrum analyzer directly to the feed cable.

5.3   Designing the Optimum Indoor Solution

The optimum indoor solution exists only in theory. All implemented indoor solutions will to some extent be a compromise. It is the main task of the indoor RF planner and the team implementing the indoor solutions to make a suitable compromise between meeting the design goals, securing the system for the future and designing and implementing the system to maximize the business case. You must also design the DAS so the antenna system utilizes the features, possibilities and limitations in the particular building to a maximum, in order to make the system applicable for practical implementation.

5.3.1   Adapt the Design to Reality

Even with theoretical knowledge on how to design the perfect indoor solution, the main task for the RF planner will often be to know when and where to compromise in order to implement the system in practice and still maintain an economical and high performing solution.

The preferred approach when designing and implementing an indoor coverage solution would be based on site survey with measurements, experience and the link budget. However, reality often dictates that the architectural and installation limitations, and cost concerns, play a major role in the final design (as shown in Figure 5.7).

5.3.2   Learn from the Mistakes of Others

An old mountaineering saying goes like this: good judgment comes from experience but experience is often a result of bad judgment! This is also true for indoor RF planning. There is no need to repeat the mistakes of others, so let us learn from some of the most common errors. Here are a few tricks to share, most of them learned the hard way.

Keep the Simple Things Correct

Indoor antenna systems are often complex and the radio services operating on these systems are getting more and more advanced, with higher data rates and multimedia applications. However, do not forget to focus on the simple and trivial things. These

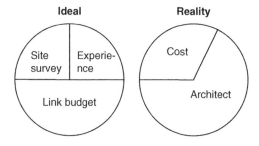

Figure 5.7   The RF planner needs to adapt to the reality of the building

simple things tend to be underestimated, but can have a major impact on the performance of the implemented system.

Remember that designing indoor distributed antenna systems it is still 'only' RF planning, and good RF design and implementation is often a matter of understanding why you need to pay close attention to RF components that might seem trivial. You must also appreciate the importance of the craftsmanship needed to install these RF components and antenna systems so they perform as planned over time. Education, craftsmanship and common sense of the installers make the difference between success and failure. Even the best designed DAS will not perform better than the weakest point in the DAS, and a small problem with a connector, for example, will have major impact on the performance.

Use Only Visible Antenna Placement

Often the RF planner is asked if it is possible to hide the antennas above the suspended ceiling. This should be avoided at all costs and you should refrain from giving any guarantee of the performance of the implemented system if forced to accept hidden antennas.

The main reason is the consequent unpredictability. You don't know what is hidden above the ceiling tiles: fire sprinklers, ventilation ducts, heating systems, electrical wiring. Worst of all, you have no post installation control of objects close to the antenna after the installation of the indoor system has been completed. Furthermore you have no control of the RF properties of the ceiling tiles themselves. What happens if the ceiling is repainted, or changed for tiles of a different material? Aluminium ceilings are in fashion in some parts of the world, and would be disastrous for the performance.

Obtain a responsible architect or designer of the building to accept and appreciate that the antennas must be 'visible' for the mobiles. After all, mobile coverage is a crucial part of the buildings utility infrastructure. If you are forced to accept installation of antennas above the ceiling or behind walls, be sure to note it in the design document and contract that the performance of the system is 'best effort' and not guaranteed due to installation restraints.

Use Only Quality Certified Components

Again, the importance of the performance of 'simple' RF components is often underestimated. It is normal to focus on the performance of the more complex components in the indoor antenna system, the base station, repeater, the active distribution system etc. However, the quality of the simpler passive components – antennas, splitters, tapers, connectors and jumpers – may have a major degrading impact on the RF service performance if these components do not perform to specification.

Use Only High-quality Antennas

One example of how important the performance of passive components can be is the 'standard' 2 dBi omni antenna. This antenna is used in 99% of installations, and therefore will play a major role on the performance of all the indoor systems in the network.

This 'simple' antenna can be obtained from various suppliers but the different antennas are markedly similar looking – if you compare the data sheets they will most likely also be similar in terms of mechanical design. Even during an RF test the antennas perform the same, but how about 5 years from now?

Is there a reason for one of the 'identical' antennas to be more costly? Could some antennas be copies, where one manufacture has not taken into account passive inter-modulation effects, due to incompatible metals inside the antenna? Could there be degrading galvanic corrosion over time? This will cause bad connections, and passive inter-modulation (see Chapter 5.7.3), with serious degrading performance of the indoor system.

Small Things Matter

The same goes for other 'simple' components like an RF connector. Could there be a reason why one type of connector comes at a cost of 1 euro more a piece? Could the more expensive connector be a better option?

Please do not misunderstand me – there is not always a 1:1 relation between price and quality. The statement is only general; be careful, and choose a manufacturer that you trust. Look, feel, measure, test and judge if you trust the quality of the component to hand. Disassemble the antenna, have a look at the internal materials: does it appear to be good craftsmanship? It is not always easy to estimate, but in many cases you will get a feeling for the quality.

Use Only Educated Installers

The craftsmanship needed to install RF systems is often underestimated. Use only educated installers who are certified to work with RF components and cables. There is a huge difference in doing a quality installation of a mains power connector, and then doing a high-quality installation of a coax connector operating at 2.1 GHz. Did you know that manufacturers of coax connectors recommend a specific torque for tightening the connector to insure the performance?

Make sure that the selected installer only uses qualified and certified workers, who know what is needed to do high-quality RF installation. Also make sure that all the installation teams have the correct tools to hand, and not just one set of tools for the first team and 'knife and spanners' for the rest! This might seem obvious, but often you will have four to six parallel teams installing the same building to be able to do it overnight. It is very important that each of these teams is equipped with the correct tools, the correct connectors, etc.

Use of quality components, tools and installers will pay of in the long run. Yes, it might seem costly, but it is worth the investment. Believe me, I know how costly it is to have all connectors replaced in an operating metro system, after the metro trains are operational!

Design Documentation

This is another example of a very important issue when designing and implementing an indoor solution. For the RF planner it might seem straightforward to decide where to install what, and obvious what the correct orientation is for that directional antenna at the end of the hall. However, again, produce detailed design documentation and installation guidelines. This will also help with future upgrades, trouble-shooting and extension of the indoor system. It can be a hard task to reconstruct the documentation of an implemented DAS system four or seven years after implementation, even for the RF planner who initially designed the system. If the same RF planner is still around, that is!

The design documentation should be easily accessible on-site in hardcopy, preferably also in an electronic format that is easy to update with future changes.

5.3.3 Common Mistakes When Designing Indoor Solutions

Often new indoor RF planners will make the same initial mistakes. Let us try to avoid these.

Dimensioning Coverage on Downlink Only

It is a fact that 2G indoor systems are mostly downlink-limited, so most focus is on the downlink. However, it is important not to forget the uplink part of the design, especially on 3G/4G, depending on the service profile of the traffic on the cell. Remember that the Link Budget consists of an analysis of both the downlink and the uplink.

We will focus more on the Link Budget calculation later in this book, in Section 8.1.

Underestimating Passive Inter-modulation

Do not underestimate potential passive inter-modulation (PIM) problems. Especially with high-capacity (power) passive DAS with many radio channels in service, the concentration of power in the cables, connectors, splitters, tappers and antennas close to the base stations can be high. If you use low-cost, low-quality components, the impact of inter-modulation can be high. We will take a closer look at this in Sections 5.7.3 and 5.7.4.

Not Accounting for Coax Loss

Many good indoor DAS designs have been ruined by the reality of the installation, often due to underestimating the shear length of the coax cable, and thus underestimating the passive losses in the system. The passive losses in the DAS have a big impact on the uplink and downlink performance of the system. It is very important to use realistic distances on the cable part of the passive system, in order to be able to calculate the link budget correctly. It is recommended to have a safety margin on the coax losses, by adding 10% to the length of the individual cables. This problem is getting bigger, as the frequencies used for mobile services are getting higher.

Underestimating the Costs of the Installed System

The business case is a crucial part of any indoor design. A major part of the cost of an indoor coverage system is installation costs, especially when installing a passive system using heavy rigid cables, which are time-consuming and labor-intensive to install.

In order to estimate the DAS cost when designing passive systems it is very important that you know the exact installation challenge, cable routes etc. upfront during the initial evaluation of the project. If you do not, the cost of the project is impossible to estimate. This means that the RF planner needs to 'walk the building' before he can do the cost estimate, and it is as much installation planning as it is radio planning.

There are many examples of passive systems costing in excess of several times the initial estimated costs once implemented. This is mainly due to lack of knowledge about the installation challenges of rigid passive coax cables. Other unforeseen installation costs, such as lack of installation trays, reconstruction of fire separation barriers in the installation ducts and access to the building only at night, will increase the labor costs significantly.

Not Accounting for the Total Cost of the Project

Mistakes in the business case are often caused by not including all the costs of the total indoor DAS. An indoor solution is much more than 'just' the cost of the indoor system itself:

- DAS costs.
- Installation costs.
- Maintenance costs.
- Site-support costs (power supply, transmission, etc.).
- Upgrade costs for future services and capacity.
- Planning costs.
- Backhaul cost.

5.3.4 Planning the Antenna Locations

The Inter-antenna Distance, Theory

Once you have calculated the link budget you can establish what the limiting link is, the uplink (from mobile to base station) or the downlink (from base station to mobile). For multilayer systems, i.e. systems where several radio services are sharing the same antenna installation, this could be 2G and 3G on the same DAS. There might be some difference in the service range from the two systems, using the same DAS antenna. An example of service ranges from different mobile services could be: 2G DL = 28m, 2G UL = 78m, 3G DL = 23m, 3G UL = 21m. In this example the determining factor of the system will be the 3G UL, and the antennas should be placed accordingly.

For the theoretical example, imagine that the environment inside a building is like a uniform pulp of RF attenuation, with uniform attenuation in all directions and no 'corridor effect' or similar real-life behavior. One could think that the inter-antenna distance of omni antennas should be 21 m as calculated to fulfill the 3G UL limit above.

However, if an area is to be covered '100%' (there is no such thing as 100% coverage, but this is a theoretical example), one could believe that the inter-antenna distance must be $2 \times 21 = 42m$, but this is not the case.

In order to have coherent coverage throughout the area, the inter-antenna distance has to be shorter than twice the service radius; after all, the footprint of the ideal omni antenna is circular, not square (as shown in Figure 5.8).

In theory the correct inter-antenna distance for omni antennas is radius $\times \sqrt{2}$. This clearly shows us that there is a need for coverage overlap between the antennas in order to provide good indoor coverage.

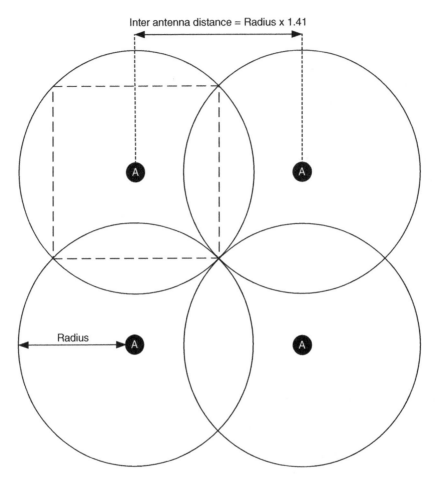

Figure 5.8   In order to provide 'full coverage', antennas need to be placed with a certain coverage overlap

Adapt the Antenna Placements to the Reality

In the real world inside a building, you will never have uniform loss in all directions from the antenna. In real buildings the antennas will typically serve many types of 'clutters' and areas, e.g. open office space, dense office areas, dense areas (stair cases), heavy dense areas (elevator shafts). Therefore you must adapt the antenna locations to these specific environments when applying the link budget-calculated service ranges from the antennas to the reality of the building.

In practice, this means that you will have a shorter range from the antenna towards the denser areas, whereas you will have longer range in the more open directions (as shown in Figure 5.9). This is where the radio planner must use his experience and knowledge of the building gained during the site survey to adapt the design to reality.

The experienced RF planner will use the calculations from the link budget. The calculated service varies from the individual antennas throughout the different clutters and areas of the building, and the antennas should be placed so that the maximum footprint is obtained for

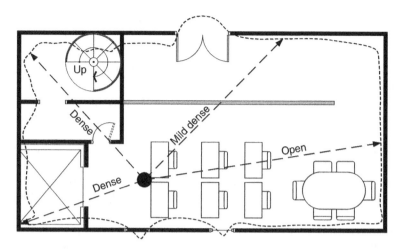

Figure 5.9   In reality the coverage from a typical indoor antenna will be uneven in different directions, due to the service of different environments

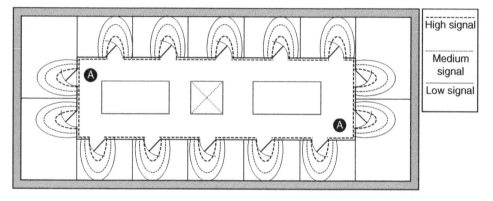

Figure 5.10   The 'corridor effect': the corridor in the building will distribute the RF signal

each antenna. The antenna locations have to be adapted to the reality of the building, taking into account the installation limitations of the individual antenna placements, cable routes and antenna overlaps.

5.3.5   The 'Corridor Effect'

One effect of placing antennas inside a building that is very important to know and utilize is the 'corridor effect'. This is the distribution of coverage when placing an antenna in one of the most typical locations, the corridor (as shown in Figure 5.10).

This antenna location will give you several advantages:

- Typically there will be easy installation access to cable conduits in the corridors of the building, which will save implementation costs.

- Corridors are often 'static' when buildings are being refurbished, so the antennas are left in place with no impact of service degradation caused by refurbishment of the internal structure of the building.
- The users of the building are less concerned about radiation when they do not have antennas installed in the ceiling above their office desk.
- You can use the corridor to distribute the signal from the antenna; there is typically line-of-sight throughout the corridor, and this is what is known as the 'corridor effect'.

The downside is that it can be hard to dominate the area along the windows at the perimeter of the building. Be sure to measure the level from any existing signals along the windows, to be able to select your RF design target level. In many modern buildings metallic-coated 'tinted' windows are installed, this will help in solving this potential problem.

The corridor effect is a strong tool, but you need to be careful when you base the design on using this method. By choosing the optimum location for the antennas, the building will help distribute the RF signal and you can utilize the corridor effect to a maximum. However, if you do not choose the antenna locations wisely, the building can limit the performance of the system. You must pay special attention when designing buildings that are planned as multicell systems, and try to minimize the areas of the handover zones.

5.3.6   Fire Cells Inside the Building

It is important to remember that most buildings are divided into several 'fire zones'. These zones within the building are separated both vertically and horizontally by the use of heavy walls and metal doors. The reason for this is to contain any potential fire inside the building, thus minimizing the damage to property and people. The heavy materials used to construct these firewalls or cells will typically also attenuate the RF signal significantly when it needs to pass between these 'fire zones'. This has a major impact on the antenna layout in the building. In reality there must always be minimum one antenna inside each of these cells, or else the firewalls will attenuate the RF signal by 30–40 dB or maybe even more. The example in Figure 5.11 shows a typical building layout including the heavy firewalls and proposed antenna locations. Note that two antennas are placed near the elevator to provide coverage inside the lift. Lift shafts are normally heavily fire- and RF-isolated from the rest of the building, although an exception to this is the 'glass' lifts used in some open atria. Two antennas are placed on the edge between two fire zones and will provide service on both sides of the firewall, thus saving one antenna placement.

5.3.7   Indoor Antenna Performance

For more detailed information about the deep theory on antennas, please refer to Reference [6].

To perform proper indoor radio planning, you must understand the basic function of the indoor antenna. The behavior and quality of the indoor antenna have major impacts on the performance of the indoor DAS system.

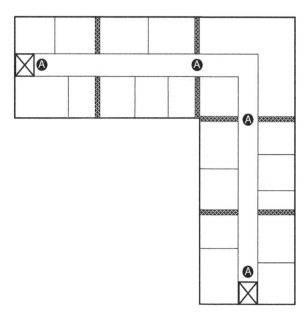

Figure 5.11 The inside of the building will be divided into 'fire cells', separated by heavy walls (as marked on the floor plan above) to contain any potential fire. These walls will attenuate the RF signal, and in most cases you will need an antenna within each of these 'fire zones'

Antennas Have No Gain

I know this statement is a bit controversial, but nevertheless it is actually true. For the indoor radio planner it is very important to realize that antennas do not have any gain! How could the antenna have any gain – it is just a piece of metal? The gain of any system is defined by the difference between the power you feed to the system and the output power; the difference is the gain. Antennas have no gain; the antenna consists of no active elements. No power to the antenna is supplied besides the feed RF signal.

Antenna Efficiency

This is shown in Figure 5.12. You feed the antenna an RF signal power on the input, and, owing to the losses of the passive elements inside the antenna, some of this RF power is dissipated as heat. Some (hopefully most) of the power will radiate from the antenna in the form of an electromagnetic RF field. So the higher the antenna efficiency, the greater the signal power. The antenna efficiency is often underestimated and parameters such as 'antenna gain' get all the attention. But antennas have no gain – they have directivity and efficiency. However, in all datasheets, you will find data relating to the 'antenna gain'. In reality, this is not the gain of the antenna, but the directivity of the antenna, relative to the 'isotopic' antenna.

Let's take a closer look at the antenna basics, which are very important and quite often misunderstood by radio planners.

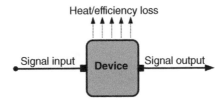

Figure 5.12 A device and its efficiency (power loss). Antennas also have loss and will lose some power as a result of this

The Isotopic Omni Antenna

The 'perfect' theoretical omnidirectional antenna is often used as the reference for this 'gain' when evaluating the performance of real antennas. This isotopic omnidirectional antenna radiates its power equally in every direction, X, Y, Z, as a perfect RF sphere with 0 dBi 'gain' in all directions (as shown in Figure 5.5 and Figure 5.13 top), hence the name 'omnidirectional' – but this antenna exists only in theory.

Where Does the 'Gain' Come In?

In reality the term 'antenna gain' refers to the directivity of the antenna, in the main beam direction of the radiation of the antenna. In practice the term 'gain' refers to the fact that the power in the main beam direction will be higher compared with an isotopic omni antenna.

This is evident in Figure 5.13 where we compare the "'gain"' (or the power density difference at a certain position) from the theoretical omni directional that re-radiates the electromagnetic field uniformly in a perfect 360° sphere – this antenna only exists in theory.

The power will be lower in other directions. This relative difference in performance compared with the isotopic omni antenna is the gain data that the manufacturer refers to as 'antenna gain', hence the name dBi (the 'i' stands for isotopic). It is used as a measure for stating the gain of an antenna. However the 'cost' of this 'antenna gain' is less directivity in other directions; the antenna is actually less sensitive in other directions compared with the isotopic omni antenna. Sometimes the antenna gain is stated in dBd, where the reference is a dipole antenna. The relative difference between dBi and dBd is 2.1 dB.

Therefore, the typical omni antenna we use inside buildings will have a 'gain' value stated in the data sheet (directivity) of 2.14 dBi, because this is actually a 1/2 λ omni directional antenna, and as shown in Figure 5.13, the directivity in the horizontal (X) plane will be 2.14 dB better, and this will 'cost' directivity in the Vvertical (Y) plane. Also keep in mind the cost, in terms of efficiency, of the antenna on the actual radiated power as described earlier.

This principle is shown in Figure 5.14, where it is clear that the perfect theoretical omni antenna distributes the signal equally in all directions, and forms a perfect sphere. The +3 dBi gain antenna is directive, more sensitive in the main direction, at the cost of less sensitivity in other directions.

Equivalent Isotropic Radiated Power (EIRP)

Equivalent isotropic radiated power is used to express the radiated power from an antenna, taking into account the passive losses in the DAS and the directivity of the antenna, in reference to an omni directional antenna with 0 dBi of gain as shown in Figure 5.13. The EIRP is the RF

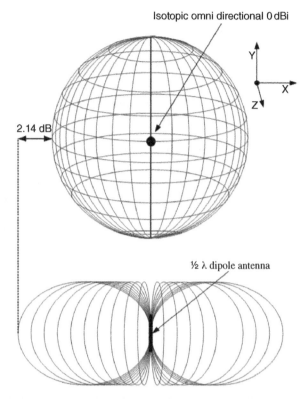

Figure 5.13   Isotopic omni antenna directivity vs. directivity of a 1/2 λ dipole antenna

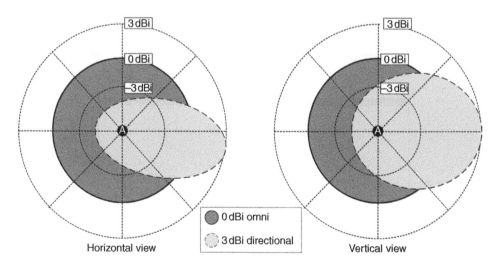

Figure 5.14   Example of horizontal and vertical directivity plots of omni and directional antennas

power transmitted in the main lope of the antenna, the direction of the antenna where the radiated RF power field is most concentrated, giving maximum field strength of RF power.

$$\text{EIRP} = \left(\text{RF power from the RF source}\right) - \left(\text{loss in the system}\right) + \left(\text{directivity of the antenna}\right)$$

So, for example, if we have a DAS remote unit with 10 dBm of transmit power, connected to a 2.14 dBi omni antenna via a coax cable with 1 dB of loss, we can calculate the transmitted EIRP:

$$\text{EIRP} = 10\,\text{dBm} - 1\,\text{dB} + 2.14\,\text{dBi} = 11.14\,\text{dBm}$$

The EIRP is used as the base RF power reference value for estimating the coverage range from the antenna, as we do in more detail in Chapter 8 when we construct the actual detailed RF link calculation of the DAS, also known as the 'link budget' for the system. Obviously the RF link has to be estimated on both the downlink (DL) from the base station to the mobile and the uplink (UL) from the mobile to the base station, taking into account all the elements of the system.

The Installation of the Antenna Plays a Role

The radiation and directivity of any antenna installed inside a building will be affected by the actual installation and local environment close to the antenna. Objects inside the building, walls and other objects will attenuate, reflect and diffract the radio waves radiated from the antenna.

In a large open hall the antenna radiation pattern from the omni antenna might look like that in Figure 5.15. In this example three different omni directional antennas are plotted: the isotopic omni, a medium-gain omni and an omnidirectional antenna with +7 dBi gain. The 'low-gain' antenna is actually more sensitive below the antenna, and the high-gain antenna is more sensitive to the sides and will give less coverage in the large area underneath the antenna.

However, the radiation pattern of the antennas will be influenced by the size and shape of the room in which the antenna is installed due to reflections from the walls. These reflections will to some degree decrease the relative difference between the antennas. However, in large open areas with high ceilings low-gain omni antennas should normally be considered, giving

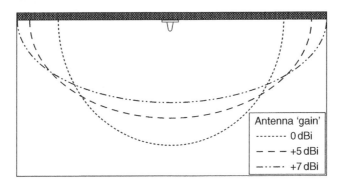

Figure 5.15   Omni antennas with different gains (directivity) in a large open space

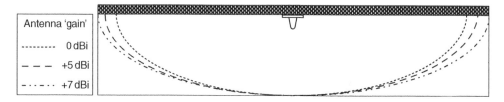

Figure 5.16   The building will 'shape' the directivity of the antennas, masking out most of the 'gain'

more direct power to the mobile users. This is important to remember when designing large convention halls, sport arenas, production facilities, etc.

Another typical example of influence on the antenna by the local environment can be seen in Figure 5.16, which shows three types of omni antennas, with three different gains in the most typical placement: a low ceiling corridor in a building. The directivity of the antenna is highly influenced by the physical constraints and shape of the corridor. The corridor masks out most of the antenna gain or directivity due to the reflections caused by the walls. Therefore, once the antennas are installed inside a building, it is often hard to measure the differences in gain of various types of low/medium/high-gain antennas that are stated in the data sheets.

Owing to installation restraints, it often makes a lot of sense to use a 'directional' antenna at the end of a long corridor. This is to save on installation costs by beaming the coverage down the corridor.

There is Only a Minor Effect of the Gain of Indoor Antennas in Reality

The previous examples show that, in a typical indoor installation (as shown in Figures 5.15 and 5.16), the shape and size of the room where the antenna is installed will have a big impact on the directivity. The installation masks out most of the 'gain' of different types of antennas; often there is not a major difference in level when comparing a '0 dBi' omni antenna with a '5 dBi' omni antenna, but installation might be less challenging for the small '0 dBi' type, due to the smaller size and less visual impact.

Directional Antennas Inside a Building

We will often use omni directional antennas in up to 95% of the antenna locations inside a building for the typical DAS deployment. There will be antenna locations where directional antennas would be preferred, both from an installation and a performance perspective. The directivity of the directional antennas could help us 'steer' the RF coverage concentration towards a specific area of the building if needed, or perhaps shield us from receiving or generating unwanted signals (interference) from other cells inside or outside the building. One example of the directivity of a typical indoor directional antenna can be seen in Figure 5.17; in the horizontal plane we have an 'opening' (directivity) of 90°, and in this example the vertical 'opening' (directivity) is 65°. We often refer to this type of antenna as a '90° antenna'. The '90°' is the point at which the RF signal power from the antenna is 3 dB less than the 'main lope' in the 0° direction.

As you can see in Figure 5.17, there will still be RF signal levels after passing the 90°, but the RF power level will just be less than 3 dB compared with the main lope, and typically it

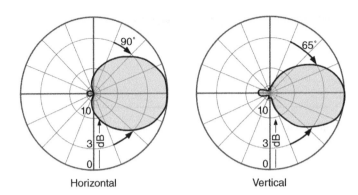

Horizontal   Vertical

Figure 5.17   Horizontal/vertical directivity plots of a typical directional indoor DAS antenna

will get lower and lower the more you move away from the main direction of the antenna. You can also see in Figure 5.17 the 'back lobe', even minor 'side lobes' where the RF power peaks lightly again. The size of the 'back lobe' is important when using directional antennas to isolate cells from each other (see Chapter 14 for more actual examples).

The Building Will Impact the Directivity of DAS Antennas

When you deploy antennas inside a building, most of the directivity will be dictated by the building structure itself, rather than the actual antenna footprint. A directional antenna inside a building is still very useful – one application of a directional antenna such as that in Figure 5.17 is to install it in one of the ends of the narrow corridor shown in Figure 5.16. This will give us close to the same RF power at the end of the corridor as the omni antenna in the center, but might be a more attractive installation option, saving cable length, etc.

Another application of the directional antenna in Figure 5.17 would be to minimize the leakage from a high-rise building such as that shown in Figure 3.8. The directional antennas could also be used to clearly define the HO zone inside large structures such as a shopping mall and a stadium, where you need multiple sectors with well-defined borders between multiple cells to minimize the HO zones, by mounting the directional antennas 'back to back' so one cell is transmitted in the exact opposite direction to the other. The principle of back to back mounted directional antennas is used in the application examples in Chapter 14.

Selecting DAS Antennas

Obviously you need to decide whether you want to use omni or more directional antennas for your DAS designs and deployments. Keep in mind that you also need to watch out for antenna efficiency, PIM performance, bandwidth, VSWR, and ease of installation. There is much more to selecting antennas than just comparing the 'gain' and the price. If you are deploying many DAS solutions with multiple frequencies and high power concentration, I recommend that you carefully test the antenna types to benchmark against the datasheets, PIM, gain performance, etc. Also, 'accelerated lifetime' testing and the RF data test should be considered to ensure a stable performance over many years.

Trust me, I have seen very bad performance from antennas that looked good on paper, and which apparently could perform as well as a 'known brand', but which, when tested, were exposed as not being very good.

Performing these tests will normally require quite expensive third-party laboratory time, so it is an investment but one that is clearly worthwhile.

5.3.8   The 'Corner Office Problem'

When designing a typical office building, it can be tempting to install a centrally placed antenna in a typical small office floor (as shown in Figure 5.18). This will often save on installation and project costs, and the solution might even fulfill the RF design levels, but be careful as this approach has an inherent issue and potential problem: lack of dominance in the corner offices.

This lack of dominance in the total area inside the building will degrade the quality, cause handover loads on the network and even increase the risk of dropped calls. The centrally placed antennas will often compromise the isolation and quality in the corner areas of the building. In many cases the corner offices are occupied by the most important users in the building, those with the highest use of mobile services and highest data speeds. These are the high-revenue users, and often these will be the users who will decide to shift to another operator if the indoor service is not accommodating their needs in terms of performance and quality. The indoor radio designer should strive to design for perfect service inside the building, especially in these important corner offices.

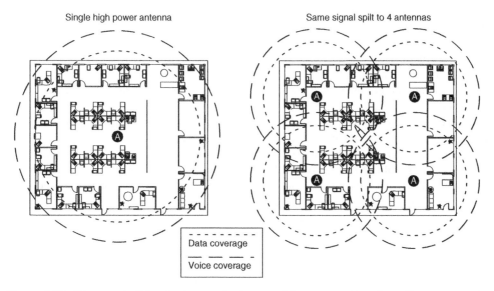

Figure 5.18   The same building implemented using two different strategies. The uniform coverage in the right example with perfect indoor dominance is to be preferred

Solving the Corner Office Problem

The solution to the problem is easy and inexpensive, especially when the extra cost is compared against the gain in performance of the indoor system, and the solution has several advantages. In order to provide a uniform, dominant indoor signal in the total area, the signal is split into more antennas, as shown in the example in Figure 5.18 to four antennas. The signal from each of these antennas will be attenuated by the 1:4 splitter by 6 dB and cable attenuation, but the gain by distributing the antennas is actually greater; they are now closer to the users with less free space loss, and the distribution is providing a more uniform coverage level and much better radio link performance. This strategy will provide:

- A more uniform signal.
- An improved RF link.
- Better data services on 2G, 3G and 4G.
- A more future-proof solution.
- Dominance, even in the corner offices of the building.
- Limited soft handover load on 3G to the macro network.
- Lower radiated power from the downlink power close to the antenna, and so less EMR.
- Lower uplink transmit power from the mobile, lowering the EMR exposure of the mobile users.
- Extended battery life of the mobiles.

5.3.9   Interleaving Antennas In-between Floors

If the floor attenuation in the building allows it (this must be confirmed by an RF survey measurement), one viable alternative to the solution shown in Figure 5.18 could be considered. The principle is to rely on coverage penetrating the floor separations, and thus only to use half the antennas per floor compared with the solution shown in Figure 5.18. In this example from Figure 5.19 it will take two antennas on every floor, and by using offset of the antenna placement on adjacent floors, you can interleave the coverage from the antennas in-between the floors. One antenna on one floor will leak signal to the adjacent floors above and below, and vice versa. The result of applying this method is that the antennas, in addition to the primary service area, will also be servicing the adjacent floors, as shown in Figure 5.20.

When considering this interleaving design strategy, it is very important to measure the exact attenuation of the floor separation. You need to take this measurement in order to estimate if this solution is viable at all. Only based on the actual measured attenuation through the floors are you able to evaluate this approach. However, even after analyzing these measurement results, it is highly recommended to add extra margin to the design, to be on the safe side.

Normally I would recommend covering everything from the antennas installed on the same floor as the intended primary service area, and then only using the coverage from the offset antennas installed on adjacent floors as 'extra' added coverage and dominance. Be sure that the antennas from adjacent floors that might service users on other floors are on the same logical cell to avoid handover load and decreased 3G/4G performance.

Do not be tempted to crank up the radiated power from the antennas to boost the effect of fill-in coverage from adjacent floors; this might result in unattended leakage of signal from the building.

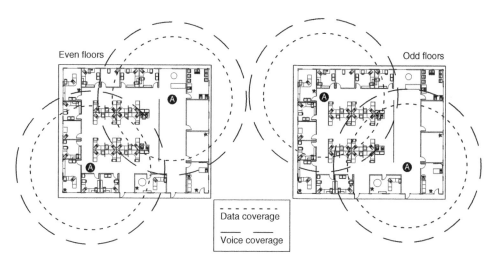

Figure 5.19 Often it is possible to interleave the layout of the antennas, in order to utilize the leakage between adjacent floors and to fill in the 'dead spots' Designing Indoor DAS Solutions 231

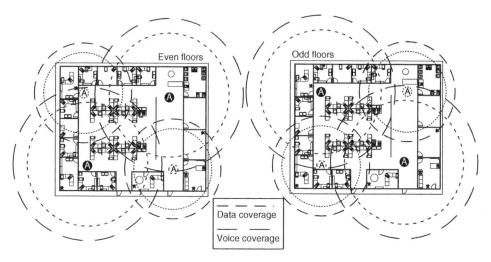

Figure 5.20 The results of the interleaving coverage are often sufficient to provide full dominance in the building

Example of a Measurement of the Floor-to-floor attenuation

Utilization of the coverage that might leak in-between floors is highly dependent on the attenuation of the individual floor separations; it is therefore highly recommended to measure the leakage from floor to floor, in order to evaluate this design approach. You must pay special attention to the layout of any handover zones between internal cells in the building. This is to avoid large soft handover zones between adjacent cells on adjacent floors and degraded 4G service.

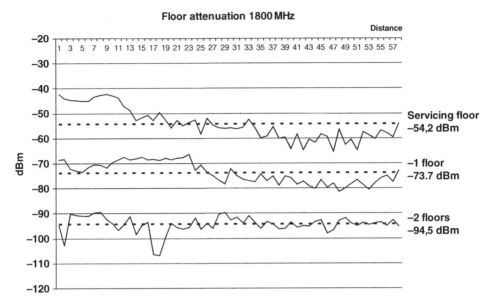

Figure 5.21   Measurement of floor attenuation on 1800 MHz

Example

One example of measurement results of isolation between floors can be seen in Figure 5.18. This is a 2G-1800 indoor system in a typical modern office building with floor separation of 30 cm concrete slaps and suspended ceilings 50 cm under the slap with 3 m floor height. In this RF survey measurement the surveyed omni antenna is located in the leftmost part of the measurement route in Figure 5.21. Three measurements were taken, all using the same route on three different levels; one on the serving floor, one on the floor below and one two floors below.

The measurement samples were averaged and it was calculated that the attenuation of the floor separation was about 20 dB. This is valuable information when evaluating if you can use the effect when designing the solution by interleaving the antenna locations in-between floors, thus saving costs and increasing performance.

It is important to measure on all the frequency bands that are supposed to be supported by the indoor system. The penetration loss can be frequency-dependant, so make sure that you measure both 2G-900 and 3G/4G if more services are to be used in the building.

Measure on all the Frequencies that are to be Used

If you only consider the free space loss (Section 5.2.8), it is a fact that, the lower the radio frequency, the less the link loss will be. Normally, when you analyze the loss through walls in a building, this effect would also be true, that the higher the frequency the higher the penetration loss. However, be careful with concrete walls and especially floor separations. The internal structure of the wall consists of a metal grid that has the purpose of reinforcing the structure of the wall or floor separation. Depending on the mask size of this internal grid structure, it will attenuate the RF signal in a nonlinear relation to the RF frequency. The reason for this is that a metal grid with a mask size of less than a quarter of the wavelength of the radio frequency will heavily attenuate the radio signal. Therefore it is very likely that a higher frequency like 2600MHz will penetrate floor separations and some types of wall better than 800MHz. In practice, always measure all bands when performing this measurement.

5.3.10   Planning for Full Indoor Coverage

The behavior of radio waves and unknown factors inside buildings will to some degree result in unpredictable effects on the radio signals. Therefore radio planning is not 100% predictable. As we have seen in Section 3.2.2, the radio signal will have a fading pattern. This exact fading behavior depends on the environment and the speed of the mobile. The consequence is that, even with a very tight layout of antennas inside a building, it is impossible to guarantee '100%' coverage. You could get close to 100%, but it would be very costly.

The standard term for full mobile coverage indoors is often referred to as '98% coverage' defined at a given RF level. With this definition, it is possible to do a RF survey measurement post implementation to verify and confirm the design level. Based on these measurements, you are able to produce statistical evaluation of the measurements that hopefully backs up the 98% requirement.

It is very important to define whether the 98% is defined as area or time, or a combination. It is equally important to differentiate between the requirements in the different areas of the building. It might be alright to have 90% coverage in the basement of the building and then have 'full' 98% coverage in the offices floors.

Coverage of 98% Might Be Perfect, or it Might Not Be

The term 98% coverage is after all just statistics, and it is important to realize that even an indoor DAS design that fulfills the 98% level can be a bad design. Depending on where the last 2% not covered is located (Figure 5.23), it might leave the mobile users inside the building quite unhappy (as shown in Figure 5.22) and with a perception of far less than 98% coverage. It might be that most of the traffic is located in the last 2% of the building with low service (as shown in Figure 5.22).

The trick is to be very careful about where the last 2% is located. You can have perfect coverage throughout the building, but if the last 2% not covered is located in the executive area where all the senior management of the building is located, high-profile users with high airtime and high expectations of high-speed services (as shown in Figure 5.22), there will be a problem even if your statistics show 99.2% area coverage.

Therefore, always be sure to note on the floor plan where the heaviest users are located, in order to make sure they have perfect coverage and capacity. That one extra antenna installed from day one on the executive floor just might be money well spent and save you a lot of worries in the long run!

Also, in indoor radio planning, where statistics and reality can be different issues, always make sure you meet the demands of the users. Pay attention to 'service areas' like the storage room and the IT server room.

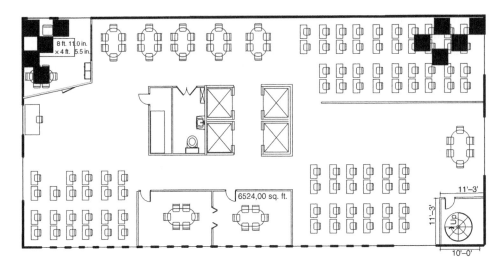

Figure 5.22   This building is covered 98%. This is verified by measurements, but the problem is that the part of the building with low signal (marked black) is located in areas with heavy users. The users are not very happy with the performance

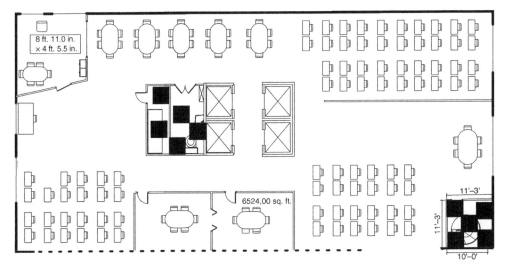

Figure 5.23   This building is also covered 98%. The RF designer has made sure that the areas with heavy users are covered and has placed the 'dead spots' in low traffic areas, thus providing perception of a quality solution

Achieving 100% Coverage

It is not possible to design 100% coverage, but if the users of the building can use the mobile service everywhere they expect, then the user perception can be 100% coverage. Therefore, a thorough and detailed input of the user expectations is very important.

Table 5.2 Example of coverage level vs solution cost

10 000 m² dense office facility, 1800 MHz 2G 14 dBm EiRP			
DL level	Number of antennas	Coverage radius	Coverage area
–70 dBm	10	21 m	1385 m²
–75 dBm	5	29 m	2642 m²
–80 dBm	3	41 m	5281 m²
–85 dBm	2	56 m	9852 m²

This input should preferably be a floor plan with clear markings of where the coverage is important, and where it is nice to have. This simple input can help in making a perfect '100%-covered' design, at a reasonable cost.

5.3.11 The Cost of Indoor Design Levels

Indoor coverage levels come at a cost – the higher the design level the higher the cost. It is crucial for the radio planner to select realistic indoor RF design levels when designing indoor DAS solutions, in order to make sure that the solution can service the needs of the mobile users. It is always wise to include some extra design margins to be on the safe side. It is also a good idea to look forward to the near future of the solution to make sure that the design can handle expected upgrades to 3G, higher data speeds or HSPA. This is a fine balance and the RF designer will need to make sure the business case is still positive. The cost of excessive design levels can be high. The example in Table 5.2 clearly shows that the radio planner has a big impact on the business case when selecting the design levels.

Do Not Under-design

Surely one should not under-design the solution, but it is recommended to use realistic design levels according to the specific building and the specific area in that building. One example could be that a 16-floor building designed for −70 dBm could be cost-optimized by using −85 dBm as the planning level in the basement and parking area. In these areas the interference from the outside network is much lower. You could have perfect quality using a 15 dB lower planning level in these areas, saving 10–20% of the system cost, just by selecting realistic design levels in these low-demand areas. Use 'the zone system' in Section 3.5.6 as a guideline, but always verify with measurements.

You can gain a lot in terms of performance boost and cost savings by pinpointing the high-use areas, the indoor hotspots. Make sure you place the antennas in these areas where the users need higher-speed services and good quality, and plan the rest of the antenna placements with the hotspot antennas as the base. By using that strategy, these areas might get really good service, −60 dBm or better when serviced by the nearby antenna. You can then without additional cost plan the rest of the DAS system so that you still keep an overall design level at −75 dBm in the rest of the area. The concept of 'hotspot' planning known from planning the macro network also applies indoors. See the example in the Section 5.4.1, where we provide really high data speeds on HSPA that can compete with Wi-Fi.

5.4 Indoor Design Strategy

When all the measurement results have been analyzed and link budget calculations have been completed, you still need to place the antennas in the building. It is one thing to calculate the antenna service radius (see Figure 5.24) and the antenna overlap, but how do you actually use this information to implement the final design?

5.4.1 Hotspot Planning Inside Buildings

The term 'hotspot' is often used when planning the macro layer. A 'hotspot' is a place of high traffic density of mobile users, needing special attention in terms of coverage, quality, capacity and data rates.

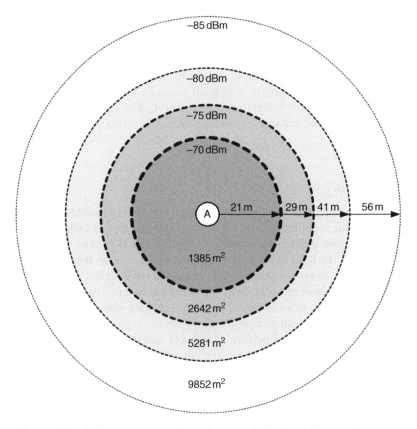

Figure 5.24   Indoor coverage radius and area vs design level from omni antenna

Hotspots also exist inside buildings and, like in the macro network, these hotspots can produce the major portion of the traffic in the cell. Indoor hotspots will typically be areas where users sit down and work using their PC/tablets and mobile for an extended time. Most highspeed mobiles are actually data cards in PCs and tablets, so naturally the user will use a convenient sitting area where he can work in relative quietness. Examples of indoor hotspots with high traffic density and high requirements for data speed performance could be business lounges in an airport, the food court in a shopping mall, the press area in a sports arena, the conference area at a hotel and executive and meeting areas in a corporate building.

It is highly recommended to place antennas in these areas, thus securing a good design margin for future data service. This strategy does not have to be costly – often it is just a matter of using the hotspot areas as the base for the antenna placements, and aligning the placement of the remaining antennas accordingly.

3G/4G Can Easily Compete with Wi-Fi

Hotspots in public buildings will often already be covered by Wi-Fi service, due to the high concentration of data users in these areas. In many cases the 2G/3G/4G indoor DAS implemented by the mobile operator will be a direct competitor to Wi-Fi service. Therefore the radio designer must be sure that he places the DAS antenna in these hotspot areas. By doing so, the 4G data service and even the 3G can compete with the speed of Wi-Fi, in many cases.

In theory, the data speed on Wi-Fi is quite high, in many cases stated as better than 54 Mbps. In reality the speed servicing the user will be limited by the ADSL backhaul from the Wi-Fi access point to the internet, not the radio speed on the air interface. Therefore in reality the speed is often less than 1 Mbps. Sometimes, however, it is up to more than 10 Mbps.

Mobile data services are also more user-friendly. There is no need for individual charging in the local Wi-Fi hotspot and all charging and roaming on mobile data are settled over the normal billing system. The mobile 2G/3G/4G data service will also provide the user with total mobility, with handovers of data service providing a global coherent data service.

Where are the 'HotSpots' in the Building?

This is a very important question to ask yourself before you start designing any DAS solution: where are the 'hotspots' in the building, the areas where users are most likely to require the best data services, and the highest capacity? It is absolutely essential that you know where the concentrations of users are the highest, and where the users expect maximum performance, before doing any design work on the DAS. Use the hotspots in the building as the base stat point when you decide the location of the DAS antennas, and focus the locations most of all on the hotspots, as well as the minimum required service level. This strategy will have a huge impact on the overall performance of the DAS; it will ensure the highest possible RF service levels and the best data speeds for most users and will have a direct positive effect on RF performance, perceived data services and the business case of the DAS. You may ask, 'How can this affect the business case?' Well, if you get higher throughput in the building on the data

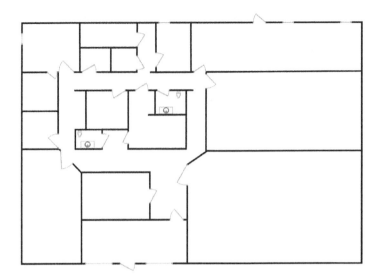

Figure 5.25 Floor plan with no detailed information except for the outline of the rooms and their locations

service, in the areas with the highest concentration of users, you will be able to service more Mb/m2 overall in the design, thus producing more revenue for the investment. This is very important for concentric deployments of multiple mobile services, like mixing 3G and 4G on the same DAS. You might want an overall 3G service to take care of voice and slow-speed data, and to have 4G in the areas that require the highest data capacity: the hotspots

If we look at a simple design in Figure 5.25 – we have received design inputs of this floor plan and been informed that this is a typical office environment and that we need to provide –75 dBm in most of the area.

The design we produce based on these inputs is shown in Figure 5.26, and appears as a solid design. We do comply with –75 dBm in most of the area, save for a few spots in the hallway and a few spots inside the rooms at the corners.

Know the Building You Are Designing

If we had taken the time and obtained the details of the function of each room in the building, as shown in Figure 5.27, instead of basing our design solely on the layout of the walls from Figure 5.25, we would have discovered that these rooms are the conference areas of the building and that the users of the building are highly reliant on wireless services for their training and presentations. Several VIP areas and meeting rooms are also on this floor – areas where the key decision-makers are located, using the latest wireless services and devices.

Had we known about the hotspots in the building, we might have done a design like the one in Figure 5.28, with a focus on the conference and VIP areas, also making sure that both the CEO and CIO had excellent service simply by adding one more antenna and focusing all the antennas on the hotspot areas.

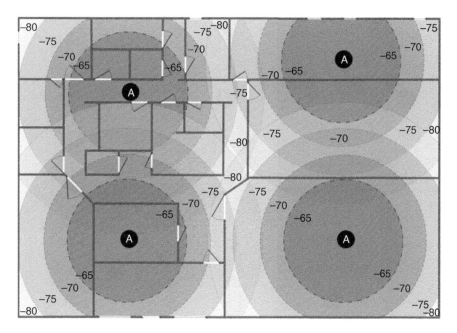

Figure 5.26   Floor plan with 'prediction plot' of the downlink RF signal level, well within the required –75 to –85 dBm minimum signal level

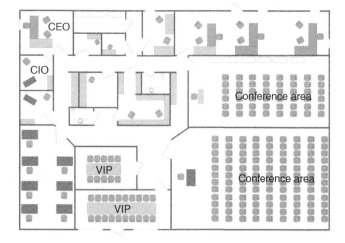

Figure 5.27   This version of the floor plan contains actual information about the usage of the individual offices and expected hotspots, unlike Figure 5.25

Power and Capacity Optimization with 'Hotspot Planning'

Having the antennas close to the hotspots ensures the most efficient utilization of the network resources, thus increasing the capacity of the cell. It also ensures that most mobiles will transmit less power to reach the DAS. This improves the interference scenario of

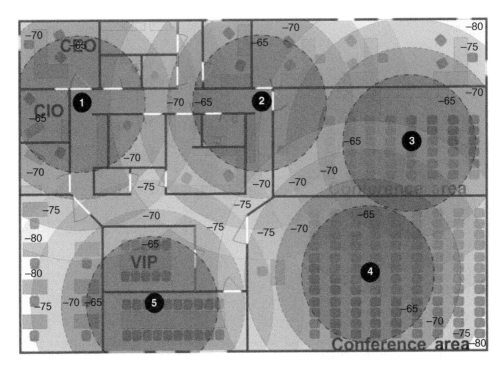

Figure 5.28   Floor plan with the DAS design, knowing the details of the hotspots in the building

neighboring cells both inside the building and possibly in the surrounding macro network, thereby increasing the overall network performance and capacity in the area. Less transmit power from the mobiles also means longer battery life and less power consumption – a more 'green' approach.

Typical Hotspots inside Buildings

Most building types will have typical hotspot areas that require focus. Below you will find a few examples – but always make sure you base your DAS design on detailed inputs from the actual building / application you are designing. Pleas also refer to Chapter 14 for more details on individual building and structure types.

Typical Hotspot Locations

- Office buildings
  - Conference areas
  - Meeting rooms
  - Training facility
  - IT Department
  - Executive/management areas

- Shopping mall
  - Food court
  - Cafe areas
  - Staff 'back office'
  - Facility management area

- Airport
  - VIP lounges
  - International gates
  - Seating areas
  - 'Back office' areas for staff

- Factory
  - Conference area
  - Meeting rooms
  - Management
  - Canteen

- Stadium
  - Conference rooms
  - VIP lounges
  - Press lounges and areas
  - Management office

Remember the hotspots. By using the information about the location of these high-traffic areas, a small adjustment in the design can have a big impact on the performance, perceived data speed and quality of the services. It might be that you meet your KPI of say −75 dBm in the building, but if you can have a 'hotspot strategy' that gives −65 dBm in locations where most people are using their wireless devices, it will have major impact on user experience, performance and the overall network economy.

5.4.2   Special Design Considerations

Even though most of the design methods and considerations are the same no matter what type of building you are designing, special attention needs to be addressed to the type of building you are designing the DAS for. These are some of the points we need to address, in addition to all the standard RF considerations.

- Make sure you prepare for more capacity or sectors for future upgrades.
- Make sure you cover the executive floor 100%.
- Is there a need for elevator coverage?
- Are there special installation challenges (e.g. fire proofing)?
- Pay attention to the service rooms or areas (e.g. IT server rooms).
- Are there special EMR concerns (like in a hospital)?
- What type of services might be needed in the future – 3G, 3,5G?
- Are there any hotspots in the building that need special attention?

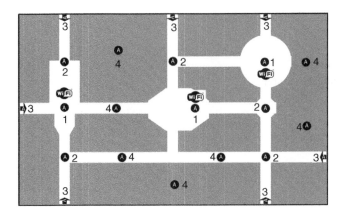

Figure 5.29   Antenna placements in a shopping mall

5.4.3   The Design Flow

When designing large projects, like campus areas, airports or shopping malls, it is worthwhile structuring how you do the design. Done correctly, you will get the most performance of the system at the lowest cost and will be able to do a rapid, future-proof design.

Let us do an example of a shopping mall (as shown in Figure 5.29), assuming that the site survey has been done and the link budget has been analyzed and calculated. Now you have an idea of the coverage range of the antennas in this particular environment and can start placing the antennas on the floor plan. First consider the characteristics of the particular building. In this example, being a shopping mall there will be some basic characteristics:

- All the shops are likely to have 100% glass facades facing the internal streets; therefore there is close to no wall attenuation from the street into the shops.
- Be careful about open skylights with regards to leakage and isolation to the macro layer. There might be nearby macro base stations leaking inside the mall in this area.
- Often you are only allowed to place antennas in the internal 'streets' of the mall.
- Capacity load can be extreme at peak shopping time, e.g. during sales.
- Prepare the system for capacity upgrades by dividing it into more sectors in the future.
- Remember the parking area, often subground level, and do not forget overlap to the outdoor network for the HO zone where the users enter and exit the mall.
- Try to avoid placing antennas near the PA speaker system to avoid interference.
- Is it necessary to cover the areas with no public access, e.g. storage rooms or offices? Those areas can generate quite high revenue.

5.4.4   Placing the Indoor Antennas

The recommended breakdown of the actual antenna placements in the shopping mall is given in Figure 5.29.

1. *Place the hotspot antennas and maximize data performance.* Locate the hotspots for data users; they will be the most demanding areas with regards to good design margin

and level. In this case of a shopping mall, the hotspots for data and voice are typically the food court, internet cafes and sitting areas. If there is a Wi-Fi operator present in these areas, you can compete against Wi-Fi on data by placing hotspot antennas in these areas. Therefore all the first placements of DAS antennas, marked '1' on the floor plan, are placed in these areas, forming the reference for the next step in the design phase.

2. *Place the 'cost-cutting' antennas.* After the hotspot antennas are placed you must place all the antennas that will maximize the coverage per antenna, marked '2' on the floor plan. By using the 'corridor effect', you can maximize the coverage of each of these antennas. You place these antennas in all the intersections of the internal streets of the mall so that the coverage will be spread in the directions of the internal streets, which will give good value for money.

3. *Isolate the building.* Dominance and well-defined HO zones along the perimeter and the entrances of the shopping mall are secured by the placement of the antennas marked '3'. These antennas will often be directional antennas pointing towards the center of the building as the preferred solution. This will isolate the building from even very close outdoor sites.

4. *Fill in the gaps.* The last placement of antennas will be 'filling the gaps' between the antennas just placed. It is often necessary to place an antenna in big shops inside the mall, if you are allowed that is. Normally installation of antennas in shopping malls is restricted to the internal streets. Therefore it can be hard to cover deep inside the larger shops; try to cover them as much as possible by having more antennas in the internal streets of the shopping mall.

5.5   Handover Considerations Inside Buildings

It is important to make sure that the indoor DAS system implemented in the building is prepared for future traffic growth. The best way to prepare this is to have a sector plan for future sectorization of the system. Even if the system is implemented as one sector, you need to look ahead, especially for 3G, in order to prepare for more sectors.

Well-defined HO zones are important for 2G and 3G/4G to avoid 'ping-pong' HO on 2G, extensive soft HO zones on 3G and degraded 4G performance. The focus should be on well-defined and controlled handover zones, preferably placed in areas with low traffic in the building. When the DAS is designed and implemented correctly, the dominance and isolation of the indoor system will insure well-defined handover zones to the outdoor macro network. However, in extreme cases, like a rooftop site on the neighboring building, you will sometimes have to deal with some signal leaking inside the building, even with high signal level from the indoor system.

As a general rule you must try to avoid having the handover zones in large open areas inside the building. Here it can be difficult to design and control the handover zone. Try to take advantage of the natural isolation provided by the building to separate the different sectors or cells. This can be done by using the floor separations as the handover border, or the fire separation zones inside the building (Section 5.3.6). Typically these fire zones will be divided by heavy walls with high RF isolation, which are perfect for giving a well-defined handover zone.

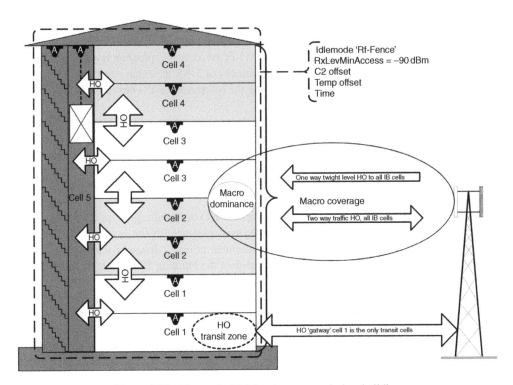

Figure 5.30 The typical 2G handover scenario in a building

5.5.1 Indoor 2G Handover Planning

It is important to realize that the handover control parameters are a crucial part of the indoor design and, in order to optimize the implemented solution, these parameters must be tuned. The typical handover scenario in a 2G multicell building is shown in Figure 5.30. This is a multicell indoor system with a total of five cells. Well-defined handover zones between the internal cells in the building are a must, and in this case the primary handover between the internal cells are done via cell-5, which also serves the elevator, so no handovers will occur when using the elevator. Even with cell-5 as the normal internal handover cell, there are defined handovers to the internal adjacent cells, just in case cell-5 is full and cannot tender for the handover between adjacent cells.

Limit the Number of Macro Handover Candidates

Normally the internal cells in the building are limited to only having defined handovers to nearby adjacent indoor cells within the building. This to prevent those mobiles typical in the topmost part of the building starting to make handovers to distant macro sites via nearby macro sites. This is a common problem. The mobile might finally end up on a distant macro cell that has no neighbor relations back to the indoor cell. The consequence is often dropped calls when the users move from the perimeter at the windows back to the center of the building.

Keep Idle Mode Traffic on the Indoor Cells

In 2G idle mode the mobile can be controlled with special cell offset parameters. You must make sure that the mobile will camp on the indoor cell, even if nearby macro sites are leaking high-level signal inside the building. In practice, you use a combination of the lowest allowed Rx access level on the mobile (RxLevAccessMin); in this example in Figure 5.30, it is set to −90 dBm. Then by applying a cell offset (C2), for example, of 40 dB, the mobile will add 40 dB to the evaluation of the cell once the cell is received more strongly than −90 dBm. The result is that the mobile evaluates the cell to be −90 + 40 = −50 dBm receive level.

This is very useful in keeping the mobiles on the indoor cell in idle mode; you can thus dominate in areas with really high macro level leaking in the building. In addition, you can use temporary offsets and time penalty periods; this also makes it a very useful tool to control the traffic if the indoor cell leaks out onto the nearby street.

Define 'Emergency' HO Candidates

It will always be a good idea to make sure you define a 'one-way' connection back to the indoor cell if mobiles should camp unintended on the nearby outdoor cell. In extreme traffic situations where the indoor cell might be congested, it can be a good idea to have traffic-controlled HO to the nearby macro site, in order to unload the indoor cell and avoid capacity blocking of calls.

Handover Zone to the Macro Network

It is very important to have a well-defined transit handover zone to and from the building. Normally this zone is around the main entrance and the entrance to the indoor parking area.

Only the cell(s) covering the entrance or exit of the building should have normal HO connections to the outdoor network and preferably to a very limited number of outdoor cells in order to ease the optimization of the handover zone.

For 2G applications it is recommended to make sure you do not pick up any outdoor traffic on the indoor system, by outdoor mobile users close to the building. The handover parameter should be tuned in order to insure that the handover takes place just inside the building.

5.5.2   Indoor 3G Handover Planning

In 2G there is some margin of offsets that can be applied in order to tune the HO zones and areas. Adjacent 3G cells will typically use the same frequency; the mobile has to be in soft HO as soon as it is able to decode more cells. If not, the adjacent cells will cause interference, degraded performance and dropped calls. Therefore there is not a large margin for tuning the 3G handover and only good radio planning will do the job.

The typical handover scenario in a 3G multicell building is shown in Figure 5.31. This is a multicell indoor system with a total of five cells. Well-defined handover zones between the internal cells in the building are a must to limit the soft handover load; in this case the primary handover between the internal cells is done via cell-5. Cell-5 also serves the elevator as the only cell, avoiding soft handovers in the elevator.

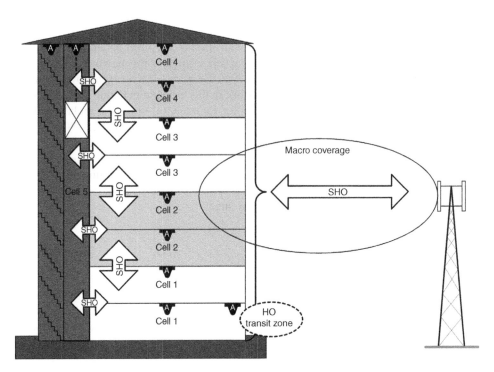

Figure 5.31   The typical 3G soft handover scenario in a building

However, you must also be sure to define handovers to the internal adjacent cells, in order to perform soft handover in the limited areas where two cells might cover the same area. Keep those areas to a minimum.

Good 3G/4G RF Planning is the Solution

The best way to optimize the handover zones on 3G/4G is by good RF planning, correct antenna placement and careful attention to the handover zone planning from the initial design phase. The solution is once again to provide dominance of the indoor cells and isolation between the cells. This is the correct way to control the handover performance and to limit soft handover zones.

To some extent the 3G CPICH (pilot channel) power can be used in order to offset the cell sizes, thereby moving the soft handover zone slightly. However, the effect is fairly limited, compared with the freedom of parameter settings for tuning the handovers on 2G.

The use of separate frequencies for the outdoor and indoor network can be a solution in some extreme cases, but these hard handovers are often a challenge for the network as well as many mobiles. Operators are typical assigned two or three RF channels for 3G, and a fixed spectrum for 4G, so spectrum- and capacity-wise this is a very expensive solution and should only be considered as a last temporary resort, until the real problem can be solved or the macro sector removed.

Furthermore you might not be able to allocate a new RF channel for 4G, and 4G might take a big hit from co-channel interference from the macro cell, degrading the speed and performance efficiency.

The HO Zone to the Outdoor Network

Obviously there needs to be a handover zone to the outdoor network designed in the entrance areas of the building. A well-defined HO zone to the outdoor network is important, and on 3G/4G it can be wise to consider having the HO just outside the entrance of the building. The reason for this is that, if the HO occurs inside the building, it is likely the mobile will be running on high transmit power in order to reach the outdoor cell. This 'noise' generated by the high power mobile will raise the noise level on the UL of the indoor cell, thus cannibalizing the capacity of the indoor cell. This can be a big issue when designing high-capacity indoor solutions with many users at the entrance of the building or shopping mall.

5.5.3 Handover Zone Size

When designing handover zones it is important that you take the speed of the traffic into account. Furthermore, you need to make sure that you include a safety margin in order to make several handover attempts, should the first attempt fail. The handover zone must accommodate this safety margin, and you must make sure that you keep the coverage design level throughout the full size of the handover zone. The last thing you want to encounter is problems with handover retries once you are below design level. That is a sure recipe for disaster and dropped calls.

Example, HO Zone Size

If you use a HO time of 4 s as an example, you can evaluate the size of the needed HO zone. If a car moves at 30 km/h out of the underground parking area (8.33 m/s), you would need a HO zone of 33 m.

Example, Handover Zone Safety Margin

As pointed out earlier in this chapter, it is not enough to design the handover zone to the minimum size. You must provide the mobile enough time to perform cell decoding, measurement, evaluation and handover execution of both cells at the same time.

The signaling load of the cells can also be an issue. This is normally not a concern when designing indoor solutions: indoor cells will typically only have to perform a few handovers simultaneously. However, there are cases that might demand consideration of the signaling load of many simultaneous handovers; consider a tunnel scenario where a train is moving at relatively high speed, and inside the train there might be 40 users in traffic performing handover at the same time. In this case you must certainly include a safety margin in the handover zone size. In addition 2G offers some presynchronization functionalities or 'chained' cells. On 2G, 3G and 4G it is recommended to have the two cells in the critical handover zone serviced from the same base station. This will increase the handover success rate and load the network far less for these critical handover types.

5.6   Elevator Coverage

Mobile users inside a building expect a good quality coherent service level throughout the building, the elevators included. It is a major challenge to service elevators with RF coverage, and requires special consideration. Most elevator lift-cars are a virtual metal enclosure with very high RF attenuation, often exceeding 60 dB. The speed of the elevator adds to the challenge, especially in very high buildings with high-speed express elevators. However, the demand for mobile coverage inside elevators is growing, primarily motivated by the normal requirement for voice and data coverage everywhere. The mobile is also considered as an extra security line by the users, even though there is an emergency phone installed inside most lifts. Therefore mobile coverage inside elevators is a must.

There are several options on how to provide mobile coverage in the elevator; the optimum approach depends on the individual solution and the constraints of the actual elevator installation. The best mobile performance is achieved if the RF design is done so that there will only be one dominant cell covering the elevators throughout the building, thereby avoiding handovers inside the lift. However providing sufficient coverage inside the lift-car can be a challenge and sometimes close to impossible, due to the metallic enclosure of the lift-car.

In small to medium-sized buildings it might be preferable not to have a dedicated cell for the elevator only, but to use one of the existing cells in the building. In that case it is recommended to use the topmost cell in the building to service the elevator shaft as well as the indoor area.

5.6.1   Elevator Installation Challenges

To obtain approval and permission to place any equipment in the shaft that is not related to the operation of the elevator itself can be difficult. If you want to install antennas inside the elevator shaft or in the elevator car itself, you will often need special approval of the cable types and all other equipment used, and issues related to fire rating and mechanical stress need to be cleared and certified. Always check and make sure you follow the installation rules and guidelines in the specific elevator in the specific building.

5.6.2   The Most Common Coverage Elevator Solution

The most used design when covering elevators is to place antennas close to the elevator shaft on each floor, preferably in the lift lobby with the antenna 1–2 m from the lift door [as shown in Figure 5.32(a)]. Often this design will work fine, especially in small buildings with only one sector/cell covering the building. For larger buildings with more sectors you must be careful using this approach, and be sure not to create a handover zone inside the lift.

The situation is shown in Figure 5.33(a), where the potential problem is evident. The handover zone must be designed just outside the lift shaft [as shown in Figure 5.32(b)], in order to avoid handovers inside the lift-car when moving at high speed.

5.6.3   Antenna Inside the Shaft

This design is shown in Figure 5.33(a), with the antenna mounted in the lift shaft, most commonly at the top, but it is also possible to have the antenna in the bottom and beam upwards. Sometimes a bidirectional antenna mounted in the middle of the shaft can be

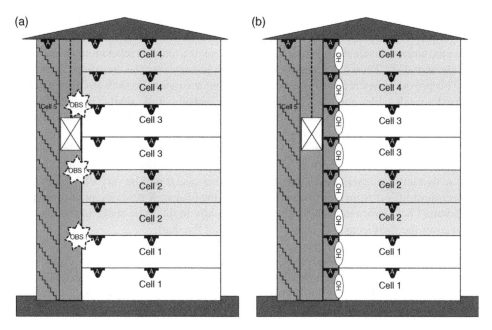

Figure 5.32 The typical way to provide elevator coverage

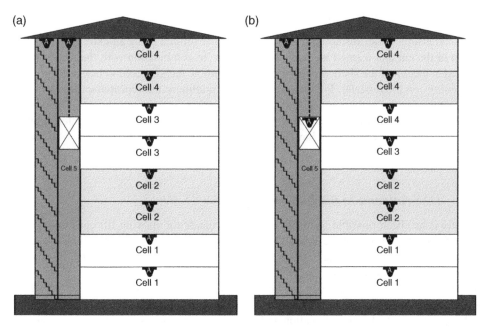

Figure 5.33 Two options for covering the elevator

considered; it all depends on penetration losses into the lift-car and the height of the building. For larger buildings, a combination of these antenna placements can be considered.

Often the antenna inside the lift shaft has to be placed on a small bracket mounted on a door to the elevator shaft. Thus you access the antenna by just opening the door from outside the shaft. This provides easy access to the antenna without accessing the shaft itself; in that case it is easy to install cables and other equipment on the outside of the shaft. This will limit the equipment inside the shaft to only the antenna. This approach might make it easier to achieve the approval for implementing coverage in the shaft.

Be careful in very high buildings, as sometimes elevators come in two levels! One elevator-car might actually be serving both even- and uneven-numbered floors at the same time; you will have two levels of users inside the elevator. In that case it will be very difficult to cover the lower-most lift-car from an antenna in the top of the lift-car. A new variant of this type of twin elevators is to have two cars moving independently in the same shaft. In this case you need to cover the shaft from both the top and bottom of the shaft.

Special consideration needs to be given to the other cells in the building, and dedicated attention to the layout of the handover zones is advisable. Often the antennas in elevator shafts will be designed to radiate high power levels in order to penetrate the elevator-car. However, be careful that this high level of signal inside the lift shaft does not leak out to the office area nearby the elevator lobby.

I have seen problems with this type of design, where all users within 20 m of the elevator lobby handed over to the elevator cell once the elevator doors opened, and were then dropped once the doors closed again, due to the abrupt loss of signal and lack of time to handover back to the office cell.

5.6.4   Repeater in the Lift-car

You could also consider installing a RF-repeater on the lift-car, using an external antenna on the top of the car and a small antenna inside the car. In combination with the shaft antenna installed in the top of the shaft, this can be a good solution, but you need to be careful with the gain settings on the repeater. Preferably you should configure the repeater settings so it just compensates for the penetration loss of the lift-car. If you use too high a gain you might saturate the mobile and the base station when the lift-car is close to the shaft antenna on the topmost floors. On 3G this might have a serious impact on all the traffic in the cell, so caution is advised.

5.6.5   DAS Antenna in the Lift-car

Recently it has been possible to install an antenna inside the moving lift-car [as shown in Figure 5.26(b)], as a part of the active DAS (Section 4.4.2). This antenna will be connected to the active DAS system of the building like all the other antennas. This option, however, is only possible when using pure active DAS. The active DAS relies on thin cabling to the remote antenna unit that services the antenna. This cable will typically be CAT5 (LAN cables) or CATV (cable TV cable), similar to the cables that are already being used to service the lift-car, for control, communication and CCTV. By using similar cables to those already installed to service the elevator, it is easier to get permission to install the DAS antenna in the lift-car. The big advantage of this approach is full control of the RF environment inside the lift and control of the zones.

The DAS antennas must be set to a very low output signal level, and you must be sure to use a version that can handle mobiles not in power control close to the antenna radiating at high power. Remember that some mobiles inside the lift are not being serviced by the elevator antenna, but by the nearby macro site of the competing operator. Hence these users are likely to transmit at full power and saturate the antenna unit in the active DAS installed in the elevator. Therefore it is recommended to install an attenuator between the antenna unit and the DAS antenna inside the lift. This attenuator (20–30 dB) will add sufficient link loss and prevent this potential problem.

5.6.6   Passive Repeaters in Elevators

Sometimes you get the question, 'why not just mount two antennas back-to-back just connected via the cable, one antenna on top of the lift-car, the other inside the lift?' Let us have a look at the set-up of this concept in Figure 5.34. First of all, we know that the free space loss in this example of 2G 1800 MHz will be 38 dB at 1 m from an antenna. Applying the free space loss, we can calculate the system gain of the passive repeater system. From 1 m in front of the donor antenna to 1 m in front of the service antenna on the other side of the passive repeater, you will have a 'system gain' of 2×-38dB + antenna gain – feeder loss.

Example

You can try to do a link calculation of a 30 m-high building. According to the free space loss (Section 5.2.8), the free space loss would be about 67 dB at 30 m distance, and then 38 dB from the antenna in the lift-car to the user.

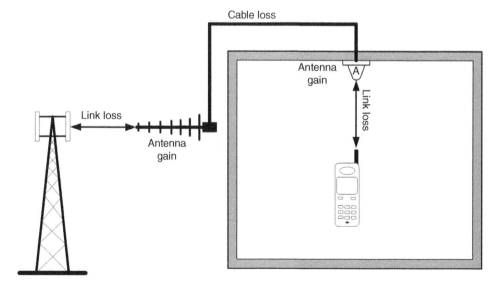

Figure 5.34   Two antennas 'back to back' might work as a passive repeater, but it takes a high donor signal

$$\text{Total free space loss} = 67\,dB + 38\,dB = 105\,dB$$

If we use a donor antenna at the top of the shaft with 7 dBi gain beaming down the elevator shaft, a 7 dBi pick-up antenna, 2 dB cable loss and 2 dBi antenna gain for the antenna in the lift, then we can calculate how much power we need to feed this antenna in order to achieve the _85 dBm inside the lift-car when it is in the bottom of the shaft:

$$\text{Total link loss} = \text{total free space loss} + \text{cable loss} - \text{antenna gain}$$
$$\text{Total link loss} = 105\,dB + 20\,dB - 2\,dB - 7\,dB - 7\,dB = 91\,dB$$
$$\text{Required power at the transmit antenna} = 91\,dB - 85\,dBm = 65\,dBm$$

5.6.7   Real-life Example of a Passive Repeater in an Elevator

The concept of passive repeaters actually works fine for elevator coverage in many cases. In a real-life elevator shaft we placed a DAS antenna, with 7 dBi pointing down in the top of the lift shaft, a feed to the antenna of 10 dBm and using a similar 7 dBi pick-up antenna at the top of the lift-car and one 2 dBi antenna inside the lift-car. The signal from the passive repeater inside the lift was better than −80 dBm at distances of up to about 80 m.

We can try to verify this by doing a fast link budget for the downlink:

$$\text{free space loss(dB)} = 32.44 + 20(\log F) + 20(\log D)$$

We can calculate the free space loss in the shaft at 75 m:

$$\text{free space loss(dB)} = 32.44\,dB + 20(\log 1850) + 20(\log 0.080)$$
$$\text{free space loss(dB)} = 32.44\,dB + 65.3\,dB - 22\,dB = 75.74\,dB$$

We need to add the free space loss at the inside of the lift (38 dB) and the antenna gains [7 dB + 7 dBi + 2 dBi + 2 dBi (mobile antenna gain)] and the cable loss (−2dB). The system gain between the connector at the DAS antenna and the mobile user inside the lift can then be calculated:

$$\text{system loss} = \text{free space loss shaft} + \text{free space loss lift} + \text{cable loss} - \text{antenna gain}$$
$$\text{system loss} = 75.44\,dB + 38\,dB + 2\,dB - 18\,dBi = 97.74\,dB$$
$$\text{calculated user level inside the lift} = \text{DAS antenna power} - \text{system loss}$$
$$\text{calculated user level inside lift} = 10\,dBm - 97:74\,dBm = -87:74\,dBm$$

You can see from the calculations that the measured level inside the lift is actually about 8 dB higher than the calculated level – why? Actually what happens here is caused by the 'tunnel effect' on the radio signal. Both the service antenna on top of the lift shaft and the donor antenna on top of the lift-car are 7 dBi gain antennas. According to the data sheet, the radiation pattern of the antenna is about 70° beam-width (opening angle), but the 'gain' of any antenna is related to the directivity of the antenna – the more 'narrow' the beam-width, the higher the gain.

The narrow elevator shaft has a big impact on the directivity; the radio waves are forced to stay inside the elevator shaft, focusing all the energy down inside the shaft. Therefore the 'gain' (directivity) of the antenna is no longer 7 dBi, but slightly higher. The mobile used inside the lift may also have been slightly closer to the service antenna, adding to the signal.

Using a passive repeater solution is definitely possible for small to medium-sized buildings for servicing the elevator, provided you can get a DAS antenna installed at the top of the lift shaft as the donor antenna to the system.

5.6.8   Control the Elevator HO Zone

The best performance is achieved if the HO zone can be placed outside the elevator, i.e. in the lobby in front of the elevator. This can only be achieved if we make sure that the same cell is covering the elevator from top to bottom, as in Figure 5.32(b). Often this cell will be one of the existing cells in the building (the topmost cell).

In order to perform a successful handover, there must be some overlap of coverage from the two cells. In a situation where you do not have antennas installed in the lift shaft and are covering the elevator only from antennas in the elevator lobby for the different floors, the RF signal has to overcome both the penetration loss into the lift-car and the penetration loss through the floor separation in order to provide cell overlap for the handover to succeed. Therefore, the speed of the elevator is an important parameter of concern when designing this type of elevator coverage for multicell buildings. You need the overlap of the two cells in order for the handover to succeed. Typically you need a minimum of 3 s to perform handover measurement, decoding, evaluation processing and signaling time in order to make the handover a success. In addition, you need to add an extra margin of time in case the first handover attempt fails, and time to perform a new handover attempt. This makes this approach in multicell buildings almost impossible in practice for use in high-rise buildings with fast-moving elevators. The only solution is to dedicate a cell to cover the elevator shaft and sometimes include the elevator lobby [Figure 5.32(b)].

5.6.9   Elevator HO Zone Size

When designing handover zones with relatively fast-moving indoor users, elevators present the most challenging case for indoor coverage solutions and paying attention to the size of the HO zone is important. When designing indoor solutions, there are typical scenarios where the speed of the users plays a major role in the size of the HO zone; for example elevators used in the building and cars driving to and from a parking area inside the building. It is recommended to have minimum of three to four seconds to allow for decoding, measurement evaluation and signaling for the handover, especially for 2G. The handover might fail; therefore you must include a safety margin for at least one handover retry to cope with extreme cases where many mobile users will perform handovers between the same two cells at the same time.

Example, Elevator HO Zone Size

You can calculate the size of the required HO zone. If an elevator moves at 5 m/s, you would need an HO zone of 20 m. If the floor separation in the building is 3 m, you would have to provide an overlap of about seven floors! This clearly underlines the potential problem in multicell buildings as shown in Figure 5.32(a).

Let us elaborate on how we can accommodate this problem.

5.6.10   Challenges with Elevator Repeaters for Large Shafts

We now have some of the basics regarding designing passive repeaters for elevator shafts under control; however, we still face some challenges on the design side, especially for large buildings and elevator coverage solutions. We will try to address some of the most common problems and challenges, and will propose some remedies for some of the most common challenges.

Larger Elevator Shafts

In a case where one cannot service the full length of the elevator shaft by simply deploying antennas in the elevator lobby and in the shaft itself, one needs to design and implement a solution that can cover the full extent of the shaft and that can provide coverage for the users inside the elevator car.

Let us look at one example, starting with a relatively short elevator shaft (60 m) like the one illustrated in Figure 5.35. Using this approach we will apply the concept of a 'Passive Repeater', a passive repeater is simply two antennas mounted on the lift car: one on the outside directed towards a donor antenna at the top of the elevator shaft (the pickup antenna) and one inside the lift car (service antenna) that extends the service to the mobile users inside the elevator car. These two antennas are interconnected via a short coaxial cable. As discussed in Section 5.6.6 we can perform the basic link calculation for this setup (Figure 5.35).

We can perform a simple link loss calculation and calculate the expected service level inside the elevator car, using 2G-900 as an example of the solution illustrated in Figure 5.35.

Elevator Shaft Example 2G 900, from Figure 5.35

In this example we have these assumptions:

- **Data on the system:**
- *Desired DL signal level in elevator car = −75 dBm*
- *Shaft & pick up antenna = 7 dBi, 1 dB feeder loss*
- *Elevator Car service antenna = 2 dBi, 1 dB of feeder loss*
- *Elevator shaft = 60 m*
- *Service distance in Lift Car = 1 m*
- *Power from the DAS = 10 dBm*
- *Power from the Mobile = 33 dBm*

Air Link Calculation

Using the basic formula for Free Space Loss (FSL) we can calculate the two Free Space Link loss components, the free space in the shaft and the free space loss from the service antenna inside the lift car to the mobile (see Section 5.2.8 for more detail). It is important that we divide the FSL calculation into two components, due to the fact that we have two separate links in series,

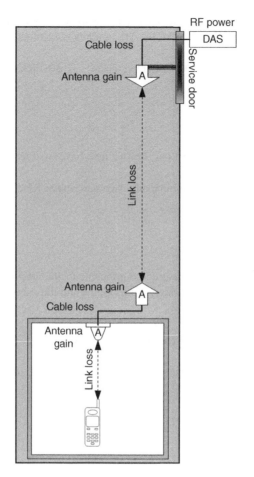

Figure 5.35   Basic elevator link calculation

- Basic Free Space Loss: FSL[dB] = 32.44 + 20 Log(Freq. [MHz]) + 20 Log(Dist.[Km])
- FSL Calculation in the Elevator Shaft @ 60 m distance = 32.44 + 59.55 − 24.43 = 67.56 dB
- FSL Calculation in Elevator Car @ 1 m distance = 32.44 + 59.55 − 60 = 32 dB
- Total FSL, Shaft + Elevator car = 67.56 + 32 = 99.56 dB

Now that we have established the total FSL, we need to consider the losses and gains of the used components, the cables, antennas, etc.

Gains in the System

In terms of gain in the solution we have the three antennas, the donor antenna (on top of the shaft), the pick-up antenna (on top of the elevator car) and the service antenna (inside the elevator car, in the ceiling).

One could argue that the mobile antenna also needs to be assessed in the calculation, this is correct – but for this example the gain of the MS antenna is assumed to be 0 dB.

- Total antenna gain in the system = 7 dB(Donor Ant) + 7 dB(Pick up Ant.) + 2 dB (Service Ant) = 16 dB

Cable Losses in the System

We have established the Free Space Loss, but we also have to consider the passive losses of the cables used in the installation.

In this example it is quite easy, we have only two sections of RF cable, each with 1dB loss, so the total cable losses are 2 dB.

Total Link Calculation

Now that we have established the required components of the total link between the system and the mobile, the Air Link Losses, the Antenna Gains and the cable losses we can calculate the total link.

- Total link loss = Air Link Loss – (antenna gain) + cable loss = 99.56 – 16 + 2 = 85.56 dB

(Note that the directivity of the tunnel of the elevator shaft will give a lower Air Link Loss in practice, see Section 5.6.7).

Calculating the Signal Levels

Knowing the transit power from the DAS and the mobile we can easily calculate the signal level at the mobile and at the DAS, for example the 'DAS' illustrated in Figure 5.35 is Active (from Section 4.4.4), the DAS is assumed to have 0 dB loss/gain back to the base station.

Signal level calculation:

- DL Level @ mobile = DAS power – total loss = 10 –85.56 = –75.56 dBm*
- UL Level @ DAS Remote Unit = MS power – total loss = 33 – 85.56 = –52.56 dBm*

These calculations assume that no power control is enabled on the Downlink or the Uplink; typically the mobile will be in Power Control and be powered by 23 dB down if the target UL level is set to –75 dBm, so the mobile will be regulated down to a lower transmit level.

Conclusion

From this example we may conclude that we can indeed reach the desired DL signal level inside the elevator car by utilizing passive repeaters in the elevator car and the use of donor antenna in the elevator shafts up to 60 m distance, under the given conditions in this example.

The Passive Repeater for Large Elevator Shafts

From these basic RF link calculations we conclude that we could use this approach for providing coverage for even large elevator solutions in taller buildings, by 'stacking' this basic application of a passive elevator repeater system. This will help us to provide sufficient RF service inside elevator cars that operate over longer distances than what may be serviced from a single antenna in the shaft. The principle is illustrated in Figure 5.36; this solution is based on the principle of servicing the passive elevator repeater in the elevator car with several donor antennas feeding the elevator shaft from both over and underneath the lift car.

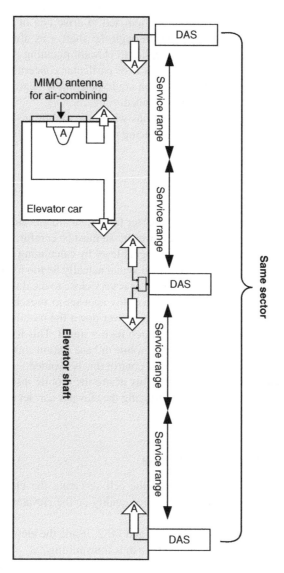

Figure 5.36 Passive Repeater System for larger elevator shafts

For obvious reasons the lift car must have pick-up antennas on both the top and bottom – these antennas are combined via a MIMO antenna inside the lift car.

As in the previous example in Figure 5.35 we will need to perform a calculation of the total link in order to establish the maximum distances of the service antennas in the shaft. As described in Section 5.6.8, it is very important to have one and only one cell covering the elevator shafts so as to avoid any handover problems – simply by applying single cell dominance throughout the whole elevator shaft. This is especially important for longer shafts where the elevator will move at high speed (the fastest elevator known to the author moves at 65 Km/h!). So by limiting the servicing cells inside the shaft to only one eliminates HO problems inside the shaft. At the same time we must be sure that the elevator cell does not leak too far into the building, so one must not be tempted to drive one or two antennas at high power to create dominance by blasting RF through the shaft – as always, a good uniform signal level is the best approach, this also applies to elevator planning.And ensure that we do not block the link with too high a signal. This is of particular concern for 3G and 4G – any receiver blocking on the mobile or base station could potentially degrade the service on the system, but thanks to power control this is avoided.

However, there could potentially be situations where the mobile inside the elevator car is not in power control by means of the cell servicing the elevator car; let us look at an example in Figure 5.28.

Receiver Blocking

Now we know how to calculate a passive repeater for a standard elevator solution and even a large shaft with multiple service antennas; however, we must be careful – at this point we have only made sure to keep clear of a minimum signal level by calculating the lowest acceptable RF level/longest service distance, but could the signal actually be too high?

What about the situation where the elevator car is very close to the donor antenna inside the shaft, would that be a problem? At first glance this appears to present no problems as the power control in the network will make sure to power down the mobile and the base station and ensure that we do not block the link with too high a signal. This is of particular concern for 3G and 4G – any receiver blocking on the mobile or base station could potentially degrade the service on the system, but thanks to power control this is avoided.

However, there could potentially be situations where the mobile inside the elevator car is not in power control by means of the cell servicing the elevator car; let us look at an example in Figure 5.35.

Problem with Mobiles Not in Power Control

Mobiles not in power control by means of the cell servicing the elevator repeater could potentially have a negative impact on the service quality on the elevator cell. Let us consider the example in Figure 5.37.

In this example we have two mobiles, MS1 and MS2, inside the elevator car, both mobiles are in traffic mode with an ongoing call and are thus transmitting.

Network Operator A provides service for MS1 and has a DAS installed in the building, utilizing a passive repeater system to provide coverage in the elevator designed as we have just

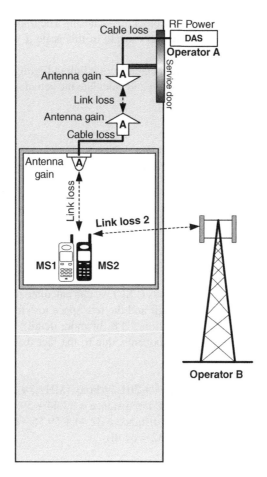

Figure 5.37   An elevator solution with two mobiles, one serviced by the elevator system, the other by a macro base station

calculated in the previous example. However, this DAS design only provides coverage for one of the two operators (Operator A). The mobile MS2 connected to Operator B is being serviced by a nearby macro base station across the street from this building. Even though the link loss to the nearby mast is high due to the solid concrete elevator shaft and the metallic enclosure of the lift car itself, the link loss is just low enough to maintain a call for MS2, provided that MS2 is powered up to full transmit level.

This scenario presents a potential problem, the fact is that when the elevator car is at the top of the lift shaft, thus close to the donor antenna, the link loss between the mobiles and the donor system is quite low – and here lies the root of the problem. MS1 is serviced by Operator A, connected to the DAS – therefore power control will ensure that the mobile transmit power from MS1 is powered down automatically when the link loss becomes lower and the elevator car is close to the donor antenna. However, MS2 will use full transmit power, thus blasting away at full power in order to overcome the high link loss to the outdoor base station. This is a potential problem for Operator A, who might suffer from the high signal level de-sensing of the base station receiver – or even receiver blocking – causing a degraded service on the uplink of Operator A or even dropped calls. Be very aware of this issue if you deploy "wideband" DAS solutions.

Let us consider a practical example and calculate the link in Figure 5.37 when the elevator car is just one meter from the donor antenna and calculate the actual signal levels:

Example 2G-900:

- *Shaft and pick up antenna = 7 dBi, 1 dB feeder loss*
- *Elevator Car antenna = 2 dBi, 1 dB of feeder loss*
- *Elevator shaft = 1 m*
- *Service distance in Lift Car = 1 m*
- *Power from the DAS = 10 dBm*
- *Power from the Mobile = 33 dBm*

Air Link Calculation

Using the basic formula for Free Space Loss (FSL) we can calculate the two Free Space Link loss components, the free space in the shaft and the free space loss from the service antenna inside the lift car to the mobile (see Section 5.2.8 for more detail). It is important that we divide the FSL calculation into two components, due to the fact that we have two separate links in series,

- Basic Free Space Loss: FSL[dB] = 32.44 + 20 Log(Freq. [MHz]) + 20 Log(Dist.[Km])
- FSL Calculation in the Elevator Shaft @ 1 m distance = 32.44 + 59.55 – 60 = 32 dB
- FSL Calculation in Elevator Car @ 1 m distance = 32.44 + 59.55 – 60 = 32 dB
- Total FSL, Shaft + Elevator car = 32 + 32 = 64 dB

Now that we have established the total FSL, we need to consider the losses and gains of the used components, the cables, antennas etc.

Gains in the System

In terms of gain in the solution we have the three antennas, the donor antenna (on top of the shaft), the pick-up antenna (on top of the elevator car) and the service antenna (inside the elevator car, in the ceiling).

One could argue that the mobile antenna also needs to be assessed in the calculation, this is correct – but for this example the gain of the MS antenna is assumed to be 0 dB.

- Total antenna gain in the system = 7 dB(Donor Ant) + 7 dB(Pick up Ant.) + 2 dB (Service Ant) = 16 dB

Cable Losses in the System

We have established the Free Space Loss, but we also have to consider the passive losses of the cables used in the installation.

In this example it is quite easy, we have only two sections of RF cable, each with 1 dB loss, so the total cable losses are 2 dB.

Total Link Calculation

Now that we have established the needed components of the total link between the system and the mobile, the Air Link Losses, the Antenna Gains and the cable losses we can calculate the total link.

• Total link loss = Air Link Loss – (antenna gain) + cable loss = 64 – 16 + 2 = 50 dB

Calculating the Signal Levels

Knowing the transit power from the DAS and the mobile we can easily calculate the signal level at the mobile and at the DAS; in the example the 'DAS' illustrated in Figure 5.35 is Active (from Section 4.4.4) and the DAS is assumed to have 0 dB loss/gain back to the base station.

Signal Levels for MS1 and MS2

• DL Level @ MS1&MS2 = DAS power – total loss = 10 – 50 dB = –40 dBm
• UL Level @ DAS Remote Unit = MS1&MS2 power – total loss = 33 – 50 = –17 dBm

Power Control will power down MS1 to reach –75 dBm UL, since MS1 is serviced by the indoor system and base station A, whereas MS2 has a high path loss to outdoor base station B and thus will be radiating at a maximum power level, giving rise to potential problems in the indoor DAS receiver system/base station.

Conclusion

MS2 serviced by Operator B will potentially saturate and block the UL of Operator A, and cause degraded quality or even dropped calls for all of Operator A's mobiles in the cell that services the elevator shaft.

So if you drive the elevator system of a passive DAS you must be careful not to have too low a loss between the donor antenna and the base station.

This is of particular importance in noise sensitive systems like WCDMA/3G – here a rise in noise on the UL will degrade UL capacity; this is of special concern for operators using adjacent channels – see Section 5.9.2 for more detail.

For active DAS the Remote Unit feeding the donor antenna might have an ALC (Automatic Limiter Circuit) in the receive path that prevents signals above a certain level – this might to some extent limit the potential problem.

The best way of preventing these issues is to agree to implement a common solution shared by all mobile operators, but this comes with a whole set of other challenges and concerns – see Section 5.7 for more detail.

Final Note on Elevator DAS and Repeater Systems

Now that we have established some basic facts when designing elevator systems, we may conclude that the best elevator DAS will make sure:

1. To cover the extent of the whole elevator shaft and car with single cell dominance.
2. We provide a sufficient DL signal level.
3. We provide sufficient UL levels without creating potential receiver blocking.

On top of this, the main challenge will often be the practical implementation of the DAS in the shaft and elevator car. One must be sure to follow all of the appropriate installation codes and guidelines, in order to ensure that all of the installation is done exactly as required.

After all, the purpose of the DAS is 'only' to provide RF service and NOT in any way to jeopardize the safety of the elevator operation or impact upon the operation of the elevator, etc.

5.7 Multioperator Systems

Often there is a need for the indoor DAS to support multioperator configurations, where more than one operator or band is to share the same DAS. This will typically be the case in large public buildings, airports, convention centers and tunnels, where there is a need for high capacity and in many cases more than one type of mobile system on air over the same DAS, 2G, 3G + 4G.

From an economic perspective there is much to be saved by the operators if they share the DAS. Seen from the building owner's perspective, it is also preferable to have only one DAS, one installation, one set of antennas, one project to coordinate and one equipment room.

In order to connect more operators to the same DAS, you need to combine several base stations, repeaters and bands into the same system. Combining the operators and bands into the same DAS is not a trivial issue; you need to pay close attention to many potential problems that might degrade the service if you are not careful.

The principle of the active DAS, where DL and UL are separated at the interface to the main unit, as well as the low power levels required at the input (typically less than 10 dBm), will ease the requirements and the design of the combiner. If considering a passive DAS for a multioperator solution, then the combiner becomes an issue. Be careful when combining several operators or bands at high power using many carriers on air. This has the potential for some really big problems. However, if you select high quality components and a careful design, preferably using a cavity filter system tuned to the individual bands, you succeed.

In many cases multioperator solutions will be installed in large buildings, airports, hotels and shopping malls. Therefore active DAS is often the preferred choice due to the improved performance on data services and ease of installation and supervision.

5.7.1 Multioperator DAS Solutions Compatibility

There are important RF parameters to take into account when designing multioperator DAS, but if you pay attention to the most critical parameters, then you will succeed.

Rx/Tx Isolation

The Rx/Tx signal from each individual base station must be separated according to the specification of the base station supplier. When using the combined Rx/Tx port on the base stations, this will normally not be a problem, due to the internal duplex filter in the base station, which will separate the uplink and downlink. However you must also make sure that the downlink signal from one base station will not reach any of the other base stations receivers, according to the isolation specification.

When using an active DAS where the main unit needs separate uplink and downlink signals, it is preferable to use the Rx diversity port for the uplink from the DAS to the base station, and use the combined Tx/Rx port for the downlink only. This will help in designing a combiner that can provide excellent isolation of the UL and DL signals. However, you must be aware that most base stations will trigger a 'main receiver fault' alarm, due to the lack of uplink signal on the main receiver port, the combined Rx/Tx. As long as you cancel this alarm trigger in the network surveillance system, then you can avoid the alarm problems. A typical value for Rx/Tx isolation requirements is 25–30 dB (without including the internal filter and combiner in the base station), but refer to the specifications and guidelines for the specific base station used.

Return Loss

This is the reverse power, the reflected power from the DAS–combiner system. In order not to trigger any VSWR alarm on the base station, you need to keep the reflected power below the trigger value. A typical value is >10 dB (the difference between forward and reverse power).

Inter-band Isolation

This is the isolation between 2G, 3G and 4G. In order not to de-sensitize the receiver in the BS it is important to achieve maximum out-of-band rejection. The basic specification on this issue is specified by ETSI. The exact value depends on the base station manufacture, how much you will allow the base station to be desensitized and the band configuration. Look up the value in the specification of the base station. A typical value for inter-band isolation is better than 50 dB.

Passive Inter-modulation

PIM is one of the biggest potential problems in combiner systems. It is very important to select a combiner system with a low PIM figure (−155 dBc at 2 × 20 W). PIM is a major concern for combiner systems, especially when combining many high-power (typical 20 W/43 dBm) carriers in multioperator systems. The power density of several operators operating with many carriers can be high, as we shall calculate later in this chapter.

The root of PIM is non-linearity and it is generated when two signals are mixed, resulting in unwanted harmonic signal products due to the non-linear behavior of a passive device/component. We know about and use this non-linear effect in the standard passive 'mixer' component, when we need to generate wanted products based on two input signals. This is typical in a diode (four diodes in a 'bridge' configuration) that is used as a 'mixer' – the non-linear behavior of the diode causes distortion and thereby generates harmonic products of the two base signals that we feed to the mixer. Passive components are not the same as a diode, so how can we get non-linear behavior from simple devices such as a connector, a coax cable, an attenuator or similar simple passive components? The PIM is mainly caused by the junction between two dissimilar materials inside the passive component. One example could be aluminum, which connects to copper (a very bad combination) inside the passive device. Initially, there might be very good contact between the two materials, but over time the galvanic correction due to the dissimilar metals will cause a non-linear connection and possibly (likely) generate a high level of unwanted signals: PIM.

4G (64QAM) modulation, in particular, is a challenge, due to the many sub-carriers and a very demanding RF signal to 'handle' at high power for any type of component. Hence there is a need for high linearity throughout the signal path to maintain performance at a maximum, and to service high modulation accuracy and thereby high data speeds. There is no point in generating a high signal level in the building if the quality of the signal has low performance.

The main PIM products are shown in Figure 5.38. Note that the bandwidth of the generated PIM signal increases with the higher order of the unwanted PIM. This can hit hard in a wideband carrier, say several 5 MHz carriers on 3G that hits a carrier using a 4G spectrum. The PIM can also desensitize a whole section of the spectrum.

It is important to note that PIM is often under-estimated, and, if/when you encounter PIM challenges, you will need to understand these basics. Obviously the best solution to PIM is to avoid the issue in the first place. Once PIM is generated and hits your service, there is no way

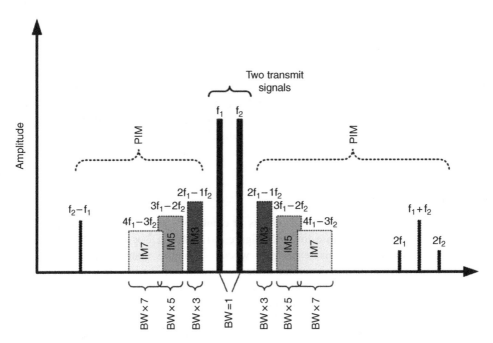

Figure 5.38 PIM power in relation to the two fundamental frequencies. Note the bandwidth of the PIM

to cure it with filters and the like, as it hits you 'in-band'. The only solution is to solve the root of the problem, the component or components that are causing PIM.

Even Simple Components Can Create PIM

Sometimes the simplest components can be a source of severe PIM performance issues – and, trust me, PIM issues can be very hard to troubleshoot. PIM is a huge concern and even more so at the 'hot' end of the DAS where high power levels are present. This will involve the RF components in the parts of the DAS system that are close to the base station: the point of interface (POI), the high-power remote unit, the remote radio head (RRH), attenuators, cables, connectors, etc.

The problem with PIM should not be ignored and it is a real concern for multi-frequency high-power systems. The impact in terms of degraded performance, low data speeds, dropped calls and other degradation of the actual or other radio services can be quite severe. And this will not only affect the quality and user experience, but also hit your business case.

Passive inter-modulation will often appear to be related to time of day or the day of the week – or, rather, the actual load on the serving base station and DAS system. This is because PIM is often related to the traffic load, so the more traffic the system is carrying at a certain time of day, and the higher the RF power density, the more likely it is that PIM will be generated. If the PIM is really bad, it might even cause the cell to drop the traffic due to quality issues. The PIM then disappears, due to the decrease of power concentration caused by the drop in traffic, at which point the cell starts to pick up traffic again, until PIM is once again triggered by power concentration due to new call set-ups and increased traffic and power load. This vicious cycle will continue until the users stops generating more calls and traffic. This is a very big concern; obviously we want the DAS system always to perform with excellence, to produce the highest possible data speeds on both downlink and uplink, in particular when the cell is carrying a lot of traffic and generating a lot of revenue. PIM will affect your modulation accuracy, your error vector magnitude (EVM; see Section 2.5.5). So even if the PIM is not bad enough to cause the cell to drop traffic or to result in a very clear degradation in performance and quality, it might degrade the data throughput in the system due to degraded EVM, preventing your service from utilizing the highest modulation rates. This will degrade the average data throughput of the cell, and thus result in less efficiency of the cell; fewer Mb of average data production in the cell will mean less revenue for the cell, and hit the business case of the deployed system directly. High-quality (low-PIM) components come at a cost, but it is definitely worthwhile investing in quality DAS components, both on the passive side and on high-specification active components and systems.

PIM Power

The PIM power is closely related to the RF power transported in the system. We can see the principle illustrated in Figure 5.39 and it is obvious that the generation of the unwanted PIM signals is non-linear behavior. In the example (Figure 5.39) the PIM starts to rise at an input power exceeding 38 dBm, and then the rise in PIM signal power is greater than the rise in input power, producing the expected non-linear behavior between the input power and the PIM in the system.

One could argue that the PIM in Figure 5.39 is of no concern as long as it stays below the noise floor of the system, in this case −117 dBm. However, it is always recommended to design using components that ensure the PIM power will be as low as possible, while keeping costs reasonable.

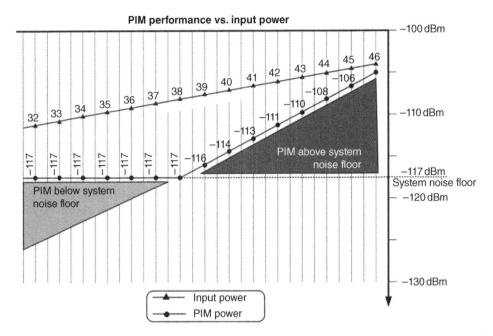

Figure 5.39 PIM power in relation to input power. PIM signals below the noise floor of a system will not be a big concern – therefore it is recommended to calculate the PIM to stay below the noise figure of the system

Selecting Passive Components to Minimize PIM

To minimize the impact of PIM, we must carefully select a combination of components with sufficiently low PIM performance to ensure that the PIM that is present remains well below the noise floor/sensitivity of the system.

$$\text{Non-linearity} + \text{multiple carriers} + \text{high RF power} = \text{PIM}$$

Connectors

Surely, a connector is a standard component and any one N-type will perform just like any other N-type from another manufacturer, the only difference being the price? Well, it is easy to conclude this initially, but the reality is not what we might expect.

In Figure 5.40 we can see a typical performance example of some standard RF connectors we use when deploying DAS solutions. It is clear that there is a huge performance difference between the different connector types. In the figure it is clear that one N-type connector will perform as badly as –120 dBc [so if there is a power level of 30 dBm passing through the connector, the potential inter-modulation power of –90 dBm, (30 dBm – 120 dB) this is clearly an issue for concern]. If instead we use a PIM-rated silver-plated N-type with a PIM performance of –165 dBc, then, with 30 dBm of power, we will potentially risk having a PIM of –135 dBm (30 dBm – 165 dB), i.e. 45 dB less potential PIM power compared with the other type of connector. As always, price and performance are closely related.

Therefore it is very important to select the right type of connectors with the right PIM performance throughout the DAS, and to pay special attention to the 'hot' end of the system.

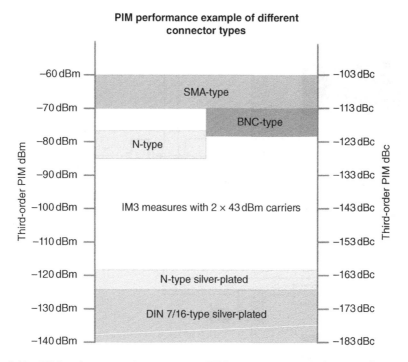

Figure 5.40   PIM performance of connectors vs. PIM product power relative to carrier power, dBc

Hence, at the POI, the attenuators, circulators etc. between the base station and the main unit of the DAS (Figures 5.41 and 5.42), it is always highly recommended to use low-PIM connectors such as DIN 7/16 due to the high RF power.

Obviously the connectors should be installed, like any other component in the DAS, in complete accordance with the manufacturer's instructions, only fitting the precise connector to the precise cable it is designed to be used with. We also need to focus carefully on the quality throughout the installation process, making sure to use the right tools, to clean, swipe, and fit all connectors and terminations as per the manufacturer's instructions. This is key: instruction, training, certification and verification.

Make sure to clean the connectors and sockets before you connect them, swipe with alcohol, and use compressed air – and then connect and tighten the connectors using the correct torque, no more, no less. I know that some of you may well be rolling your eyes at this point, as you may not have experienced any issues. Are you sure? It may well be that some of the unexplained dropped calls, the lack of maximum performance or the 'odd' statistics you have been struggling with have come about because of a simple lack of discipline during the installation phase.

Minimize the PIM Issues
Here are a few hints to help you avoid some of the issues. Obviously you need to select low-PIM components, but also make sure to focus on the following:

• Torque the connectors
   Yes, I know it sound tedious but it will make the difference between bad PIM performance and excellent PIM performance if the connectors are tightened at the specified torque to

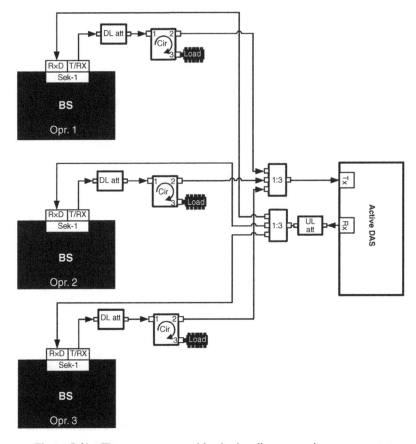

Figure 5.41 Three operators combined using discrete passive components

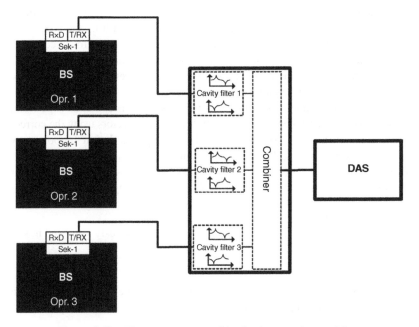

Figure 5.42 Three operators combined using a cavity combiner

ensure the correct contact surface, and thereby the lowest possible PIM. Most manufacturers of high-quality, low-PIM connectors can also supply torque wrenches to fit each connector type. These will 'release' with a specific 'click' once the correct torque has been applied. If you ever get the chance to see a PIM tester/instrument connected to a low-PIM reference, and to undo the connectors and tighten them with your fingers, you will see how dramatically the PIM product decreases once you use the right tool with the right torque instead of your fingers. I did, and that is why I am very insistent about this.

- The coax cable
  Even a 'simple' component like the coax cable is a concern when it comes to PIM performance. The heavier coax with a solid shield is not normally a concern, but flexible, braided cable is most definitely a concern in many cases. Therefore, at the 'hot' (high power) end of the DAS, typically from the base stations to the POI, attenuator, etc. where the power levels are high and you don't want to use a heavy coax, make sure to use low-PIM-type cables.
- Attenuators
  Like any other component in the RF link, the attenuators also play a big role. A normal attenuator is traditionally made of carbon and has very bad PIM performance, due to the construction, which is basically a lot of carbon particles squeezed together to get the right value of attenuation. All these particles are essentially millions of junctions between individual 'resistors', and these junctions will struggle to behave in a linear fashion. Never use this type of standard attenuator in a DAS, especially at the 'hot' end of the DAS, which,in an active DAS, would typically be between the base stations and the POI. In this application we make sure to use a low-PIM attenuator, a so-called 'cable absorber'. This type of attenuator is actually a low-PIM cable that is reeled up in a spool inside a box. You should always use low-PIM-type attenuators between the base stations and the active DAS hub (see Figure 5.41). This will minimize PIM and be below −150 dBc or better. (Check the specifications from the manufacturer of the low-PIM attenuator.)

Testing for PIM

In order to test the PIM, one needs specialized high-precision RF equipment and deep knowledge. I highly recommend participating in specialized training program or hiring the right team with the right equipment if PIM tests are to be conducted correctly.

You could try to use the equipment you have to hand, and feed dual carriers to a system, trace the inter-modulation products to see the effect of PIM and inter-modulation distortion (IMD); however, this should not be considered a reference test, though it can be helpful to clarify the basic performance.

If PIM is suspected, a nice trick is to step back the RF input power level to the system, and see if there is a non-linear drop of the unwanted products. If so, it is likely that there is be a PIM/IMD problem.

5.7.2   The Combiner System

There are many ways to combine operators and bands into the same DAS, from constructing your own combiner using discrete broadband components (as shown in Figure 5.41) to cavity filter combiners with filters tuned to the specific UL/DL band of each of the operators (as shown in Figure 5.42).

The broadband system in Figure 5.41 is ideal for use when combining low power signals into an active DAS. Note that the downlink attenuator, used for leveling the downlink signal to the correct value of input to the active DAS, is installed prior to the circulator. Standard attenuators have rather poor PIM performance 120–140 dBc; therefore it is recommended to use a low-PIM cable attenuator. This will keep power levels in the circulator to a minimum, thus minimizing the potential PIM in the circulator. Note that this example is used for an active DAS, where the downlink power requirement at the input at the main unit is about 5 dBm. Therefore low-power base stations are used with about 30 dBm output power.

You can keep the uplink and downlink separated by using circulators; this will make sure that one base station's transmitters downlink signal is not feed to the other base station's combined Tx/Rx port, via the 1:3 splitter–combiner, thus securing isolation. Note the specification of the circulator; it might be an idea to use a double circulator with high reverse attenuation.

The custom tuned filter combiner in Figure 5.42 will typically perform much better. This combiner must have low PIM and high isolation, even at high power, which is typically used when combining high power systems into a passive DAS. The combiner is tuned individually to the specific frequency bands of each operator, and there is good inter-operator and inter-band isolation. Only high-quality components are recommended for use in any combiner system.

5.7.3   Inter-modulation Distortion

When two or more signals are mixed in a nonlinear component, passive or active, there will be generated other signals as a product of the two or more original input signals. Inter-modulation distortion (IMD) will occur in amplifiers when operating in nonlinear mode. Therefore it is very important to always operate amplifiers according to their specification, and to keep below a certain limit, e.g. ETSI −36 dBm at 2G-900. Make sure you stay within the linear operating window of the amplifier, and the amplifier will have linear performance with low distortion. If you overdrive the amplifier, you might get few dB more output power, but the side effect will be IMD problems that will generate inter-modulation interference.

What is Inter-modulation Interference?

There are three basic categories of inter-modulation distortion:

1. *Receiver produced IMD*: when two or more transmitter signals are mixed in the receivers RF amplifier.
2. *Transmitter produced IMD*: when one or more transmitted signals are mixed in a nonlinear component, in the transmitter.
3. *Passive IMD*: normally radio planners are only concerned about inter-modulation problems caused by active components like transmitters, amplifiers and receivers. However, you must realize that passive components like cables, splitters, antennas can also produce inter-modulation – passive inter-modulation. Typically the source of PIM is the junction between different types of materials, an example could be where the cable connects to the connector. The connection between any internal or external point of contact in a passive component can generate PIM.

The PIM performance is often the big difference between quality and low-cost passive components. Normally good quality passive components, antennas included, will be constructed using the same base material, with only a few or no internal connections or assembled parts and a high level of craftsmanship.

The PIM problem is often underestimated, especially when designing passive systems for large, high-capacity passive DAS, multioperator solutions and multiband solutions. The concentration of the high power in the passive DAS, especially close to the base stations, can be high.

Example
Let us look at a typical example of a multioperator solution in an airport, with this configuration: four 2G operators are each using eight carriers, transmitting 40 dBm (10 W) from their base stations. This is a total of 32 carriers. The total composite power is up to 55 dBm (320 W) in the first part of the passive DAS. Even if the first splitter in the DAS can handle 700 W, you might have a problem. How high is the generated inter-modulation power?

A typical value for a 'standard' splitter is − 120 dBc, meaning that the IM3 is 120 dB below the c (carrier). With 40 dBm of power for one carrier to the splitter, the IM3 could be up to 40 −120 = −80 dBm.

Depending on the frequencies used, this will generate a problem: it could hit the uplink of one of the radio services on the DAS. If you exceed the power rating of the components, or there is a bad connection somewhere in the passive DAS, then the PIM will increase dramatically.

5.7.4   How to Minimize PIM

Passive inter-modulation occurs when two or more signals are present in a passive device (cable, connector, isolator, switch, antenna, etc.) and this device exhibits a nonlinear response. The nonlinearity is typically caused by dissimilar metals, dirty interconnections or other anodic or corrosion effects. Bad connections are also a typical source, and often the effect does not appear at low power levels, but increases exponentially at higher power levels. Then the passive device starts acting like a frequency mixer with a local oscillator and an RF input, generating its own unwanted signals.

There are some rules of the thumb on how to design for low PIM:

- All passive components must fulfill a minimum specification of −155 dBc at 2 × 20 W.
- Cable absorbers should be used as terminations.
- The 7/16 type of connectors should be used.
- All connectors should be tightened according to the specifications, using the correct torque and tools.
- Low PIM cables should be used, with all connectors fully soldered.
- It is vital to maintain disciplined fitting of connectors to the cable, craftsmanship, use of correct tool and all metal cleaned before fitting the connector.
- Tools, cables and connectors that match and are from the same manufacture should be used.
- Proper installation means:
  no loose cables, everything strapped with cable binders;
  no mechanical stress on RF parts;
  all RF interconnections cleaned as specified by the manufacturer.

Table 5.3 The inter-modulation components and results

Inter-modulation class	Result
Second order 2 CH	$f_1 + f_2$ $f_1 - f_2$ $2f_1$ $2f_2$
Third order 2 CH	$2f_1 + f_2$ $2f_1 - f_2$ $f_1 + 2f_2$ $2f_2 - f_1$
Third order 3 CH	$f_1 + f_2 - f_3$ $f_1 + f_3 - f_2$ $f_2 + f_3 - f_1$
Fifth order 2 CH	$3f_2 - 2f_1$ $3f_1 - 2f_2$
Fifth order 3 CH	$2f_1 + f_2 - 2f_3$ $f_1 + 2f_2 - 2f_3$ $2f_1 + f_3 - 2f_2$ $f_1 + 2f_3 - 2f_2$ $2f_2 + f_3 - 2f_1$ $2C + f_2 - 2f_1$
Seventh order 2 CH	$4f_1 - 3f_2$ $4f_2 - 3f_1$
7th order 3 CH	$3f_1 + f_2 - 3f_3$ $f_1 + 3f_2 - 3f_3$ $3f_2 + f_3 - 3f_1$ $3f_3 + f_2 - 3f_1$

5.7.5 IMD Products

The frequencies that are generated by inter-modulation can be found by mathematical calculations of the performance of nonlinear circuits (as shown in Table 5.3). The terminology used to define inter-modulation products classifies their order as second-order, third-order, fourth-order, etc.

The frequencies generated are calculated as the sum or differences between these inter-modulation products (as shown in Figure 5.43). In theory there are no limits to the number of them; however, typically just a few can result in serious consequences.

The main concern in indoor DAS solutions is the IM3 product,

$$2f_1 + f_2, 2f_1 - f_2$$

and you must strive to minimize that problem. The powers of products of higher orders, such as fifth to ninth orders and even higher, are usually so low that they do not create serious problems, but be aware of PIM and check.

The products of even second, fourth or sixth orders usually have limited impact due to the fact that the resulting PIM frequency is attenuated by the filters the base station.

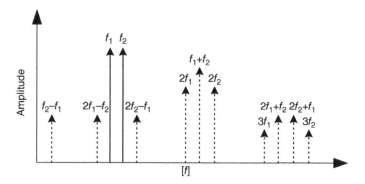

Figure 5.43   Example of second- and third-order IMD products

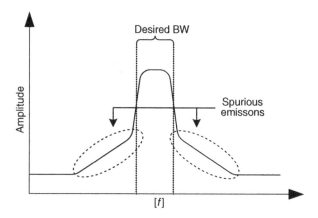

Figure 5.44   Spurious emissions from a transmitter

5.8   Co-existence Issues for 2G/3G

When deploying different radio systems and frequency bands using the same DAS for distribution, there are concerns that need special attention. You need to make sure that one system does not cause IM problems that will degrade the performance on its own or on other bands.

The problem rises exponentially with the power you feed to the DAS. The more power, the bigger the issue, mainly due to PIM (see Sections 4.3 and 5.7.3).

5.8.1   Spurious Emissions

A major source of co-existence problems is spurious emissions from the 2G transmitters (as shown in Figure 5.44). Therefore it is highly recommended *always* to use high-quality pass band filters on all transmitters, in a multioperator or multisystem DAS solution, when combining 2G and 3G. Those filters will normally be a part of the base station, the combiner, but check the specifications of the base station to be sure.

Spurious emissions from 2G are restricted in the UL 3G band (1920–1980 MHz) to a maximum –96dBm measured in 100 kHz bandwidth (equivalent to –80dBm in a 3.84 MHz channel). Any multioperator or multisystem DAS should be designed accordingly, in order to minimize the impact on the 3G UL.

Big Impact on 3G

As we know, the 3G system is noise-limited; any noise increase on the UL of the 3G system will severely impact the performance. Just a slight increase in the UL noise load will offset admission control; high noise increase will collapse the cell and cause the admission control to block for any traffic.

The number of 2G carriers in the system plays a significant role: when doubling the number of 2G carriers (with same level), the spurious emission is increased by 3 dB.

5.8.2 Combined DAS for 2G-900 and 3G

When combining 2G-900 and 3G on the same DAS, you must pay attention to the second harmonic from 2G-900 that might fall into the uplink band of 3G (TDD) (as shown in Figure 5.45).

5.8.3 Combined DAS for 2G-1800 and 3G

When combining 2G-1800 and 3G into the same DAS, you need to pay attention to the third-order inter-modulation products. The third-order IMD from 2G-1800 can fall into the UL band of 3G, if you do not pay attention to the frequency allocation on 2G-1800 (as shown in Figure 5.46). When doing the frequency planning on 2G-1800, attention to this problem is advised in order to minimize it. This can be a challenge in multioperator solutions, where frequency coordination between competing operators must be carried out.

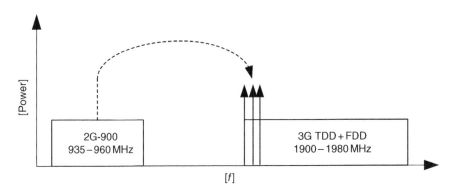

Figure 5.45 Example of second-order IMD products from 2G-900 hitting 3G UL

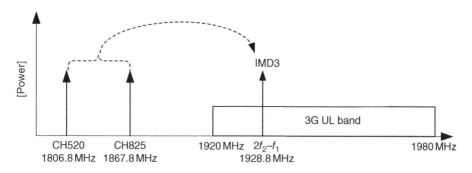

Figure 5.46   Example of third-order IMD products from 2G-1800 hitting the 3G uplink

Example
Let us look at one example with 1800 MHz third-order IM. Using CH520 (1806.8 MHz) and CH825 (1867.8 MHz) will produce this IMD3 (refer to Section 5.7.3):

$$2f_1 + f_2 = 5481.4\,\text{MHz}$$
$$2f_1 - f_2 = 1745.8\,\text{MHz}\,(\text{will hit 2G-1800\,CH690\,UL})$$
$$f_1 - f_2 = 5542.4\,\text{MHz}$$
$$2f_2 - f_1 = 1928.8\,\text{MHz}\,(\text{will hit 3G UL})$$

It is highly recommended to plan the frequencies on 2G-1800 so they do not fall in the UL band of any 3G channels that are in use. The fact is that 3G has good narrow-band interference immunity, but a high narrowband signal on the UL of the 3G base station can cause receiver blocking.

Knowing that the first 2G channel used is CH520 = 1806.8 MHz and that the problem is the third IMD ($2f_2 - f_1$), we can calculate what the highest frequency can be, in order to make sure that the third IMD will 'hit' just below the 3G FDD frequency band 1919.8 MHz, and not cause any problems (if 3G TDD is not used). The highest allowable frequency is:

$$1919.8\,\text{MHz} = 2*f_2 - 1806.88\,\text{MHz} = (1919.88\,\text{MHz} + 1806.88\,\text{MHz})/2$$
$$= 1863.38\,\text{MHz} = \text{CH802}\,(1863.2\,\text{MHz})$$

With CH520 as one of the frequencies in operation, we must make sure that we do not select frequencies above CH802 as the other channel.

5.9   Co-existence Issues for 3G/3G

In an indoor DAS system it is very likely that operators will have adjacent channels active on the same DAS. Because of the limited selectivity of the filters used in the mobile and base station, power from one channel will leak into the adjacent channels.

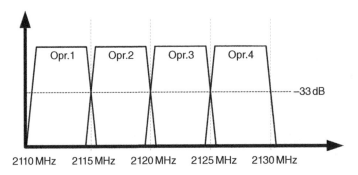

Figure 5.47   Channel allocation on 3G

5.9.1   Adjacent Channel Interference Power Ratio

This is the major issue when co-locating 3G operators in the same building. It is specified that the adjacent channel suppression must be better than 33 dB (as shown in Figure 5.47), but this is in many cases not enough, especially indoors where you typically have users close to the antennas. A user being serviced by another operator (and therefore not in power control of the cell) is likely to cause adjacent channel interference power ratio (ACIR) when this mobile is close to the antenna, having low path loss. Conversely, the adjacent operator might impact your user when he is close to the other operator's antenna.

Noise Increase on Node B

The adjacent channel interference will cause noise increase on the UL of node B, degrading the capacity with noise load.

Example

We can calculate how much noise increase a mobile will generate and the noise load on our cell, when this mobile is being serviced by an adjacent operator and when the mobile is close to our antenna. The mobile on the adjacent channel is transmitting 21 dBm. The mobile (UE) is 10 meters from our antenna (free space loss = 59 dB and then we have 10 dB loss from the node B to the antenna and 2 dBi antenna gain; the path loss is $59 + 10 - 2 = 67$ dB:

$$P_{RxAdj} = P_{TxUE} - ACIR - PL$$

Where P_{RxAdj} is the adjacent channel power level; P_{TxUE} is the mobile transmit power on adjacent channel; ACIR is the adjacent channel interference rejection; PL is the path loss from the adjacent mobile to our node B.

$$P_{RxAdj} = 21\,dBm - 33\,dB - 67\,dB = -79\,dBm$$

Now we can calculate the new noise level (assuming a noise floor on node B of -105 dBm). We need to sum up the noise, by adding the power contribution from each different source (the P_{RxAdj}+noise floor on node B of -105 dBm):

$$\text{power total} = \text{power1} + \text{power2} \ldots \text{power}(n)$$

We now need to convert the dBm to mW:

$$P(\text{mW}) = 10^{\text{dBm}/10}$$
$$-105\,\text{dBm} = 31.6 \times 10^{-12}\,(\text{mW})$$
$$-79\,\text{dBm} = 12.58 \times 10^{-9}\,(\text{mW})$$

Then we can add the two powers:

$$\text{combined noise power} = 31.6 \times 10^{-12} + 12.58 \times 10^{-9} = 12.62 \times 10^{-9}\,(\text{mW})dBm$$

We then convert the noise power back to dBm:

$$\text{noise floor (dBm)} = 10\log(12.62 \times 10^{-9}) = -78.98\,\text{dBm}$$

The $-105\,\text{dBm}$ from the BS plays only a minor role; it is after all 26 dB lower than the power from the adjacent mobile. This adjacent channel noise causes the uplink of the cell to collapse, disabling the cell from carrying any more traffic, and dropping all ongoing calls!

5.9.2 The ACIR Problem with Indoor DAS

ACIR is a major concern inside buildings, when the operator uses adjacent frequencies and separate DAS systems. The problem is demonstrated in Figures 5.48 and 5.49. Two separate DAS systems are deployed within the same building; one for operator 1 (OPR1) and one for operator 2 (OPR2).

When we look at the signal power as a function of the location relative to the antennas (as shown in Figure 5.49), it is clear that, when you are close to an antenna from the other operator and have low signal from the serving cell, there is a real potential for ACIR.

The Potential ACIR Problem in the Future

There are many of these potential ACIR problems in real buildings around the world, but the problem is not evident until the operators deploy more channels. This is due to the fact that most 3G operators have two or three 3G channels allocated and if they deploy the first

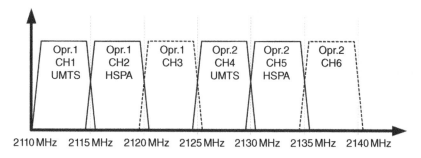

Figure 5.48 Typical channel usage of two operators with 3G/HSPA deploys, and a third channel for future use

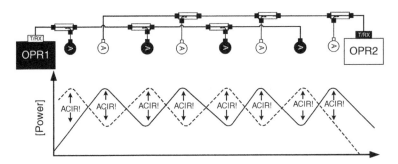

Figure 5.49 Adjacent interference problems in a building with two operators on separate DAS systems

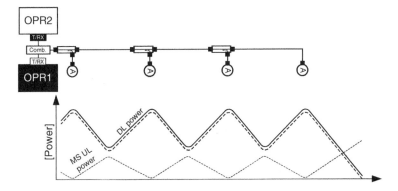

Figure 5.50 No adjacent interference problem in a building with two operators on the same DAS systems

carrier only, there is still 10–15 MHz distance to the next channel from the other operator, minimizing the ACIR problem. Most operators deploy 3G (R99) traffic on the first carrier, HSPA on the second carrier and leave the third channel for 'future upgrades'. This usage of the 'future' channel could be a potential problem in many buildings (as shown Figure 5.38).

5.9.3 Solving the ACIR Problem Inside Buildings

There are means to solve the ACIR issue; the best way is for the operators to agree between themselves to use the same DAS and antenna locations, and coordinate the roll-out in these buildings.

It is always a good idea to place the indoor antenna so that the user cannot get too close to it; however this is not always possible inside a building. After all, if the ceiling height is 3 m, and the user is 180 cm tall, the mobile will come as close as 120 cm to the antenna (with a path loss of about 40 dB).

Multioperator Systems will Help

When using the same DAS, the scenario can actually be controlled by the fact that both the wanted and the unwanted signals track the same relative level. If both operators use the same antenna locations (as shown in Figure 5.50), then the two signal levels will track and

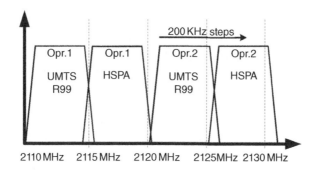

Figure 5.51   Operators might offset their WCDMA frequency to solver the ACIR problem

have the same relative difference in level throughout the building, thus avoiding any potential ACIR problems, on both UL and DL.

Offsetting the 3G/WCDMA Frequency

It is possible for the mobile operator to offset the WCDMA frequency in steps of 200 kHz, thus minimizing the problem. If CH1 is used for 3G, CH2 for HSPA and CH3 is unused, then the two channels can be offset 1 MHz, improving the adjacent channel isolation in the building (as shown in Figure 5.51).

5.10   Multioperator Requirements

The complexity of designing a multioperator solution is often underestimated. To design, implement and operate a multioperator solution is much more than 'just' the RF design issues. It is highly recommended that all the mobile operators in each country or region work towards a mutual accepted document that defines all the parameters, interfaces and issues with regards to multioperator DAS solutions.

Often RF planners focus on the RF design only and go ahead with the project. However costly mistakes and performance degrading problems can be avoided if you plan more than just one step ahead and pay attention to the whole process. It is recommended to prepare, develop and agree on a multioperator agreement before the first multioperator project is decided. This agreement takes time to develop and obtain agreement on; therefore it is highly recommended that you complete the multioperator agreement before the first multioperator DAS solution design is actually needed. If you wait until the first project, it will probably be delayed.

There are many issues, technical, political and legal, that must be clarified in the process of developing the multioperator agreement. In most cases it is worthwhile hiring a neutral consultant to help the operators with this agreement. Compromises sometimes need to be made, and it can be much easier for an independent consultant to make ends meet and come to a mutual agreement and compromise.

5.10.1   Multioperator Agreement

The multioperator agreement (MOA) document must be agreed and must cover all aspects of the process and lifetime of the DAS system. There must be clear definitions of all aspects of the project, including technical, logistical, installation and legal aspects.

The MOA should ideally be a universal document that can be used for all projects in the future and constantly be updated and adapted. Clear responsibilities must be defined for all aspects of the project. It needs to be precise and accurate, preferably containing document templates to handle and control the process. The best way to handle any multioperator project is to get the MOA in place before anything else is done. Even if you currently do not have any multioperator projects in mind, start now to define the MOA, so it will be ready when the first multioperator project is initiated.

In many aspects we consider other operators as competitors, but it is important to agree on a common MOA to handle these projects.

5.10.2   Parties Involved in the Indoor Project

We, RF planners, like to think that the whole indoor project revolves around us. In fact, many other important persons and departments are involved, internally at each operator but also quite a few external parties. Here are a few examples of parties involved in the typical project:

- The building owner, and also the architect of the building.
- The local IT consultant or personnel responsible for the site.
- Radio planners from all operators.
- Co-location managers from all operators.
- Site hunters.
- Transmission planners.
- Parameter optimizers.
- Equipment procurement departments.
- Commissioning teams.
- Contractors and installers.
- Optimizing team.
- Operations and maintenance team.

5.10.3   The Most Important Aspects to Cover in the MOA

Indoor DAS projects are complex and many parameters and aspects need to be taken into account. Obviously there is much focus on all the radio parameters, but many other parameters and issues need to be taken into account when designing, implementing and operating a multioperator system. Here are some examples of the most important aspects.

The Radio Design Specification

This is the basis of the design. It is very important to absolutely crystal clear on the RF specifications:

- RF design levels – UL and DL level and quality or BER levels.
- Data service levels – DL and UL speeds 3G/4G.

- Noise levels – UL noise power and DL noise level.
- Delay – maximum allowable delay of the system, end to end.
- Differentiated design goals – special design levels in specific areas.
- Link budgets – complete link budget for all services, all bands must be documented.
- General design guidelines – other, non-RF design parameters.
- Performance merits of the used DAS components – PIM specifications, power rating and type approvals.
- Handover zones – HO zone size and traffic speed, and 2G/3G or 4G zones.
- Capacity – number of channels needed for voice and data requirements.
- Maximum downlink transmit power at the DAS antenna – to avoid near–far problems, especially on 3G and 4G.

Co-existence Parameters

It is important to insure compatibility between all the operators and radio services, to make sure that the radio services can co-exist on the same DAS without any service degradation of any of the radio services.

- RF interface specification – type of connectors, power levels and isolation between bands and between operators.
- Inter-modulation optimization – frequency restrictions, spurious emissions, and ETSI and other requirements.
- Public radiation safety guidelines (EMR) – WHO/ICNIRP and local limits.
- Co-existence issues with other equipment – EMC and RFI compliance.
- Leakage from the building.

Future of the Indoor DAS

We need to look into the expected future of the building and DAS project. We must be sure that we define how to react on any changes in DAS:

- Future upgrades – more sectors, capacity, how and when new services are added to the DAS, 3G, 4G, etc.
- Connecting new operators to the DAS in the future – legal issues and headroom in the link budget.
- Disconnecting from the DAS – how this is handled if an operator wants to disconnect.
- Discontinuing the DAS – how this is handled when the DAS has to be removed.

Logistics

Clear definition of who is responsible for what in the project is very important. Gentlemen's agreements might be nice, but clearly defined contracts and agreements in writing are a must. These documents well help resolve any misunderstandings about responsibility in the future:

- Responsibility matrix – all parties included.
- Selection of DAS system – vendor and supplier, and request for quotation (RFQ) process.
- Selection and certification of installers – education, tool requirements and certification.

- Legal agreement and contracts – between operators, between operators and building owner, between operators and installer or DAS supplier, on who owns the DAS and insurance issues.
- Documentation – specify function, in- and output of all documents, documentation control and tracing updates, and design and as-built documentation.
- Installation guidelines – building code and fire retardants.
- Implementation plan.
- Acceptance testing and measurements – pass or levels, test method specification and operations.
- Be sure to define how and who takes care of maintenance of the DAS in the future – site access, alarm monitoring, spare parts, operation and maintenance service level agreement.

6

Traffic Dimensioning

One of the most important design parameters for indoor DAS solutions is to define how many speech channels you must provide for the users of the building in order to cater for their need for voice traffic channels. Voice traffic is time-critical. Many data services can be degraded in speed, offset slightly or delayed in time without any degradation of the user experience. It does not matter if your mail or SMS is delayed by 30 s, but the voice connection must be maintained 1:1 in time; one voice call needs one permanent traffic channel data services however, is dictating more and more the system capacity of the DAS especially multimedia services is a challenge if they need 1>1 support in real time.

How do you calculate the number of channels needed to cater for the capacity requirements of the users for a specific DAS project? And when you know the number of users and the type of users, how do you then use this information to calculate the number of channels/sectors? In the early days of telephony over fixed lines, a Danish mathematician faced similar challenges when dimensioning how many telephone lines were needed to interconnect fixed telephone switches. This principle. although used for voice is very useful to understand when designing for capacity in a DAS project. Basic understanding of Erlang and trunking efficiency, will also help you understand the benefits of shared resources when designing for high data capacity.

6.1 Erlang, the Traffic Measurement

Agner Krarup Erlang (Figure 6.1) was born in Denmark in 1878 (close to where I live). He was a pioneer in the theory of telecommunications traffic and he proposed a formula to calculate the percentages of users served by a telephone exchange that would have to wait when attempting to place a call. He could calculate how many lines were needed to service a specific number of users, with a specific availability of lines for the users.

Indoor Radio Planning: A Practical Guide for 2G, 3G and 4G, Third Edition. Morten Tolstrup.
© 2015 John Wiley & Sons, Ltd. Published 2015 by John Wiley & Sons, Ltd.

Figure 6.1   The Danish statistician Agner Krarup Erlang

A. K. Erlang published the result of his study in 1909: *The Theory of Probabilities and Telephone Conversations*. Erlang is now recognized worldwide for this work, and his formula is accepted as the standard reference for calculating telecommunication traffic load. The unit we use for telephony traffic load is 'Erlang' or E. Erlang worked for the Copenhagen Telephone Company (KTAS) for many years, until his death in 1929.

6.1.1   What is One Erlang?

An *Erlang* is a unit of telecommunications traffic measurement. One Erlang is the continuous use of one voice channel. In call minutes, one Erlang is 60 min/h, 1440 call min/24 h. In practice, when doing mobile capacity calculations, an Erlang is used to describe the total traffic volume of 1 h, for a specific cell.

Erlang Example

If a group of 20 users makes 60 calls in 1 h, and each call had an average duration of 3 min, then we can calculate the traffic in Erlangs:

$$\text{total minutes of traffic in 1 h} = \text{duration} \times \text{number of calls}$$
$$\text{total minutes of traffic in 1 h} = 3 \times 60$$
$$\text{total minutes of traffic in 1 h} = 180 \, \text{min}$$

The Erlangs are defined as traffic (minutes) per hour:

$$\text{Erlangs} = 180/60 = 3\,\text{E}$$

Knowing the number of users (20), we can calculate the load per user:

$$\text{user load} = \text{total load} = \text{number of users}$$
$$\text{user load} = 3/20 = 0.115\,\text{E} = 150\,\text{mE per user}$$

Then, if we have the same type of users inside a building with 350 mobile users, we can calculate what capacity we need:

$$\text{total load} = \text{number of users} \times \text{load per user}$$
$$\text{total load} = 350 \times 0.150 = 52.5\,\text{E}$$

Yet what do these 52.5 Erlangs mean in terms of number of voice channels needed? Do we then need 53 channels in order to service the users? Not exactly. It depends on the user behavior: how the calls are distributed and how many calls are allowed to be rejected in the cell. In theory – if you do not want any rejected calls due to lack of capacity, you should provide 350 channels for the 350 users, and then you would have 100% capacity for all users at any given time. However, it is not likely that all users will call at the same time, and users will accept a certain level of call rejection due to lack of capacity. To estimate the number of channels needed, we need to look at maximum call blocking we allow in the busiest hour during the day.

6.1.2   Call Blocking, Grade of Service

We need to calculate how many lines, or in our case how many traffic channels, we need to provide in order to carry the needed traffic. In this example we calculated that 350 users, each of whom load the system with 150 mE, would produce a total load of 52.5 E. What if all the users call at the same time? How are the calls distributed? How many call rejects due to lack of channels can we accept? This was exactly the basis of Erlang's work; he calculated how many channels you would need to carry a specific traffic load, with a specific service rate or blocking rate, for a given distribution of the traffic arrival.

The blocking rate (grade of service or GOS) is defined as the percentage of calls that are rejected due to lack of channels. If the users makes 100 calls, and one call is rejected due to lack of channels (capacity) the blocking rate is 1 in 100, or 1%. This is referred to as 1% GOS. Operators might differentiate the GOS target for different indoor solutions, with a strict GOS of 0.5% in an office building but allowing a GOS of 2% in shopping malls.

6.1.3   The Erlang B Table

Provided that the calls are Erlang-distributed, you can use the Erlang B formula to calculate the required number of channels at a given load rate, and a given grade of service. The Erlang distribution is a special case of gamma distribution used to model the total interval associated with multiple Poisson events. The Erlang B formula is complex; it is more practical to use an Erlang B table based on the Erlang B formula (see Table 6.1).

Using the Erlang Table

First of all we need to decide the quality of service in terms of how many calls we will allow the system to reject in the busiest hour. This is expressed as the grade of service. If we design for a GOS of 1%, a maximum of 1 call out of 100 will be rejected due to lack of traffic channels. Using the Erlang table is straightforward; you simply select the GOS you want to offer the users, and a typical number for mobile systems is 1%. Then you look up in the table how many voice traffic channels (N) you need to support this traffic.

Example
If we have calculated that the traffic is 5 E, and we will offer 1% GOS, we need 11 voice channels (5.1599 E) to support the traffic. We can also use the Erlang to work in reverse, and conclude that having only 10 traffic channels to support 5 E will give a GOS (blocking) of about 2%.

Table 6.1 Erlang B table, 1–50 channels, 0.01–5% grade of service

Number of channels	8*pc* Blocking (GOS)					Number of channels
	0.1%	0.5%	1%	2%	5%	
1	0.001	0.005	0.010	0.020	0.052	1
2	0.045	0.105	0.152	0.223	0.381	2
3	0.193	0.349	0.455	0.602	0.899	3
4	0.439	0.701	0.869	1.092	1.524	4
5	0.762	1.132	1.360	1.657	2.218	5
6	1.145	1.621	1.909	2.275	2.960	6
7	1.578	2.157	2.500	2.935	3.737	7
8	2.051	2.729	3.127	3.627	4.543	8
9	2.557	3.332	3.782	4.344	5.370	9
10	3.092	3.960	4.461	5.084	6.215	10
11	3.651	4.610	5.159	5.841	7.076	11
12	4.231	5.278	5.876	6.614	7.950	12
13	4.830	5.963	6.607	7.401	8.834	13
14	5.446	6.663	7.351	8.200	9.729	14
15	6.077	7.375	8.108	9.009	10.633	15
16	6.721	8.099	8.875	9.828	11.544	16
17	7.378	8.834	9.651	10.656	12.461	17
18	8.045	9.578	10.437	11.491	13.385	18
19	8.723	10.331	11.230	12.333	14.315	19
20	9.411	11.092	12.031	13.182	15.249	20
21	10.108	11.860	12.838	14.036	16.189	21
22	10.812	12.635	13.651	14.896	17.132	22
23	11.524	13.416	14.470	15.761	18.080	23
24	12.243	14.204	15.295	16.631	19.031	24
25	12.969	14.997	16.125	17.505	19.985	25
26	13.701	15.795	16.959	18.383	20.943	26
27	14.439	16.598	17.797	19.265	21.904	27
28	15.182	17.406	18.640	20.150	22.867	28
29	15.930	18.218	19.487	21.039	23.833	29
30	16.684	19.034	20.337	21.932	24.802	30
31	17.442	19.854	21.191	22.827	25.773	31
32	18.205	20.678	22.048	23.725	26.746	32
33	18.972	21.505	22.909	24.626	27.721	33
34	19.743	22.336	23.772	25.529	28.698	34
35	20.517	23.169	24.638	26.435	29.677	35
36	21.296	24.006	25.507	27.343	30.657	36
37	22.078	24.846	26.378	28.254	31.640	37
38	22.864	25.689	27.252	29.166	32.624	38
39	23.652	26.534	28.129	30.081	33.609	39
40	24.444	27.382	29.007	30.997	34.596	40
41	25.239	28.232	29.888	31.916	35.584	41
42	26.037	29.085	30.771	32.836	36.574	42
43	26.837	29.940	31.656	33.758	37.565	43
44	27.641	30.797	32.543	34.682	38.557	44
45	28.447	31.656	33.432	35.607	39.550	45
46	29.255	32.517	34.322	36.534	40.545	46
47	30.066	33.381	35.215	37.462	41.540	47
48	30.879	34.246	36.109	38.392	42.537	48
49	31.694	35.113	37.004	39.323	43.534	49
50	32.512	35.982	37.901	40.255	44.533	50

How do we determine what load in Erlang an indoor system will give? First of all we need to know the traffic profile of the users in the building.

6.1.4   User Types, User Traffic Profile

Each user in the network will have a specific load profile; the profile may vary and some days one user will load the mobile system more than other days. You need to use average load numbers, based on the type of user, in order to make some assumptions to help us design the capacity need (see Table 6.2). They can only be assumptions, but these assumptions will be adjusted over time when analyzing the post traffic on the implemented solutions. Thus you will build up a knowledge database to help you be more accurate in future designs. As a general guideline the typical numbers shown in Table 6.2 could be used.

If you are designing an indoor system for a building with 250 users and the users comprise 40 heavy users, 190 normal office users and 20 private users, you will need a capacity of:

$$(40 \times 0.1) + (190 \times 0.05) + (20 \times 0.02) = 4 + 9.5 + 0.4 = 13.9\,E$$

However we must remember that this load is only present when all users are in the building at the same time. It may be more likely that on average 80% of the users are in the building, and thus the traffic load would be $13.9 \times 0.8 = 11.12\,E$.

If we design for a GOS of maximum 1%, we can look up in the Erlang table the necessary number of traffic channels. In this case we will need 19 channels in order to keep the GOS below 1% (11.23 E).

Mobile-to-mobile voice Calls in the Same Cell, Intracell Mobile Calls

In a totally wireless environment where users generate a lot of internal traffic in the same cell, you must remember that to support one call where a user is calling another user in the same cell will actually take up two traffic channels. Therefore the 'internal call factor' must be applied. This is an important parameter to take into account, when analyzing the capacity need for the specific building.

In practice, if 10% of the load is internal mobile-to-mobile traffic, the load per user can be multiplied by a factor of 1.10; a heavy user that normally loads the network with 100 mE must be calculated as 110 mE if the internal traffic is 10%.

Table 6.2   Typical voice user load in Erlang

User type	Traffic load per user
Extreme user	200 mE
Heavy user	100 mE
Normal office user	50 mE
Private user	20 mE

6.1.5   Save on Cost, Use the Erlang Table

If it is an indoor 2G cell you are designing, then typically the number of traffic channels is seven per transceiver (radio channel). The exact number of traffic channels per transceiver depends on the configuration of the logical channels in the cell, but seven is often the average.

To calculate how much capacity you need, take the calculated total traffic in the cell, in this case 13.9 E, and multiply it by the 'intracell' factor, 1.10 in this case. To calculate the total capacity requirement:

$$13.9E \times 1.10 = 15.29E$$

Using the Erlang table, you can determine that you would need a total of 24 traffic channels to make sure you can keep the voice service below 1% GOS.

You know that each radio channel comes with seven traffic channels, so you would need to deploy four TRX with a total of 28 traffic channels – in fact an overcapacity of four traffic channels. The four TRX would be able to carry 18.64 E, or an overcapacity of 3.35 E, or 30 heavy users. Thus some extra margin is integrated for capacity growth in the solution. However, there is also another option.

Instead of deploying four TRX, you could save the cost of one TRX and only deploy three TRX with a total of 21 traffic channels. You can now look up in the Erlang table the GOS of a load of 15.29 E on 21 channels and see that you would have a GOS of 2–5%. Depending on the solution, you might decide to go with three radio channels and save the cost of one transceiver unit, and then analyze the call blocking performance once the system was on air, over a period of 4–6 weeks, to make sure that the capacity was acceptable. In fact, the Erlang table and traffic calculation can help you cost-optimize the design.

You must remember that the cost of an extra TRX is not only limited to the HW price of the TRX itself, there might be combiner and filter costs, BSC costs, A-bis interface and transcoder costs, as well as annual license fees to the network supplier.

6.1.6   When Not to Use Erlang

There are cases where we cannot use Erlang for voice calls; the most common example is when calculating the capacity for a metro train tunnel system. Even though the traffic inside the trains is Erlang-distributed and has normal telephone user behavior, you must realize that the trains moving the users are not Erlang-distributed! In practice, the arrival of traffic in cells in a metro train system is distributed by the train schedule, not by Erlang distribution, thus Erlang cannot be applied.

In practice, these trains are 'capacity bombs' that drift around the network, so be careful. In practice you might not even discover the real blocking of calls in a tunnel system; most network statistics will not be able to detect the call blocking in this case due to the fact that the blocking occurs only in a very short moment in time, maybe a few seconds when all the users inside a train hand over (or try to hand over) to the next cell when traveling in the train. To make sure there is no blocking, you might have to analyze the specific network counter indicating blocked calls for the specific cell, not the Erlang statistics, which might be averaged out over 15 min. Signaling capacity in this case is also a major concern when a 'train-load' of calls are to handover within the same 2–5 s.

6.1.7   2G Radio Channels and Erlang

Now you know how to use the Erlang table, you can easily produce a small table that can help in the estimation of how many voice users you are able to support for a given 2G configuration of radio channels in the cell (see Table 6.3).

Bear in mind that you also need to be able to support the required data services, and in some cases you will dedicate specific time slots (traffic channels) to data traffic.

It is interesting to note that the capacity of two radio channels is more than double the capacity of one! This effect is called trunking gain, and is explained in detail later in Section 6.1.9.

Half-rate on 2G

Note that 2G can utilize 'half-rate' where two voice users share the same TSL, with half the data rate. This will result in double the voice capacity for each radio channel. In cells with high peak loads this is normally implemented as 'dynamic half-rate', where the voice calls in normal circumstances gets one normal TSL per call. However, if the load on the cell rises to a preset trigger level, the system starts to convert existing calls to half-rate, and assigns new calls as half-rate, until the load fall below another trigger level.

The use of half-rate voice coding will to some extent degrade the perceived voice quality, but in normal circumstances the voice quality degradation is barely noticeable, and is to be preferred considering that the alternative is call rejection due to lack of voice channels.

6.1.8   3G Channels and Erlang

The voice capacity on a 3G cell is related to many factors. How much noise increase will we allow in the cell? How many data users and at what speed do we need to support these users? How much processing power does the base station have (channel elements) to support the

Table 6.3   Typical user load in Erlang and number of users vs TCH

Number of radio channels	Number of traffic channels (at seven TCH/TRX)	Number of office users at 50 mE/1% GOS	Number of private users at 20 mE/2% GOS
1	7	50	147
2	14	147	410
3	21	256	702
4	28	372	1007
5	35	492	1322
6	42	615	1642

traffic? Typically one channel element can support one voice connection. It also depends on how much power is assigned to the CPICH and the other common channels.

Many factors interact, but assume that the only traffic on the cell is voice calls and the other parameters are set to:

- Sum of power assigned to CPICH and other common channels is 16% (−8.1 dB).
- Maximum noise rise is 3 dB (load 50%).
- Voice activity factor is 67%.

Then one 3G carrier can support 39 voice channels of 12.2 kbps. This corresponds to 28.129 E at 1% GOS, enabling the cell to support 563 office users at 50 mE per user, or 33.609 E at 2% GOS, enabling the cell to support 1680 private users at 20 mE per user.

However we must remember that this is pure voice traffic, and data calls will take up capacity. Using the same conditions, the same cell will only be able to support one 384 kps data call and 27 voice channels, or three voice channels and three 384 kbps data calls. This is provided that there are channel elements to support the traffic installed in the base station.

6.1.9 Trunking Gain, Resource Sharing

Trunking gain is a term used in telecom traffic calculations to describe the gain from combining the load into the same resource (trunk). The effect of the trunking gain can also be seen in the Erlang table, and in Table 6.3. If you look up in the Erlang table the traffic capability of a 2G cell using one radio channel with seven traffic channels and 1% GOS, you can see that the capacity for the cell is 2.50 E. Then two separate cells with this configuration of seven traffic channels will each be able to carry $2 \times 2.50 = 5$ E.

However if we combine the resources, all of the 14 traffic channels into the same cell, we can see in the Erlang table that 14 channels can carry 7.35 E. This is 2.35 E more with the same resources, 14 channels. This is a gain of 47% with the same resources, but combined into the same cell; this effect is called 'trunking gain' (see Figure 6.2).

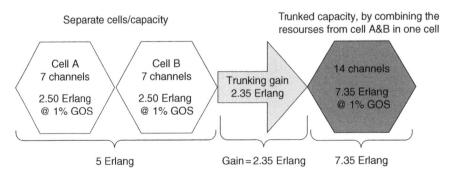

Figure 6.2 Trunking gain when combining the resources into the same cell

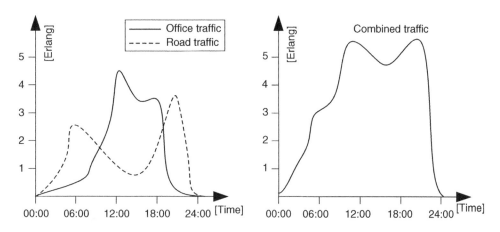

Figure 6.3   Trunking gain when combining different daily traffic profiles into the same cell

Traffic Load Profile of the Cell and Shared Resources

The traffic in different cells will often have different traffic profiles, the profile being the distribution of the load throughout the day. A shopping mall will have low traffic until opening hours and almost no traffic at night time. An airport might have a relatively high load constantly. This offset in load profile can be used to improve the business case of indoor solutions.

Example
If we take two traffic profiles, one from an outdoor cell, and the traffic from the cell covering inside an office building and combine the load from the two areas (see Figure 6.3), we can actually get 'free' capacity. This is the case for the solution shown in Section 4.6.3, where we are tapping into the capacity of the outdoor macro cell, distributing the same capacity resource inside the building. By combining the two load profiles, a high trunking gain is obtained due to the offset in the traffic profile between the two cells.

Sharing Load on a Weekly Basis

The offset in traffic profile should not only be evaluated on a daily basis, but also over a week (see Figure 6.4) or even longer periods. It might be that the capacity of a big office building next to a stadium could be shared if the game in the arena occurred outside shopping hours. Thus you can save a lot of roll-out cost if you can share the load from different areas with noncorrelated traffic profiles.

6.1.10   Cell Configuration in Indoor Projects

Trunking gain has a direct effect on the business case of the indoor system. The more traffic you allow to be serviced by the same cell, the better the trunking gain is, and the lower the production cost per call is. Why then do you need to divide large buildings into more sectors?

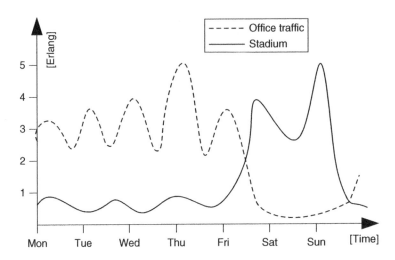

Figure 6.4 Load sharing by combining different weekly traffic profiles into one cell

3G/4G

When designing 3G indoor solutions it is evident that you cannot just deploy a new scrambling code (sector) in the same cell. This would be an overlap of the two cells in 100% of the area, and all the traffic would permanently be in soft handover, using resources from both cells, so actually nothing would be gained.

Dividing the building in to more sectors is the only way to add capacity in the building, but 3G needs physical isolation between the sectors in order to minimize the soft handover zones and decrease cannibalization of the capacity of the cells. Deploying 4G you need to evaluate the bandwidth you want to deploy in the DAS Cell. Like for 3G you must avoid extensive overlap of cells, especially in "hotspot" areas with high traffic.

2G

On 2G you can add new radio channels to the cell; the limit is the number of channels the operator has in the spectrum and restrictions with regards to frequency coordination with other adjacent cells. Also, hardware restrictions on the base station will demand a channel separation of typically three channels in the same cell. However, in theory you could deploy 16–24 channels or even more in the same cell. In practice it is preferable to keep the number limited to a maximum of about 12 for an indoor project. In a high-rise building you should use a maximum of four to six carriers in each of the topmost sectors, to avoid 'leaking' too many carriers to the nearby macro area. However, there are other reasons – for the passive part of the DAS it is a matter of keeping the composite power at a relative level when combining several high power carriers to minimize inter-modulation problems. For active DAS it can be a matter of a too low a power per carrier, due to the limited resources of the composite amplifier, where all carriers must share the same power resource, and this will limit the downlink service range from the antennas.

6.1.11 Busy Hour and Return on Investment Calculations

Busy Hour

You must bear in mind that the capacity calculation for estimating the needed traffic channels must be based on the busiest hour in the week, whereas the business case calculation is based on the total production of traffic, or call minutes in the building. The number of traffic channels needed will be dependent on the GOS you want to offer in the busy hour.

Return on Investment

Return on investment (ROI) is mostly the motivation for implementing any indoor solution. Usually an operator will know what the revenue is per call minute, but how do you calculate the expected traffic production for the system? It is actually quite easy: $1\,E$ is, as we know, defined as a constant call. Therefore, you can calculate that $1\,E$ is $24 \times 60 = 1440\,\text{min/day}$. This will mean that a system with a traffic load of $11\,E$ will produce $11 \times 1440 = 15840$ call min/day. Knowing the revenue per call minute, you can then calculate the revenue for the system, and do the business case evaluation.

Example, Shopping Mall

Sometimes it is a little more complex to calculate the traffic and the needed traffic channels. If you have a case with a multioperator solution, you will need to divide the traffic according to the market share.

Inputs

The input is the statistics from the shopping mall: an average of 18 000 shoppers per day each in the mall for an average time of 2.5 h. There are six shopping days; opening hours are Monday to Friday 9.00 to 19.00 and Saturday 10.00 to 14.00. The busiest day of the week is Saturday, when 25 000 shoppers are in the mall. The mobile penetration in the country where the shopping mall is located is 85%. The market share for the operator we are designing for is 55%. The operator allows 2% GOS in the shopping area.

Calculations

First of all we need to understand that there actually are two calculations of the traffic: one for the business case and one for the traffic design. The business case should be based on the total call minutes, but the traffic design should be done to accommodate the busy hour.

Busy Hour Calculation

From the data we can see that the busiest day is Saturday, with 25 000 shoppers in the mall, and we assume that in the busy hour there are 20 000 shoppers present. We can then calculate the traffic load in the busy hour:

$$BHL = users \times LPU \times MP \times MS$$

Where BHL is the busy hour load; users is the number of users; LPU is the load per user (assumed to be 20 mE); MP is the mobile penetration; and MS is the market share for the operator.

$$BHL = 20000 \, shoppers \times 0.02 \, E \times 0.85 \, mobile \, penetration \times 0:55 \, market \, share$$
$$BHL = 18:7 \, Erlang \, busy \, hour \, traffic$$

The GOS was 2%, so using the Erlang table we can see that we need 27 traffic channels in order to support this traffic.

The Business Case Calculation
In order to evaluate the business case, we need to sum up all the traffic for the week. On weekdays the mall is open for 10 h, and on Saturday for 4 h. This is an average of 1800 shoppers per hour on weekdays and 6250 on Saturdays. Using the average of 20 mE per shopper, we can calculate the traffic production per hour: we know that 1 E produces 60 call min/h; 20 mE will then produce $60 \times 0.02 = 1.2$ call min/h on average.

We need to incorporate the market share of 55% and the mobile penetration of 85%, so for the operator we have an average of $1800 \times 0.55 \times 0.85 = 841.5$ users/h on weekdays, and $6250 \times 0.55 \times 0.85 = 2921.8$ users/h on Saturdays. Using the 1.2 call min/h, we can calculate the production:

$$daily \, production = users \times call \, min \, per \, user \times opening \, hours$$
$$total \, all \, weekdays = 842.5 \times 1.2 \times 10 \times 5 = 50550 \, call \, min \, for \, all \, five \, days$$
$$Saturdays = 2921.8 \times 1.2 \times 4 = 14024.6 \, call \, min \, per \, Saturday$$

The total production per week is then $= 50550 + 14024.6 = 64574$ call min.
We could also have done the calculation in Erlang:

$$Erlang = users \times E \, per \, user \times opening \, hours$$

$$All \, five \, weekdays = 842.5 \times 0.02 \times 10 \times 5 = 842.5 \, E \, per \, week.$$

$$Saturdays = 2921.8 \times 0.02 \times 4 = 233.7 \, E \, per \, Saturday.$$

$$One \, week \, is \, a \, traffic \, production \, of \, 842.5 + 233.7 = 1076.2 \, E$$

We know that one E is 60 call min, so we can easily convert back to call minutes:

$$call \, min = E \times 60$$
$$call \, min = 1076.2 \times 60 = 64574 \, call \, min$$

Sector Strategy for Larger Indoor Solutions

Example 2: Multi Operator DAS in a Shopping Mall
Let's have a look at how the total capacity and trunking gain are related in a real example, looking at the example of a 2G design for a multiple cell system in a shopping mall.

It is a relatively big shopping mall with a good potential for traffic; we have several options – in this example we are designing for a 2G solution, and our budget is limited to using six TRX for total capacity. So how do we utilize these six TRX to ensure that we achieve the greatest traffic production for our investment? In this example we assume the load per user to be 25 mErlang (0.025 E).

Option 1, Six Sectors

One option is to divide the shopping mall into as many sectors as we have TRX; this is illustrated in Figure 6.5 implementing three sectors on each floor, a total of six sectors.

At first glance this might seem to be a reasonable configuration, but let us have a look at the capacity per cell and the total capacity of the system.

First, we can establish the capacity per cell using the Erlang table (Table 6.1); the capacity is 2.5 Erlang per cell (7 TCH per TRX@1% GOS).

Then we can add up the capacity for all six cells; total capacity 6 × 2.5 Erlang = 15 Erlang.

If we estimate one standard user in the shopping centre to be 25 mE per user we can easily calculate the number of users we can support in the shopping centre:

$$\text{Number of users} = \text{Total capacity}/(\text{load per user})$$
$$\text{Number of users} = 15/0.025 = 600 \,\text{users}$$

So with a configuration of six cells (Figure 6.5), each using 1 TRX we can support 600 users/shoppers in the centre.

Option 2, Two Sectors

A more logical option could be to divide the shopping mall into two sectors, as illustrated in Figure 6.6, implementing one sector per floor, each with a capacity of three TRX.

Let us have a look at the capacity per cell and the total capacity of this configuration.

First, we can establish the capacity per cell using the Erlang table; the capacity is 14.47 Erlang per cell (23 TCH per TRX@1% GOS, the first TRX has 7 TCH – the next two 8 TCH per TRX).

Then we can add up the capacity for the two cells; total capacity 2 × 14.47 Erlang = 28.84 Erlang in total capacity.

If we estimate one standard user in the shopping centre to be 25 mE per user we can easily calculate the number of users we can support in the shopping centre:

$$\text{Number of users} = \text{Total capacity}/(\text{load per user})$$
$$\text{Number of users} = 28:84 = 0:025 = 1153 \,\text{users}$$

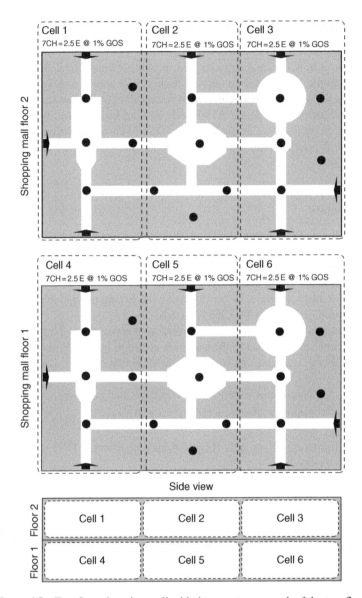

Figure 6.5   Two floor shopping mall with three sectors on each of the two floors

So with a configuration of two cells (Figure 6.6), each using three TRX we can support 1153 users/shoppers in the centre – this is almost at a factor of two compared with option 1(Figure 6.5) – so with the same base capacity resource and investment we can almost double the capacity! This is a very important parameter when considering the efficiency of the solution and the business case which has a great effect on the payback time for the solution and the production cost for the mobile operator.

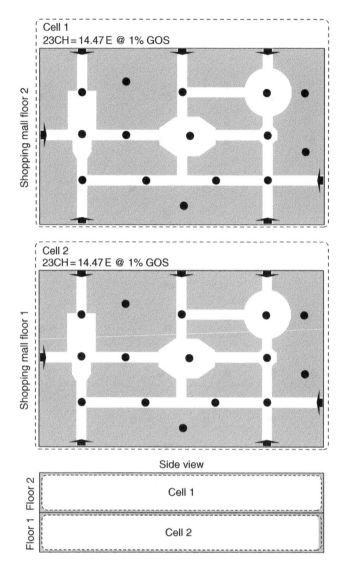

Cell 1
23CH = 14.47 E @ 1% GOS

Shopping mall floor 2

Cell 2
23CH = 14.47 E @ 1% GOS

Shopping mall floor 1

Side view

Cell 1

Floor 2

Cell 2

Floor 1

Figure 6.6 Two floor shopping mall with two sectors, one on each of the two floors

Option 3, One Large Sector

Let us have a look at the third option; to have all six TRX utilized as one shared resource, in one cell.

If we configure the cell to have 7 TCH for the first TRX and 8 TCH for the next five, we have a total of 47 TCH in the cell. Using the Erlang table we can see that the capacity is 35.21 Erlang for the cell with six TRX.

As in options 1 and 2 we may estimate one standard user in the shopping centre to be 25 mE per user and we can easily calculate the number of users we can support in the shopping centre:

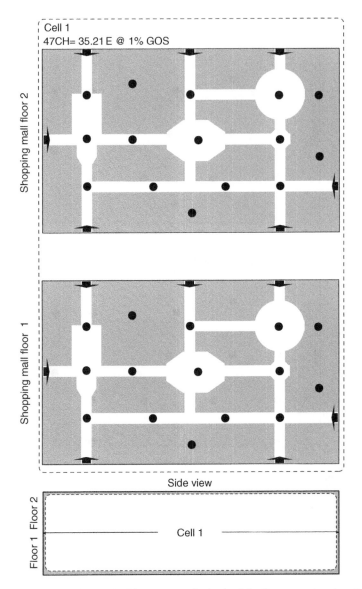

Figure 6.7 Two floor shopping mall with one sector for both of the floors, serving the whole mall with the capacity of one cell

Table 6.4 Showing the capacity of the three different configurations of the shopping mall

TRX	# of Cells	Total Erlang (1% GOS)	Total number of users	Users per TRX
6	6	15	600	100
6	2	28.84	1153	192
6	1	35.21	1408	234

$$\text{Number of users} = \text{Total capacity}/(\text{load per user})$$
$$\text{Number of users} = 35.21/0.025 = 1408 \text{ users}$$

So with a configuration of one cell (Figure 6.7), using six TRX we can support 1408 users/shoppers in the centre. This is almost at a factor of three compared with option 1 – so with the same base resource we can almost double the capacity! Again, this is a very important parameter when considering the cost efficiency of the solution and the business case.

Boost the Business Case

The examples that we have just presented; using six TRX in a two floor shopping mall with different configurations of capacity/cells ilustrates the point of trunking gain.

We summarize the results from the three options in Table 6.4:

In Table 6.4 it is evident that there is a large capacity gain when one combines the same six TRX into a single shared resource, rather than dividing the TRX into more cells.

This is a very important point to keep in mind when deciding on sector layout and configurations inside buildings.

Attention to the Handover Zones Between Multiple Cells

There are other parameters that also affect the decision; we must be careful with the sector lay-out on 3G and make sure that we do not have any large Soft Handover zones in high capacity areas in the shopping mall. This is a considerable challenge inside buildings with large open areas like shopping centres and airports that have a high concentration of traffic.

Make sure to utilize the internal walls to divide the sectors, and most of all make sure not to place extensive HO zones in areas with a high data load, in the food court, for example.

6.1.12   Base Station Hotels

Load sharing can be used on a large scale, and be a type of 'Base Station Hotel', where a central resource of base stations distributes coverage and capacity to many buildings in the same area (see Figure 6.8), typically a campus environment. This has several advantages but the main points are:

- very high trunking gain;
- low cost rollout – only needs one base station and site support;
- only needs one equipment room;
- easy upgrades in capacity.

Base Station Hotels are designed by deploying optical transmission between remote located active DAS solutions. One central location will house the base station and main units for the active DAS and from this central location coverage and capacity are distributed to nearby buildings. One must be careful with the total delay introduced by the distribution

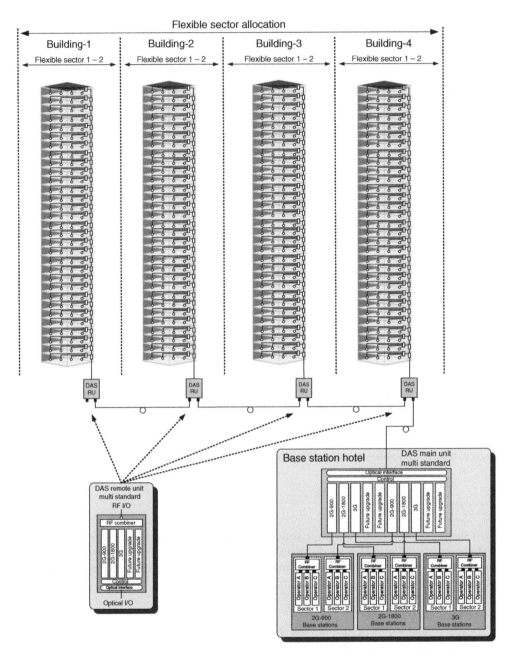

Figure 6.8 A base station concept used to provide multiple RF services in four buildings from one central location

system, and not exceed the maximum timing advance in 2G, and be watchful for other timing issues.

The Base Station Hotel and Remote Units

In this example, the central Base Station Hotel (Figure 6.8) supports four buildings. The Base Station Hotel could be located in a central equipment room in one of the buildings – typically an IT room, taking advantage of easy access for service and upgrades, battery back up, transmission, air-conditioning, etc. This could resolve a major issue to do with lack of space in remote locations when deploying indoor DAS solutions – especially if there is a requirement to support multiple mobile standards and sometime multiple operators. This topology is possible due to the fiber distribution of the DAS, the DAS will be able to multiplex the optical transport and extend the reach to the Remotes for more than 15–20 km between the central Base Station Hotel and the remote units.

One will not need space in each building to accommodate all of the base station and site support equipment. Trunking gain, OPEX and CAPEX savings all have a major impact on the business case of the implemented solution; one simply has the structure in place with the Base Station Hotel and remote units. Come future services and standards it will be relatively easy to upgrade, as compared to the traditional solution of upgrading the base stations in each and every building.

Different Sector Plans for Different Services

The possibility of implementing a flexible cell cluster and adaptive sector plans, using different simulcast schemes greatly improves the business case and lowers production costs for the mobile operator. Depending on the type of DAS that one is deploying the system could even be able to provide a single sector plan on 2G, a separate sector plan for 3G and a third for 4G. Some types of active DAS will even give one the opportunity to implement different sector plans for different mobile operators using the same frequency band and sharing the DAS infrastructure. This makes the Base Station Hotel a very attractive, versatile and future proof concept, and this flexibility can have a very positive effect on the business case and be a strong tool for handling capacity.

You will find further detail and examples of the advantages of utilizing Base Station Hotels in Section 12.3.1. Some DAS solutions will even "power down" certain Remote Units when they are not needed (low traffic) to save on power load.

6.2 Data Capacity

In many countries the user volume of mobile data downloads has surpassed the volume of data load on the fixed line networks. This drive to enable more and more data capacity in the mobile networks demands high data speed capabilities throughout the network from end to end. The networks supporting this ever-rising demand for data obviously have to be designed with this challenge in mind, and be constantly upgraded to keep up with this never-ending challenge. This also applies to the wireless part of our DAS designs. I know from experience that, although you might try to prepare your system for the expected rise in data load, you will often be surprised at how quickly the demand for more data capacity catches up with your design. What might appear to be a good margin to cope with future capacity demands at the time of the design will easily be cannibalized by the change in demand, and much sooner than anticipated.

6.2.1   Application-driven Data Load

There is no doubt that the many smart mobile devices with easy access to applications relying on high-speed mobile data bandwidths have been the game changer in the wireless industry and in the network design strategy and challenges. As you can see in Figure 6.9, the timing of the abrupt increase in mobile data consumption coincides with the advent of smart mobile devices. Easy to use and with easy access to social media, music and TV streaming services, as well as video and image uploads and downloads, these have been the main driver behind the accelerated growth of and constant increase in mobile data requirements and expectations.

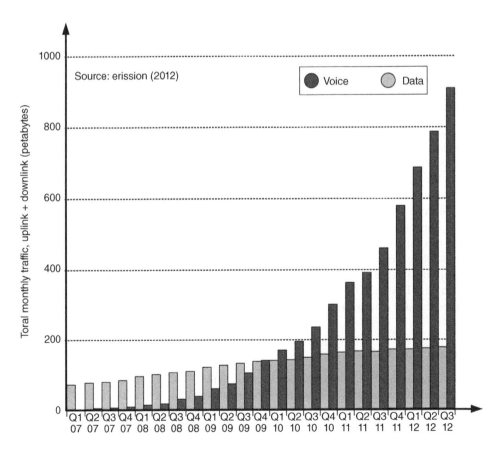

Figure 6.9   The accelerated increase in mobile data consumption is mainly driven by smart devices and applications. Source: Ericsson (2012).

The consumption of these streaming services can be very unpredictable and a big challenge for us as radio and capacity planners designing the DAS solutions to support these high-speed data services. The 'hockey stick effect' (exponential growth over time) is very obvious in Figure 6.9 and it appears that the need for ever-increasing mobile data speeds and capacity will present a never-ending challenge.

Data Services Are Often Unlike the Voice Service

Unlike the voice service, which needs 1:1 real-time support ensuring communication between the two parties on the call, data traffic is normally not that time-sensitive: short delays of a text message, SMS, instant messaging (IM), or email or slight degradation of data speed will normally not have a big impact upon the user's perception of the quality of service. Even music and video streaming can be maintained with buffer functionality when there are short interruptions and temporary delays of what appears to be an 'online' live streaming service. It is very important that you take this into account in your designs, and you need to keep this in mind when you are designing capacity for data services. There are greater dynamics in play with regard to user behavior and requirements in terms of acceptable 'blocking' levels, depending on the actual service.

One will typically not use Erlang to estimate the capacity requirements of data services; however, Erlang could be used as an indicator – a guideline as to when to expect that users will have a need for data sessions in some cases – although one must realize that the data traffic will take up capacity that could have been used for voice services. Sometimes you will assign 2G and/or 3G in the cell for voice only, and then leave the 4G service for pure high-speed data support.

Normally, mobile operators will design data capacity per user in kbps or Mbps, and then sum up the real-time requirements separate to the total estimated volume, as the real-time services will be the most critical. You need to understand the load profile for the type of users and services you are supporting in your design.

Advanced traffic control parameter settings in the network allow us to adjust various margins so as to control the amount of data traffic, and the remaining capacity for voice service. However, most data-driven mobile networks try to use 4G for data only, and then use 3G for voice and slow-speed data. 2G will often be a voice-only service, and in many cases the former 2G spectrum is re-farmed to 3G or 4G services to keep up with the high demands of data, and maximize the utilization of the sparse RF frequency spectrum the mobile operators have acquired with their license to operate the network. Design targets for indoor data services are normally expressed as maximum peak data speeds and the maximum numbers of users for a particular data speed and load profile, and guarantees for quality of service (QOS) are normally pretty relaxed for non-real-time service. The increasing need for high-quality data services imposes special demands on delay and round-trip delay of the total end-to-end network. The specific design requirement is directly related to the specific service; therefore it is hard to provide any detailed and concrete guidelines here. Tables 6.5 and 6.6 show some typical values. Be sure to confirm what applies in your actual design.

One of the main challenges we face is the unpredictability of the many new data services that are enabled when the data bandwidth is available for the users. Although you might try very hard to calculate the requirements (and you should), it is often very difficult to predict the actual load and type of services that new applications and services will require once the traffic

Table 6.5 Typical mobile data load sessions

Activity	Estimated average data load per activity
One hour of text-based chat service, IM, SMS etc.	10–20 kB
Single text-based email, via web-based mail service (no attachments)	200 kB
Loading a photo (1024 × 768 jpeg)	1 MB
Music download of 4-minute song	5–8 MB, depending on encoding grade
Video service (e.g. YouTube etc.)	Average of 1 MB/min (non-HD format)
Skype video call for 10 minutes	2–11 MB
Streaming movie (480 × 640)	1.5–3 GB
Streaming music online	60 MB @ 128kbps service

Table 6.6 Typical data rates for mobile services

Application	Real-time?	Typical bandwidth guidelines	Latency/jitter tolerance
VoIP	Yes	4–64 kbps	<150 ms/< 20 ms
Video calling	Yes	32–384 kbps	<150 ms/< 20 ms
Online gaming	Yes	40–100 kbps	<150 ms/NA
Music streaming/voice	Yes (buffered)	3–128 kbps	NA/< 100ms
Video streaming	Yes (buffered)	20–384 kbps	NA/< 100ms
Movie (non-HD)	Yes (buffered)	384 kbps – 2 Mbps	NA/< 100ms
Instant messaging	No	10 kbps	NA
Email	No	>500 kbps	NA
Web browsing	No	>500 kbps	NA
Media service download	No	>1 Mbps	NA

is active. Recently I had the opportunity to learn about some of the ideas and applications that might be implemented in the near future: during a live game in a sports arena, for example, you will be able to video stream live from other matches, follow your favorite players live on your phone's screen from various angles via the tracking cameras deployed in the same arena where you are watching the game live, enable instant replay, and pull up live video from a small camera mounted on one of the players, or the referee, and so on. I am not much of a sports fan, but I do see the benefits these types of service will bring, and I can also appreciate the challenges that we as DAS designers will face in supporting these applications and data requirements in our designs.

Backhaul Issues

On 3G and 4G, the data service required in the indoor cell will also demand the needed bandwidth on the backhaul transmission line from the base station to the network, to support the services over the air interface. This is a major issue for mobile operators; it does not make sense if the indoor system can support in excess of 100 Mbps over the air interface while the backhaul transmission is limited to 2 Mbps. The users will only experience the 2 Mbps no matter how well designed the DAS is. One can conclude that wireless data services will need a lot of wires to work at the required data speeds throughout the network. Often you will need fiber transport to the base station in order to support the high-speed wireless data rate needed from the base station to the core network.

6.2.2 Data offload to Wi-Fi and Small Cells

Sometimes the wireless data load and data requirements in a venue for which you are designing a DAS solution can be extremely high, and you might even struggle to deploy enough resources and cells in your mobile network to support these requirements, even using both 3G and 4G in the venue. One of the ways to support these high-speed data capacity requirements for high data load on your mobile network cells is to provide mobile data offloading to Wi-Fi systems (see Section 13.2 for details) and other hotspot small cells, such as those shown in Figure 6.10. This approach will normally be used in capacity hotspots (see Section 5.4.1 for more details) such as shopping malls, airports, sports venues (see design details for these applications in Chapter 14).

Sometimes Wi-Fi will be more strategically used to provide indoor high-speed services by the mobile operators, and the mobile network, typically 3G, will be used to provide coverage and capacity mainly for voice. One of the main challenges with Wi-Fi is the use of the open, non-regulated spectrum that is out of your control. So providing a specific quality of service metrics can be a challenge, as you have no control to restrict others deploying transmitting Wi-Fi access points on the same frequency, thus degrading your services. On the other hand, Wi-Fi can support high data speeds close to the access points, which are in- fact small cells for Wi-Fi (currently not supporting mobility, handovers). The limited coverage range of the Wi-Fi access points could be used inside the buildings to support location-based services, as the location will be limited to the specific area of the individual SSID (service set identifier) that is unique to each Wi-Fi cell (AP)

6.2.3 Future-proof Your DAS to Handle More Data Load

Most of the DAS systems you design and implement will benefit from a clear strategy on how to prepare your solutions to handle future upgrades, adding new frequency bands (typically higher frequency bands) and new generations of mobile services. One example is to ensure the DAS can evolve from 2G and 3G to 4G to overcome and keep up with data requirements. Strategies for adding more sectors to your solution in order to upgrade and boost the capacity in various DAS solutions can be seen in Chapter 14, with several applications, such as high rises, shopping malls, sports arenas, airports and other applications, analyzed in more detail. One simple way to prepare for more capacity is to define a clear design strategy that allows you to design and implement the DAS in 'sections' that can be combined into different sector (cell) layouts and changed in the future to accommodate both the current and expected growth in capacity needs. One typical example would be a high-rise office building like the one in Figure 6.11, showing the same building divided into 16 sections (areas); each of these DAS sections provides coverage and capacity for two floors, to service the 32 floors in total. In the figure we can see the principle of preparing for more capacity, with three different configurations of the same DAS in the same building.

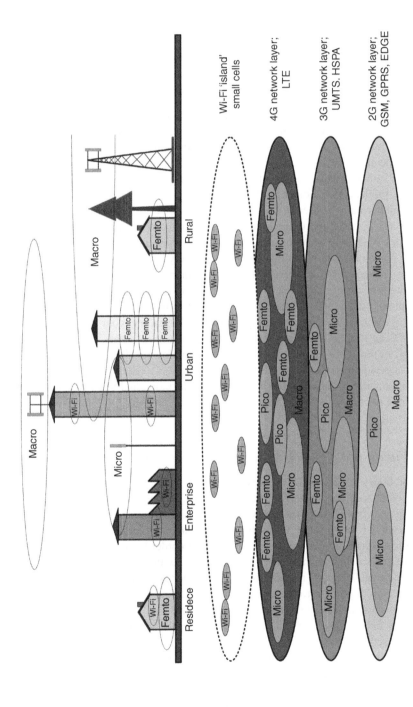

Figure 6.10 Wi-Fi and small cell offload is very useful to provide increased network capacity and is sometimes vital in order to cope with high data loads

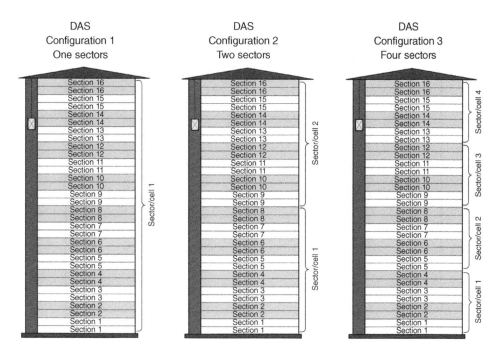

Figure 6.11 Large office building designed in coverage areas that can be combined into various combinations of sectors, is a good strategy so you can divide the DAS into more sectors in the future

Initially all of these 'sections' are combined into one large sector as shown ('simulcasting' the same sector from all sections/antennas, as shown in Figure 6.11, left). Then over time we split the sections into two sectors (cells) to double the capacity (Figure 6.11, center), and later divide them into four sectors (Figure 6.11 right) to quadruple the initial capacity. This strategy, if designed and implemented correctly from day one, will give you the flexibility in your solutions, in order to cope with future needs with a minimum of impact and cost to the implemented DAS solutions in the buildings. So your office building in Figure 6.11 could support any configuration from one to 16 sectors (cells). You could even prepare the building to have individual sector (cell) configurations for 2G, 3G and 4G.

Sports venues such as sport arenas/stadiums are among the biggest capacity challenges you are likely to face. One example of preparing a stadium design for extreme capacity in the future is the plan of a sports arena shown in Figure 6.12, where the DAS is divided into 36 sections. These 36 sections can be configured as 36 sectors (cells), or, alternatively, if there is less need for capacity at the time of deployment you can start with 12 sectors (three DAS sections per sector; see Figure 6.12) or another configuration that makes sense for your actual design and future plans. Some radio planners use a rule of thumb – based on initial calculations and experience – of deploying one sector/cell per 3000 seats in a sports arena, but it will depend on the services applied in the network, the capacity in the cell, the load profile of the users. Do you use 2G, 3G and 4G or only 3G and 4G, etc.? You may also have to prepare your DAS to cover the field, in case there are going to be events like concerts with many mobile users in that area. Even Wi-Fi offload could be needed as indicated in Figure 6.12, and you

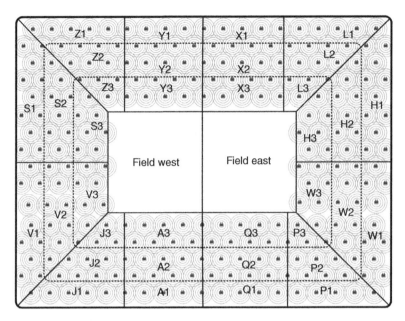

Figure 6.12 The different areas in this capacity layout are now divided into sectors (cells) by joining (simulcasting) the cell over multiple DAS areas. We are using Wi-Fi offload on a separate layer, and are observant of the handover zones, especially in the field

have to use a heterogeneous network strategy (Figure 6.10) such as that described in Section 6.2.2. Figures 6.11 and 6.12 illustrate the principle of designing a flexible DAS solution – for more detailed examples, please refer to Chapter 14 where various high capacity applications are described in more details.

Where Are the Users?

One pitfall you need to watch out for is the geographical distribution of the users in the building structure, and thus in your deployed DAS. Let us take the office building in Figure 6.11 as an example, and assume you have combined all sections into one large cell. You experience, resource blocking of the cell due to capacity load and therefore need to prepare an upgrade of the system. First you will upgrade the base station feeding the cell to a maximum, and then, if this is still not enough, you might decide to divide the building into more vertical separated sectors, perhaps two sectors (cells). However, this strategy will only work if the users are evenly distributed over the floors, and this might well not be the case. For example, if you have one floor in the building (perhaps a stockbroker) with extreme load, responsible for 90% of the usage in the building located on floor 6, they would still experience blocking in sector 1 (cell 1) after upgrading to two sectors. And even adding more sectors (as shown in Figure 6.11), up to 16, might still not be enough to support the users on floor 6. The only solution would be to 'split' that specific floor into more sectors horizontally in order to cope with the demand. The challange will be to maintain good isolation among these cells on the same floor though.

6.2.4   Event-driven Data Load

It is relatively 'easy', although a challenge, to design capacity to support the voice and data requirements of typical buildings and structures. It is even more of a challenge to take into account the non-typical load in your system. This could be holiday shopping in shopping malls, large exhibitions in conference centers and venues – I bet most of us who participate each year in the Mobile World Congress are still amazed that the mobile operators are able to support the capacity demands of 80,000 high-demanding mobile industry users with a high-quality user experience and only limited access problems.

Sometimes the biggest challenges are the less predictable data loads; this can be launch of a new service or 'unusual users'. A real-life example of an 'unusual user' was encountered by one of my friends after going live with a new 4G 2 × 2 MIMO DAS in a large office building. Within a few days of concluding that everything was okay, and that data throughput was verified to support 80 Mbps+ in most of the areas, the data capacity became heavily loaded over a period of a few hours almost every day. After some detailed troubleshooting, the mobile operator was able to narrow the extreme load down to a single user. It turned out that this user was working for the local IT support team. This person was responsible for configuring new PCs to new users in this large company, and he had noticed that it was more convenient to use a 4G data card and download the total software package via the indoor DAS, rather than using the wired ethernet, and that the 4G speed was fast enough to do so! This just proves the saying, 'if it's there, we will use it', so once you provide users with the option of better data services, they will very well use it, even before you expected them to do so.

Therefore you must have a strategy in the expected 'high load' buildings and areas, such as that described in Section 6.2.3 and in more detail in Chapter 14, where you can add more sectors/cells. In addition, make sure to prepare for future generations of mobile technologies, such as MIMO. Also please refer to the application examples in Chapter 14.

6.2.5   Calculating the Data Load

We covered the basics of traffic calculations of voice services earlier in this chapter, using Erlang. Now let us have a look at some basic considerations of data load, and designing for data capacity.

First of all when designing for data-driven services, you need to understand the load and data profile of the users and of the various types of data-service you are supporting in your DAS design. You need to know the location of the hotspots in the area (see Section 5.4.1) and have a clear strategy on the deployment of 2G, 3G and/or 4G services (see Figure 6.13). You also need to understand based on the data-traffic type and load profile in the cell, how you want to utilize the different services in the cell for the different services in order to support the traffic needs, as shown in Figure 6.13. Often you will be deploying a mixed technology – for new designs this means mostly 3G in order to support the voice service and also to provide a relatively good footprint for slower-speed data services, whereas the 4G deployed in the same DAS is intended to support the high-speed pure data services. Therefore you need to ensure you are targeting the data-heavy areas when deploying your DAS antennas in order to achieve high data services for the highest concentrated areas, the 'hotspots'.

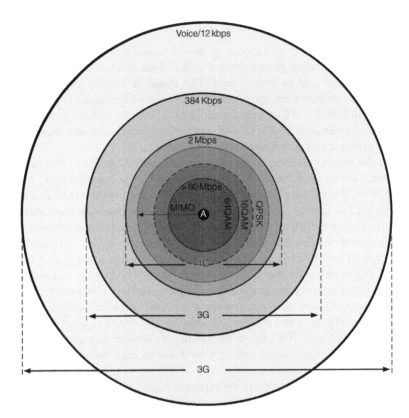

Figure 6.13 When designing a multi-technology system, 3G–4G, you need to appreciate the different data speeds and the different footprints of the various services. You need to have a clear strategy regarding what service you want 'push' to a specific layer of technology

The Load Profile

Understanding the behaviour of the data users you intend to support in the DAS is absolutely key to the capacity design phase. You need to estimate the total data requirements of all the data services you need to maintain in real time, such as voice over IP and live video conferences (see Table 6.6) – it is critical to provision for these services, because of the real-time 1:1 requirement. You may want to aim for a maximum expected load of 80% in the cell, in order to provision for sufficient extra overhead for additional traffic, as shown in Figure 6.14.

Other data services are not critical to maintain in 1:1 real time, such as IM, email, and web browsing, where slight delays of a few seconds are not that critical. If the DAS you are designing is mainly supporting these types of services for data, then you can load up the cell to a much greater degree, closer to 90–95% (peak), rather than the 80% shown in Figure 6.14. One example would be a mixed technology cell of 3G and 4G (as in Figure 6.13), with the critical 1:1 real-time services such as voice and video calls supported by 3G, using Erlang as the design merit, and the non-real-time data services carried by the 4G part of the cell.

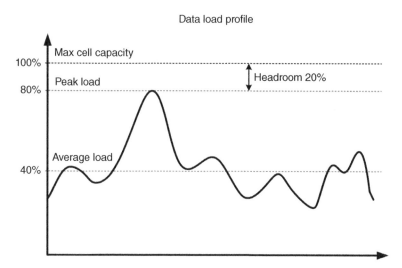

Figure 6.14 Designing for data capacity is very dependent on both the average and expected peak load, and on whether you need to support 1:1 real-time applications

Complex algorithms are available of moving the traffic between the different 'layers' (2G, 3G and 4G) and you need to understand how to utilize this, as well as the limitations when performing the data design.

The Real-time Services Will Dictate the Requirements

Using the typical profiles of data load in Tables 6.5 and 6.6, you can estimate the number of sessions of various services a specific cell can support. Your design requirements could be dictated by the real-time services, with voice calls obviously the main concern. I highly recommend getting detailed, local, actual information about the data load and user profiles from your own network and users. Table 6.5 is merely a typical example for the purpose of calculating the load of various sessions. Table 6.6 gives the typical services requirements to support the various sessions. Please use your own local up to date inputs, for most accurate calculations.

Erlang and Data Services?

Erlang is, as we know, used to estimate the voice load distribution, and although this is primarily for voice, one could user Erlang distribution to estimate the load distribution of typical data services – i.e. the user load pattern of the requests for data services. However, this is only for typically distributed traffic load, and does not consider the 'event-driven load profiles', such as sports events, half-time, when a goal is cored and so on.

7

Noise

Noise performance and noise calculations are the most important aspects in radio planning. The quality of any radio link is not defined by the absolute signal level, but by the signal-to-noise ratio. The lower the noise is relative to the signal level, the better the radio performance. Radio quality and mobile data rates are dependent on a certain signal-to-noise ratio. The better the SNR, the better the quality, the higher the data rate on the radio link will be.

Many RF planners consider the issue regarding noise in DAS systems, on the UL of their link budget to be 'black magic'. However, if you do not overcomplicate the subject of noise, the concept is really pretty straight forward and easy to understand. Understanding noise and noise calculations is crucial for the radio planner in designing any RF system. Radio planners need to understand that passive loss in a DAS system will also increase the total system noise figure. Noise power generated by amplifiers and the base station also has a big impact on the performance of the radio link.

Over the following pages, I will present the basics of noise, by presenting examples of amplifier configurations and analyzing the performance step by step. All the examples on the next pages have been calculated on a standard scientific pocket calculator (the one on the cover on this book). You do not need advanced computer programs to be in control of the noise calculations when designing passive or active DAS solutions. However, the creation of spreadsheets in Excel for doing the trivial part of the work with these calculations is recommended. It saves a lot of time, and you can limit the mistakes.

7.1 Noise Fundamentals

Any object capable of allowing the flow of any electric current will generate noise. The noise is generated by random vibrations of electrons in the material. The vibrations, and hence the noise power, are proportional to the physical temperature in the material. This noise is referred to as 'thermal noise'.

Indoor Radio Planning: A Practical Guide for 2G, 3G and 4G, Third Edition. Morten Tolstrup.
© 2015 John Wiley & Sons, Ltd. Published 2015 by John Wiley & Sons, Ltd.

7.1.1 Thermal Noise

Thermal noise is 'white noise'; white noise has its power distributed equally throughout the total RF spectrum, from the lowest frequency all the way to the highest microwave frequency.

Spectral Noise Density

In other words, the power spectral density of white noise is constant over the RF frequency spectrum; hence the noise power is proportional to the bandwidth. If the bandwidth of the RF channel is doubled, the thermal noise power will also double (+3 dB).

Thermal Noise Level

In order to be able to calculate the noise power of a given bandwidth, you need to establish the 'base noise', that is, the noise in 1 Hz bandwidth. Knowing this level, you can simply multiply the noise power in 1Hz by the bandwidth in Hz. Thermal noise power is defined as

$$P = 10 \log(KTB)$$

Where P is the noise power at the output of the thermal noise source (dB W), K is Boltzmann's constant $= 1.380 \times 10^{-23}$ (J/K), T is temperature (K) and B is the bandwidth (Hz).

Reference Noise Level per 1 Hz

At room temperature (17 °C/290 K), in a 1 Hz bandwidth we can calculate the power:

$$p = 10 \log(1.380 \times 10^{-23} \times 290 \times 1 = -204 (\text{dB W})$$

where dBm takes its reference as 1 mW. The relation between 1 W and 1 mW is 1/0.001 = 1000 = 30 dBm, i.e. 1 W = 30 dBm. Therefore you can calculate that the thermal noise power at (17 °C/290 K) in a 1 Hz bandwidth in dBm:

$$\text{thermal noise power} = -204 + 30 = -174 \, \text{dBm/Hz}$$

where -174 dBm/Hz in noise power is the reference for any noise power calculation when designing radio systems. Relative to the bandwidth, you can use the reference level of the -174 dBm/Hz and simply multiply it by the actual bandwidth of the radio channel. It does not matter if it is a radio service that operates on 450 MHz or at 2150 MHz, the noise power/Hz will be the same if the radio channel bandwidth is the same.

Reference Noise Level for 2G GSM/DCS

The noise power density is frequency-independent: the noise power remains the same no matter what the radio frequency. The noise power is equally distributed throughout the spectrum and is related to the bandwidth of the RF channel.

2G

Using -174 dBm/Hz, we can calculate the thermal noise floor for the 200 kHz channel as used for 2G. We just calculate the thermal noise in that bandwidth:

$$KTB \text{ for } 2G(200KHz) = -174 \text{ dBm/Hz} + 10\log(200.000\text{ Hz}) = -121 \text{ dBm}$$

The -121 dBm is therefore the absolute lowest noise power we will have in a 200 kHz 2G channel.

3G

The 3G radio channel is 3.84 MHz. This makes the thermal noise floor for 3G:

$$KTB \text{ for } 3G(3.84\text{ MHz}) = -174 \text{ dBm/Hz} + 10\log(3.840.000\text{ Hz}) = -108 \text{ dBm}$$

7.1.2 Noise Factor

The noise factor (F) is defined as the input signal-to-noise ratio (S/N or SNR) divided by the output signal-to-noise ratio. In other words the noise factor is the amount of noise introduced by the amplifier itself, on top of the input noise. For a passive system such as a cable, the SNR would also be degraded, by attenuating the signal nearer to the thermal noise floor would equally degrade the SNR.

The noise factor is a linear value:

$$\text{noise factor}(F) = \frac{\text{SNR}_{(input)}}{\text{SNR}_{(output)}}$$

7.1.3 Noise Figure

The noise figure (NF) is noise factor described in dB, and is the most important figure to note on the uplink of any DAS or amplifier system. The NF will affect the DAS sensitivity on the uplink, and will determine the performance of the uplink. This will be the limiting factor for the highest possible data rate the radio link can carry, no matter whether it is 2G, 3G and 4G or any other service. The lower the NF, the better the performance.

$$\text{noise figure (NF)} = 10\log(F)$$

7.1.4 Noise Floor

The noise floor is the noise power at a given noise figure at a given bandwidth; you can calculate the noise power or noise level for a receiver, amplifier or any other active component.

$$\text{noise power (noise floor)} = KTB + NF + \text{gain of the device}$$

A typical case for a 2G BTS is shown in Figure 7.1.

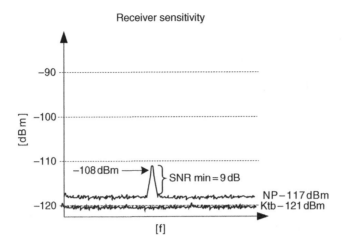

Figure 7.1   2G receiver with 4 dB NF

7.1.5   The Receiver Sensitivity

Knowing the NF and bandwidth of the receiver in the BTS, you can calculate the receiver sensitivity. However, first you need to calculate the noise floor of the receiver.

Noise Floor

As an example, we can take a 2G receiver as shown in Figure 7.1 with an NF of 4 dB, then the noise floor for that receiver can be calculated as:

$$\text{receiver noise floor} = KTB + \text{NF}$$
$$\text{receiver noise floor} = -121\,\text{dBm} + 4\,\text{dB} = -117\,\text{dBm}$$

This is the base noise power, the absolute lowest signal level present at the receiver input, a noise power of −117 dBm, a noise level generated by the receiver itself.

Sensitivity

With reference to the noise floor, you can calculate the minimum signal level for any given service that the receiver is able to detect; you just need to reference the required SNR for the service, and make sure the service is the required level above the noise floor of the receiver.

If, for example, you want to provide a 2G voice service, you need a 9 dB SNR. You can calculate the minimum required level the receiver will be able to detect to produce the quality needed for voice (without adding any fading margins):

$$\text{receiver sensitivity} = \text{receiver noise floor} + \text{service SNR requirement}$$
$$\text{receiver sensitivity} = -117 + 9 = -108\,\text{dBm}$$

Thus the lowest detectable signal for a voice call will be −108 dBm. Therefore the receiver sensitivity for 2G voice is −108 dBm in this example, excluding any fading margins.

7.1.6 Noise Figure of Amplifiers

As in the example with the receiver above, you can calculate the noise increase in a single amplifier, knowing the bandwidth of the signal, the NF and the gain of the amplifier.

Example: 30 dB Amplifier with 10 dB NF

Using an amplifier for 2G with a 200 kHz bandwidth with 30 dB gain and an NF of 10 dB (as shown in Figure 7.2), you can calculate what happens if you feed the amplifier a signal of −90 dBm (as shown in Figure 7.3). Some would consider this a relatively low signal, but the quality, the SNR of the input signal, is actually good because in this example there is no interference from any source present at the input. The only noise power present on the input of the amplifier is the *KTB* (−121 dBm/200 kHz). Therefore we can calculate that the signal has an SNR of 31 dB [−90 dBm − (−121 dBm)] – this is in fact a really good quality signal.

Nevertheless, you use the amplifier to boost the signal level, and you can now calculate the consequences for the signal, the noise and the SNR at the output of the amplifier.

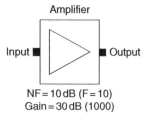

Figure 7.2   2G 30 dB/10 dB NF amplifier

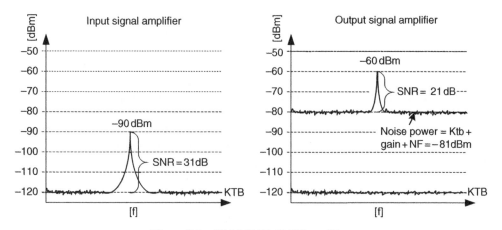

Figure 7.3   2G 30 dB/10 dB NF amplifier

New Noise Floor

The input signal from Figure 7.3 is fed into the amplifier shown in Figure 7.2 with the specified 30 dB gain and 10 dB NF. Naturally the amplifier will amplify the signal from −90 to −60 dBm, but it will also amplify the noise at the input (*KTB* in this example) from −121 to −91 dBm. In addition to the 30 dB amplification of the noise, the amplifier will add an additional 10 dB to the noise power due to the 10 dB NF (as shown in Figure 7.3), so the output noise level from the amplifier is:

$$\text{NFloorOut} = \text{NFloorIn} + \text{Gain} + \text{NF}$$
$$\text{NFloorOut} = -121\,\text{dBm} + 30\,\text{dB} + 10\,\text{dB} = -81\,\text{dBm}$$

The Noise Figure Degrades the SNR

The result is amplification of the wanted signal by 30 dB, from −90 dBm at the input to −60 dBm at the output – so we have a higher signal level. However, what happened to the quality of the signal, the SNR?

We actually degraded the SNR of the signal from 31 to 21 dB. In fact, we degraded the performance and data rate for the base station on the uplink due to the 10 dB NF degradation of the SNR. We can conclude that an amplifier will always degrade the SNR, if you consider the amplifier as a single block.

Why Do We Use Amplifiers?

We just calculated that the amplifier will degrade the SNR, due to the NF of the amplifier. Then why do we use amplifiers at all, if they degrade the SNR of the signal; after all, the performance of the signal is not the level but the SNR? To understand this we need to analyze the impact of loss of passive cable in the system (Sections 7.2.2 and 7.2.3) and the impact of loss in cascaded amplifier systems (Section 7.2.4).

7.1.7 Noise Factor of Coax Cables

To understand how and when you can use amplifiers, to maximize the performance of the DAS system, you need to understand the impact of the attenuation of a passive coax cable or component on the total system noise figure performance. You need to realize that any loss of a passive cable or any other passive component prior to the base station will also impact and degrade the SNR, like the path loss on the radio link or the NF of an amplifier will.

This may sound a bit odd; after all the passive cable or component will not produce any noise power, but it will degrade the SNR performance of the system just the same. The passive losses will diminish the signal, the 'S' in the SNR, whereas active elements will raise the 'N' in the SNR. Both will result in degraded SNR.

We know from the previous section that the noise factor (*F*) is defined as the input SNR divided by the output SNR:

$$\text{noise factor } (F) = \frac{\text{SNR}_{(\text{input})}}{\text{SNR}_{(\text{output})}}$$

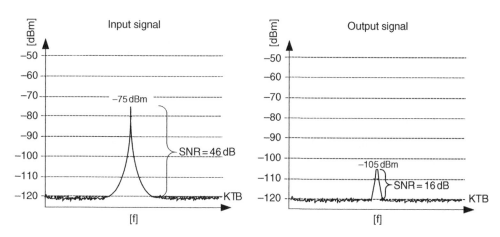

Figure 7.4   A coax cable with 30 dB loss

Example of NF of a Passive Cable with 30 dB Loss

Let us analyze the noise performance of a passive coax cable with 30 dB loss, a typical attenuation for a passive DAS. In the example we feed the input of the cable a 2G signal level of −75 dBm, as shown in Figure 7.4. We can calculate the SNR at the input of the cable to be:

$$SNR = -75\,dBm - (-121\,dBm) = 46\,dB\,(factor = 39811)$$

At the output of the cable the signal is now attenuated with the loss of the cable by 30 dB, and the signal level is now:

$$output\ signal\ level = signal\ input - attenuation$$
$$output\ signal\ level = -75\,dbM - 30\,dB = -105\,dBm$$

The SNR at the output of the cable is:

$$noise\ is\ still\ the\ KTB,\ so\ the\ SNR = -105\,dBm - (-121\,dB) = 16\,dB\,(factor = 39.8)$$

A value of 16 dB will be acceptable for most services, but might impact the performance on HSPA.

Now that we have the SNR for the input signal and the output signal we can calculate the noise factor:

$$noise\ factor\ (F) = \frac{39811}{39.8} = 1000$$

In dB:

$$noise\ figure = 10\log(F)$$
$$noise\ figure = 10\log(1000) = 30\,dB$$

The noise figure of the cable is 30 dB (noise factor 1000). We can then conclude that the noise figure of a passive component is the same as the loss or attenuation of the component.

The SNR of the signal is now degraded by 30 dB due to the loss of the cable. Like any NF degradation this will severely impact the performance of the RF link.

The Good SNR is Lost Forever

Amplifying the signal at the end of the cable would not help; we would also amplify the noise (the *KTB*) and degrade the SNR with the NF of the amplifier.

Conclusion, Loss on Passive Cables

Any loss or attenuation of a passive cable or passive component will degrade the NF with the attenuation in dB – dB per dB. Once the SNR is degraded cable, you are getting closer to the thermal noise floor, the *KTB*. There is no way back, not even with an amplifier at the output of the cable; this amplifier will only degrade the SNR even more – you simply cannot retrieve the SNR once it is lost.

7.2   Cascaded Noise

Components rarely work on a stand-alone basis. You will need to connect the base station to the DAS via cables. The passive DAS consists of many passive cables and many passive components, adding up the losses from the base station to the antenna and vice versa. Using active DAS you still need to use a passive cable to connect to the main unit, and the active DAS consists of amplifiers and other active components. Hybrid DAS is a mixture of a large portion of passive components and cables, with active amplifiers and distribution units.

When you chain amplifiers and passive components or cables in a system, it is called a 'cascaded' system. Most RF systems will be cascaded systems. The amplifier will amplify the signal and the noise of the preceding amplifier or passive component or cable, and therefore the NF builds up.

7.2.1   The Friis Formula

We can calculate the noise factor (F) of any number of chained amplifiers or passive components (stages) using the cascaded noise formula known as the Friis formula. The Friis formula can be used for any number of cascaded components, from 1 to n. F_1/G_1 is the noise factor and gain of stage 1, F_2/G_2 is the noise factor and gain of stage 2, etc.

$$F_S = F_1 + \left[\frac{F_2 - 1}{G_1}\right] + \left[\frac{F_3 - 1}{G_1 \times G_2}\right] + \cdots \left[\frac{F_n - 1}{G_1 \times G_2 \times G_3 \dots G_{(n-1)}}\right]$$

The values for gain and NF have to be input as linear values (factors).

Linear Gain (G)

To calculate the linear gain to use in the Friis formula, or gain factor, we use this formula:

$$\text{linear gain} = 10^{[\text{gain (dB)}]/10}$$

Example: Linear Gain

$$6\,\text{dB gain will be } G = 3.98$$
$$6\,\text{dB attenuation} (-6\,\text{dB gain}) G = 0.251$$

Linear Noise Factor (F)

We calculate the linear noise factor to use in the Friis formula, using the same formula:

$$\text{linear noise factor } (F) = 10^{[\text{NF (dB)}]/10}$$

Example

$$20\,\text{dB NF (amplifier) will be } F = 100$$
$$20\,\text{dB loss (cable} = \text{NF} = 20\,\text{dB}) F = 100$$

Calculation of Cascaded Noise

Knowing the gain factor and noise factor of the elements in a cascaded system, we are now able to calculate the performance of the whole system.

Include the Passive Losses

We need to include NF and loss (attenuation = negative gain) of the cables interconnecting the amplifiers in the system as stages in the cascaded system, in order to accommodate the noise factor and attenuation of these losses.

The gain factor of a passive cable is of course less than zero due to the attenuation of the cable. When a passive component or cable is one of the stages, we need to inset the noise and factor of the cable, as just calculated in the example above.

7.2.2 Amplifier After the Cable Loss

In Section 7.1 we calculated the degradation of the signal level and SNR caused by a passive cable or DAS, so why do we not just use an amplifier to boost the signal again? Let us have a look; in this case the amplifier is connected to the end of the coax cable to compensate for the attenuation of the cable loss of the signal, as shown in Figure 7.5. However, is this the best approach? Let us analyze the system performance end-to-end.

Example

This is a two-stage system. The passive cable is stage 1 and the amplifier stage 2.

We can use the Friis formula to calculate the cascaded performance of the system: the first stage, as shown in Figure 7.5 (always counting the stages from the input of the system), is the

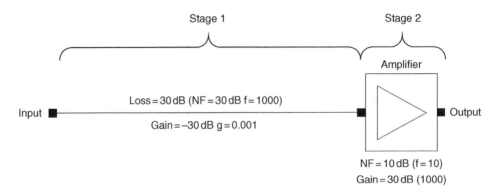

Figure 7.5   Cascaded system, with the passive cable as the first stage

coax cable; the second stage is the active amplifier. Inserting the values in the Friis formula of this two-stage system we get the cascaded noise calculation

$$F_S = F_1 + \left[\frac{F_{2-1}}{G_1} \right]$$

$$F_S = 1000 + \left[\frac{10-1}{0.001} \right]$$

$$F_S = 1000 + [9000]$$

$$F_S = 10\,000$$

$$NF = 10\log(10\,000) = 40\,dB$$

We can clearly see in the formula that stage 2 is highly affected by the attenuation of stage 1, with a noise factor of 1000 and a gain factor of 0.001.

Signal Power and Signal-to-noise Performance

We know now that the cascaded NF of the system is 40 dB, but what happens with the signal and the SNR?

Input Signal
We input −70 dBm at the input of the cable.

Signal Power
We can calculate the signal level throughout the system:

$$\text{Signal input} = -70\,dBm$$
$$\text{Signal at the output of the cable} = -100\,dBm$$
$$\text{Gain of amplifier} = 30\,dB$$
$$\text{Signal at the output of the amplifier} = -70\,dBm$$

The amplifier actually compensates for the loss of the cable. We have restored the signal level, but what about the quality of the signal, what about the SNR? We can start by calculating the noise power at the amplifier output. We already know the signal level to be $-70\,$dBm.

Noise Power

After we have verified that the system maintains the signal level, we must have a look at the impact of the different stages on the noise power: noise input = $KTB = -121\,$dBm; noise at the output of the cable = $-121\,$dBm (it cannot be lower than KTB, and the cable does not generate noise power); noise at the input of the amplifier = $-121\,$dBm; noise output from the amplifier = noise input + gain + NF, = $-121\,$dBm + 30dB + 10dB = -81dBm. The cascaded noise formula gave an NF of 40dB for the system.

We could also have used this formula for the system:

$$N\,\text{Floor Out} = N\,\text{Floor In} + \text{gain} + \text{NF}$$
$$N\,\text{Floor Out} = -121 + 0 + 40 = -81\,\text{dBm}$$

Signal-to-noise Calculation

Now that we have calculated the signal level as well as the noise level on the output, the SNR is then easy to calculate:

$$\text{SNR} = \text{signal power} - \text{noise power} = -70\,\text{dBm} - (-81\,\text{dBm}) = 11\,\text{dB}$$

Knowing the cascaded NF of the system, we could also have calculated the SNR on the output more easily using this formula:

$$\text{SNR output} = \text{SNR input} - \text{NF}$$
$$\text{SNR output} = -70\,\text{dBm} - KTB = -70\,\text{dBm} - (-121\,\text{dBm}) = 51\,\text{dB}$$
$$\text{SNR output} = 51 - 40\,\text{dB} = 11\,\text{dB}$$

Conclusion: Passive Loss will Degrade the System

According to the cascaded noise formula, it is clear that the first stage of the system is the dominant factor on the system performance. In this case, the passive coax cable has the predominant role. The 30dB factor of the cable plays a dominant role in the cascaded noise performance, owing to the high NF caused by the loss of the cable.

The example clearly shows that all passive loss (loss = NF) prior to the first amplifier can be added to the amplifier's NF to calculate the system NF. We can conclude that, to obtain the best performance in a cascaded system, we need to have a low NF in stage 1, and a gain better than 0dB.

7.2.3 Amplifier Prior to the Cable Loss

As we just documented in Section 7.1.1, having the amplifier placed after the cable did not perform that well. We concluded that this was mainly due to the high NF of stage 1, the passive coax cable. We need as low an NF as possible in stage 1 to optimize the NF of the

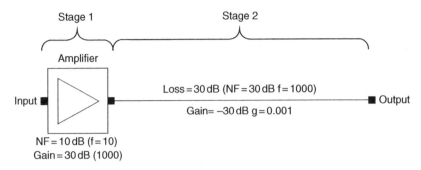

Figure 7.6 Cascaded system, with the passive cable as the last stage

total cascaded system. Let us try to swap the cable and the amplifier, making the amplifier stage 1 and the cable stage 2, and evaluate the performance to see if we can confirm this strategy.

Example
The first stage in Figure 7.6 (always counting from the input) is now the amplifier, and the second stage is the passive cable. Inserting the values of this two-stage system in the cascaded noise formula we get the cascaded noise calculation:

$$F_S = F_1 + \left[\frac{F_2 - 1}{G_1} \right]$$

$$F_S = 10 + \left[\frac{1000 - 1}{1000} \right]$$

$$F_S = 10 + [0.99]$$

$$F_S = 10.99$$

$$NF = 10\log(10.99) = 10.41\,dB$$

Better NF

We can see that now the NF is improved close to 30 dB, and the first configuration with the amplifier after the cable the NF is 40 dB. The NF in the new configuration is only 10.41 dB. This is almost the only contribution of the 10 dB NF of the amplifier in stage 1. It is evident in the formula that the passive cable, now stage 2, has only a minor impact on the total NF of the system.

Let us have a look at what happens with the signal and noise level through the system, by applying the same signal at the input as in the previous example.

Input Signal
We input −70 dBm at the input of the cable.

Signal Power

We can calculate the signal level throughout the system:

$$\text{signal input} = -70\,\text{dBm}$$
$$\text{gain of amplifier} = 30\,\text{dB}$$
$$\text{signal at the output of the amplifier} = -40\,\text{dBm}$$
$$\text{loss of cable} = 30\,\text{dBm}$$

As in the example from Section 7.1.1, the amplifier still compensates for the loss of the cable, with regards to the signal – the system is 'gain transparent': $-70\,\text{dBm}$ input and $-70\,\text{dBm}$ output.

Noise Power

We have verified that the system maintains the signal level; we need to analyze the noise performance and the noise power throughout the system:

$$\text{noise input} = KTB = -121\,\text{dBm}$$
$$\text{noise output amplifier} = KTB + \text{gain} + \text{NF} = -121\,\text{dBm} + 30\,\text{dB} + 10\,\text{dB} = -81\,\text{dBm}$$
$$\text{noise output cable} = \text{noise input cable} - \text{attenuation} = -81\,\text{dBm} - 30\,\text{dB} = -111\,\text{dBm}$$

Signal-to-noise Calculation

The SNR is easy to calculate:

$$\text{SNR} = \text{signal power} - \text{noise power} = -70\,\text{dBm} - (-111\,\text{dBm}) = 41\,\text{dB}$$

The SNR is now 30 dB better (compared with Figure 7.5), simply due to the amplifier being stage 1, prior to the passive cable (stage 2).

Conclusions: Amplifiers Prior to Passive Loss can Improve Performance

From the two examples in Sections 7.1.1 and 7.1.2 it is evident that, by placing the amplifier at the input of the system and preferably having minimum a gain equal to the subsequent cable loss, you can improve the link. In this example this is an improvement of 30 dB. It is evident in the cascaded noise formula that F1 plays the major role.

In the example the noise figure of the link is improved from 40 to 10.41 dB; just by moving the amplifier to the first stage of the signal chain you gain 29.58 dB in NF. This will 'clean' the UL signal and improve the SNR of the link dramatically.

7.2.4   Problems with Passive Cables and Passive DAS

In a passive DAS the first amplifier in the system is in the base station receiver. The loss of the passive DAS will impact the NF of the system as we just calculated from Section 7.2.2. The loss of the passive DAS will severely impact the NF and the SNR on the UL, limiting the UL data service performance. One decibel of passive loss prior to the base station will degrade the system NF by 1 dB; in passive systems the attenuation is typically in the range 25–40 dB. Every decibel lost on the passive cable is to be subtracted from the maximum allowed path loss for the given link budget. The passive losses of the coax cable do not have to be a problem, but they will degrade the uplink service range of the DAS antenna, as with any other loss on the radio link.

Doing the link budget calculation for a passive DAS, you will need to calculate every antenna individually if there is different loss to each antenna from the base station. However we also must have a pragmatic approach: if the radio service we are designing the DAS to requires an SNR of 14 dB, we should not worry much about whether the link is performing at 20 or 34 dB in SNR.

This is How a Low Noise Amplifier Works

This is the exact reason for deploying mast-mounted amplifiers (LNAs) in mobile systems, which are frequently used in 3G deployment. A small low noise amplifier is mounted close to the antenna, in order to compensate for the loss of the feeder cable running to the base station. Often the local LNA inside the base station will be turned off, and the tower mounted LNA will compensate for the gain in the system. By doing so, the loss of the feeder is compensated and the gain in the system is applied where it should be, as close to the receive antenna as possible, before any passive losses degrade the SNR.

The LNA boosts the performance and raises the data rate on the uplink. The UL is often the limiting factor on 3G, especially at the higher data rates; this will impact the more demanding uplink data services like 4G most. The LNA will improve the cell range on the uplink, and thus save on roll-out cost of the network operator, or obtain better performance for the same roll-out cost.

This Also Applies for Active DAS Systems

The exact same principle is used by some manufacturers of active DAS systems; the best performing active DAS will actually have the remote unit placed close to the antenna. Inside this remote unit is the first UL amplifier, the LNA. However, this only applies for pure active systems; hybrid active DAS will, as with any other cascaded system, suffer on the NF from the passive losses prior to the remote unit. But surely be better than a full Passive DAS, depending on the balance between loss of the passive, NF and gain of the amplifier. For DAS systems where the UL data performance plays a major role, especially 3G and 4G systems, you need to be very careful. Do a detailed analysis to see if the uplink data service you are designing can cope with the NF of the DAS, passive, hybrid or active.

Uniform Performance

The pure active DAS system in Figure 7.7 will have the same NF at each antenna. Therefore there will be the same uniform uplink performance throughout the DAS system, making link budget calculation and planning fairly easy. Downside is the high number of active elements arround the building (hidden in ceilings) that might need service and maintanance.

For hybrid systems the loss of the passive prior to the remote unit will impact the performance of the NF, and might limit the uplink data speed. Do check with noise calculations of the total system from antenna to base station in order to be sure. In the case of deploying the hybrid DAS, the active remote unit can often be in the vertical shaft, using passive only for the horisontal part of the building. This is often a good balance between performance, investment and services access.

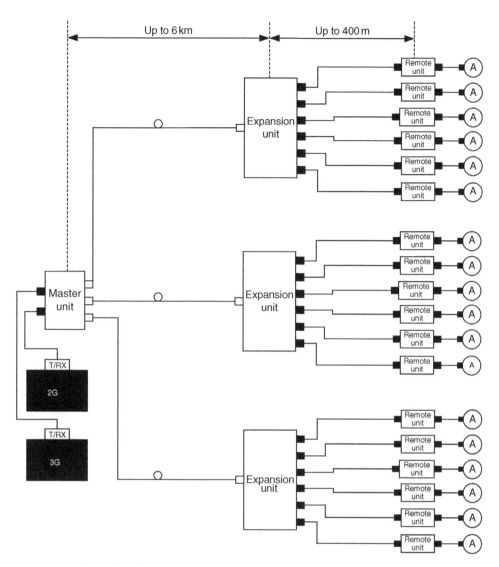

Figure 7.7   The pure active DAS: amplifiers located close to the antennas

Conclusion: Higher Data Rates with Low NF

The impact from the NF on the UL is evident in the link budget, especially when the radio channel is servicing higher data rates, for 3G and for HSPA. Refer to Chapter 8 for more details of the link budget.

7.3   Noise Power

From the example in Section 7.2, it is clear that the NF could be improved by installing the amplifier prior to the coax cable. Place the amplifier as near as possible to the signal, the antenna. However, it is also a matter of having control of the noise power, especially in noise-sensitive systems like 3G/4G. In these systems the SNR of the system plays a major role for the high data rate services.

Let us have a closer look at what happens with the input signal as shown in Figure 7.9 (−90 dBm) and the noise in the system. Backing up the calculations with a 'spectrum analyzer look', it all becomes more obvious, compared with 'just' looking at formulas.

7.3.1 Calculating the Noise Power of a System

An example of a pre-amplifier 30 dB gain with 10 dB NF and 30 dB loss on cable is shown in Figure 7.8. The spectrum in Figure 7.9 looks at the input signal.

Signal-to-noise Ratio at the Input of the Amplifier

This is a 2G system operating at 200 kHz bandwidth; the *KTB* is −121 dBm (noise floor) and we can calculate the signal-to-noise ratio:

$$\text{SNR} = \text{signal/noise (linear)}$$

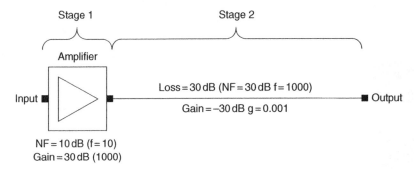

Figure 7.8 Pre-amplifier solution, −90 dBm input signal

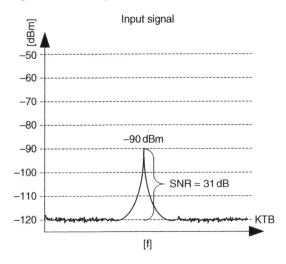

Figure 7.9 Signal at the input of the pre-amplifier solution

or in dB,

$$SNR = \text{signal power} - \text{noise (dB)}$$
$$SNR = -90\,dBm - (-121\,dBm) = 31\,dB$$

Output Signal from Amplifier

Analyzing the output of the amplifier (Figure 7.8) as shown in Figure 7.10, we can see the gain of the signal, as well as the noise increase caused by the gain and the NF.

Amplifier Output, Signal Level

The amplifier has 30 dB gain and a NF of 10 dB, so the signal is now:

$$\text{output signal} = \text{input signal} + \text{gain}$$
$$\text{output signal} = -90\,dBm + 30\,dB = -60\,dBm$$

Amplifier Output, Noise Level

The amplifier also amplifies the input noise, the *KTB* (−121 dBm) and adds 10 dB of noise due to the NF of the amplifier; we can calculate the new noise power:

$$NFloorOut = NFloorIn + \text{gain} + NF$$
$$\text{output noise floor} = \text{input noise floor} + NF + \text{gain}$$
$$\text{output noise floor} = -121\,dBm + 10\,dB + 30\,dB = -81\,dBm$$

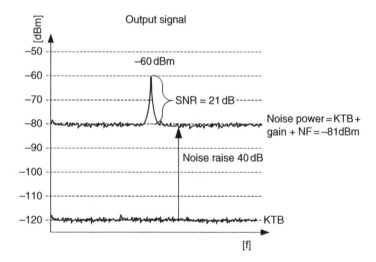

Figure 7.10 The output of the amplifier, signal and raised noise floor

Amplifier, SNR Degradation

The NF of the amplifier will degrade the signal-to-noise ratio:

$$\text{SNR output} = \text{SNR output} - \text{NF}$$
$$\text{SNR output} = 31\,\text{dB} - 10\,\text{dB} = 21\,\text{dB}$$

Output Signal from the Coax

Present at the output of the passive cable will be the attenuated signal level from the amplifier, as well as the noise power out of the amplifier. Both are attenuated by the cable loss (30 dB), as shown in Figure 7.11.

Output, Coax Cable

The coax cable has 30 dB of attenuation, the same as the 30 dB gain of the amplifier. We can calculate the output of both the signal and noise power of the cable:

$$\text{output signal} = \text{input signal} - \text{cable loss}$$
$$\text{output signal} = -60\,\text{dBm} = 30\,\text{dB} = -90\,\text{dBm}$$
$$\text{output noise floor} = \text{system input noise floor} + \text{system NF} + \text{system gain}$$
$$\text{output noise floor} = -121\,\text{dBm} + 11\,\text{dB} + 0\,\text{dB}$$
$$\text{output noise floor} = -121\,\text{dBm}$$
$$\text{SNR} = -90\,\text{dBm} - (-111\,\text{dBm}) - 21\,\text{dB}$$

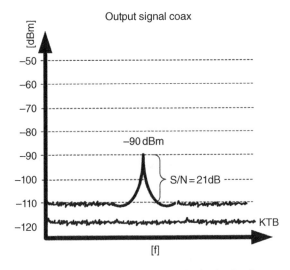

Figure 7.11 Signal at the output of the cable; the signal and raised noise floor are attenuated by the loss of the cable

SNR is Maintained

Obviously the cable will attenuate both the wanted signal and the noise, but as long as we make sure that the signals are not attenuated down to a level where the noise power from the amplifier will reach the KTB(−121dBm) (no other noise is present than the KTB in this example), the SNR of the system will remain intact. In this example, −81dBm − (−121dBm) = 40dB, we can keep the impact of the SNR minimal. However, if the attenuation of the cable is higher than the input noise power − KTB, the SNR will be degraded by the difference between the attenuation and the margin to the KTB. This is due to the fact that the KTB, −121 dBm (200 kHz 2G), is the absolute lowest level. Once the noise power 'hits' the KTB the SNR will be degraded and we will never be able to restore the lost SNR.

High SNR Means Higher Data Rates

The performance on any radio link is not related to the absolute signal power, but to the SNR of the radio channel. This is demonstrated in Figure 7.12; signal 1 is 'only' −90dBm, whereas signal 2 is 30dB higher. However, the SNR in signal 1 is 10dB better; therefore this link will perform much better even if the signal is 30dB lower.

Far too often all focus is on level, but keep in mind that the quality of the radio service is not defined by the signal level, but the SNR. The better the SNR, the higher the data rates on the radio link.

When designing 3G and 4G DAS solutions, the loss of passive cables has a big impact, due to the higher losses on higher frequencies. The stricter SNR requirements for high data rates will also demand good radio link quality, especially when designing 3G/4G DAS solutions.

Conclusion

From the examples just covered, we can conclude that, to some extent, we can compensate for the loss of a coax cable with the use of pre-amplifiers. For maximum performance we must have the lowest possible NF of the amplifier, and a gain that can compensate for the loss of the passive cable following the amplifier.

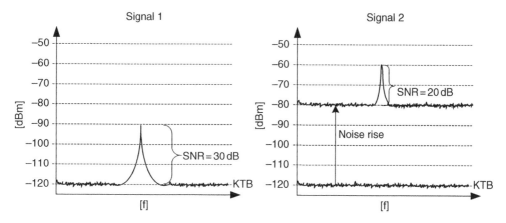

Figure 7.12   It is the SNR of the signal that is important for the performance, not the absolute signal level

7.4   Noise Power from Parallel Systems

Often there will be more than one noise signal (noise power sources) present at the input of a system. This could be two parts of an active DAS, using two repeaters, that needs to be combined to a base station receiver, like the 2G (200 kHz) example in Figure 7.13.

7.4.1   Calculating Noise Power from Parallel Sources

Random Power Signals

Noise is random (white noise), and two noise signals will be noncorrelated. Thus the noise powers (in Watts) of two sources can be added in order to calculate the total noise power from the system.

First we need calculate the noise floor for each of the two amplifiers (2G):

$$NFloorOut = NFloorIn + gain + NF$$
$$NFloorOut = KTB + gain + NF$$
$$NFloorOut = -121\,dBm + 10\,dB + 16\,dB = -95\,dBm$$
$$NFloorOut = -121\,dBm + 20\,dB + 22\,dB = -79\,dBm$$

Watts

We need to sum up the noise, by adding the power contribution from each source.

$$power\ total = power\,1 + power\,2 \ldots power\,(n)$$

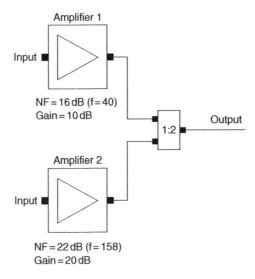

Figure 7.13   Two parallel noise sources combined into one output

Now we convert the noise powers from dBm to mW:

$$P(\text{mW}) = 10^{\text{dBm}/10}$$
$$-95\,\text{dBm} = 136 \times 10^{-12}\,(\text{mW})$$
$$-79\,\text{dBm} = 12.58 \times 10^{-9}\,(\text{mW})$$

Then we can sum up the two powers:
combined noise power $= 136 \times 10^{-12} + 12.58 \times 10^{-9} = 12.89 \times 10^{-9\,(\text{mW})}$

Convert to dBm
We convert the noise power back to dBm:

$$\text{noise floor (dBm)} = 10\log(\text{mW})$$
$$\text{noise floor (dBm)} = 10\log(12.89 \times 10^{-9}) = -78.89\,\text{dBm}$$

The Combiner
The noise power is combined in a splitter with $3\,\text{dB}$ loss ($F = 0.5$); the output noise floor will be:

$$\text{noise floor output} = -78.89\text{dBm} - 3 = -81.89\text{dBm}$$

OBS!
Note that for the ease of the example, the loss of the cables and the insertion loss of the splitter have not been accounted for.

7.5 Noise Control

Noise control is important, especially in cascaded systems with multiple amplifiers and interconnecting cable. Gains have to be adjusted according to noise figures and losses of all the individual system components in order to optimize the performance of the whole system, end to end.

7.5.1 Noise Load on Base Stations

When an amplifier or an active DAS injects a noise power signal on the uplink port of the base station, it may potentially cause problems that will affect the performance of the base station. The main issue is that the injected noise will desensitize the receiver in the base station, limiting the performance of the uplink. This issue needs special attention in applications where an active DAS is connected to a macro cell, or if the base station is also serving a passive DAS. In these applications we need to insure that we configure the active part of the DAS with regards to gain and attenuation to minimize the effect of the uplink degradation. To select the correct UL attenuation in order to minimize the noise power, refer to Chapter 10. Furthermore, the raised noise floor has additional side effects for 2G and 3G systems.

7.5.2   Noise and 2G Base Stations

Injected Noise Might give Interference Alarms

2G base stations will monitor the noise level in idle mode time slots to estimate any potential interference problems from mobiles in traffic using the same frequency, in nearby or distant cells, PIM products, etc. This enables the base station to provide warnings about potential interference issues that might degrade the performance.

When the noise floor on the 2G base station increases due to any external noise source, like the active DAS, which might inject noise power on the uplink of the base station, this will cause the BSS system to generate an 'idle mode interference alarm'. The trigger level for this alarm can be changed, but the solution is to attenuate the noise to a level below the alarm value (see Chapter 10 for more details).

7.5.3   Noise and 3G Base Stations

3G Systems: Noise will Affect Admission Control

Noise control is paramount for the performance of any WCDMA system, like 3G. This is the motivation for very strict power and noise control in 3G. 3G base stations will evaluate (measure) the noise rise on the uplink in order to evaluate the load in the cell (in 3G traffic increase is equal to the noise increase).

The 3G base station will have a preset reference for the base level of the noise floor with no traffic, i.e. the NF of the base station itself, and with this as a reference the base station can calculate the current load (noise) in the cell. The purpose of this noise control is not to admit new traffic if the cell exceeds the pre-set maximum noise increase or load rate. A load rate of 50% in a 3G cell is equal to a noise rise of 3 dB.

If we inject noise into a 3G base station that increases the noise floor by 3 dB, then the base station might be tricked into evaluating the current traffic load as 50%, even with no traffic. The base station will then only admit another 10% traffic if the maximum load is preset to 60%.

The result will be that this base station will only be capable of picking up 10% traffic and will be heavily underutilized if we do not minimize the noise load from the external source.

Solution

First of all, the uplink gain of any in-line amplifier in a DAS system should only be adjusted to a level where it compensates for the losses of the subsequent passive network. This will insure that the noise build-up is kept to a minimum, and the noise power injected into the base station or amplifier is at the level of the *KTB*, minimizing the ramped-up noise of the system.

The solution is to attenuate the noise (and signal), in order to lower the noise load of the base station. As we have just clarified, we can do that without affecting the SNR as long as we keep clear of the *KTB* (see Section 7.3.1).

It is impossible to completely remove the noise power on the input without degrading the SNR of the system. Luckily for 3G, where the noise control is crucial for the performance of admission control, you can offset the base station noise reference level. For further details refer to Chapter 10.

7.6   Updating a Passive DAS from 2G to 3G/4G

In recent years RF designers have tried to prepare their passive DAS designs for 3G/4G to some extent. Often the planning of these passive DAS was done years prior to the launch of their 3G/4G service. Thus, it lacks a 3G/4G link budget, guidelines and knowledge as to how the NF caused by the attenuation of the passive system would impact the 3G/4G uplink performance once implemented.

Many of these 2G systems were 'future proofed' to support 3G, normally by using passive components that covered the 3G (2100MHz)/4G (2600MHz) band, and sometimes a slightly heavier coax cable.

For some of these projects it has become clear that, in some areas of the building, especially in the areas serviced by the most remote antennas, the antennas with the highest loss have problems with UL data speed on 3G, and in particularly with HSUPA.

7.6.1   The 3G/4G Challenge

The reason for this is often the increased loss, the higher noise figure on the passive DAS when the frequencies get higher, but also the much stricter SNR requirements for the high data rate on 3G/4G. These requirements are covered in more detail in Chapter 8.

Solution

As we have just explored in the first sections of Chapter 7, we can use amplifiers to improve the performance in the part of the passive DAS that is having difficulties with the loss/noise figure.

Often antennas close to the base station and those with relatively low loss will perform acceptable. Typically it will be just the more distant part of the passive DAS that needs an upgrade to perform on 3G/4G.

Example

This is a real-life practical example on why and how we can upgrade a passive DAS from 2G to 2G + 3G. This passive DAS, shown in Figure 7.14, was originally designed for 2G-900. All components, splitters, tappers, etc., were prepared for 1800 and 2100MHz, so the system was to some extent prepared for the 3G upgrade, which was expected within 5–7 years after installing the original DAS system.

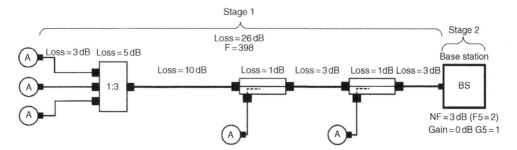

Figure 7.14   2G passive DAS with 26 dB loss to the three most remote antennas

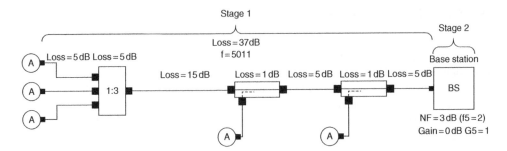

Figure 7.15   3G on the passive DAS designed for 2G, now with excessive (37 dB) loss to the three most remote antennas

7.6.2   The 3G Problem

However, after the deployment of 3G, it turns out that there are issues for the performance on the uplink data service in the areas covered by the three most remote antennas. The radio planner clarified that the problem was due to the high loss on the three remote antennas, as shown in Figure 7.15.

Cascaded noise calculations showed that the losses on the 3G band (2100 MHz) were now 37 dB, and that this increased the noise figure for the system performance for the three antennas to 40 dB and degraded the data service in that area of the building.

The UL Problem on 3G

The problem is related to the high NF that impacts the UL data performance. As we can see in the calculation, stage 1 plays a major role in the degradation, being the passive DAS.

Cascaded Noise Calculation

We can calculate the NF of this two-stage system:

$$F_S = F_1 + \left[\frac{F_2 - 1}{G_1} \right]$$

$$F_S = 5011 + \left[\frac{2 - 1}{0.0001995} \right]$$

$$F_S = 10\,022$$

$$NF = 10\log(10012) = 40\,dB$$

Conclusion on the Existing Passive DAS

The problem is related to the high NF, and impacts the UL data performance. As we can see in the calculation, stage 1 plays a major role in the degradation, stage 1 being the passive DAS.

3G UL Data Coverage at 40 dB NF

When analyzing the link budget (Chapter 8), it becomes clear that the coverage range for 384 kbps is only about 11 m in the dense office environment. The practical experience with degraded uplink 384 kbps data performance is confirmed by the link budget calculation. Even the data services with lower data rates and less strict requirements are limited. For 3G UL Data service:

- 384 Kbps = 11 m.
- 128 Kbps = 14 m.
- 64 Kbps = 16 m.

Several solutions were considered in order to solve the problem.

7.6.3   Solution 1, In-line BDA

The initial idea was to install one amplifier after the last tapper, in order to boost all three antennas as shown in Figure 7.16, thus only needing to install one active element in the system, and thereby saving on hardware and installation costs.

Cascaded Noise Calculation

We can calculate the NF of this, now four-stage, system:

$$F_S = F_1 + \left[\frac{F_2 - 1}{G_1}\right] + \left[\frac{F_3 - 1}{G_1 \times G_2}\right] + \left[\frac{F_4 - 1}{G_1 \times G_2 \times G_3}\right]$$

$$F_S = 316 + \left[\frac{10 - 1}{0.00316}\right] + \left[\frac{15.8 - 1}{0.00316 \times 1000}\right] + \left[\frac{2 - 1}{0.00316 \times 1000 \times 0.63}\right]$$

$$F_S = 316 + [2848] + [4.68] + [0.05]$$

$$F_S = 3168$$

$$NF = 10 \log(3168) = 35 \, dB$$

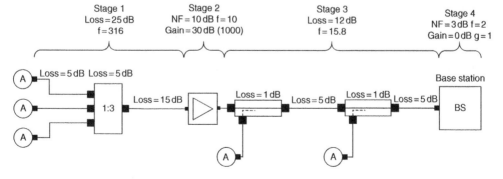

Figure 7.16   3G upgrade by deploying an amplifier at a central point in the system

Conclusion on Upgrade with Central BDA

This system with one in-line bi-directional amplifier BDA and repeater is only a 5 dB improvement on the uplink performance. The link budget calculation shows that this is clearly not enough to get the noise figure down to a level that will improve the performance to the required service level on higher data rates. The NF was only improved by 5 dB.

From the cascaded noise formula we can see that stage 2 takes a big 'hit' from the 25 dB loss of the passive stage 1. Thus we conclude that, in order to increase the uplink performance of the system to a desired level, we must try to limit the impact of the passive loss prior to stage 1 of the system. We must deploy the pre-amplifier as close to the antennas as possible to minimize the impact of the coax loss.

Solution 2, Active DAS Overlay

In order to minimize the impact of any passive loss prior to the amplifier, we can now analyze the performance if we install the amplifier right next to each of the three antennas as shown in Figure 7.17. By using the strategy, we can eliminate the passive losses in stage 1 of the cascaded system. Note that the diagram only shows one amplifier, but since the three antennas have the exact same loss back to the base station, only one link is analyzed. The performance for all three antennas with three amplifiers will be the same.

Cascaded Noise Calculation

We can calculate the NF of this, now three-stage, system:

$$F_S = F_1 + \left[\frac{F_2 - 1}{G_1}\right] + \left[\frac{F_3 - 1}{G_1 \times G_2}\right]$$

$$F_S = 10 + \left[\frac{5011 - 1}{1000}\right] + \left[\frac{2 - 1}{1000 \times 0.0001995}\right]$$

$$F_S = 20.02$$

$$\text{NF} = 10\log(20.02) = 13.01 \text{ dB}$$

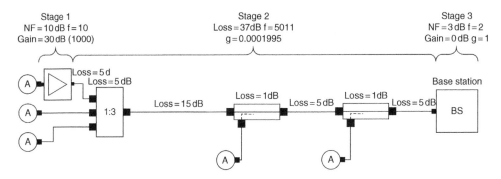

Figure 7.17 3G upgrade by deploying the amplifier at the antenna

Conclusion with BDAs Installed Close to the Antennas

The cascaded noise analysis shows the result; there is a stunning 27 dB improvement of the noise figure. This will boost the uplink performance on the desired 3G data rates, and even perform well on HSUPA. Furthermore, we can actually reuse most of the old 2G installation by upgrading it with amplifiers close to the antennas. This will insure a fast deployment and limited interruption of the mobile service in the building during the installation of the upgrade.

The part of the existing passive system with relative low loss we can leave intact; here the performance is acceptable. We only need to pinpoint the problem areas, and upgrade these antennas.

Be careful about the noise power injected in the base stations uplink. In this case the attenuation of the passive DAS components, from the BDA to the base station, will attenuate the noise power, and minimize the noise load on the base station. See Chapter 10 for more details on how to minimize the noise load on 3G base stations using attenuators.

3G UL Data Coverage at 13 dB NF

The link budget analysis shows the boost in the uplink performance. The cure is clear: install the amplifier as close to the antenna as possible. This is actually how pure active DAS works (see Section 4.4.2):

- 384 Kbps = 39 m.
- 128 Kbps = 51 m.
- 64 Kbps = 59 m.

Note that ideally we should have calculated the combined NF of the three parallel BDAs, but it turns out that the DL is now the limiting factor, and we must use another approach.

The Downlink

We just analyzed how to boost the performance on the uplink 3G data performance, on a passive DAS designed for 2G, but what about the performance of the downlink coverage? After all we have 37 dB of loss on the DL from the base station room in the basement to the topmost floors with the problem area.

In this case we are using a +40 dBm base station (CPICH at 10% = +30 dBm), and the CPICH power into the antenna will only be −3 dBm.

The Problem is in the Top of the Building

The problem area is actually located in 'zone C' of the building (see Section 3.5.6), on the highest floors (the executive management floor), where we need a good dominant signal from the in-building DAS system and preferably HSDPA/4G performance.

Lack of DL Signal, CPICH Dominance

According to the free space loss formula (see Section 5.2.8), we have a path loss from the antenna of about 67 dB at 25 m from the antenna, giving us a CPICH level of about −3 dBm − 67 dB = −70 dBm.

The design level for this floor is actually −65 dBm CPICH, so we need to boost the DL power at the antenna in order to get dominance and good SNR and maintain pilot dominance inside the building.

When analyzing the link budget, we see that for the data service for DL, we have less than 12 m of service radius.

UL/DL Service Balance

Often the highest data load is on the DL, and we have just improved the UL with the pre-amplifier to these service radiuses:

UL 3G Service (Dense Environnent)

- 384 Kbps = 39 m.
- 128 Kbps = 51 m.
- 64 Kbps = 59 m.

There is an imbalance between the uplink and downlink performance, making the system downlink-limited. We will need to boost the downlink level in order to balance the performance and coverage on UL/DL.

Boost both UL and DL on 3G

We need to boost the DL as well as the UL in order to balance the links; the solution would be to install two-way (UL/DL) amplifiers close to all the distant antennas. These types of amplifiers are often referred to as bi-directional amplifies, but by installing remote BDAs in the building, other considerations arise.

Installation and Performance Issues with Remote BDAs
Often it will be a major challenge to install all the required in-line 3G BDAs, in these typical, large indoor projects where a passive DAS designed for 2G needs an update to perform on 3G. The challenge is more than installing the BDAs.

Power Supply
You are frequently not allowed to tap into the existing AC power supply but will need to install a new 'AC group' for the BDA cluster. Installing new AC power groups for this throughout a large building is time-consuming, expensive and has a heavy installation impact on the building. You also need to consider if power supply back-up will be needed for these distributed BDAs. This distribution no-break power will further add to the cost and complication of the system.

2G Coverage

Make sure that the selected BDAs will not degrade the performance on the existing 2G system. The BDAs must have a 'bypass' function that makes sure that any 2G (and 3G) signals will be bypassed through the amplifier in case of malfunction or AC power loss.

Commissioning

The distributed BDAs frequently do not have a centralized control point, so setting and tuning them must be done locally for each individual BDA. They are also often installed in locations that are not easily accessible, such as above ceilings (they have to be close to the antennas to perform).

Operations and Maintenance and Surveillance

In case of a fault on the system, it is challenging to find the cause. A cluster of BDAs often lacks a common alarm structure, and a common external alarm interface to the base station.

7.6.4   Solution 2: Active DAS Overlay

There is actually a way to upgrade the passive DAS with remote BDAs located at the antenna, having a common alarm infrastructure with the alarm interface to the base station and a common centralized power supply, if we use the system from Section 4.4.3 (shown in Figure 7.18). If you combine this small active DAS with the passive DAS in the problem areas, the problem will be solved.

The Existing Passive DAS, Designed for 2G

This 14-floor office building is designed for 2G (the passive DAS is shown in Figure 7.19), and the performance on 2G is perfect. After several years of good 2G operation on the DAS, 3G was deployed. Initially the 3G base station was connected to the DAS, expecting full 3G performance on the existing DAS. However it turned out that there were problems similar to those described in the previous section: lack of pilot dominance and high noise figure on

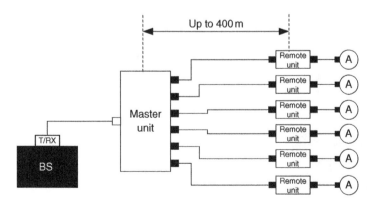

Figure 7.18   Small active DAS, with amplifiers (RU) close to the antennas and no loss

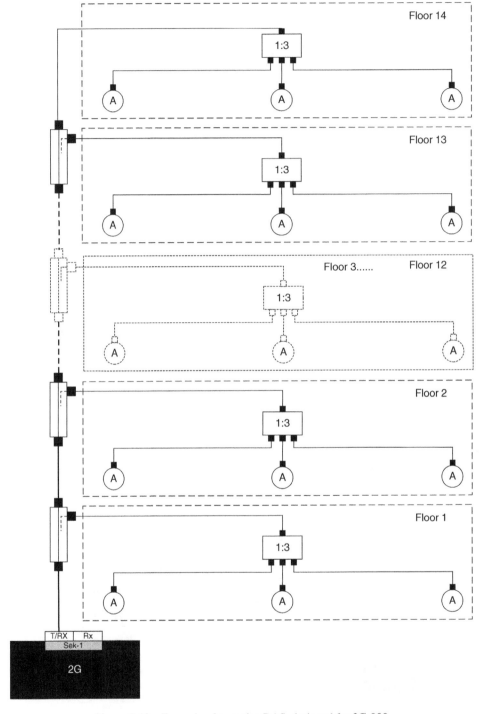

Figure 7.19 Example of a passive DAS, designed for 2G-900

floors 13 and 14. The users on these executive floors suffered from degraded 3G data performance, dropped calls and no HSPA performance. Measurements on-site confirmed the problem as lack of dominance in 'zone C' the most distant from the base station, and with high signal levels from the outside network present inside the building. As in the previous example, the cascaded noise figure of the antennas covering the topmost floors was calculated, and link budget calculations confirmed the problem.

The users on these executive floors needed an upgrade fast; the problem was solved by deploying small local uplink and downlink amplifiers close to the antennas in the problem, as described below.

The Upgrade to 3G

By deploying the small active system from Figure 7.18, you can increase the DL power and improve the noise figure of the DAS on the two problem floors 13 and 14 (see Figure 7.20). The downlink power from the active DAS is 15 dBm power per carrier (PPC), so the DL could be boosted; 15 dB uplink gain and an NF of 16 dB could improve the performance of the uplink.

You use filters installed in the base station room (S), and at the antenna locations on the two topmost floors 13 and 14 (F) to inject the 3G signal and to separate the 2G.

Noise Control

When connecting the active DAS system to the existing passive DAS, you must be aware that the noise power from the active DAS can desensitize the base station and degrade the uplink performance on the passive section of the DAS. In order to avoid this, we need to insert an appropriate attenuator to attenuate the injected noise power. Refer to Chapter 10 for calculation of the optimum attenuator.

The Performance

The link budget calculation shows that, after the upgrade, the DL will have more power, producing higher downlink levels, and the uplink will perform much better due to the lower noise figure of the system.

The system (in Figure 7.20) was implemented, and the link budget calculation was confirmed by measurements. A summary of real-life measurement results from office buildings using this upgrade approach can be seen in Table 7.1.

These measurements confirm the theory, and results from other real-life upgrades and implementations also show similar results. This approach is applicable on existing passive DAS, boosting the DL signal and raising the UL data speed.

Still Full 2G Service

The 2G service is still maintained. The only 'cost' on 2G is the insertion loss of the filters, less than 1 dB. In practice there will be no performance degradation, and the 2G service in the building is maintained at the current level.

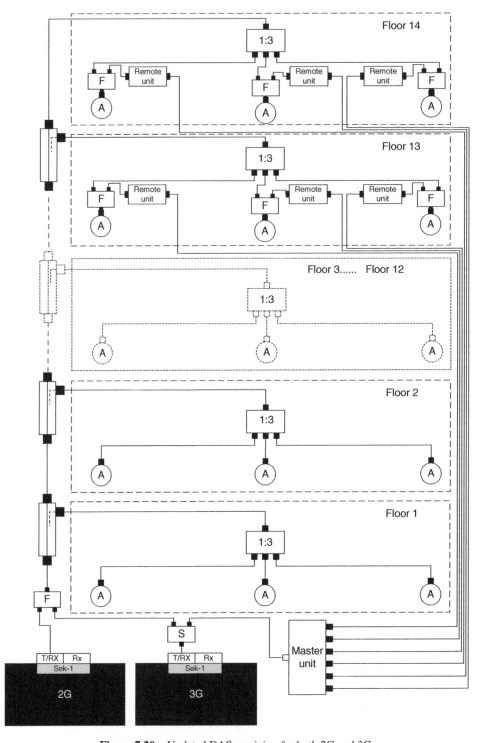

Figure 7.20   Updated DAS servicing for both 2G and 3G

Table 7.1 3G UL/DL performance improvement after upgrade of the passive DAS

Passive DAS loss	NF improvement	DL gain	E_c/I_0 gain
30 dB	13 dB	5 dB	2 dB
35 dB	14 dB	12 dB	8 dB
40 dB	15 dB	17 dB	14 dB

Conclusion: Upgrading Passive DAS to 3G Performance

The use of remote located BDAs close to the existing antennas can increase the UL and DL performance on an existing DAS originally designed for 2G. Using the small version of an active DAS is applicable in practice, and use of thin IT-type cabling between the master unit and the remote units ensures a rapid deployment of the system. In some cases you might be able to reuse some of the existing IT cables, minimizing the need to install new ones. The small active DAS uses central power, distributed to the remote units from the master unit, so there will be no need for local power supply at the antenna locations. The master unit has external alarm outputs that can be connected to the base station in order to maintain full visibility of the status of the active elements (the MU and RUs).

7.6.5   Conclusions on Noise and Noise Control

All electronic components generate noise. The density of the noise is constant over the spectrum, and the power is −174 dBm/Hz. This base noise is referred to as the *KTB*. By multiplying the −174 dBm/Hz by the bandwidth of the radio channel, we can calculate the absolute minimum noise level for 2G (200 kHz) as −121 dBm, and 3G (3.84 MH$_z$) = −108 dBm.

The performance of a radio link is defined as the signal-to-noise ratio, not the absolute signal level. Any noise figure of an amplifier or loss of a passive cable will degrade the radio link. The loss or attenuation of a passive cable or passive component will degrade the NF with the attenuation in decibels. Once the SNR is degraded by the passive cable, you are getting closer to the thermal noise floor, the *KTB*. Then there is no way back, not even with an amplifier at the output of the cable. This amplifier would only degrade the SNR even more – you simply cannot retrieve the SNR once it is lost. Placing an uplink amplifier at the end of the passive DAS, close to the base station, will not improve the SNR, but degrade the SNR with the noise figure of the amplifier.

From the two examples in Sections 7.2.2 and 7.2.3 it is evident that you must place the amplifier at the input of the system. It is preferable to have a low noise figure in the amplifier and minimum gain as the subsequent cable has loss. With this strategy you can improve the link, and compensate for most of the impact of the NF from the passive loss in the following cable. In the example the link was improved by almost 30 dB. If you make sure you keep below the maximum loss, SNR is maintained, and the performance on the RF link is maintained.

When using amplifiers prior to the base station, you must be careful with the noise you inject on the uplink of the base station. The uplink gain of, any in-line amplifier in a DAS system should only be adjusted to a level where it compensates for the losses of the following passive network. This will insure that the noise build-up is kept to a minimum, and the noise power injected into the base station or amplifier is at a level of the *KTB*, minimizing the ramped-up noise of the system. You can decrease the noise power down close to the *KTB* with only minimum noise impact of the link.

It was also demonstrated how this basic knowledge about noise and noise control enables us to upgrade an existing passive DAS designed for 2G to perform high data rates on 3G. By deploying an active DAS overlay on the part of passive DAS that are designed for 2G, you could boost the performance on the uplink to service for 3G/HSPA. This strategy will boost the uplink as well as the downlink. Understanding noise and noise performance and calculations is the basis of all radio planning.

8

The Link Budget

The link budget (LB) is the fundamental calculation for planning of any RF link between a transmitter (Tx) and a receiver (Rx). In two-way calculations we actually have two LB calculations, one for the DL and one for the UL.

The result of the link budget calculations is the maximum allowable path loss (MAPL) from the base station to the mobile in the downlink and the maximum allowable link loss on the reverse link, from the mobile to the base station the uplink. You need to include all the attenuation and gains of the end-to-end signal path from the Tx to the Rx, the attenuation due to the distance, adding the clutter loss of the environment, cable attenuation and antenna gains. You will also need to include a safety margin to provide a given probability of the desired signal, accounting for fading margins and body losses.

Depending on the type of distribution system you design, there are various parameters to take into account when calculating the link budget. Based on these parameters for the DAS, the radio service requirement, and the impact of noise from existing signal sources operating on the same frequency or channel, you must calculate the link budget for both links, the uplink and the downlink, to determine the service range of the system in both directions. The link budget for the particular radio service range can be either UL- or DL-limited; this will depend on the parameters affecting the links. One cell can be downlink-limited for one type of service, and the same cell can be uplink-limited for another service (service profile).

Indoor Radio Planning: A Practical Guide for 2G, 3G and 4G, Third Edition. Morten Tolstrup.
© 2015 John Wiley & Sons, Ltd. Published 2015 by John Wiley & Sons, Ltd.

8.1   The Components and Calculations of the RF Link

Let us try to break down the link budget and have a look on the most important parameters in the calculation. The simplest LB calculation looks, in principle, like this:

$$\text{Rx-Level (dBm)} = Tx \text{ power (dBm)} - \text{cable attenuation (dB)} - \text{propagation losses} + \text{antenna gain (dB)}$$

However, the real-life 2G and 3G LB are more complex, so you will need to break down the link budget calculation into more detail.

8.1.1   The Maximum Allowable Path Loss

The essential purpose of making the link budget is to calculate the maximum allowable path loss, as shown in Figure 8.1. Once you have calculated the MAPL, you can calculate the service area and service radius to use for antenna placement in the building.

Note that the example in Figure 8.1 is rather simplified, and merely shows a few of the factors you need to take into account when calculating the link budget. This simplified example shows only the downlink, but both the downlink and the uplink APL need to be calculated, for all services on the DAS.

8.1.2   The Components in the Link Budget

In order to understand the details in the link budget for indoor DAS design, you need to break it down to the all various components of gains and losses. Then you can understand how the different parameters interact and affect the link budget calculation.

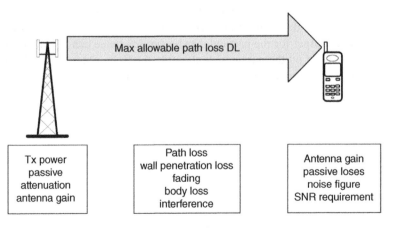

Figure 8.1   Principles of the link budget (DL)

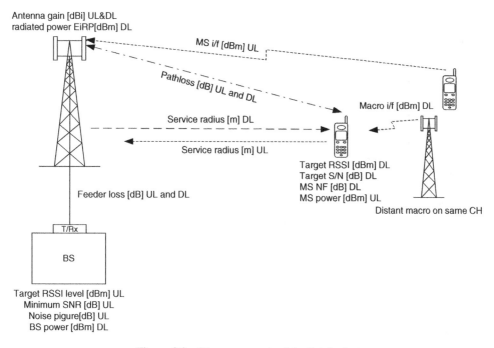

Figure 8.2   The components of the link budget

The components of the basic link budget are shown in Figure 8.2 and Figure 8.3. To be able to do a detailed analysis, you need to take a look at the different components of the link budget calculation shown in basic form in Table 8.1, one by one (the symbols from Table 8.1 are defined below).

- a, BS power (dBm): this is the generated RF power from the base station, at the antenna connector of the output of the base station rack.
- b, Feeder loss (dB): this is the attenuation of the coax cable from the BS to the antenna. The loss is symmetrical for the UL and DL.
- c, BS antenna gain (dBi): this is the antenna gain (directivity) of the BS antenna.
- d, EiRP (dBm): effective isotropic radiated power – this is the radiated power from the base station antenna. It is BS power – feeder loss + antenna gain.
- e, MS antenna gain (dBi): the mobile terminal antenna has a gain that we need to include in the LB. This antenna gain may in fact be negative! There are many measurements available that show the gain for various types of mobiles. Some of these measurements show mobiles with antenna gain down below $-7\,$dBi. That is important to realize this when doing the LB calculation.
- f, Mobile station noise figure, dB: the amplifiers and electronics inside a receiver will generate noise. The relative power of that noise is defined as the noise figure. This NF together with the thermal noise floor will define the reference for the noise floor at the MS; refer to Chapter 7 for more details.

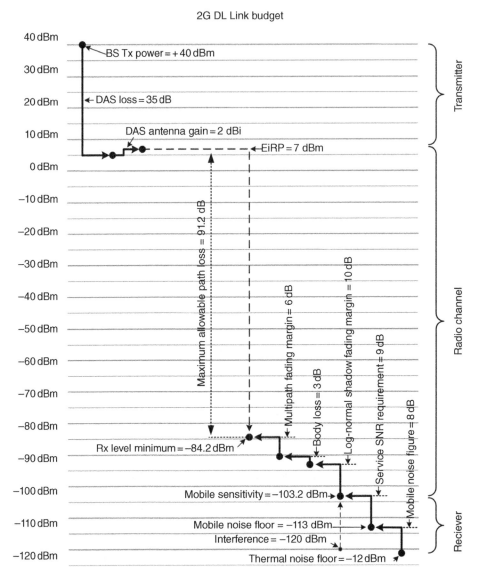

Figure 8.3 Graphical example of 2G DL link budget for voice service. EiRP, equivalent isotopic radiated power; SNR, signal-to-noise ratio.

- g, MS noise floor: as an example consider a typical 2G MS with an NF of 8 dB. You simply add the NF to the thermal noise to calculate the noise floor in the MS

$$\text{noise floor} = \text{noise power at 1 Hz} + \text{NF} + 10\log\ (\text{bandwidth})$$
$$\text{noise floor} = -174\,\text{dBm / Hz} + \text{NF} + 10\log\ (\text{bandwidth})$$
$$\text{noise floor} = -174\,\text{dBm / Hz} + 8\,\text{dB} + 10\log(200000\,\text{Hz}) = -113\,\text{dBm}$$

Table 8.1 Link budget example, 2G downlink (2G-900)

BS transmitter BS Tx power	40	dBm	*a*	Input
Feeder loss	35	dB	*b*	Input
BS antenna gain	2	dBi	*c*	Input
EiRP	**7**	dBm	*d*	a − b + c
MS receiver				
MS antenna gain	0	dBi	*e*	
MS noise figure	8	dB	*f*	2G type, 8 dB; 3G type, 7 dB
MS noise floor	**−113**	dBm	*g*	
Thermal noise floor	−121	dBm	*h*	2G = −121 dBm, 3G = −108 dBm
Interference	−120	dBm	*i*	
Service SNR requirement	9	dB	*j*	Signal-to-noise demand
Mobile sensitivity	**−103.2**	dBm	*k*	
The RF channel				
Log–normal shadow fading	10	dB	*l*	
Multipath fading margin	6	dB	*m*	
Body loss	3	dB	*n*	
Total margin	**19**	dB	*o*	l + m + n
Rx minimum level	**−84.2**	dBm	*p*	k + o
Maximum allowable path loss	**91.2**	dB	*q*	d − p
Service radius from antenna	**35**	m	*r*	Example based on PLS 38.5

For explanation of *a* to *r*, see text.

- *h*, Thermal noise floor (dBm): depending on the operating bandwidth of the radio channel, there will be certain thermal noise floor. This is a physical constant. At room temperature (17°C), the thermal noise floor is defined as:

$$\text{noise floor} = -174 + 10\log(\text{bandwidth})$$

So for 2G (200 kHz) the thermal noise floor will be:

$$2G = -174\,\text{dBm}\,/\,\text{Hz} + 10\log(200000\,\text{Hz}) = -121\,\text{dBm}$$

For 3G (3.84Mc) it will be

$$3G = -174\,\text{dBm}\,/\,\text{Hz} + 10\log(3840000\,\text{Hz}) = -108\,\text{dBm}$$

- *i*, Interference: interfering base stations transmitting on the same carrier must be taken into account. This interference will add to the noise floor, increasing the signal requirement to fulfill the SNR. It is very important to take this interference into account when doing the LB. It is highly recommended always to perform a measurement (see Section 5.2.2) of the interference level present in the building, and to use this input when calculating the link budget. In this example the interference is set to −120 dBm (very low).

We can also look at the 2G DL link budget calculation presented in Table 8.1 in a graphical representation, as shown in Figure 8.3. For me at least, this helps to clarify the calculations in a logical way. We are trying to establish the maximum allowable path loss over the radio channel, knowing the performance of the transmitter and the receiver. This is the same basic principle for all radio systems. Let us have a look at the three main components.

The Transmitting Part of the System

As we can see in Figure 8.3, we have a base station transmitting 40 dBm of power. This is then fed to a passive DAS with 35 dB of loss. The remaining power is now +40 dBm – 35 dB = 5 dBm.

This power (5 dBm) is fed to a 2 dBi omni directional gain, and we can now calculate the equivalent isotopic radiated power (EiRP) to be 5 dBm + 2 dB = 7 dBm EiRP.

So our transmit power from the antenna is 7 dBm – this is our 'launch power' in the system shown in Figure 8.3:

$$\text{Transmit power} = \text{base station power}\left(40\,\text{dBm}\right) - \text{losses}\left(35\,\text{dB}\right) + \text{antenna gain}\left(2\,\text{dB}\right) 7\,\text{dBm}$$

The Receiving Part of the System

This being a downlink example (Figure 8.3), the receiving part of the system is a 2G mobile in a 200 kHz radio bandwidth. The thermal noise power in a 200 kHz radio channel is –121 dBm. In our example, the noise figure of the mobile receiver is 8 dB so the noise floor is –121 dBm – 8 dB = 113 dBm.

We are aiming for a voice-only service, and as the signal-to-noise requirement for 2G is 9 dB, we can calculate the signal requirement to be 113 dBm – 9 dB = –104 dBm. In this example, almost no interference is received on the same channel, only –120 dBm offsetting the required receive level to maintain the 9 dB signal to noise at –103.2 dBm (–120 dBm + (–)104 dBm).

We also need to include fading margins, a log-normal shadow fading of 10 dB, multipath fading of 6 dB and 3 dB of body loss, giving us a total fading margin of 19 dB.

Now we are able to calculate the minimum required signal level:

$$\text{Rx-Level minimun} = \text{mobile sensitivity for service } (-103.2\,\text{dBm}) + \text{fading margin (19\,dB)}$$
$$= -84.2\,\text{dBm}$$

I am sure that a lot of you with 2G experience will recognize the design level above as the 'standard design level of –85 dBm.

The Maximum Allowable Path Loss (MAPL)

Now that we have established the transmit power at the antenna, and the signal level required at the mobile to maintain the service requirement, we can calculate the maximum allowable path loss. This is straightforward, as it will be the difference between the transmit power and the required signal level at the receiver.

MAPL = transmitted power (7 dBm) − required signal level (−84.2 dBm) = 91.2 dB

Both Table 8.1 and Figure 8.3 show the same 2G downlink calculation. I believe the illustration in the figure is a very good way of showing the link estimation in a straightforward and logical way.

Typical 2G Design Levels

The actual design level on 2G obviously depends on the types of service we are designing for. These requirements are reflected in Table 8.3, where you can see the requirements for each of the different 2G services. Increasingly, 2G will service voice only, with 3G and 4G then typically being used as a separate layer in the DAS to service data much more efficiently than 2G is capable of. Like any other radio system, the performance of the DAS (loss/noise figure), the transmitted power levels of the transmitters, the noise figure of the receivers and the existing co-channel interference have huge impact. If we assume voice services, most systems for 2G would be designed for:

- Basement and low interference areas: −85 dBm Rx-Level (BCCH).
- Office and high-use environment, with limited interference: −75 dBm Rx-Level (BCCH).
- High rises, high-use critical areas with some interference: −65 dBm Rx-Level (BCCH).

3G Noise Increase

On 3G we also need to take into account the noise increase due to the traffic load in the cell. Therefore, with 50% of load in the cell we need to add 3 dB in noise increase; refer to Section 2.4.3 for more detail on how to calculate the noise increase.

- j, Service SNR (Signal to Noise Ratio) requirements (dB): for the specific service we need to support over the radio link there will be a quality requirement, in terms of SNR. This is a definition of how strong the signal receive level has to be above the noise floor in the channel in order for the RF service to work. If we have a service that needs 9 dB SNR (2G voice) in order to work, the lowest acceptable signal would be:

$$\text{Required signal level} = -113\,\text{dBm (MS noise floor)} + 9\,\text{dB (service requirement)}$$
$$= -104\,\text{dBm}$$

The SNR Requirement is Related to the Data Rate

The critical parameter that affects the service level on the UL and DL is not the absolute signal level on the link but the quality of the RF link, the SNR. For different services there will be different demands on the quality of the RF link. The higher the service demands (data rates), the better the RF link needs to be. This is why the higher data rates on 3G and HSPA are more sensitive to any degradation of the indoor DAS noise figure or attenuation. Therefore we must insure the lowest possible noise figure and attenuation of the DAS, in order to perform at the higher data rates, and cater for future data services.

3G design margin

For 3G design we need to define the desired E_b/N_o and also apply the processing gain of the radio service.

$$\text{Processing gaing} = \text{chip rate / user data rate}$$
$$\text{Processing gaing} = 3.84 \text{ M / user rate (linear)}$$
$$\text{Processing gaing} = 10\log(3.84\,\text{M / user rate}) \text{ (dB)}$$

Using these formulas we can calculate the processing gain of the specific 3G data service as shown in Table 8.2.

Examples

Processing gain for high − speed data, 384 kbps : $10\log(3.84\,\text{Mcps} / 384\,\text{kb}) = 10\,\text{dB}$

Processing gain for speed 12.2 kbps : $10\log(3.84\,\text{Mcps} / 12.2\,\text{kbps}) = 25\,\text{dB}$

The use of orthogonal spreading codes and processing gain is the main feature of 3G/WCDMA, giving the system robustness against self-interference. The main reason is the frequency reuse factor of 1, due to the rejection of noise from other cells/users.

When we know the required bit power density, E_b/N_o (energy per bit/noise) for the specific service (voice 12.2 kbps + 5 dB, data + 20 dB) we are able to calculate the required signal-to-interference ratio.

Voice at 12.2 kbps needs approximately 5 dB wideband signal-to-interference ratio minus the processing gain, $5 - 25\text{dB} = -20\text{dB}$ In practice, the processing gain means that the signal for the voice call can be 20 dB lower than the interference/noise, but still be decoded.

Soft Handover Gain (3G)

In traditional macro link budgets you will include a soft handover gain, typical using 3 dB. This is because the users on the edge of the cell are in soft handover, and using macro diversity improves the link. However, for indoor 3G planning I recommend not using the soft handover gain in the link budget. You should minimize the areas of soft handover, to limit the load on the node B and the network.

Table 8.2   Processing gain

User	Processing rate gain
12.2 kbps	25 dB
64 kbps	18 dB
128 kbps	15 dB
384 kbps	10 dB

Table 8.3 Examples of 2G coverage indoors on different service requirements; note that the signal level in this example is with no interference

LB example, 2G-900 at 12 dBm EiRP, moderate dense office				
Service	Minimum *C/I*	Maximum APL	Radius	DL level
2G voice	9 dB	97 dB	55 m	−85.0 dBm
EDGE-MCS1 8.8 kbps	9.5 dB	96 dB	53 m	−84.5 dBm
EDGE-MCS2 11.2 kbps	12 dB	94 dB	46 m	−82.0 dBm
EDGE-MCS3 14.8 kbps	16.5 dB	89 dB	35 m	−77.5 dBm
EDGE-MCS4 17.6 kbps	21.5 dB	84 dB	26 m	−72.5 dBm
EDGE-MCS5 22.4 kbps	14.5 dB	91 dB	39 m	−79.5 dBm
EDGE-MCS6 29.6 kbps	17 dB	89 dB	34 m	−77.0 dBm
EDGE-MCS7 44.8 kbps	23.5 dB	82 dB	23 m	−70.5 dBm
EDGE-MCS8 54.4 kbps	29 dB	77 dB	16 m	−65.0 dBm
EDGE-MCS9 59.2 kbps	32 dB	74 dB	14 m	−62.0 dBm

2G Example

The impact on the data service offerings in a typical indoor 2G-900 system can be seen in Table 8.3. This example shows first and foremost the demand on the SNR and the *C/I* (channel/interference).

For 2G voice the demand is a minimum of 9 dB *C/I*. In this example (the downlink budget in Table 8.1), the maximum allowable link loss on the DL in order to provide voice service is 97 dB. Therefore, the total loss from the base station to the mobile must not exceed 97 dB. In this environment (moderate dense office) that corresponds to about 55 m (note that other parameters and margins also play a major role – the sensitivity of the mobile, the noise figure of the mobile, the fading margin, co-channel interference etc.; all these parameters will be elaborated later on in this chapter).

In this case the DL level needed to provide sufficient signal level is −85 dBm. Whereas one can see that the service requirement for EDGE-MCS9 (59.2 kbps) is 23 dB stricter on *C/I*, with a maximum allowable path loss of 74 dB. This results in a service level of −62 dBm, or a reduction of the service range from the indoor DAS antenna from 55 m to support voice to only 14 m to support EDGE-MCS9.

- *k*, Mobile sensitivity: interference, i.e. signals transmitting on the same frequency as the supported service, will desensitize the receiver in the mobile.
  Knowing the interference coming from other base stations using the same DL frequency (measured with a test receiver as described in Section 5.2.2), and knowing the mobile noise figure, the mobile antenna gain and the SNR requirement for the specific service, you can then calculate the mobile receiver sensitivity for the specific service:

$$\text{RxSensitivity} = 10\text{Log}\left(\left(10^{\left(\frac{ilf+MSantgain}{10}\right)}\right)+\left(10^{\left(\frac{TH_{noise+MSNF}}{10}\right)}\right)\right)-\text{MsAntGain}+\text{ServiceREQ}$$

Example

$$\text{RxSensitivity} = 10\text{Log}\left(\left(10^{\left(\frac{-120+0}{10}\right)}\right)+\left(10^{\left(\frac{-121+8}{10}\right)}\right)\right) - 0 + 9$$

$$\text{RxSensitivity} = 10\text{Log}\left(\left(10^{(-12)}\right)+\left(10^{(-11.3)}\right)\right) - 0 + 9$$

$$\text{RxSensitivity} = -121.20 - 0 + 9$$

$$\text{RxSensitivity} = -103.2\,\text{dBm}$$

In this example, shown in Table 8.1, the MS sensitivity is calculated to $-103.2\,\text{dBm}$.

The Radio Channel

- $l + m$, Fading margins: in any RF environment there will be fading due to reflections and diffractions of the RF signal. Several books and reports have been written about fading margins and RF planning; this is beyond the scope of this book. Typically, in indoor environments you should use a total fading margin of around 16–18 dB to obtain 95% area probability of the desired coverage (Refer to [6]).

RF DL Planning Level

If we continue to calculate using the example from Figure 8.1, we can calculate the planning level for the RF signal:

- n, Body loss (dB): the MS will be affected by the user, who will act as 'clutter' between the MS and the BS antenna. To take this into account in the LB, you will need to apply 'body loss' to the calculation. A typical number used for body loss is 3 dB.
- o, Total design margin: in this example, for the total design margin i,

    $$\text{total fading margin} = \log-\text{normal shadow fading} + \text{multipath fading} + \text{body loss}$$
    $$\text{total fading margin} = 10\,\text{dB} + 6\,\text{dB} + 3\,dB = 19\,\text{dB}$$

- p, Minimum level at cell edge (RxMin): knowing the mobile sensitivity and the total fading margin, we can now calculate the design goal at the cell edge:

    $$RxMin = \text{mobile sensitivity} + \text{design margin}$$
    $$RxMin = -103.2\,\text{dBm} + 19\,\text{dB} = -84.2\,\text{dBm}$$

This is 0.8 dB higher than the signal requirement shown in Table 8.3, and is due to the $-120\,\text{dBm}$ interference on the channel, as shown in Table 8.1.

- q, Maximum allowable path loss: finally, we can calculate the maximum link loss that can fulfill the design criteria. We know the radiated power EiRP and the minimum signal level at the cell edge; therefore we can calculate the MAPL

    $$MAPL = EiRP - RxMin$$
    $$MAPL = 7\,dBm - (-84.2\,dBm) = 91.2\,dB$$

- *r*, Calculating the antenna service radius: now we have established the MAPL, we can calculate the service range of the antenna in the particular environment:

$$\text{coverage radius}(m) = 10^{(\text{APL}-\text{PLlm})/\text{PLS}}$$

Remember the Uplink!

We have only done half the link budget at this point. We need to perform the exact same calculation, in reverse, for the uplink (as shown in Table 8.4; see also Figure 8.4 for a graphical representation). For 2G indoor planning, the LB will often be limited by the DL power. However, it is important to confirm this in the LB, so always perform a two-way link budget calculation for 2G as well. For 3G/4G you could well be limited by the UL, depending on the load profile of the cell.

Balance the Link

You will have different service distance on the DL compared with the UL. Try to balance your LB, maybe with parameter adjustments on the cell, once it is in service. It makes no sense to blast to high power on any link UL or DL if you are dependent on the other link limiting the service radius of the cell. Then you might as well power down the dominant link and balance the LB, thus limiting the interference from that particular link to adjacent cells or mobiles.

Table 8.4 Link budget example, 2G uplink

MS Tx power	33	dBm	a	Input
MS antenna gain	0	dBi	b	Input (can be negative!)
EiRP	**33**	dBm	c	$a+b$
DAS receiver				
BS noise figure	3	dB	d	2G type, 3 dB; 3G type, 4 dB
DAS noise figure	0	dB	e	The NF of the active DAS
DAS passive loss	35	dB	f	The loss of the passive DAS
System noise figure	**38**	dB	g	$(d+f)$ or $(d+e)$
Thermal noise floor	−121	dBm	h	2G = −121 dBm 3G = −108 dBm
Interference	−120	dBm	i	
Service SNR requirement	9	dB	j	Signal-to-noise demand
DAS antenna gain	2	dBi	k	
BS sensitivity	**−76**	dBm	l	
The RF channel				
Log–normal shadow fading	10	dB	m	
Multipath fading margin	6	dB	n	
Body loss	3	dB	o	
Total margin	**19**	dB	p	$m+n+o$
Rx minimum level	**−57.0**	dBm	q	$l+p$
Maximum allowable path loss	**91.2**	dB	r	$c-q$
Service radius from antenna	**23**	m	s	Example based on PLS 38.5
				MS transmitter

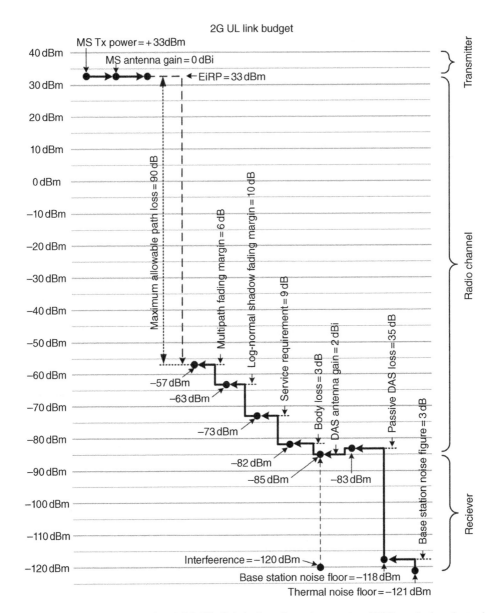

Figure 8.4 Graphical example of 2G UL link budget for voice service. EiRP, equivalent isotopic radiated power.

If we consider the example we calculated in Figures 8.3 and 8.4 (Tables 8.2 and 8.4), we conclude that the maximum allowable path loss on the downlink (from base station to mobile) is 91.2 dB, while on the uplink (reverse link) the maximum allowable path loss is 90 dB. This link is very well balanced and should have no issues.

Passive Loss Will Attenuate the UL Interference

Also note that the uplink interference we receive at the DAS antenna in Figure 8.4 (−120 dBm) will never actually reach the base station receiver due the loss of the passive DAS.

The UL interference will only start to be an issue once it is not attenuated enough by the loss of the passive DAS. I would recommend a margin of about 3 dB more than the loss of the passive DAS. In our case, with a passive DAS with 35 dB of loss and a thermal noise floor on 2G of −121 dBm due to the 200 kHz bandwidth, we can calculate the UL interference 'problem level' at the DAS antenna as $35 − 3 + (−121) = −89$ dBm. As long as we receive less than −89 dBm of interference, then this interference signal power will never reach the UL port of the base station due to the attenuation of the passive DAS

3G DL CPICH Calculation Example

In Figure 8.5 we can see a simple downlink calculation of common pilot channel (CPICH) requirements in a passive DAS; as the CPICH has an activity factor of 100% (it is constantly transmitting, typically at about 10 dB less than full power), it is a simple link calculation. One could argue that the fading margins used in Figure 8.5 are too conservative for a wideband signal, inside a building, and often a 'standard' design level for CPICH power would be in the order of −90 to −85 dBm.

For the actual services, you will need to do a link budget for the specific data service, including the processing gain (see Table 8.2). You also need to take into account the power transmitted for the specific service, activity factor, etc. Most people, however, will design purely on the CPICH DL power, but remember that you might be uplink-limited for the higher data services in 3G.

You also need to decide how much noise increase you will allow in the cell. Typically it will be a 3 dB noise increase (50% cell load) to avoid too much impact of noise load on the surrounding cells. However, sometimes for well isolated indoor cells, you could consider allowing more capacity load, and use less allocated power for the CPICH, allowing more capacity to be carried on the cell.

Typical 3G Design Levels

The actual design level on 3G obviously depends on the type of services being designed for. More and more 3G will service only voice and slow-speed data (SMS and IM), and 4G will then typically be used as a separate layer in the DAS to service high data rates. Like any other radio system, the performance of the DAS (loss/noise figure), the transmitted power levels of the transmitters, the noise figure of the receivers and existing co-channel interference have huge impact. If we assume voice and slow-speed data services, most systems for 3G would be designed for:

- Basement and low-interference areas: −90 dBm CPICH.
- Office and high-use environment, with limited interference: −85 dBm CPICH.
- High rises, high-use critical areas with some interference: −75 dBm CPICH.
- High-interference areas even higher than −65 dBm CPICH design level.

Isolation Requirement for 3G

To avoid extensive soft handover zones, it is recommended to design for a 6–10 dB dominance of your DAS over other cells covering the building to provide sufficient isolation. There will be 'transition' areas where you need to have overlap, so make sure to plan these handover areas in places where there is limited capacity load.

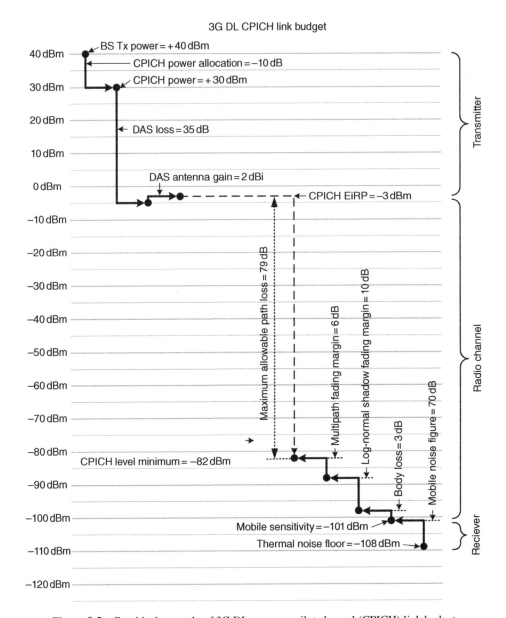

Figure 8.5 Graphical example of 3G DL common pilot channel (CPICH) link budget

8.1.3 Link Budgets for Indoor Systems

When designing indoor systems, there are some additional parameters you need to take into account in the link budget calculation. Different parameters need to be considered, depending on whether it is a passive or an active DAS you are designing (see Figure 8.6).

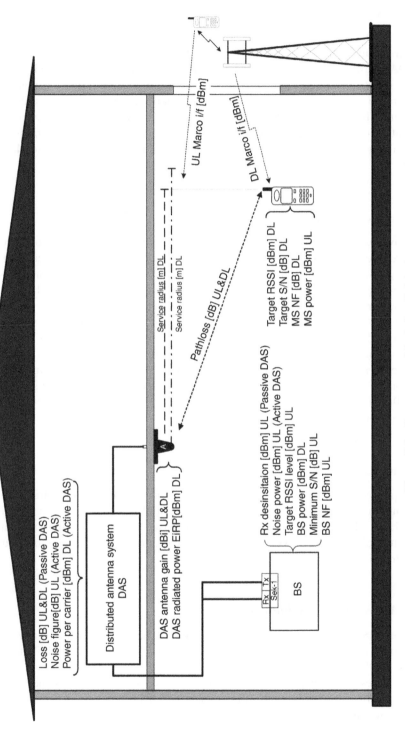

Figure 8.6 Components of the indoor link budget

8.1.4   Passive DAS Link Budget

The passive DAS is based on coaxial cables and splitters, as described in Section 4.3. From a link budget calculation it is fairly easy to design a passive DAS. The attenuation of the DAS will degrade both UL and DL with the same loss, corresponding to added link loss due to distance. The challenge is to get the exact information regarding the attenuation to each antenna in the system. This depends on the exact installation and cable route. Therefore detailed information on the actual installation is needed, and this is upfront before any system is installed. This can be a challenge for the radio planner, often making passive DAS design more installation planning exercise than radio planning.

The Downlink Calculation

The passive DAS will attenuate the power from the signal source, base station or repeater, so be sure to offset the DL power with the loss of the passive DAS when calculating the link budget. Be aware that you need to evaluate the DL link budget for all antennas in the system, owing to the fact that the individual antennas in the system will have different losses, and hence different radiated power levels and service ranges. Be very careful not to overdrive the passive DAS, especially when operating a multicarrier or multioperator system. It is often forgotten how high the power on the passive DAS actually is, especially on the components close to the base stations. One example is a system with four 2G base stations, each feeding 43 dBm to eight carriers, a total of 24 carriers; 43 dBm is 20 W, and the composite power will be 58 dBm, or 630 W. Even using good PIM with specifications of 120 dBc, you might have severe inter-modulation products (refer to Section 5.7.3 for more details on PIM).

The Uplink Calculation

The UL part of the link budget calculation of the passive DAS is also straightforward; the attenuation of the DAS will impact the NF of the base station, as described in Section 7.1.7. The different attenuation from the base station to each individual antenna will result in different UL coverage ranges from each antenna, so be sure to calculate all antenna locations.

8.1.5   Active DAS Link Budget

The active DAS is described in detail in Section 4.4. An active DAS will typically be 'transparent', so the downlink signal you feed into the DAS you will have out of the antennas, and vice versa for the uplink plus or minus the gain you set in the system. This makes active DAS very easy to design and implement since you do not have to take cable distances and losses into concern. However, there are other things you need to pay attention to when calculating the link budget for an active DAS.

The Downlink Calculation

The limitation on the downlink power in an active DAS is usually the composite power capabilities of the DL amplifier in the 'remote unit'. The RU is located close to the antenna and will have one shared amplifier that the entire spectrum supported by the DAS is sharing. For multiband DAS you would typical have dedicated amplifiers for each supported band. Typically you would have a power level at one carrier of, for example, 15 dBm, and then every time you doubled the number of carriers you would back off about 3–3.5 dB in order not to overdrive the system and cause distortion. Be very careful only to operate the active DAS system at the specified PPC (see Section 4.12.2 for more details). Otherwise you might cause interference problems for own and other services.

Hybrid DAS Link Budget

The hybrid DAS as described in Section 4.5 is a mix of passive and active DAS; you need to incorporate both strategies from above when doing the link budget calculation.

8.1.6 The Free Space Loss

The Path Loss

A major component in the LB is the path loss between the Tx and Rx antenna. This loss depends on the distance between the Tx and Rx antenna, and the environment that the RF signal has to pass. Free space loss (Figure 8.7) is a physical constant. This simple RF formula is valid up to about 50 m from the antenna when in line-of-sight indoors. The free space loss does not take into account any additional clutter loss or reflections, hence the name. The free space loss formula is:

$$\text{free space loss } (\text{dB}) = 32.44 + 20(\log F) + 20(\log D)$$

where f = frequency (MHz) and D = distance (km).

8.1.7 The Modified Indoor Model

In addition to the free space loss, we have the loss due to the environment. The main component from the indoor environment is penetration losses through walls and floor separations. The complete 'model' for indoor propagation loss could look like this:

$$\text{path loss} = \text{free space loss} + \text{wall loss}$$

This simple model gives a fair accuracy, provided that you have sufficient data for wall losses for the particular building (see Figure 8.8).

It is straightforward to create your own model and also to calculate the free space loss, and add the losses of the various walls. Obviously this method is not 100% accurate – no prediction

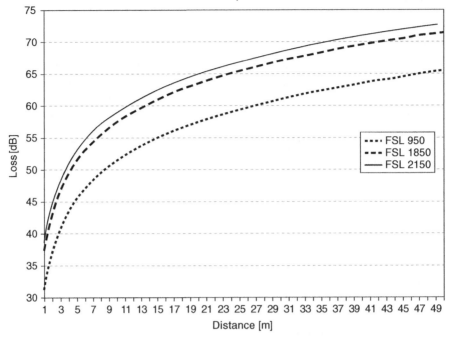

Figure 8.7 Free space losses 1–50 m

<table>
<tr><td colspan="5" align="center">Indoor calculator MTO. V.1.0.1</td></tr>
<tr><td></td><td colspan="3" align="center">Inputs</td><td></td></tr>
<tr><td></td><td>No. of walls</td><td>Att-900</td><td>Att-1800</td><td></td></tr>
<tr><td>Wall Att. wall-type 1</td><td>1</td><td>4</td><td>8</td><td>[dB]</td></tr>
<tr><td>Wall Att. wall-type 2</td><td>1</td><td>12</td><td>16</td><td>[dB]</td></tr>
<tr><td>Wall Att. wall-type 3</td><td>0</td><td>14</td><td>18</td><td>[dB]</td></tr>
<tr><td>Distance to antenna</td><td></td><td></td><td>25</td><td>[m]</td></tr>
<tr><td></td><td></td><td>900 MHz</td><td>1800 MHz</td><td></td></tr>
<tr><td>BTS RX sensitivity</td><td></td><td>−107</td><td>−104</td><td>[dBm]</td></tr>
<tr><td>BTS max power</td><td></td><td>33</td><td>33</td><td>[dBm]</td></tr>
<tr><td>MS RX sensitivity</td><td></td><td>−102</td><td>−100</td><td>[dBm]</td></tr>
<tr><td>MS max power</td><td></td><td>33</td><td>33</td><td>[dBm]</td></tr>
<tr><td>DAS coax system loss</td><td></td><td>35</td><td>35</td><td>[dB]</td></tr>
<tr><td colspan="5" align="center">Results</td></tr>
<tr><td></td><td></td><td>900 MHz</td><td>1800 MHz</td><td></td></tr>
<tr><td>Pathloss</td><td></td><td>75.46</td><td>89.56</td><td>[dB]</td></tr>
<tr><td>RX signal UL @ BTS</td><td></td><td>−77.46</td><td>−91.56</td><td>[dBm]</td></tr>
<tr><td>RX signal DL @ MS</td><td></td><td>−77.46</td><td>−91.56</td><td>[dBm]</td></tr>
<tr><td>C/N margin DL @ MS excel. fading</td><td></td><td>−13.46</td><td>−29.56</td><td>[dB]</td></tr>
<tr><td>C/N margin UL @ BTS excel. fading</td><td></td><td>−8.46</td><td>−25.56</td><td>[dB]</td></tr>
</table>

IB-Model
900 :L(dB) = 91.5 + 20 log d + p·W(k)
1800:L(dB) = 97.6 + 20 log d + p·W(k)

L(dB) = 32.5 + 20g f + 20 log d + k * F(k) + p*W(k) + D(d–db)
 L = path loss (dB)
 f = fequency (MHz)
 d = distance MS / DAS antenna
 k = no. of floor separations, the signal has to pass
 F = floor separation loss.
 P = no. of walls the signal has to cross
 W = wall attenuation
 D = linear attenuations factor dB/m
 db = 50 m 'breakpoint' additional 0.2 dB/m

Figure 8.8 Indoor 'model' calculator in Excel, the first homemade tool I ever made for indoor designs

tools actually are. However, it gives you a fair estimate of the possible path. The tool shown in Figure 8.8 is my first indoor link calculator; I have done about 75 indoor 2G designs using this calculator.

8.1.8   The PLS Model

A widely used model for calculating the path loss relies on 'path loss slopes' (PLS). These PLS are different attenuation slopes for different types of environments and frequencies. A general model is based on empirical analysis of a vast number of measurement samples in these different types of environments.

How is the PLS Established?

The PLS is derived from measurement samples in different environments (see Figure 8.9). The path loss is measured at different distances from the antenna, and it is possible to average out these thousands of measurements and calculate a path loss slope.

It is important to note that the PLS is an average value, and that there will be both positive and negative variations from the PLS. It is my experience that these variations will be covered by the fading margins suggested in Section 8.1.2.

It is very important to use the right PLS constant, according to the environment. The model based on the PLS values from Table 8.5 looks like this:

$$\text{path loss } (\text{dB}) = \text{PL at 1m } (\text{dB}) + \text{PLS} \times \log(\text{distance, m})$$

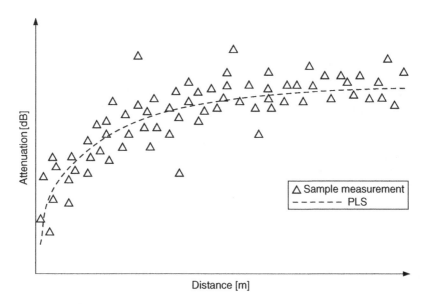

Figure 8.9   Example of a path loss slope, based on measurement samples

Table 8.5 PLS constants for different environments

Type of environment	900 MHz PLS	1800/2100 MHz PLS
Open environment, few RF obstacles Parking garage, convention center 33.7	30.1	
Moderately open environment, low to medium number of RF obstacles Factory, airport, warehouse	35	32
Slightly dense environment, medium to large number of RF obstacles Shopping mall, office that is 80% cubicle and	36.1	33.120% hard wall
Moderately dense environment, medium to large number of RF obstacles Office that is 50% cubicle and 50% hard wall	37.6	34.8
Dense environment, large number of RF obstacles	39.4	38.1Hospital, office that is 20% cubicle and 80% hard wall

You can calculate the free space loss at 1 m, using the free space formula from Section 8.1.6:

- 800MHz = 31dB
- 950MHz = 32dB
- 1850MHz = 38dB
- 2150MHz = 39dB
- 2600MHz = 41dB

You will find different PLS values from various sources; the internet is a good place to start. You might even develop your own PLS based on indoor RF survey measurements (see Section 5.2.5) in the various environment types and on the frequencies you use. Or you can use Table 8.5 as a start, and fine tune these PLS to match your needs. Do understand that these types of RF models are just guidelines, and can never be considered 100% accurate. Use these models accordingly, and apply some common sense based on your experience. If in doubt, then always do an RF survey measurement to verify the model used (see Section 5.2.5). Thus it would also be possible to establish the correct fading margin for the particular environment measured.

In Figure 8.10 you can see an example of the suggested PLS in Table 8.5 for a dense office environment, for 2G, 3G and 4G frequencies. There is considerable difference between the PLS attenuation and the free space loss, as shown in Figure 8.11. Note that these PLS models assume that you use common sense when placing the antennas, so only use visible antenna placement below the ceiling (refer to Section 5.3 for more details).

8.1.9 Calculating the Antenna Service Radius

When you have completed the link budget calculation and have established the APL, you can calculate the service range of the antenna in the particular environment:

$$\text{coverage radius (m)} = 10^{(APL-PL1m)/PLS}$$

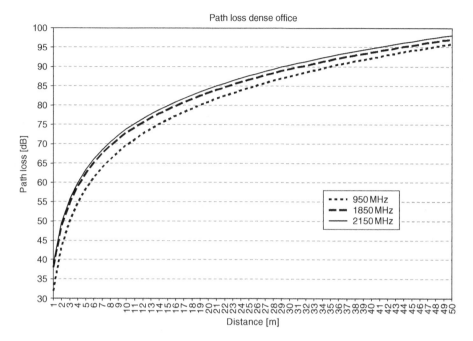

Figure 8.10   Path loss based on PLS for a dense office, 1–50 m

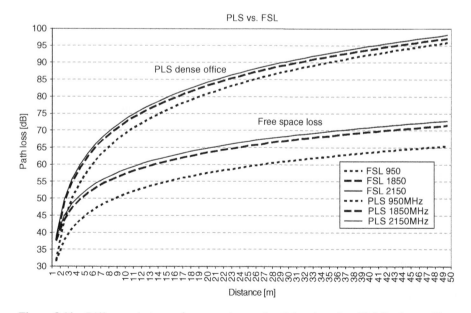

Figure 8.11   Difference between free space loss and path loss based on PLS for dense office

Table 8.6 2G 1800 example of DL service range from 2 dBi omni antenna

2G 1800 service radius at four different DL levels	DL EiRP from antenna							
	7 dBm		12 dBm		17 dBm		22 dBm	
DL target level	−65	−70	−65	−70	−65	−70	−65	−70
	−75	−80	−75	−80	−75	−80	−75	−80
Open	14 m	20 m	20 m	30 m	30 m	44 m	44 m	64 m
Coverage radius	30 m	44 m	44 m	65 m	65 m	94 m	94 m	138 m
Moderately open	12 m	17 m	17 m	25 m	25 m	35 m	35 m	50 m
Coverage radius	25 m	35 m	35 m	50 m	50 m	72 m	72 m	103 m
Slightly dense	11 m	16 m	16 m	22 m	22 m	31 m	31 m	44 m
Coverage radius	22 m	31 m	31 m	44 m	44 m	63 m	63 m	89 m
Moderately dense	10 m	14 m	14 m	19 m	19 m	26 m	26 m	37 m
Coverage radius	19 m	26 m	26 m	37 m	37 m	51 m	51 m	71 m
Dense	8 m	11 m	11 m	15 m	15 m	20 m	20 m	27 m
Coverage radius	15 m	20 m	20 m	27 m	27 m	36 m	36 m	50 m

Log–normal shadowing fading, 10 dB; multipath fading margin, 6 dB; body loss, 3 dB; no interference; 8 dB MS NF.

Examples of DL Service Ranges for Indoor Antennas

This is just an example on the downlink service ranges from an indoor DAS antenna (Table 8.6). Make your own link budget calculation to establish what the actual service range is for both uplink and downlink for your particular design.

8.2 4G Link Budget

We covered some of the 4G system basics in Section 2.7, but let us take a look at the link calculation and estimation for 4G.

Isolation Requirements

Like other radio systems we need to maintain a good 'clean' serving cell signal, with good isolation to any other cells on same frequency. In 4G we have a frequency reuse of 1, as described in Section 2.7 (all cells using same frequency). For well-isolated, high-capacity cells in a DAS, it is recommended to be about 10–15 dB more dominant than other cells in most of the coverage area. Obviously there will be 'transition' areas where you need to have an overlap – make sure to plan these handover areas in places where there is limited capacity load.

Receiver Blocking on 4G

To avoid saturation of the receivers in 4G, it is highly recommended not to exceed –25 dBm reference symbol received power (RSRP).

8.2.1   4G Design Levels

In 2G we designed for Rx-Level, and in 3G we designed for CPICH level; in 4G, RSRP is the reference and a design target level of –95 to –85 dBm seems to be standard among most mobile operators, with –80 dBm in high-demand areas, and even –75 to –70 dBm RSRP in VIP areas. Like any other RF system, the actual design target level will depend on the interference level in the building, and you have to adjust the design level to overcome this interference (see section Section 3.5.6 for more details):

- Basement and low-interference areas: –100 to –95dBm RSRP.
- Office and high-use environment, with limited interference: –90 dBm RSRP.
- High rises, high-use critical areas with some interference: –85 dBm RSRP.
- High-interference areas: even higher than –85 dBm RSRP design level.

We must keep in mind that the total composite power of the 4G base station must be shaed by the sub-carriers in service. The number of sub-carriers relates to the RF channel bandwidth (see Figure 8.12 for an example of a 5 MHz 4G channel). The 5 MHz bandwidth allows 25 physiscal resource blocks (PRBs). Each PRB comprises 12 sub-carriers so there is a total of

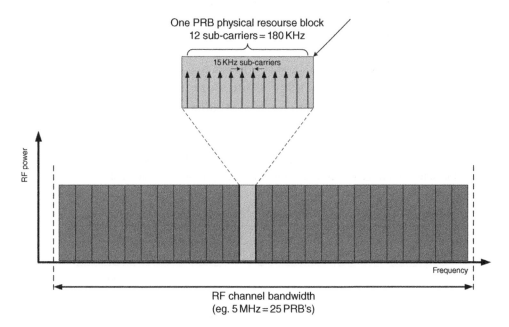

Figure 8.12   Physical resource blocks (PRBs) vs. 4G CH bandwidth

$25 \times 12 = 300$ sub-carriers to share the composite power resource. So the more carriers, the less power there is per sub-carrier, and the more DAS equipment would need to be deployed to maintain the required design-level RSRP.

8.2.2 RSRP, Reference Symbol Transmit Power

The RSRP is used to measure the RF level of the cell for mobility management. The RS is also used to identify the individual MIMO paths. The RSRP is the linear-averaged RS signal level over the six RS in each resource block (RB). In order to calculate the link budget, we first need to be able to calculate the transmitted power of the RS. The RS transmit power relates to the bandwidth of the 4G channel (the number of sub-carriers).

This is also evident in Figure 8.12, which shows a 5 MHz 4G channel with 25 PRBs, each with 12 sub-carriers, giving a total of $25 \times 12 = 300$ sub-carriers.

We can calculate the power per sub-carrier in relation to the total power of the transmitter.

Example

Let us calculate an example of a 15 MHz 4G channel and a composite power of the transmitter of 43 dBm. In Table 8.7 we can see that a 15 MHz channel has 75 RBs. Each RB comprises 12 sub-carriers, giving a total of $75 \times 12 = 900$ sub-carriers.

Therefore each of the 900 sub-carriers gets assigned 1/900 of the total power. We will now be able to calculate the power assignment to each sub-carrier in relation to the total power:

$$\text{Power drop per sub-carrier of 900 sub-carrier} = 10 \times \log(900) = 29.5 \text{ dB}$$

Now we can calculate the transmitted power (TxPWR) per sub-carrier:

$$\text{TxPWR RSRP} = \text{total power} - (\text{power drop per sub-carrier})$$
$$\text{TxPWR RSRP} = 43 \text{dBm} - 29.5 \text{dB}$$
$$\text{TxPWR RSRP} = 13.5 \text{dBm}$$

In Table 8.7 we can see the same calculation results, of each 4G channel bandwidth from 1.4MHz to 20MHz. Like any other shared resource, the PPC Power per Carrier drops the more carriers the same amplifier will have to support. Therefore we need to back off from the full

Table 8.7 RF bandwith (BW)/ resource blocks showing RS transmit power

4G channel BW (MHz)	No. of resource blocks	No. of sub-carriers	Total power / RS power (dB)
1.4	6	72	18.6
3	15	180	22.6
5	25	300	24.8
10	50	600	27.8
15	75	900	29.5
20	100	1200	30.8

composite power according to the Table 8.7 to assure we do overload the power amplifier, and drives it into a nonlinear performance area and degrades the quality of the transmitted 4G service.

8.2.3   4G RSSI Signal Power

The received signal strength indicator (RSSI) is the total composite received strength of a radio channel, a 'raw' RF measurement over the full channel bandwidth. In case of 4G, the RSSI is the sum of the power of all active sub-carriers. Therefore the RSSI is dependent on the bandwidth of the carrier; the wider the bandwidth, the more sub-carriers will be transmitted.

When designing 4G systems, we take as our reference the RSRP. We can calculate the expected RSSI of the full 4G channel when we know the RSRP level. This is very useful when measuring signal strength and power levels in 4G systems, e.g. if you do not have the equipment that can decode and measure the RSRP directly.

Let's try to calculate the RSSI when we know the RSRP we are designing for, using the same approach as before and using the values of Table 8.7, on a 5 MHz 4G carrier.

Example
We can see in Table 8.7 that a 5 MHz 4G carrier has 25 RBs, each with 12 sub-carriers, giving a total of $25 \times 12 = 300$ sub carriers.

This is a power offset of: $10 \times \log(300) = 24.77\,dB$. This means that the RSRP is 24.77 dB lower than the RSSI.

If we were designing for −85 dBm RSRP then the expected RSSI would be 24.77 dB higher:

$$RSSI = RSRP + offset$$
$$RSSI = -85\,dBm + 24.77\,dB = -60.23\,dBm\,(in\,full\,5\,MHz)\,bandwidth$$

Typical 4G Design Levels

Obviously, the exact RSRP design target level is dependent on the actual interference in the building, noise power from any active elements, and losses from passive elements. Typical design levels will often range from −87 dBm RSRP in low-interference areas up to more than −65 dBm for high-interference areas.

8.2.4   4G Coverage vs. Capacity

As we can see in Table 8.7, there is a difference in the assigned RS transmit power, depending on the transmitted 4G channel bandwidth, according to the number of sub-carriers supported by the actual channel bandwidth. If we consider a DAS remote unit with a composite power capability of 40 dBm, we can calculate the transmitted RS power level in Table 8.8 according to the actual 4G channel bandwidth by compensating with the RS power offset shown in Table 8.7.

Table 8.8 RF bandwidth (BW)/resource blocks showing RS transmit power

4G channel BW (MHz)	DAS remote unit composite power (dBm)	No. of sub-carriers	RS transmit power (dBm)
1.4	40	72	21.4
3	40	180	17.4
5	40	300	15.2
10	40	600	12.2
15	40	900	10.5
20	40	1200	9.2

We can see in Table 8.8 that, for example, a 5 MHz 4G channel would transmit an RS power level of 15.2 dBm, whereas a 20 MHz channel would have a power drop of 6 dB relative to the 5MHz channel (four times the bandwidth, a 6 dB difference) and transmit 9.2 dBm. In other words, if you are receiving a 5 MHz 4G channel at a given distance from the 40 dB remote unit from Table 8.8 of –85 dBm RSRP, then a 20 MHz 4G channel from the same remote unit would receive 6 dB lower at –91 dBm.

Therefore you can decide to design for more coverage-driven areas (sectors) in your DAS for the low-capacity part of the system, and other cells in the same project could be more focused on capacity with a wider bandwidth, needing more DAS equipment to support the wider bandwidth, and hence higher power levels in the building. This could be a large airport, where you would design for more coverage than capacity in the parking areas, and need more bandwidth in the hotspot areas in the terminal buildings (see Section 14.4 for more details on airport design.)

The principle of different coverage footprints of a 4G antenna in a DAS with different bandwidths can be seen in Figure 8.13. It is evident that the extra power of using a narrower bandwidth gives you a lot of footprint when aiming for same RSRP design level, in the example –85 dBm RSRP.

8.2.5 4G DL RS Link Budget Example

Let us make a brief RS DL simple link budget for a 20 MHz carrier. We assume a base station with 43 dBm of composite power output, feeding a passive DAS with 20 dB of loss; we are aiming for an RSRP level of –85 dBm:

Example
Calculating the RS transmit power from the base station

$$\text{Base station composite power} = 43\,\text{dBm}$$

Reference signal power drop (see Table 8.8) = 30.8 dBm
Reference signal output power at base station = 43 dBm – 30.8 dB = 12.2 dBm

Calculating the RS transmit power from the DAS antenna
Passive DAS loss = 25 dB, DAS antenna gain = 2 dBi
  RS EiIRP = BS RS power – DAS loss + DAS antenna gain
  RS EiRP = 12.2 dBm – 20 dB + 2 dB = –5.8 dBm EiRP

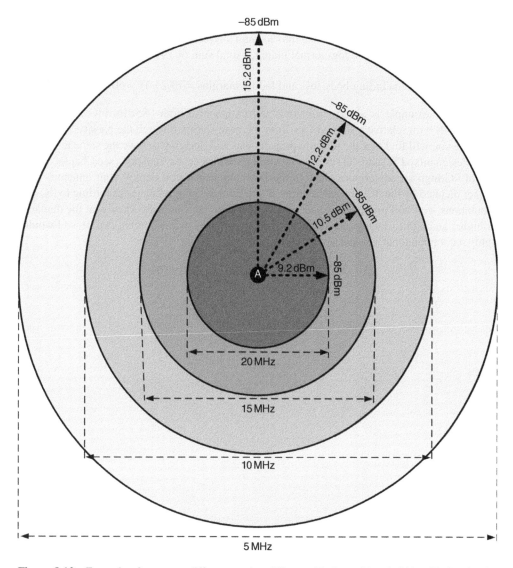

Figure 8.13 Example of coverage difference using different 4G channel bandwidths. Aiming for the same RSRP DL level will give different coverage radii due to the different transmitted RSRP

Calculating the MAPL

The MAPL is the difference between the transmitted power and the desired receive power including fading margins and body losses. From the above we have:

$$RS\,EiRP = -5.8\,dBm$$
$$RS\,target\,level\ = 85\,dBm$$

So without any margins, the maximum loss we can allow is -5.8 dBm $-(-85$ dBm$)=$ 79.2 dB. However, we also need to take into account a typical 'body loss' of 3 dB and an additional 16 dB of multipath and log-normal fading, a total sum of 19 dB.

MAPL including body loss and fading margins $= 79.2 - 19 = 60.2$ dB.

The above example is a very strict link calculation of a high -85 dBm 4G MAPL in a passive DAS with relatively limited loss from the base station through the passive DAS. But I am sure you will find that the real-life projects you will design will have the same tendency. If you design mixed 3G and 4G systems, you will have to accept a 'hotspot' (see Chapter 5.4.1) type of 4G high-speed services close to the DAS antenna if you deploy your antennas in a manner dictated by the 3G link calculation. If you allocate your antennas according to the 4G calculations, you will probably over-design the 3G part. However, we know that the demand for higher and higher data speeds is catching up on us faster than we might expect, so I would highly recommend that you consider the 4G approach.

9

Tools for Indoor Radio Planning

The trade of indoor radio planning relies on several types of tools to aid the radio planner with the process. These tools help in accurately predicting and verifying radio coverage, capacity and quality. More logistical tools may be also needed to calculate the BOM (Bill of Material), component count, equipment lists and project costs. Obviously tools for design documentation, diagrams, etc., are needed as well. There are many different types of tools, and each operator has a tool box that relies on different types of tools for the various aspects of the process.

The choice of the tools you should select is must be relevant to your individual needs. If you only design three to five indoor projects per year, it might be alright to use standard spreadsheets for link budget calculations and simple diagram drawing programs for documentation. However, if you are dedicated to indoor radio planning on a full-time basis and design 20 or more indoor systems per year, it would be worthwhile investing in automatic tools that can make your design process more cost-efficient and improve the quality of your work. Remember that the design documentation should also be used by other parties in the project, so all documentation must be clear and precise.

In the following, examples of tools are presented. These are examples, and other tools from various manufacturers are available, each with their own pros and cons.

9.1 Live and Learn

Whatever nice tools you might have on your PC, it is highly recommended to leave your desk behind and get out there in the building, walk the turf, conduct measurements, take part in the optimization campaign and try to participate in a couple of late-night installations of one of your designs. That will enable you to get a feel for the whole process. Indoor planning is much more than 'just' desktop work on the PC. Trust me there is a lot to learn, outhere in the real world!

Indoor Radio Planning: A Practical Guide for 2G, 3G and 4G, Third Edition. Morten Tolstrup.
© 2015 John Wiley & Sons, Ltd. Published 2015 by John Wiley & Sons, Ltd.

An essential part of the learning process for the radio planner is also to gain experience from the indoor solutions and installations that are live with real traffic. Conduct the post installation RF verification surveys yourself; do these measurements enough times to gain experience so that you can trust your 'RF sense'. Real-life feedback from the designs you have done is vital, and the launch pad for improvement.

Another valuable input for the radio planner is feedback on network performance statistics of the indoor cells implemented. Monitor the call drop rate, quality, traffic RF performance and handovers to learn where the weak point in the design is.

9.2   Diagram Tools

All indoor radio designs need to be well documented, including diagrams. It is essential to use an electronic tool that enables you to store and update the diagram documentation in the future.

Always leave a set of diagrams of the DAS on-site, preferably with all components and their location in the building. It can be very helpful if you draw the diagram, using the floor plans as the backdrop.

9.2.1   Simple or Advanced?

The simplest tools can take you a long way. Most diagrams and an illustration in this book were done using Microsoft Visio (as the example in Figure 9.1). Visio is very easy to use, and you can easily create your own stencil containing the DAS components you use. You can also

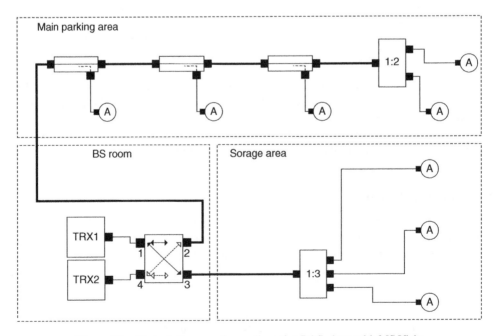

Figure 9.1   Typical diagram of a small passive DAS, done with MS Visio

import a picture format of the floor plan you are working on, and draw components on top. This is a very easy approach and can be useful if you do only a few DAS designs a month. If you are designing DAS on a larger scale, it is defiantly worthwhile considering a more automatic tool that can help you with the total process. Many designers are also using AutoCAD for DAS design; this can be advantageous if you can get the floor plans in AutoCAD format.

You are welcome to contact me if you want a copy of my Visio stencil I used for this book. If you are doing a lot of indoor designs, the optimum solution would be to get a tool similar to the one described in Section 9.7, which is what I use on a daily basis.

9.3   Radio Survey Tools

In some cases it is advisable to perform an RF survey in the building, to verify the antenna locations and the model you are using for simulation of the coverage in the building (refer to Section 5.2.5).

9.3.1   Use Only Calibrated Equipment

You will need a calibrated transmitter, preferably battery operated, and a pole or tripod that enables you to place the survey transmit antenna at the same location as the planned antenna. The transmitter needs to be able to have adjustable output power so you can simulate the same power as the final antenna.

Log the Measurement Result

You will also need calibrated measurement tools that are able to log the measurement, and preferably save the data on a PC. The best system enables you to place the measurement on a floor plan that is imported to the system (see Section 5.2.6). If you are using, or plan to use, a design program that can do indoor modeling and RF prediction, it is important to chose a radio survey tool that can export the measurement data to that tool, and thereby measurements from the actual building can be used to fine tune or calibrate the prediction model. Needless to say, a radio planner never leaves home without a 'test mobile' that is able to display, sometimes even log, the basic radio levels and parameters.

9.4   The Simple Tools and Tips

I am a big fan of simple, straightforward tools. In radio planning, as in so many other cases, some of the best tool and tips are often the simplest ones.

9.4.1   Use a Digital Camera

A digital camera is a must for the radio survey, but also for installation documentation. It is recommended to document all antenna and equipment locations with digital pictures. Name the files accordingly, so the picture of the proposed installation location of antenna 'A1' is saved as 'A-1.jpeg', etc. It is also useful to use a laser pointer to pinpoint the exact antenna

Table 9.1   Traffic profile calculator for calculating the traffic load of an indoor cell

User profile/type	mE/user	Number of users	Total E
Normal private user	15	25	0.375
Business user, not 100% WO	30	200	6.000
SOHO, 100% WO	50	0	0.000
Normal business user, 100% WO	70	40	2.800
Office user, 100% WO	100	100	10.000
Heavy business/telecomm, 100% WO	140	5	0.700
User def.-1	0	0	0.000
User def.-2	0	0	0.000
Total			*19.875*

location when recording the picture; then there will be no debate with the installer as to where the agreed antenna location is.

9.4.2   Use the World Wide Web

Prior to the survey it can be useful to visit the homepage of the company who owns the building where you are implementing the DAS. Often you can find site plans, photos and other useful information that can help you prepare the draft design concept.

9.4.3   Traffic Calculations

Obviously the Erlang table is one of the main tools when calculating voice capacity (see Section 6.1 & 6.2), and you will find a copy of the Erlang table in the Appendix. Excel is a useful tool for the basic traffic calculations. It is very easy to produce a dedicated calculator for a specific need. The example in Table 9.1 is a typical easily made tool for calculating the traffic profile of an indoor cell, by adding the various users.

9.5   Tools for Link Budget Calculations

Any science pocket calculator will do fine, but it is preferable to us a spreadsheet type of link budget calculator, such as the example in Tables 9.3 and 9.4. Most of my tools are done using standard Excel for windows. Dedicated tools for link budget calculations are available, but the flexibility of a standard Excel spreadsheet makes it very easy to upgrade and maintain. You can easily integrate the results of other calculations, such as cascaded noise (as shown in Table 9.2) or power per carrier, into your link budget tool if it is all based on Excel.

9.6   Tools for Indoor Predictions

It is true that once you have done 20 or 30 indoor design, radio surveys and post installation measurements, you will have a really good feeling, and even a sixth 'RF sense'. It is important to recognize and trust that feeling and experience, but *always* do a link budget, and do use a link calculation, at least in the expected worst case areas of the building.

Table 9.2 Simple cascaded noise calculator using Excel

Cascaded Noise										
Total NF		**Stage 3**			**Stage 2**			**Stage 1**		**Passive Loss**
17.38 [dB]		BS-NF	6 [dB]		UL-Att	10 [dB]	Active Gain 0 [dB] / Active NF 12.0 [dB]			Loss 0.0 [dB]
Calculated total Noise Figure		The Noise Figure of the Basestation			The UL attenator value		Gain and Noise Figure of the Active system			Attenuation of passive losses, prior to the first stage (RAU)
					Input all the RED fields					

$$NF = NF1 + \frac{NF2 - 1}{G1} + \frac{NF3 - 1}{G1 \times G2} + P$$

$$NF = 15.85 + \frac{10 - 1}{1} + \frac{3.98 - 1}{1 \times 0.1} + 0.0$$

$$NF = 54.7 = 17.38 \ dB + 0.0 = \boxed{17.38 \ dB}$$

Table 9.3 Link budget tool in Excel, 2G uplink

MS transmitter		
MS Tx power	33	dBm
MS antenna gain	0	dBi}
EiRP	**33**	**dBm**
DAS receiver		
BS noise figure	3	dB
DAS noise figure	0	dB
DAS passive loss	35	dB
System noise figure	**38**	**dB**
Thermal noise floor	−121	dBm
Interference	−120	dBm
Service SNR requirement	9	dB
DAS antenna gain	2	dBi
BS sensitivity	**−76**	dBm
The RF channel		
Log–normal shadow fading	10	dB
Multipath fading margin	6	dB
Body loss	3	dB
Total margin	**19**	dB
Rx minimum level	**−57.0**	dBm
Maximum allowable path loss	**90**	dB
Service radius from antenna	**23**	m

If you only use your experience, you might make expensive mistakes, or expensive overdesigns of the DAS solutions. If you want to utilize your experience, the best way is to first plot the antennas on the floor plan, then re-check with link budget and propagation simulations, maybe even RF survey measurement with a test transmitter in the building, to confirm you are right. This will make your 'gut feeling' even better.

Table 9.4 Indoor 'model' calculator, the first homemade tool I ever made for indoor designs

Indoor calculator MTO. V.1.0.1				
		Inputs		
	# of Walls	ATT-900	ATT-1800	
Wall Att.Wall-Type 1	1	4	8	[dB]
Wall Att.Wall-Type 2	1	12	16	[dB]
Wall Att.Wall-Type 3	0	14	18	[dB]
Distance to antenna			25	[m]
		900 MHz	1800 MHz	
BTS RX Sensitivity		−107	−104	[dBm]
BTS Max power		33	33	[dBm]
MS RX Sensitivity		−102	−100	[dBm]
MS Max power		33	33	[dBm]
DAS Coax system loss		35	35	[dB]
Results				
		900 MHz	1800 MHz	
Pathloss		75.46	89.56	[dB]
RX signal UL @ BTS		−77.46	−91.56	[dBm]
RX signal DL @ MS		−77.46	−91.56	[dBm]
C/N Margin DL @ MS excl. Fading		−13.46	−29.56	[dB]
C/N Margin UL @ BTS excl. Fading		−8.46	−25.56	[dB]

IB-Model

900 : L(dB) = 91.5 + 20log d + p*W(k)

1800 : L(dB) = 97.6 + 20log d + p*W(k)

L(dB) = 32.5 + 20log f + 20log d + k* F(k) + p*W(k) + D(d-db)

L = Path loss (dB)

f = frequency (MHz)

d = Distance MS/DAS Antenna

k = # of floor speerations, the signal has to pass

F = floor separation loss

P = # of walls the signal has to cross

W = Wall attenuation

D = linear attenuations factor dB/m

db = 50m 'breakpoint' additional 0.2 dB/meter

9.6.1 Spreadsheets Can Do Most of the Job

Like the link budget calculator, you can use a standard Excel spreadsheet to predict the coverage in the building. This approach is limited to calculating a specific level or distance in a specific area or place in the building. Therefore you will simulate the worst case in the building, and also several location samples that represent the building.

9.6.2 The More Advanced RF Prediction Models

Some tools are able to import floor plans and to define the loss of the individual wall and floor separations. The most advanced of these prediction tools can even calculate 'ray-tracing', estimating the signals being reflected throughout the building in three dimensions.

It is very useful to use these types of tools, where you also can include the nearby macro sites in your simulation. You can produce *C/I* plots or best server plots and save a lot of investment, while keeping the planning mistakes to a minimum.

Any tool will only be good if you use it correctly; no tools output is better than the quality of the input you provide. Performing an accurate RF simulation of a building can be time-consuming; you need all the details of penetration loss through walls and floor separations, reflectivity of the surfaces, etc. It is evident that these tools cannot be 100% accurate, but they allow you to simulate the environment and to experiment by moving antennas around, thereby getting a good feeling on how and where to place the antennas. If in doubt, always do a verification survey with a test transmitter and receiver.

9.7   The Advanced Toolkit (iBwave Unity, Design, and Mobile from iBwave.com)

Only a couple of years ago, in-building wireless was a new concept that mobile operators were starting to slowly explore and address and was typically being done on a building-by-building basis. Typically buildings were equipped with indoor systems delivering no more than one or two technologies for a single mobile network operator, and the main goal of the designer was to bring signal into a building. Fast-forward to the world of today and we see that in-building wireless has not only become a nice-to-have indoors, but an expectation and requirement from an end-user perspective. Those changes are due to the amount of newly connected devices that have been brought to the market and that have changed the way users interact with wireless networks. To keep up with these devices and usage patterns, wireless protocols have also evolved to be able to deliver the capacity and data rates needed migrating from 2G, 3G to 5G (Chapter 2). Innovative solutions such as DAS solutions and sometimes small cells (Chapter 13) have become the norm, and mobile network operators now have a mandate to provide coverage where most of the traffic is being generated, namely indoors.

The design of today's in-building systems has become exponentially complex. In-building systems now consist of projects supporting multiple wireless protocols, multiple wireless operators, and multiple sectors at the same time in the same venue. Relying on spreadsheet tools to approximate signal strength link budget calculations used to be sufficient, but today's wireless protocols require special attention to metrics such as signal-to-noise ratio (SNR), data rates, electromagnetic radiation (EMR) levels and interference levels not only from the macro outdoor network but even from within the building as well.

Above and beyond these technical considerations, it is also important to note that with the exponential growth in deployments and upgrades, there are now new challenges pertaining to deploying more projects more quickly, mixing small cells with indoor DAS and outdoor networks, leveraging operations staff to streamline design and deployment processes, design revision tracking, collaboration with non-technical teams, asset tracking, and so forth.

All that said, careful consideration must be exercised when selecting a solution to tackle today's challenges. There are many types of tools on the market for the various individual tasks with regard to indoor radio planning, and in many cases you will end up with separate tools and files for each task or project. It can be challenging to handle all the different data inputs and outputs from these tools so that you are sure that each project is always updated with accurate data. There is, however, one solution that combines all the aspects of the indoor design process into one ecosystem (the solution I use on a daily basis, which is shown in Figure 9.2). This is not a sales pitch for that specific solution, but I am sure you will find the design software from iBwave.com, a collaboration solution and mobile application, very useful and cost-saving if you do indoor radio planning on a frequent basis.

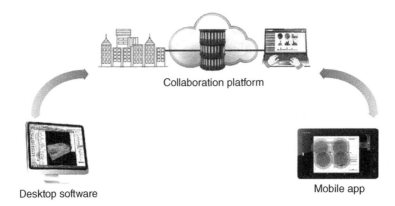

Collaboration platform

Desktop software   Mobile app

Figure 9.2   File collaboration platform and indoor design ecosystem – exchange files between desktop tool and mobile platform while keeping track of changes and project revisions

Using one efficient tool with data consistency and that supports the process from end to end will allow you to focus more on applying the most efficient solution, rather than juggling multiple tools with the risk of unconsistant data. The end-to-end process described in Chapter 15 will be covered within the same tool suite on the planning and documentation side.

9.7.1   Save Time, Keep Costs and Mistakes to a Minimum

These tools are not only very easy and intuitive to use, but they save a lot of planning costs, streamline processes, and keep mistakes to a minimum, as all data is linked across platforms and there is an ability to track design revisions. Updating a component on the schematic diagram simultaneously updates the floor plan, simulation, and equipment lists. Leveraging your smartphone or tablet device, you can also capture site information in the form of annotations, create preliminary designs and markups, and update as-built designs which will then be synced back to your desktop tool, displaying the changes that were done in the field. In other words, the tool suite embraces the full process, raises the quality of your implemented solutions and allows the designer to focus on the intelligence of the solution designed.

9.7.2   Collaboration, Visibility, and Revision Controls

Project design and documentation information is stored in one central repository (as shown in Figure 9.2). This enables the tracking of the workflow and project status as well as the collaboration with all the stakeholders on a project through different roles and permissions that have been created for each person. Project information can be accessed from any web client when on the go and also leverage the use of smartphones and tablets to access and populate information. All projects are also geo-located and can be viewed on a map. The system tracks revisions and milestones as projects evolve, with the latest wireless protocols being added or retrofitted.

9.7.3 Multisystem or Multioperator Small Cells, DAS, and Wi-Fi

The software is a powerful multi-technology, multi-band, and multioperator planning solution that displays the different wireless services using color coding to visualize link budget calculations. It's possible to define wireless services per operator, and designate multiple sites and sectors for each. Wireless systems can be defined as small cells, traditional base stations, or off-air systems. Wi-Fi access points are also available to be used in the design, either together with a traditional cellular in-building design or on their own for a Wi-Fi-only project. This allows you to consider the full heterogeneous networks (HetNet) design and interoperability in your design (see Section 13.2 for details)

9.7.4 The Site Survey Tool

You have the ability to leverage the latest mobile and tablet devices to create preliminary designs, and capture site survey information in the form of annotations (as shown in Figure 9.3) using pictures, video, audio, and text notes. It is also possible to update designs from the field in order to create project revisions, and make notes directly on the floor plans or pictures that were taken on site. So when your site walk team is out there looking at your initial design, every note and picture saved on the tool is 'pushed' to your design and saved in the same project file. This saves a lot of time and ensures both speed and consistency in the process.

9.7.5 The Mobile Planning Tool

The solution includes a tool that allows you to use a tablet device to design small cells in the field, out there on the spot (as shown in Figure 9.4) from actual manufacturer parts. Floor plans can be created from images captured in the field, by simply taking a photo of the floor plan at hand. Often you will find 'fire evacuation plans' posted in most public buildings – these are very useful and it is very simple to scale the floor plan once captured by the tablet's camera.

Geo-location data and scales can be set directly on the tablet for the tool to demonstrate signal and throughput contours as you design your system. The tool can even measure the existing coverage levels in the building to take these into account in the design process.

9.7.6 Import Floor Plans

You can easily import any standard type of floor plan, e.g. *.gif, *.jpeg, *.pdf, and even AutoCAD. Floor plans are then scaled and associated with materials from a database to match the actual type, size, and structure of the building (as shown in Figure 9.6). All the floor plans can then be aligned together in order to have a perfect three-dimensional modeling of the building. The cable lengths will also be automatically measured horizontally and vertically based on the floor-plan scales. When predicting signal quality and coverage, the materials assigned to the building are taken into account to provide very accurate results.

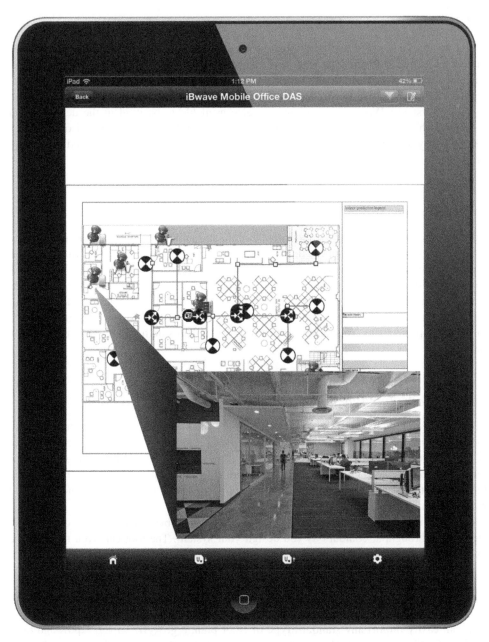

Figure 9.3 Use a tablet or smartphone to capture site survey information by attaching pictures, video, voice, and text notes to geo-located annotations and to draw mark-up notes directly onto the floor plan

9.7.7 Schematic Diagram

The schematic diagram is where the user can see all the components connected together and analyze the RF power at each single connector (as shown in Figure 9.5). Each component is drawn for precise identification and it also contains all its technical specifications in order for

(a)

(b)

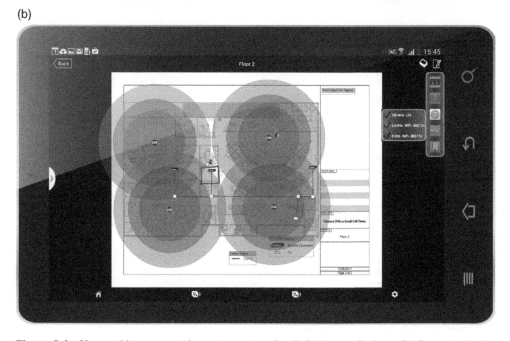

Figure 9.4 Use a tablet or smartphone to create a Small Cell or preliminary DAS remote design directly from the field

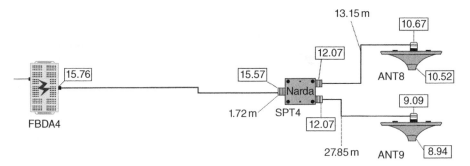

Figure 9.5 Schematic diagram of a small section of the DAS – power levels at the input and output of each component are calculated in dBm

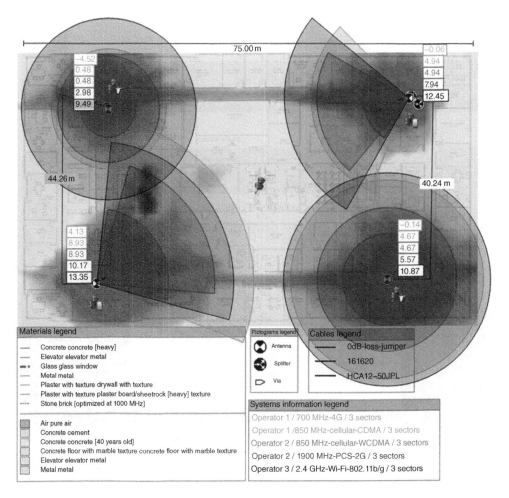

Figure 9.6 Imported floor plan, scaled to size, with the diagram of the DAS on top; it automatically calculates cable distance, losses and installation costs, path loss slope (PLS) contours, and visual link budget

the tool to simulate its actions as it would perform in real life. Most of today's in-building projects are now multiband and multi-technology where people are using the same infrastructure as a big RF pipeline to carry multiple types of signals at different frequencies. The schematic view provides the ability to analyze the interaction of all the signals going through a component and perform the specific RF calculations based on the particularity of the technology. The tool performs many downlink and uplink calculations, which is useful to increase speed and accuracy of the design. But many projects are still carried out using passive DAS, which is hard to balance without a tool, especially when there are many components. The uplink portion is even worse, as the cascaded noise figure of the system is crucial but very hard to calculate manually. Another aspect to consider in design is passive inter-modulation (PIM), which can be calculated automatically. When modifications are required due to a site visit or other factors, the tool will save a lot of time by re-doing all these calculations in a second. The link budget, noise calculations, power, and inter-modulation calculations are also done within the tool and it supports all the mobile communication protocols from around the world.

9.7.8   Floor Plan Diagram

When you place components such as antennas, couplers, amplifiers, cable and others on the floor plan, the schematic diagram is automatically drawn for the user, so the diagram will always be accurate and match the floor plan, and vice versa.

Figure 9.6 illustrates simple floor plan with several antennas. The software automatically calculates the actual cable distance and link budget due to the scaled floor plan. The cable distances and the predicted cell borders are updated in real time when moving the components on the floor plan and a prediction can then be run to simulate the actual coverage. Users can choose what type of supplementary information is needed and the diagram will present the results in a visual DL link budget; in the example in Figure 9.6, it is the power per channel level in dBm at the output of the antennas.

9.7.9   Site Documentation

It is possible to include site documentation, annotations, as-built pictures, video, audio, and text notes as well as photo mock-up pages (as shown in Figure 9.7) directly in the design file in order to help the installers avoid mistakes and to help the building owner understand the installation. This is obviously also very important when considering any future upgrades of the system, in particular the details from main and sub-equipment rooms, to estimate how much space is available for adding more active equipment in these locations.

9.7.10   Error Detection

The software has detailed configurable warnings on many different topics and will prevent users from making errors. The software will check for errors, such as connecting two outputs together, incorrect connectors (gender and type), composite RF power going through a component, overdriving an amplifier, and many more. This will also ensure that the installer

Figure 9.7   Example of installation mock-up, which is very useful for the installation process

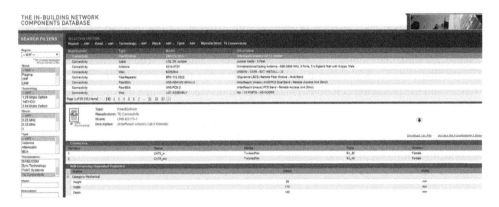

Figure 9.8   The in-building network component database

can save time in the installation process, simply by making sure all the correct components, jumpers, connectors are documented automatically in the bill of material (BOM) that is automatically generated from the tool.

9.7.11   Component Database

The tool has a large database of components from the different manufacturers (as shown in Figure 9.8), so the design calculations and documentation will show and use the real data updated online from the manufacturer. This also includes the antenna radiation diagram so you can make accurate predictions using the actual antenna you plan to implement. It is also easy to create your own components if needed, or to change the existing component data.

You can define the cost per component, per meter of cable etc. In addition, you can define the cost of installation per component, antenna and meter of cable. So not only can you generate the BOM, but also the total estimated implemented cost of the deployed DAS, early in the design phase.

9.7.12   RF Propagation

The main purpose of the tool is to estimate the RF coverage in the building so that users can select various prediction models and modes (as shown in Figures 9.6 and 9.9) such as basic PLS-based contours, to advanced COST231, and three-dimensional Fast Ray Tracing algorithms. The database contains multiple types of materials in order to define as closely as possible the building characteristics and the environment. Three-dimensional prediction can be performed and analyzed for all the bands at the same time and displayed in different formats, such as best server, link loss or signal strength, among others.

The predictions of the RF signal inside the building are obviously important for the radio designer, but are also very important to us in order to present the solution to the end-customer so they can appreciate the coverage they can expect in the building. Plotting the signal levels in an easy-to-understand color scheme on the actual floor plan is a very efficient form of communication to the end-user of a relatively complex radio design. Like any tool, the quality of the output and how the implemented performance will match the predictions obviously depend on the details and accuracy of the inputs you provide.

9.7.13   RF Optimization

There is an ability to optimize the location and transmit power of antennas and transmitters to achieve quality and throughput metrics required to take advantage of the latest wireless protocols. Users can choose from several types of output maps to optimize

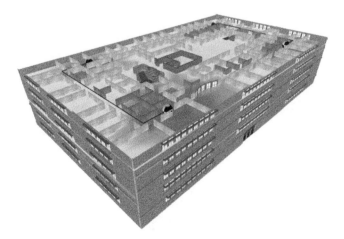

Figure 9.9   The same DAS as shown in Figure 9.6; this is simulated in three dimensions (obviously, predictions are in color in real life)

designs such as SNR, maximum achievable data rate, and interpolation of interference from the macro outdoor network. Capacity can also be optimized and analyzed using a Monte Carlo algorithm.

The optimization process is quite often very labor-intensive, so an efficient tool is an important part of the process.

9.7.14   Complex Environments

Designing a solution for a few office floors or a small building can be a challenge. Sometimes you are faced with big challenges of designing really complex solutions, where you have to consider the signals that are provided from other levels in the structure, the floor above, below, and so on, so the tool also allows for designing complex environments such as campus scenarios, stadiums (see details in Chapter 14), and train or metro stations. The three-dimensional modeler allows for the design of inclined surfaces which can be used to represent escalators, tunnels, and seating. Using the capacity analysis tool, we can validate the proper capacity and dimensioning for such environments (as shown in Figures 9.10 and 9.11).

9.7.15   Importing an RF Survey

The results file from the RF survey measurement tool can be imported and the data compared with the prediction model. This makes it very easy to find installation problems when comparing the post-installation survey with the prediction data. Survey data can also be used to calibrate and tune the prediction algorithm and material losses to attain a higher order of accuracy. This will make sure that your field experience of the actual RF signal performance is used to constantly improve the tool and ensure that it evolves over time to provide an even more accurate and more efficient design and solutions.

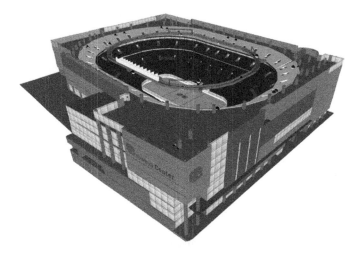

Figure 9.10   Design of an arena, including the seating area as well as concessions and office area

(a)

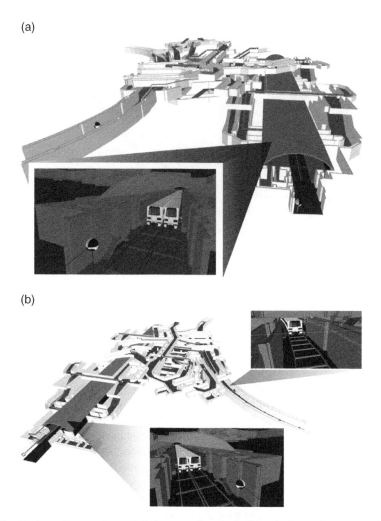

(b)

Figure 9.11 Design of a complex train/subway station, including platforms, interconnecting tunnels, and various floor heights

9.7.16 Equipment List and Project Cost Report

It is possible to define the cost per component and the installation cost per component or meter of cable, and at the push of a button you will get a complete BOM and project cost. This enables the radio planner to estimate the project cost fairly accurately and the complete business case for the project at hand.

9.7.17 RF and Installation Report

Many other reports are available to help the user to quickly analyze and optimize the project. Other reports allow the user to quickly validate the compliancy metrics of a project, look at the detailed link budget of each antenna for the different bands, analyze the average output power

Figure 9.12 Design of a building campus or city block, including a combination of indoor and outdoor transmitters and small cells

per antenna to better balance its system, and give a detailed cable routing report to installers in order to speed up the installation process.

9.7.18   Fully Integrated

The main benefit of using this solution is that all the data is centralized into the same platform and leverages the use of a mobile and desktop solution, making it is easy to update the project at any stage, from anywhere. The documentation, calculations, reports, diagrams, floor plans and all the project information are contained in one file and can easily be shared across the in-building value chain using the collaboration solution. Replacing single or multiple components requires merely a few clicks with the mouse and then all data will be linked. You can also quickly review projects by establishing compliancy metrics and the tool will automatically generate output maps and a project report detailing whether or not the project meets those required metrics. This can save a lot of project time and limit the number of mistakes.

In this way, the tool integrates the full process, even complex campus or city designs of various building structures and types such those shown in Figure 9.12, into one coordinated project.

9.7.19   Outputs from the Tool

The following list summarizes some of the types of outputs available and provides some examples of each.

- On the schematic diagram
  - ○ Trunking diagram, displays the interconnection details of all the parts using graphical representations of the actual manufacturer's components.
  - ○ Visual link budget, ability to toggle on and off the various RF calculations for the various sectors that can be analyzed visually, avoiding the need for a spreadsheet.

- On the floor plans
  - Three-dimensional floor architecture details, including all the wall and surface materials properties as well as their respective scales.
  - Cabling diagram, showing the cable lengths and routes of each cable.
  - Geo-located component locations using longitude and latitude.
  - Annotations information captured in the form of pictures, video, text, and audio notes gathered during a site survey using a smartphone or tablet.
  - Multiple coverage plots, including the expected design signal levels, such as serving sector, signal strength, signal quality, and throughput among others available depending on the technology being analyzed.

- In the reports section
  - Compliancy report, showing quickly if the design project meets given key performance indicators required for coverage, quality, and throughput.
  - Bill of materials, including the actual manufacturer part numbers and specifications.
  - Cost details, giving a total project cost estimate using the predefined cost of the individual parts, as well as implementation costs.
  - Cable routing report, details of all the cable routes specifying the 'from–to' cable connections of all parts.
  - RF link budget, a detailed tabular form displaying all the signal paths from the signal source to the output of the antennas.
  - EMR, detailing the safety limits of RF exposure for various regions of the globe.
  - Capacity report, showing the effective traffic that can be supported for each sector.
  - Debug message list, details any errors in the design such as connector mismatch, unsupported frequency bands, over powering a component, etc.

9.7.20   Team Collaboration

Designing and deploying DAS systems is a complicated process with many teams and people involved. Everyone is an important cog in the big machine, and every detail and exchange of information and documentation is essential to the success of the team, and the project.

Team collaboration between field technicians and RF professionals is facilitated by a collaboration platform that delivers the following capabilities in the collaboration platform:

- Individual project spaces for each project.
- Project revisions are tracked.
- Export project locations to mapping tool.
- Ability to track the project progress and workflow.
- Safely share files with other collaborators on a project.
- Attach additional project documentation to the assigned project spaces.

This improves both the design quality and time to service and optimizes costs, as errors and mistakes are minimized owing to the fact that the whole process, end to end, is in synchronization.

9.7.21   Make Sure to Learn the Basics

The tool described is extremely efficient and will assist you with almost all phases and tasks of designing an indoor DAS. Any tool will not give you better output than the input you provide. I highly recommend that you still learn to perform basic RF calculations, so you will get a good sense of whether the output is as you expect. If that is not the case, you will have a chance to correct the inputs to the tool in order to tune the various settings so that the actual design will match the final performance.

9.8   Tools for DAS Verification

When going live with the DAS, we need to perform some pre-testing, the actual commissioning and system verification. Let us have a look at some of the tools needed for the process. The process itself is described in Section 15.2.3. Before we perform the commissioning and set the system on-air, we need to ensure that the DAS performs to specifications on a component/ sub-system basis. It would be far too comprehensive to give all the details of the various tests, so please consult the appropriate literature from the equipment manufacturers and the test equipment manufacturers for details. All the installed equipment and components have already been subjected to a visual inspection during the installation phase. There are some tools that are absolutely essential to invest in when deploying passive and active DAS. In the following we provide a brief introduction to these and to their basic purpose and function. Please consult the documentation and instruction manuals for the specific measurement equipment, as what follows is merely a short description.

Documentation is Important

For documentation purpose, make sure that your test equipment can generate the needed output documentation of the measurements, log files, text dumps, screenshots, etc. to include in your project documentation (see Chapter 15).

Calibration

Needless to say, you must follow a strict calibration procedure and labeling process as per the instructions of the test equipment manufacturer. You need to be able to trace all measurements according to the calibration dates on the test equipment; if there is a slip, you need to re-calibrate and re-test.

- *Antenna and cable tester* You need a tester that can perform the basic VSWR sweep on some selected antennas. Perform the required sample tests at least of the antennas in the 'hot' part of the DAS where there is higher power. Often the tester can perform cable sweep, distance to fault, etc. You will be able to program start and stop frequency for the sweep, alarm triggers and the like.
- *PIM tester* Testing PIM (see details in Section 5.7.4 for details) is a complex process and it takes highly specialized (and expensive) equipment to perform. Normally this will not be standard equipment you acquire, but rather hire in an external specialized team to perform

real PIM testing, and please do so. You can, to some extent, confirm if you have PIM issues by changing the transmit power in the suspected system, and if there is a non-linear relation between the change of power and the unwanted signals, then it could very likely be a PIM issue. Then you absolutely need to get the right equipment on location and do a real PIM test.

• *Fiber test tester, optical power meter* Before the fiber plant is put into use, all fibers should be tested end to end with optical time domain reflectometer (OTDR) test. This allows you to sweep the fiber end-to-end; all patches, splices and distances will be shown in detail in terms of reflections and distance losses. (Note that if you use analog active DAS, the fiber will most probably need to be terminated with APC connectors all through the installation to avoid reflections.) You also need to make sure you have equipment that can measure the optical power levels, and make sure your optical testers support the needed wavelengths.

• *CAT5/6 test* Some types of active DAS utilize CAT5/6 ethernet type cables, and these should typically be tested against the TIA/EIA-568A standard. As a minimum you should test the 'wire map' and the loop resistance, to make sure that the DC power to the remote unit can be maintained.

• *CATV test* Some types of active DAS utilize cable TV type coax (75 Ω), which typically carries an intermediate frequency for the RF signals as well as DC power to the remote antenna units. Again, you perform the same types of test as for the standard coax: PIM, sweep over distance (reflection vs. distance), VSWR, etc. DC loop resistance is also important to ensure we can maintain the DC power to the remote antenna units.

• *DAS build in testing* The active DAS equipment will often have a quite detailed test program internally. These automatic tests will often be performed 'end-to-end', even checking signal levels on the various fiber and RF, spectrum, PIM, ports of the system etc.

• *RF verification measurement tools* The most important measurement tool for DAS RF testing is a spectrum analyzer, preferably portable and one that is able to decode various signaling and power levels for various channels in the system tested. This is crucial when leveling the signal power and gain in active systems.

In the following section, we look at some measurements of a typical spectrum analyzer that are able to decode the 3G/4G channels and provide detailed information.

9.8.1 3G Example Measurement

Figure 9.13 shows the spectrum power sweep over about 38 MHz. It is clear that there are two 3G carriers active and a composite power of about 9.4 dBm. You have several options of setting various bandwidths, start/stop scan frequencies, markers and tracers/triggers.

Figure 9.14 shows detailed results of the measurement in Figure 9.13 in a table format. One of the two 3G carriers from Figure 13 is analyzed – in this example, CH1087 with a center frequency of 2.1574 GHz. You can see the various offsets and reference levels you might have applied. Typically the most important information you will be looking for during the commissioning, power leveling/gain setting of an active DAS is the CPICH power level, marked as a 'P-CPICH' of 5.71 dBm in this measurement.

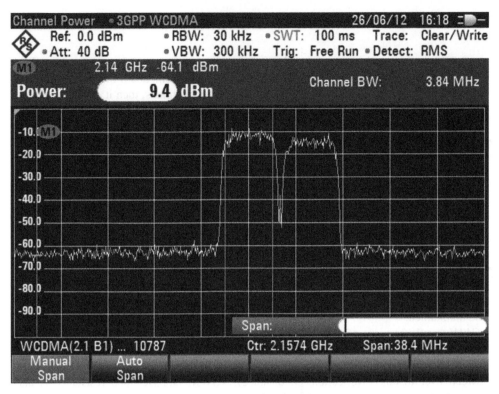

Figure 9.13 A screen dump in 'spectrum mode' of the RF analyzer, measuring two 3G channels

Channel Power ● 3GPP WCDMA 26/06/12 16:18
Ref: 0.0 dBm ● RBW: 30 kHz ● SWT: 100 ms Trace: Clear/Write
● Att: 40 dB ● VBW: 300 kHz Trig: Free Run ● Detect: RMS
2.14 GHz -64.1 dBm
Channel BW: 3.84 MHz
Power: 9.4 dBm
WCDMA(2.1 B1) ... 10787 Ctr: 2.1574 GHz Span:38.4 MHz
Manual Span Auto Span

Result Summary	3GPP WCDMA BTS		26/06/12 16:21	
Center:	2.1574 GHz	Ref Level: 0.0 dBm	Sweep:	Cont
Channel:	10787	Ref Offset: 0.0 dB	Antenna Div:	None
Band:	WCDMA(2.1 B1)	Att: ● 40.0 dB	P-CPICH Slot:	0
Transd:	---	Preamp: Off	Ch Search:	Off
		Scr Code: 176 / 0		

Global Results   SYNC OK

RF Channel Power:	9.41 dBm	Active Channels:	2
Carrier Freq Error:	258.6 Hz	Scr Code Found:	N/A

Channel Results

P-CPICH (15 ksps, Code 0)		P-CCPCH (15 ksps, Code 1)	
Power:	5.71 dBm	Power (Abs):	3.71 dBm
Ec/Io:	-3.70 dB	Ec/Io:	-5.70 dB
Symbol EVM rms:	1.35 %	Symbol EVM rms:	1.08 %
P-SCH Power (Abs):	1.07 dBm	S-SCH Power (Abs):	1.12 dBm

Result Display	Display Settings	Level Adjust		Signal Settings	Power Settings

Figure 9.14 For the same measurement as in Figure 9.13, this screen shows the decoding of one selected 3G carrier

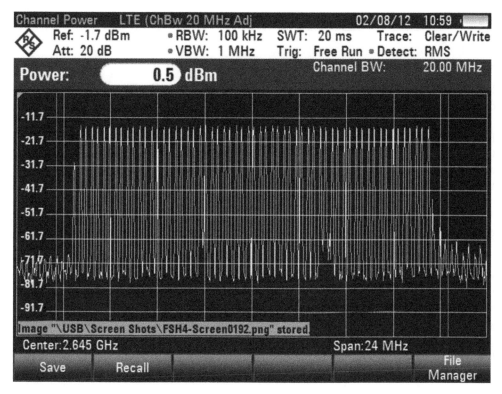

Figure 9.15 Screen dump in 'spectrum mode' of the RF analyzer, measuring a 20 MHz 4G channel

Result Summary			LTE-FDD BTS			16/09/13 11:34	
Center:	2.6625 GHz		Ref Level:	-20.0 dBm		Sweep:	Cont
Channel:	---		Ref Offset:	0.0 dB		Cell [Grp/ID]	Auto
Band:	---		Att:	20.0 dB		Cyclic Prefix:	Auto
Transd:	---		Preamp:	Off		Antenna:	M 2x2 / Tx2
Ch BW:	20 MHz (100 RB)					Subframes:	10

Global Results		SYNC OK	
RF Channel Power:	-4.86 dBm	Cell Identity [Grp/ID]:	15 [5/0]
Overall EVM:	83.95 %	Cyclic Prefix:	Normal
Carrier Freq Error:	575.0 Hz	Traffic Activity:	26.12 %
Sync Signal Power:	-24.53 dBm		
OSTP:	-52.41 dBm	IQ Offset:	-73.18 dB

Allocation Summary					
	Power:	EVM: (Std)		Power:	EVM: (Std)
Ref Signal:	-22.65 dBm	92.34 %	PSYNC:	-24.53 dBm	141.38 %
QPSK:	-77.49 dBm	194.31 %	SSYNC:	-24.54 dBm	141.42 %
16 QAM:	--- dBm	--- %	PBCH:	-24.53 dBm	1.73 %
64 QAM:	--- dBm	--- %	PCFICH:	-30.52 dBm	88.64 %

Result Display	Display Settings	Level Adjust	Antenna Settings	Signal Settings	Meas Settings

Figure 9.16 For the same measurement type as in Figure 9.15, this screen shows the decoding of the 4G carrier, power levels etc.

9.8.2 4G Example Measurement

Figure 9.15 shows the spectrum power sweep over about 24 MHz. There is a 20 MHz 4G carrier active; composite power is about 0.5 dBm.

Figure 9.16 shows the detailed results of another 4G measurement, like the one in Figure 9.15, in table format. All relevant and primary information is showed in the table, and the typical measurement we will be looking for in order to level our power and gain in an active DAS is the reference signal received power (RSRP) level. In Figure 9.16 we can see the RSRP is measured to be –22.65 dBm.

9.8.3 Final Word on Tools

'Good tools are half the work.' I am sure you will have heard that phrase before, and I agree with it wholeheartedly. Good tools throughout the process from start to finish are essential, for the design, implementation, and verification. But good tools needs to be maintained, calibrated and, most of all, operated as specified. Take your time to select the needed tools and make sure you follow the process and instructions given by the tool vendor. Good tools can save a lot of grief and money when they are maintained and used correctly; on the other hand, good tools that are not maintained or operated as specified will end up costing you time, quality and money.

10

Optimizing the Radio Resource Management Parameters on Node B When Interfacing to an Active DAS, BDA, LNA or TMA

When connecting any external uplink or downlink amplifiers to node B, you will offset the UL and DL. It will be necessary to evaluate and adjust the basic power control, system-info, noise and timing parameters in the BSS system in order to optimize the performance of an indoor active DAS system. If the parameters in the radio access network (RAN) (node B and RNC) are not tuned to cope with the offsets in DL power, noise load and timing offset, there will be a serious impact on the performance of the indoor system.

10.1 Introduction

3G is an advanced system. As it is noise- and power-limited, enhancement of the UL and DL performance can be obtained by the use of UL and DL remote amplifiers to overcome the passive losses from the base station (node B) to the antenna. This is the basic principle of an active DAS, which functions as distributed amplifiers in the system, offsetting the DL power and to some extend loading the base station with noise power on the uplink.

10.1.1 3G Radio Performance is All About Noise and Power Control

It is a fact that, in order to have optimum performance on a 3G system, we need to insure:

- *Strict power control*: all mobiles use the same UL frequency, so all mobiles have to be controlled so that the received signal strength from them on the UL of the node B is kept at the same receive level (see Section 2.4.5). The offset of the node B DL power caused by a BDA, TMA or active DAS has to be considered in the node B setting.

Indoor Radio Planning: A Practical Guide for 2G, 3G and 4G, Third Edition. Morten Tolstrup.
© 2015 John Wiley & Sons, Ltd. Published 2015 by John Wiley & Sons, Ltd.

- *Noise load is controlled in the system*: node B will measure the total UL noise (traffic) continuously, and use that measurement for admission control, to make sure that the overall system noise increase on the UL is kept below the desired threshold. The added UL noise from the active DAS, BDA or remote LNA has to be considered in the node B evaluation of noise/traffic increase.
- *Delay of the DAS (synchronization window)*: delay in the external amplifier system has to be incorporated in the timing and cell size.

10.1.2   3G RF Parameter Reference is Different from 2G

In 2G the reference for the radio parameters is the antenna connectors on top of the base station rack. However, for 3G it was foreseen that external equipment between the base station and the antenna would offset the downlink power, as well as the uplink noise load. The 3GPP-3G specification foresees that external equipment will be connected to the node B antenna connector (3GPP no's 25.215 and 25.104). This equipment will typically be an LNA (to enhance the uplink) or TMA (tower-mounted amplifier, to enhance the downlink).

10.1.3   Adjust the Parameters

The 3GPP specification takes the performance of this external equipment (see Figure 10.1) into account in the parameter set, by having the BS antenna as the reference. This is different from the 2G specification where the antenna connector of the base station is the reference. This change of reference for the node B parameter is motivated by the fact that 3G is a

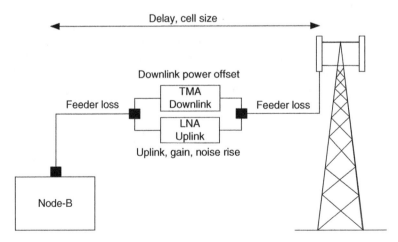

Figure 10.1   3G base station with external TMA or LNA; the noise offset on the UL and the power offset will affect the performance of the noise and power control

power- and noise-limited system, so it is crucial that the performance of any external equipment is taken into account the radio resource management parameter evaluations.

10.1.4   How to Adjust this in the RAN

Typically, there will be some offset adjustments to consider in the parameter set of the specific RAN vendor. The actual parameter name varies from vendor to vendor, but the principle is the same. In order for node B to be 3GPP-compliant, these offsets must be tunable, in order to offset the reference for the radio resource management in the BSS.

In the typical RAN you use the 'external TMA/LNA' function; here you define the DL power at the antenna, the noise figure and gain on the uplink, and then the RAN tunes all the control parameters accordingly.

10.1.5   Switch Off the LNA in Node B when Using Active DAS

When connecting an active DAS to node B, it is pretty much like connecting a TMA and a LNA to node B. Some vendors to have an easy way of taking the offset off the DL power and UL noise increase into account, by setting node B into 'external TMA' and 'external LNA' mode. Then you just define the NF and DL power of the DAS, and everything is taken into account. Note that the pure active system has a UL amplifier close to the antenna, and by using the UL gain in the pure active system, and switching off the internal LNA in node B, the UL performance increases because to the basics of cascaded noise; it is always best to have the gain closest to the source (the antenna).

10.2   Impact of DL Power Offset

When node B is connected to an active DAS, e.g. a 40 dBm, the CPICH power will typically be 30 dBm (10%/−10 dB). With one 3G carrier on the active system, the maximum power will be, for example, 15 dBm, so the CPICH power of the remote access unit will be 6–10 dB lower. However the system information transmitted by node B will (if left unchanged) still inform the mobiles that the transmitted CPICH power is 30 dBm from node B. These 25 dB in offset between the system information and the actual CPICH power will cause problems during access burst, if left unchanged.

10.2.1   Access Burst

When a mobile is in idle mode (open loop power control), it will monitor and decode the system information broadcast by node B (transmitted CPICH power). In order to enable the mobile to access the cell with the correct initial access burst level (not to overshoot the UL), node B broadcasts the CPICH Tx Power as a reference to the mobile. This CPICH Tx power level is used by the mobile to calculate the link loss. Therefore, the mobiles are able to start the initial access sequence with a power level that will insure that this access burst does not overpower all the other traffic on the UL (as shown in Figure 10.2).

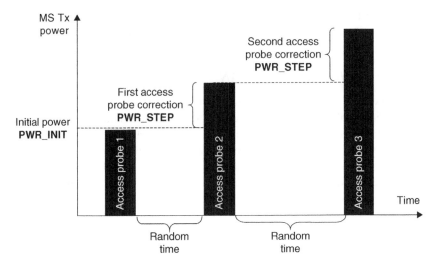

Figure 10.2 The 3G access burst principle: PWR_INIT = CPICH_Tx_Power − CPICH_RSCP + UL_ Interference + UL_Required_CI, where PWR_Init = calculated initial MS power, for the first access burst; CPICH_Tx_Power = the BS broadcasts the transmitted CPICH power, so the MS can estimate the path loss; CPICH_RSCP = received CPICH level at the MS; UL_Interfeerence = UL interference level at the BS, which the MS UL signal has to overcome; UL_Required_CI = the required CI on the UL

Example
A mobile is decoding a cell where the BS reports that the CPICH is transmitted at +33 dBm; the mobile receives the CPICH signal at −85 dBm. Thus the mobile will calculate the link loss to be:

$$\text{link loss} = \text{CPICH received} - \text{CPICH transmitted}$$
$$\text{link loss} = 85\,\text{dBm} - 33\,\text{dBm} = 118\,\text{dB}$$

Using this 118 dB of link loss, the mobile can estimate the initial power needed to overcome the link loss, for the first access burst.

10.2.2 Power Offset Between Node B and the Active DAS

The active DAS has a remote unit with an integrated DL power amplifier. The actual DL power will therefore depend on the composite power resource of that specific DL amplifier. In a typical active DAS a RU would typically transmit the CPICH at +5 dBm (one 3G carrier at full power is +15 dBm, and with the CPICH set to −10 dB = 5 dBm CPICH).

If the signal source to the active system is a +43 dBm macro node B, the CPICH is +33 dBm, and the system information broadcasts +33 dBm as the CPICH reference power (not the actual +5 dBm radiated from the RU), there is an obvious problem: using the same example as above, where the mobile is receiving the CPICH at −85 dBm, the mobile will still decode the information broadcast by node B, which the CPICH is transmitted at +33 dBm by the active DAS. The mobile will once again calculate the link loss to be 118 dBm:

$$\text{link loss} = -85\,\text{dBm} - 33\,\text{dBm} = 118\,\text{dB}$$

The mobile will then calculate the needed UL power to overcome the 118 dB link loss when accessing node B. However, in reality the CPICH is not transmitted at +33 dBm, but at +5 dBm due to the power amplifier (PA) in the RU, so the real link loss is actually:

$$\text{link loss} = -85\,\text{dBm} - 5\,\text{dBm} = 90\,\text{dB}$$

The result is that the mobile will use to high power in the access bursts. This erroneous CPICH power information will cause the mobile to access the cell with a level 28 dB too high (118–90 dB).

This overshoot on the UL will cause:

- Excessive noise increase on the serving cell and potentially in the neighboring cells.
- Degraded service for other mobiles in the serving cell and potentially in the neighboring cell.
- Potential for dropped calls in serving cell and neighboring cell.
- UL blocking of the RU (trigging the UL automatic level control).

10.2.3 Solution

Thanks to 3GPP, the solution is easy. The broadcast information on the CPICH power has to be changed to the actual level transmitted level by the RU. It is important to note that the CPICH level itself (−10 dB) is not changed, but only the information on the absolute transmitted level is adjusted to the CPICH level of the RU.

10.2.4 Impact on the UL of Node B

The noise load of the UL is a limiting factor on the UL performance, so it is crucial for 3G to control the noise increase in the cells.

10.2.5 Admission Control

Node B will constantly measure the overall noise power on the UL to evaluate the UL noise increase in the cell. The total noise power on the UL of the node B will be a result of:

$$\text{noise}_{\text{total}} = \text{own}_{\text{traffic noise}} + \text{other}_{\text{cells'noise}} + \text{node B}_{\text{noise power}}$$

So in addition to the noise generated by the traffic in the cell, there will be a contribution of noise from users in other cells and finally the NF of the node B hardware. Using this noise measurement and adding the hardware noise of node B makes it possible to evaluate if new admitted traffic will cause a noise increase that exceeds the predefined limit for the maximum UL noise increase in the cell (as shown in Figure 10.3).

10.3 Impact of Noise Power

An active DAS consists of amplifiers and, depending of the actual configuration, will have a given noise figure that adds to the noise on the UL of node B.

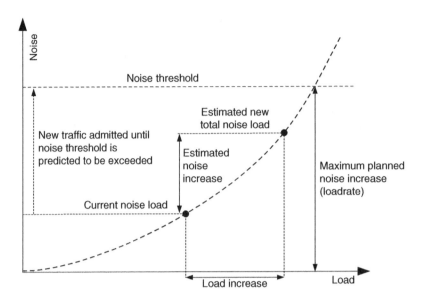

Figure 10.3   The principle of the admission control function in 3G

10.3.1   The UL Noise Increase on Node B

The active DAS needs to be configured so that the active DAS and UL attenuators are set at optimum values, minimizing the resulting noise increase on node B – without degrading the performance of the UL. However there will be a noise increase on the UL of node B. Therefore, it is crucial to tune the admission control parameter, with regards to the added noise power on the UL of the node B

Solution

$$Tune\ the:\ node\ B_{noise\ power}$$

This includes the noise increase caused by the active DAS (minimized by a correct UL attenuator). If this is not tuned, the result will be that the admission control assumes too high a UL traffic (noise level), limiting the UL capacity. In extreme cases (if no UL attenuator is used), this results in no call set-up in the cell.

10.4   Delay of the Active DAS

Connecting external equipment between node B and the antenna will cause an offset of the timing in the cell. In reality this delay will result in an increase in cell size (timing window). An active DAS connected directly to the node B will typical add 2–3 km to the cell size. This delay or offset depends on the delay in the active DAS amplifiers, but also on the length of fibers and CAT5 cables. The longer the fiber and cable runs, the longer the delay will be, in some cases exceeding 15–20 km offset of the cell size. Some operators will set a given cell size according to the environment the cell is serving.

Typically, a node B serving an indoor solution would be set to a cell size of 2 km, so that node B does not pick up any traffic distant to the cell. Therefore if one does not consider the delay on an active indoor DAS, and sets the cell size to 2 km, the cell might not be able to set up calls on the indoor system, due to the delay of the DAS.

10.4.1 Solution

Set the cell size to cope with the offset caused by the delay in the active DAS.

10.5 Impact of External Noise Power

It is crucial to select the correct value for the UL attenuator between node B and the active system. The purpose of this attenuator is to minimize the UL noise power. However the correct value of the attenuator needs to be selected in order to have the minimum noise impact and not affect the UL (SNR) performance.

The ideal attenuator will be the difference between the noise power from the DAS and the *KTB*. The 'noise power normograph' in Figure 10.4 will help you to calculate the noise power in any given bandwidth, and thereby enable you to select the optimum UL attenuator. It is the standard noise calculations from Chapter 7, although the noise normograph might be easier to use. This is how to use the normograph.

- *The left graph* – this is the reference point for the noise figure from the DAS plus the UL gain set in the system, so if your NF from the system is 12 dB, and you use 3 dB UL gain, you should select 15 dB as the reference.
- *The right graph* – this is the reference point for the RF bandwidth for the noise calculation, so for 2G you select 200 kHz for 3G 3.84 Mc (Figure 10.4).

10.5.1 To Calculate the Noise Power

You select the RF bandwidth and the NF and draw a line between the two points; then the noise power is given by the point where this line crosses the center graph. Too calculate the noise power of a 3G node B with an NF of 4 dB, the two reference points are the 3G bandwidth (3.84 Mc) on the right graph and the NF (4 dB) on the left graph. The line connecting these two points crosses the noise power graph at −104 dBm [4 dB NF+ (−108 dBm)], so this is the noise power loading the Rx port of node B, and this new RAN admission reference for zero traffic on node B.

10.5.2 To Calculate the UL Attenuator

The example shows a 3G (3.84 Mc) system, connected to an active DAS with 20 dB NF. As we can see, the noise power is −89 dBm, loading the input of the node B, if we do not use any attenuator. The ideal attenuator is calculated as the difference between this noise power and the

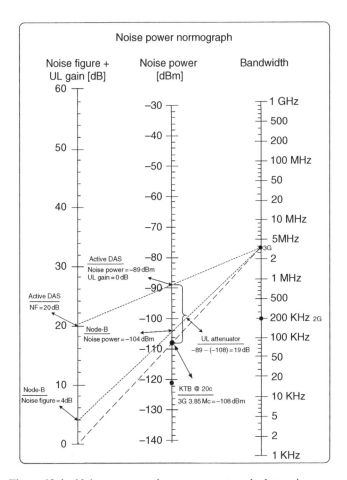

Figure 10.4   Noise normograph, an easy way to calculate noise power

KTB (the *KTB* is the noise power at 0 dB NF). In this example the ideal attenuator is 19 dB, so in practice you would select a 20 dB attenuator and adjust the UL gain of the DAS to +1 dB. Note that selecting a higher attenuator will degrade the UL dynamic range and sensitivity of the node B. However, there will still be a slight impact of the noise power from the DAS when node B has a noise power of −106 dBm and the DAS with attenuator has −108 dBm of noise power, and we will need to sum up these two noise powers by adding the powers in Watts.

First we convert the dBm to W:

$$P\,(\text{mw}) = 10^{\text{dBm}/10}$$

The idle noise power on the base station = −106 dBm = 1.58 × 10⁻¹¹ mW. The noise power from DAS = −108 dBm = 2.51× 10⁻¹¹ mW. Now we add the two noise powers:

$$\text{noise power} = P1 + P2 = 1.58 \times 10^{-11}\,\text{mW} + 2.51 \times 10^{-11}\,\text{mW} = 4.09 \times 10^{-11}\,\text{mW}$$

Then we convert back to dBm:

$$P(\text{dBm}) = 10\log(\text{mW})$$
$$P(\text{dBm}) = 10\log(4.09 \times 10^{-11}) = -103.88\,\text{dBm}$$

10.5.3   Affect on Admission Control

The resulting noise power load would be about $-104\,\text{dBm}$, and the admission control should offset to take this into account. It would be a noise load increase of $+2\,\text{dB}$ on the UL of the base station, corresponding to a UL load of 37%. If the base station admission control was set to a maximum UL load of 50% ($3\,\text{dB}$ noise increase), this 'phantom load' would have the effect that the base station would only be able to admit 13% of traffic load (50% max UL load -37% phantom load). If no UL attenuator was used, the noise power injected to the base station would saturate admission control with phantom noise, and no traffic would be admitted to the cell.

11

Tunnel Radio Planning

Tunnel Coverage Solutions

Coverage systems for tunnel solutions can often be designed using DAS in combination with radiating cable. The precise approach, however, is greatly dependent on installation limitations and challenges. Normally, we may say that tunnel radio planning is 5% radio planning and 95% other more pratical and implementation challenges.

If you plan to perform the practical work and measurement pre, during and post implementation (and this you should do, when RF designing for a tunnel DAS) you must be prepared to spend many late nights down in the dark; have your orange clothes, boots and hard hat ready ....You are going to need them, and a lot of patience, waiting for access to the tunnel in the early morning hours!.

Tunnel coverage systems pose specific challenges when compared with traditional indoor or outdoor coverage, due mainly to the physical nature of the tunnel itself. Also, the type of vehicle employing the tunnel will affect the design and sometimes the RF properties. The vehicle is a crucial design element of the total DAS solution; the role of the vehicle is often underestimated, especially when designing coverage solutions for rail tunnel systems. In addition to the specific challenge of handover zones and providing sufficient signal levels, special attention must be paid to the cell layout out and capacity design.

These challenges specific to tunnels will be addressed in this chapter, starting with the challenge of interfacing the tunnel coverage system with the normal network via the crusial and challenging handover zone. Different layouts of internal handover zones in a tunnel solution are proposed and evaluated in Section 11.7.

The type of coverage system to apply; radiating cable, DAS or combinations of these options will be explored and the pros and cons evaluated. Cell lay-out and capacity strategy for larger tunnel systems such as major metro tunnel systems will be explored and several strategies evaluated. Tunnel solutions are often a practical and project management challenge more than a RF planning challenge.

Indoor Radio Planning: A Practical Guide for 2G, 3G and 4G, Third Edition. Morten Tolstrup.
© 2015 John Wiley & Sons, Ltd. Published 2015 by John Wiley & Sons, Ltd.

As well as the theoretical and practical challenges of installing DAS in the tunnel, you will obviously also need to plan the needed RF services. In the RF planning and testing process you must pay attention to the services you design for, other RF services present in the tunnel; typically there will be emergency and other RF systems operating in the confined space of the tunnel, the co-existence of multiple radio systems will require specific attention from an early stage in the design process. You must try to maximize RF isolation of these systems, often by separating the installation of cables and antennas (this is obviously a challange, given the limited space), etc. Sometimes these systems will be transported on the same DAS when you plan for mobile services; here you must be especially careful. For more detail on some of these issues please refer to Section 5.7.

11.1   The Typical Tunnel Solution

The typical tunnel coverage system is illustrated in Figure 11.1 for a car or rail tunnel; showing a short tunnel being served inside the tunnel by yagi antennas or radiating cable. Typically, the RF signal from the base station covering the tunnel will be distributed by a fiber optical DAS from a central base station location. When using the yagi antenna option in the tunnel it will often be possible to some extent to take advantage of the back lobe footprint from the yagi antennas to cover some of the tunnel/area. This can be useful in the platform area; if you are covering the tunnel itself with yagi antennas it will often be possible to cover the platform itself with the back loop from a yagi antenna near the tunnel entrance. The tunnel system must also provide sufficient signal level in the handover zones to the external macro network, as well as any overlapping zones between cells internally in the tunnel system, typically for larger tunnel systems.

But before even starting on designing the tunnel DAS solution it is very important to obtain some basic information in order to construct a link budget, and start the actual design work.

One of the most important parameters in the RF design process is the penetration loss into the train coach; don't even think about starting the RF planning before you have ascertained that and many other specific facts.

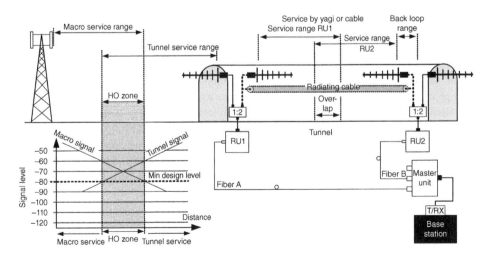

Figure 11.1   The typical tunnel coverage system

11.1.1 The Penetration Loss into the Train Coach

One of the most important issues to address when designing any RF coverage solution for wireless service of mobile users inside a train is the RF characteristics and RF performance of the train-car itself. The penetration loss of the RF into the train is a major component in the link budget and is often forgotten or assumed to be a much too low value compared with the actual loss. Trains are essentially just metal tubes seen from an RF perspective. The only way in and out of the train for the RF signal is through the windows – for modern trains the windows will often be coated with a metallic foil in order to shield passengers from sunlight, and this attenuated the RF! Sometimes it is necessary to install an onboard antenna system feed via a small repeater system installed onboard the train-car.

In order to carry out the RF design and design of HO zones, etc., it is very important to take a benchmark measurement of the exact penetration loss for the RF signal through the windows of the train, in order to establish the exact penetration loss. This measurement can be conducted as described in Section 5.2.5, but with the test transmitter outside the train. This measurement is required no matter whether you plan for an onboard DAS system, an onboard repeater or for the users to be served from a signal coming from outside the train. The measurement must be conducted on all the frequency bands that are within the scope of the coverage system and the RF planning. There can be a big difference in penetration loss from one band to another. Preferably, the measurement should be conducted with the exact same type of antenna, or radiating cable, mounted in a test section of the actual, tunnel.

Increasingly, windows with metallic coating are being used for trains; this will result in high penetration losses, and can come as a big and expensive surprise if the exact losses are not clarified at an early point in the project phase.

You must be very careful when evaluating the penetration loss because it depends on several factors: obviously the service frequency, the material of the windows, etc., but the incident angle of the service antenna/signal also plays a significant role. The presence of passengers can also play a major part, so using a typical body loss of 3 dB is far from sufficient. Just consider a fully packed train coach in a large urban area with passengers packed from wall to wall – this can easily eat up 10 dB of the coverage when compared with the normal off peak situation.

The penetration of the RF signal into the train coach is illustrated in Figure 11.2; the same power and frequency are fed to three antennas; A, A2 and B. The penetration loss will typically

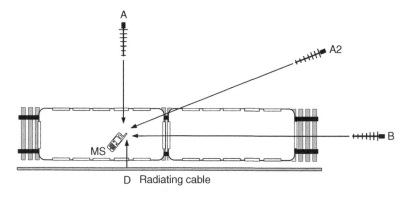

Figure 11.2 RF Penetration losses into the train, seen from the top of the train

not be the same even if you use same type of antenna in terms of distance, frequency and power. Penetration loss from antenna 'A' is typically the lowest, the service antenna 'sees' inside the train through a relatively large window opening – but seen from location 'A2' the window at this angle is relatively smaller and adding the reflection of the metallic coating of the window there will be an increased penetration loss – this can easily be 10–20 dB more than the penetration loss from antenna location 'A'.

The worst antenna location will typically be the antenna location 'B' where the antenna is could be located on top of the tunnel portal entrance (in the HO zone to and from the tunnel) – sadly this is the chosen location for 98% of the HO antennas designed to provide a sufficient HO zone to and from the tunnel service area to the outside mobile network.

The signal to/from portal antenna location 'B' will have to pass through the length of the entire train, or rely on reflections from objects near the train to reach the users inside the coach. This will greatly increase the link loss from antenna location 'B' when compared with the other options, antenna location 'A' and 'A2'.

The link loss from portal antenna 'B' when compared with antenna location 'A' can easily be 30–50 dB more and this often cause concern when optimizing the HO zone to and from the tunnel.

It is also important to note that the penetration loss will vary according to the relative location between the user inside the train and the antenna as the train moves.

The most uniform coverage and penetration loss will be option 'D' in Figure 11.2, showing a radiating cable that is installed at the same level as the windows in the train. The radiating cable is often only considered as a solution inside the tunnel itself, but might actually be a viable solution in the HO zone to/from the tunnel DAS system providing a long 'RF Corridor' 50–200 meters outside the tunnel entrance/exit, to tender for the HO Zone.

11.2   The Tunnel HO Zone

One of the most challenging components of any tunnel design are the handover zones; big and expensive mistakes are often made when designing that crucial part of the system, when considering the HO zone as any other handover zone in the network. It is very important to understand the potential challenges and pitfalls at an early stage in the project. Often, RF planning is mainly concentrated around the signal level inside the tunnel, and though this is one of the most important parameters and factors for success, there are other important parts of the design that must be considered carefully. It will be a major issue if the design and implemention solution do not ensure the smooth and seamless handover of traffic to and from the tunnel system when the vehicles moves between the outside network and the tunnel service area. The initial design of the handover zone is theoretical and typical RF desktop work, but the handover zone design is greatly dependent on installation limitations of the service antennas, so it is crucial with actual measurements in order to define the antenna locations for the tunnel HO antennas, etc. at an early stage in the project. Therefore, close cooperation with the construction team on the tunnel project itself is important at an early stage in the project. We, as radio planers would like to consider the RF coverage system to be a crucial part of the project, however the reality is often different. Remember that the radio coverage system is normally a minor part of the construction of a tunnel, therefore the specifications and limitations of cable routes, antenna, equipment and cable location must be defined at an early stage if we are to have the slightest hope of influencing the project to incorporate special requrements for locations of DAS elements.

Quite often you will simply have to adapt your RF design and your final DAS solution accord to the actual tunnel design and construction; take it or leave it. So tunnel DAS planning and design is way beyond desktop design work, you have to get out there, in the real world.

11.2.1   Establishing the HO Zone Size

The typical tunnel application can be seen in Figure 11.1 and the first step towards designing a successful handover (HO) zone is to define the size of the HO zone.

It is important to realize that several factors will have a great effect on the required size of the HO zone.

Considering the typical handover scenario for the HO, your design for the HO zone must accommodate the following steps in the HO process:

1. Allow the mobile the needed time to detect and measure the potential new serving cell; the HO candidate.
2. The system need suficcient time to process the measurement and evaluate the handover process.
3. Leave sufficient time for signaling and control of the handover.
4. Leave a time margin to ensure the required HO retries in case the initial HO is unsuccessful.
5. Provide enough time to make sure that the HO process is successful and the old connection with previous cell is closed.
6. Note that signaling load can be a bottle neck in the system; you will have a train load of passengers in handover, all at the same time for a few seconds; this will load the two hand- over cells severely.

This minimum HO zone mentioned above, size will be required whether it be the HO zone between the outside macro cell and the tunnel system, or the internal HO zones between cells inside the tunnel.

Typically the HO zone between the macro network and the tunnel will pose the greatest challenge; however, for larger metro subways HO zones between the cells in the subway can also be a considerable challenge, and there are fundamental differences between the require- ments for 2G, 3G and 4G services. This can lead to a compromise when designing a shared system for 2G, 3G and 4G in tunnels. Rail tunnels often pose the greatest challenges so the main focus in this chapter will be on that type of application; the same principles can be applied to car tunnels, but often these will be less of a challenge.

Avoid 'Ping Pong' Handovers

'Ping Pong' handover zones are handover zones that are not clearly defined, there is no clearly defined margin for hysteresis between the potential serving cells – so the traffic shifts for- wards and backwards between two or more cells during the handover until finally settling on the intended handover candidate cell. A 'good' bad example is illustrated in Section 11.7.2.

'Ping pong' handover zones are very problematic anywhere in the network and should be avoided; however, in tunnel solutions they can be lethal for the system performance. Take a train tunnel where a train with 20 people engaged in a telephone call must hand over between two cells in a matter of a few seconds. This is a significant challenge when it comes to sig- naling, etc., but a 'ping pong' HO zone will multiply the load with an increased number of obsolete handovers. This loads the system severely and increases the number of unsuccessful handovers and potentially the number of dropped calls. In some mobile systems you can tune and optimize HO parameters, hysteresis, etc. so as to cope with some potential issues, but like anything in radio planning, you will be so much more successful in your design if you do the radio planning right from the start, and then – if required – tune the network, not the other way round. Throughout this chapter (and in Section 12.5) there will be examples of how to deal with this for tunnel and rail solutions.

11.2.2   The Link Loss and the Effect on the Handover Zone Design

The link loss characteristics described above are dependant on the tunnel HO-antenna location relative to the train windows, as illustrated in Figure 11.2. The link loss also depends on the location of the macro network handover candidate from the outside network and the location of any potential interfering cells.

The typical handover scenario is shown in Figure 11.3, where a typical tunnel/macro HO-Zone is illustrated. This is a simplified example, the typical real life scenario can be a great deal more complex with more than one macro serving cell, several handover candidates at the entrance of the tunnel and more than one potential interfering cell operating on the same frequency as the macro or tunnel cell. Figure 11.3 depicts serving macro cell 'A' which is expected to provide coverage in the area just outside the tunnel and this makes cell 'A' the prime HO candidate to/from the tunnel. Cell 'B' is the tunnel system, and this HO zone is

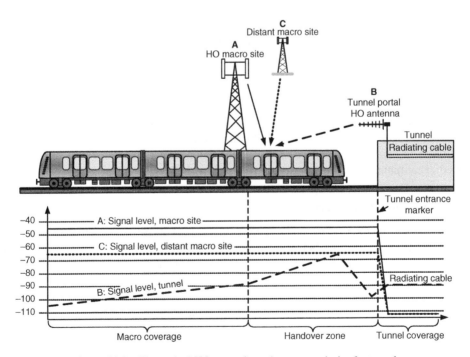

Figure 11.3   The typical HO scenario at the entrance/exit of a tunnel

covered by a directional antenna installed on top of the tunnel portal – antenna 'B': the purpose of this antenna is to ensure a sufficient overlapping signal from the tunnel system in the HO zone in order to ensure that the traffic can be handed over seamlessly to/from the tunnel and to/from the macro cell.

Inside the tunnel the users in the train will be served by the radiating cable by the same cell that is connected to antenna 'B'.

A typical design specification might provide that one requires a certain RF signal level from the tunnel system in the pre defined handover zone measured on the track. A potential problem arises due to the issues addressed in the previous section (Figure 11.2); that the link loss is highly dependent on the relative location between the mobile users inside the train and the serving antenna. The signal from antenna location 'B' might provide a strong signal on the empty track, but will not provide sufficient dominance inside the train coach and lacks margin to potential interferers; in this example cell 'C' (Figure 11.3).

The serving macro cell serves the users inside the train directly through the windows (Figure 11.3) and will only suffer from relatively low penetration loss through the window; and in this example provides a service level inside the train of –45 dBm, a powerful signal. The signal from the tunnel system, radiated from antenna location 'B', however, will suffer from high penetration loss from the longitudinal direction through the train, and rely mostly on reflections in order to reach users inside the train. So even with a very high signal level on the track, in the example in Figure 11.3 –45 dBm, the fact is that one has to consider the high link loss in reaching the users inside the train from antenna location 'B', in this example 20 dB, so inside the train the signal from antenna 'B' is only –65 dBm. This is not attractive enough for the mobile to hand over from cell 'A' to 'B' when entering the tunnel, being 20 dB less than the serving Macro cell A. The impact of the uneven penetration loss from Figure 11.2 is evident.

11.2.3   The Handover Challenge Between the Tunnel and Outside Network

At first glance you might think that it is fairly easy and straightforward to design the tunnel handover zone. After all, there are only two cells to consider: the macro handover candidate and the tunnel cell, and then the speed of the train – so from that perspective it seems pretty straightforward. However, there are several pitfalls that one must be careful to avoid – these issues, if left unattended, can cause serious problems and entail expensive changes to the system in order to make it work. Let us look at a real life problem, as illustrated in Figure 11.3.

HO Problem Entering the Tunnel

Owing to the link losses described in the section above, it is now clear that due to the high signal level from the macro cell served from antenna location 'A' the tunnel system, coverage by antenna B will not be dominant enough to trigger the handover, at least not until we reach the entrance of the tunnel. Now the problem starts, due to the fact that the macro cell signal served from antenna location 'A' will fade rapidly because of the structure of the tunnel, and leave a very short HO zone, if any, for the system to hand over the traffic successfully from cell 'A' to the tunnel system cell 'B'.

Adding to the problem will be the distant macro cells perpendicular and in line of sight to the train, in this example antenna location 'C'. Due to the low penetration loss from the direction of antenna location 'C' relative to the link loss to the HO antenna location 'B' even very distant cells can interfere with the signal from the signal from antenna location 'B' and cause dropped calls/unsuccessful handovers. If the macro cell 'A' is very close to the track, a few meters, it can even overshoot the train and not reach through the windows, and will be shielded by the metallic roof of the coach.

HO Problem When Exiting the Tunnel

The potential problem is less complex when exiting the tunnel and performing handover from the tunnel system served by antenna location 'B' and the macro handover candidate served by antenna location 'A'. The high signal level from macro (A) cell will trigger the handover from the tunnel system (B), provided that no potential interference (C) will cause problems, and there is a sufficient handover zone. The root of the problem is the uneven link loss shown in Figure 11.2.

Calculating HO Zone Size

We can easily calculate the required HO zone overlap if we know the speed of the vehicle and the time needed to perform the handover, remembering to include time for handover retry if the initial handover is not possible.

Let us look at one example:

We have a vehicle with a speed of 120 km/h, and we need six seconds for handover (once the trigger value is fulfilled), and then we will add extra margin – a total HO zone that last 18 seconds.

First, we will calculate how many meters per second the train moves:

The relation between km/t and meters per second [m/s] is: m/s = (km/h)/3.6.

So at 120 km/t this would be 120/3.6 = 33.33 m/s

For 18 seconds of valid HO zone we would then need 18 × 33.33 = 600 meters of valid overlapping HO zone. And remember that throughout the HO zone both cells (serving and candidate) would need to have sufficient level to fulfil the HO trigger and provide signaling, etc.

This is a challenge to design for, and especially for high speed rail (see Section 12.5 for more on high speed rail); however, it is a genuine requirement, even under ideal conditions, where all of the antennas are planned and installed correctly.

Please note that some (most) rail systems will often exceed the official planned speed in order to catch up on delays, so having a good margin in the handover zone size is important.

11.2.4 Possible Solutions for the Tunnel HO Problem to the Outside Network

There are several possible solutions for the potential HO problem described in Figure 11.3. The best solution is obviously to avoid any problems by taking all aspects into account in the design phase of the project.

Optimize the Location of the Tunnel Entrance Antenna

One of the possible solutions is illustrated in Figure 11.4B; the problem is that the tunnel HO antenna is located on top of the portal of the tunnel entrance (Figure 11.4A). The link loss from this antenna location (A) will be a problem when serving users inside the train due to the fact that the signal from the portal antenna is not 'visible' through the windows and this results in high path loss, as described in Figure 11.2.

One solution, and the best way to avoid the problem, is simply to move the antenna location of the tunnel HO antenna from on top of the portal down to the side of the track, thus making it more visible to the users inside the train. It is preferable to split the signal between a minimum pof two antennas (Figure 11.4B); one antenna should be located at the side of the tunnel just at the entrance, the other should be located 50–100 meters further down the track on the opposite side. If possible, the antennas should be moved off to the side of the track in level with the windows, at a distance of 5–10 meters from the track. This lay out of the antennas will raise the signal inside the train coach by 20–40 dB and provide a more uniform and well defined HO zone with good dominance of the tunnel cell in the desired area.

It is always recommended to do a trial set-up and perform a RF survey measurement to verify the exact antenna locations and signal power needed at the antennas. If possible, use the exact train type and speed for the RF survey.

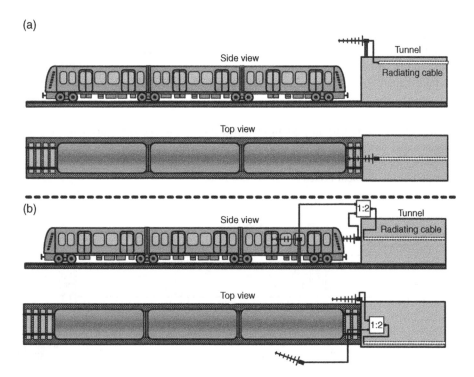

Figure 11.4 One possible solution (B) of the HO zone problem (A) at the entrance of a typical tunnel

Other Solutions for the HO Problem of the External Network

There are several other solutions that might be considered in order to resolve the issue with rapid signal fades in the HO zone. Obviously the best solution is to avoid the potential HO problem in the first place. But if you end up with problems in the HO zone some of the topics worth considering would be the following:

- Limit the number of neighbors in the neighbor list in order for the mobile to perform faster measurements.
- Use fast averaging HO evaluation algorithms to speed up the HO processing time.
- Use smaller HO margins.
- 'Chain' or pre-synchronize the cells; this is highly vendor and technology dependant.
- You could consider re-radiating the macro cell inside the tunnel on the first section of radiating cable in order to perform the tunnel/macro HO inside the tunnel in a more controlled RF environment.

11.3   Covering Tunnels with Antennas

For a short tunnel of a few hundred meters where you have line of sight from one end to the other the preferred coverage system in a tunnel will often be yagi antennas at the end of the tunnel beaming through the tunnel as well as covering the HO zone to the tunnel. This is a fast and relatively inexpensive solution (See Section 11.11.4 & Figure 11.28). Often these solutions will be operating on the nearby macro cell via a repeater system and thus not even require a dedicated base station.

Although antennas inside the tunnel, even for large tunnels, feeding sections of the tunnel to provide end to end coverage (as illustrated in Figure 11.5) are a viable solution there are some pitfalls that are to be avoided. Let us have a look on some of the common issues that one needs to consider carefully.

- Mechanical issues; it is very important to ensure that the antennas are mechanically sturdy and are actually designed to be installed in the challenging environment of the tunnel. This will mean using antennas that are 100% enclosed in a plastic radome to protect them from dust and other mechanical issues.

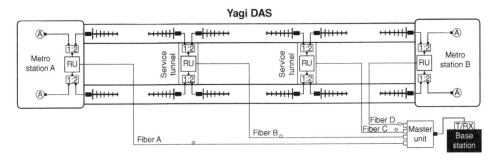

Figure 11.5   Tunnel DAS system using antennas in the tunnel sections

- Vibrations and wind load from passing traffic are not to be underestimated and special antennas that can be fixed to the wall in each end of the antenna radome are often to be preferred inside the tunnel.
- Tunnel clearance profile (the kinetic envelope; see Figure 11.6): you must make sure to stay clear of the 'traffic profile' of the tunnel – this is the defined zone of the tunnel occupied by the vehicles - in order to ensure that no cars or train coaches or payload hit any objects in the tunnel (Like your DAS equipment & antennas). This is obviously very important; the RF coverage in the tunnel must remain a secondary priority compared with the safety of the moving traffic in the tunnel.
- Building code; there are often special requirements for mechanical and electrical testing of any equipment in a tunnel. So you must make sure to document that all parts of your system will comply with these codes.
- Always verify with RF survey measurements: when using antenna distribution in a tunnel it is highly recommended to verify all the different types of tunnels, and tunnel sections, with measurements. It is impossible to simulate the coverage inside a tunnel, there are so many 'Tunnel Factors' to take into account: size of the tunnel, shape of the tunnel, wall surface material. The antenna mounting position in the tunnel is an important factor, a few cm increase in distance from the corner of the tunnel profile can mean a difference of 5–10 dB in propagation.
- The 'near-far effect' (see Section 4.12.10): using distributed antennas in a tunnel system will often give rise to 'near-far' issues. Consider the typical DAS deployment in a tunnel; a high power Remote unit feeding antennas with relative low coupling loss between the RU and the antenna. The problem is that the users are not being served by the tunnel DAS but by the outside network – if there is a signal leaking in – and at the same time are in close proximity to the antenna and operating in the same frequency band. This mobile, not within power control by the DAS, is likely to be transmitting at quite high UL power in order to reach a more distant server. This can cause UL blocking problems on the front end of the RU caused by a limited dynamic range and cause signal distortion in the front stage of the amplifier. This is one of the major problems to consider when designing DAS systems for tunnels, especially when you have a tunnel DAS system that is not shared by all mobile operators. See Section 5.7 for more specific detail regarding multi operator DAS. It is recommended that one minimizes this problem by simply installing the antennas in such a manner that users cannot get closer than 1–2 meters from them, if possible.
- Antenna placement and coupling to the tunnel: it is recommended to experiment with the antenna location of the yagi/circular antennas that feeds signal into the tunnel. Just by moving the antenna 20–50 cm around the proposed installation area can have a significant effect on how much RF energy is being beamed down the tunnel. This is especially so for antennas near the entrance of the tunnel, or antennas located at the subway station area and feeding into the tunnel. A difference in the average signal level inside the tunnel of 5–10 dB can easily occur when using the same power to the antenna, but moving it around slightly.
- 'RF blocking': one must realize that objects in the tunnel, chiefly the train, will block the RF signal on the opposite side of the vehicle to the antenna. So you must make sure, when designing a yagi DAS for a tunnel system, that there will be only one train in each antenna section at one time, or ensure that both trains can be covered, each from an antenna in the opposite direction.

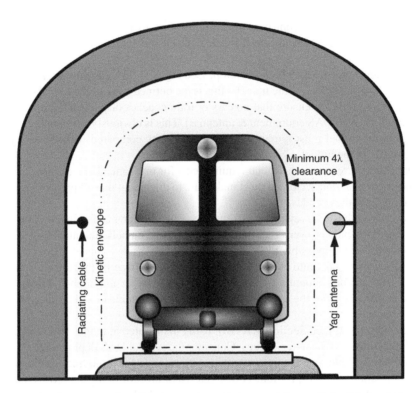

Figure 11.6 Tunnel with train – 4λ clearance is needed as a minimum if the RF signal from a yagi antenna is to get past the train. Radiating cable is best aligned at the center of the windows to maximize the signal. Remember to stay clear of the 'kinetic envelope' of the train, to avoid disaster.

- One needs to ensure that the clearance between the tunnel and the train is at least 4λ (see Figure 11.6), and then you will often be able to service even a long train through the windows, served by an antenna inside the tunnel. But it is always recommended to verify this with RF survey measurements.

11.4   Radiating Cable Solutions

Radiating cable systems inside tunnels are often to be preferred when compared with antenna distribution in the tunnel; however, in many cases a solution based on distributed Yagi antennas could work just as well and at a much lower cost. The RF environment using radiating cable is much more uniform and easier to plan. The main drawback is the challenge of installation and the price of the installation.

Traditionally, radiating cable is only used for tunnel coverage solutions. However, there are cases where a building has one or several long tunnels, i.e. interconnecting buildings in a campus, long vertical shafts, emergency staircases or elevator shafts, where a radiating cable could be considered as a possible solution.

The typical tunnel application for radiating cable can be seen in Figure 11.7 with a section of tunnel between two underground metro stations. High power remote units installed in the emergency escape tunnels feed the cable sections, and at the stations also the platform antennas. The base station is located at Station B and the optical fiber DAS distributes the signal throughout the tunnels and stations from the central base station.

11.4.1 The Radiating Cable

The radiating cable or 'leaky feeder' (as illustrated in Figure 11.8) is typically based on a traditional coaxial cable with (1) an inner conductor, (2) a dielectric, (3) an outer shield, (4) tuned slots in the outer shield and finally (5) the jacket. The size, shape, orientation and placement of the slots are optimized to tune the coverage from the cable. The cable can be optimized during the design and production with regard to coupling loss at certain distances, frequency bandwidths, insertion (longitudinal) and loss.

The cable acts as a long antenna, or many individual small antennas. The slots will radiate and pick up signal along the length of the cable. Typically the cable will be optimized to serve a specific frequency range of RF spectrum, and optimized to maximum coverage within a certain distance, within the space of the tunnel. Normally a radiating cable is not applicable for servicing users who are not in the proximity of the cable; it is typically ideal for distances of 2–10 meters perpendicular to the cable.

Initially the radiating cable was produced as a 'simple' coaxial cable where the manufacturer simply cut with a parallel saw, thus splitting the outer conductor (shield) of a standard corrugated cable. These days, however, designing and producing radiating cables is a fine art,

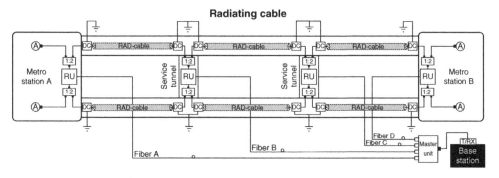

Figure 11.7   A typical section of a metro tunnel DAS, using radiating cable in the tunnels and antennas in the station area

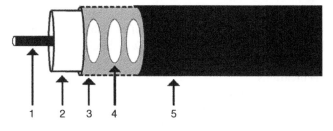

Figure 11.8   Radiating cable, principle

with the apertures (holes) finely tuned in shape, size and layout, optimizing the cable to be more broadband or fine tuned for optimized performance at a given bandwidth.

The Technical Data of the Radiating Cable

Radiating cable might initially look very simple and easy to understand, but designing and installing radiating cable coverage solutions is a challenge. To understand how to utilize the cable, you first of all need to understand the main parameters of the cable (see Figure 11.10). The most important parameters to consider when selecting, designing and installing radiating cables are listed below.

- **Frequency Range**
  The radiating cable will typically be designed and optimized to support specific frequency ranges or bands. For use in designing systems for 2G/3G/4G you can find standard radiating cable optimized to support typical band combinations such as 900, 1800 and 2100 MHz. Be careful if you plan to apply other radio services on the same cable system, and make sure that the cable will be able to support it.
- **Longitudinal Loss**
  The longitudinal loss increases with distance and will vary with frequency as with standard coaxial cable. Note that the longitudinal loss is also related to the coupling loss of the cable, so the lower the coupling loss the cable is designed for, the more signal is coupled out of the cable, and hence there will be a higher longitudinal loss.
- **Coupling Loss**
  The coupling loss is the loss between the cable and the mobile terminal, specified at a given distance and probability, typically 2 or 6 meters. The specification of coupling loss usually has a relatively high variance margin, typically 10 dB. Be sure to include this in the design overhead margin when calculating the link budget. Note that the coupling losses stated in the datasheet are related to a certain probability. A coupling loss of 80 dB at 50% probability could mean a coupling loss of 95 dB at 95% probability.
- **System Loss**
  The system loss is the sum of the longitudinal and coupling loss. This information is some-times used in data sheets. Some manufacturers of radiating cable can produce the cables to fit the exact installation, and decreasing the coupling loss as the longitudinal loss increases, thus having a more uniform system loss over the distance of the radiating cable.
- **DC Resistance**
  In many tunnel applications DC is injected onto the radiating cable at a central point, and distributed via the inner and outer copper conductor of the coaxial cable. This is used for distributing power supply to distributed active elements, amplifiers, etc., connected in the system. It is important that the DC characteristics of the cable will be able to accommodate the power load requirements of the amplifiers, BDAs (bi-directional amplifiers) and other equipment on the line.
- **Mechanical Specifications**
  As for other types of cable, mechanical specifications, size, bend radius and fire rating are important. Be sure to fulfill the installation speciation and local guidelines for the specific site where the cable is to be installed, as well as the specifications from the manufacturer of the cable.

Delays in the System

Owing to long cables, optical links, etc., delay and timing can be an issue in tunnel solutions.

Remember that the velocity of the coaxial cable and optical fiber as well as delays in active elements will cause the timing to shift. This will offset the timing advance in 2G (and for very long tunnels you might even reach the limit for maximum timing advance in the 2G system). For 3G you might need to widen your 'search window'. Refer to Section 10.3 for more detail.

For more general information on calculating delays in large DAS systems using fiber distribution and radiating cable refer to Section 11.11.1.

11.4.2   Calculating the Coverage Level

One must realize that fading in the tunnel and coupling variance from the cable, as illustrated in Figure 11.9, has to be accounted for in the link budget (a simple example is given in Table 11.1). It is fairly easy to calculate the service level in the tunnel. It is important to include fading margins; these margins will depend on several 'tunnel factors' such as tunnel size, shape, material, speed of the users and obstructions in the tunnel.

Most manufacturers of radiating cable will provide you with design support and help with tools that can help you do the radio planning of the tunnel sections.

Ideally, especially on a large tunnel project, it is recommended to verify RF calculations and the link budget with a trial installation in a section of the tunnel. It can be very expensive if mistakes are made in tunnel projects; be absolutely sure to check, re-check and double-check everything, twice. Believe me; I learned that the hard and expensive way!

Sample Link Budget for a Radiating Cable

This is a simplified link budget for a downlink calculation (Table 11.1) Please note that this is just an example, you will need to consult with the manufacturer of the radiating cable to make sure you use the appropriate numbers. Also bear in mind that you will need to make sure to include the needed overlap for handover zones, as discussed in Section 11.2.3. Needless to say but you will also need to perform the appropriate uplink calculation.

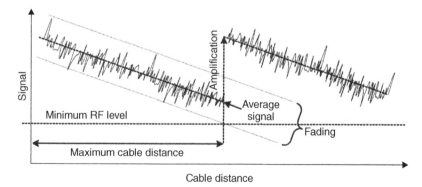

Figure 11.9   Fading and variance of the signal in a tunnel can be high, so good design margins are recommended

Table 11.1 Sample link budget (downlink)

Sample downlink budget example		
Radiating cable type	1¼′	
Inputs to the link budget		
Description	Value	Reference
Coupling loss at 2100 MHz (95% probability)	82 dB	A
Longitudinal loss per 1000 meters	78 dB	B
RF planning level	–85 dBm	C
Body loss (loaded train)	8 dB	D
Fading margin	10 dB	E
Train penetration loss	8 dB	F
Power per carrier from the DAS Remote Unit	15 dBm	G
Margin for extra loss in jumpers, connectors, etc.	3 dB	H
Calculations / Output		
Total margins (D+E+F+H)	29 dB	Q
Maximum allowable path loss (G-C)	100 dB	X
Maximum allowable longitudinal loss (X-Q)	81 dB	Y
Radiating cable longitudinal loss per meter (B/1000)	0.078 dB	Z
Maximum meters the radiating cable can service (Y/Z)	1038 meter	

Link Calculation Example: Radiating Cable

In order to demonstrate the basics of tunnel radio planning using a radiating coax, let's have a look at the example in Figure 11.10; this is a small, simplified example to show the basic calculations when designing radiating cable systems. Remember, as stated several times before, it is important to verify all calculations with field measurements. Also keep the handover zones and overlap in mind, as described in Section 11.7, as well as the delay in tunnel systems, as described in Section 11.11. In principle, however, these are the calculations you would do, regardless of whether it is a short tunnel or a large complex metro system with several sections of radiating cable, several remote units, and several sectors. You divide the tunnel DAS into individual sections, and then try to verify the performance at the expected worst-case location for each section.

We have the following data for a radiating cable (1 1/4") and we want to do a simple link calculation at 2100 MHz. According to the datasheet:

$$\text{Longitudinal loss} = 8.2 \, \text{dB} \, @ \, 100 \, \text{m} \, @ \, 2100 \, \text{MHz}$$

Coupling loss = 71 dB @ 95% @ 2 m (95% probability at a distance of 2 m)
    Other design inputs are as follows:

$$\text{Designtarget level} \left(\text{DL} \right) = -85 \, \text{dBm CPICH}$$

$$\text{Fading margin is assumed} = 10 \, \text{dB}$$

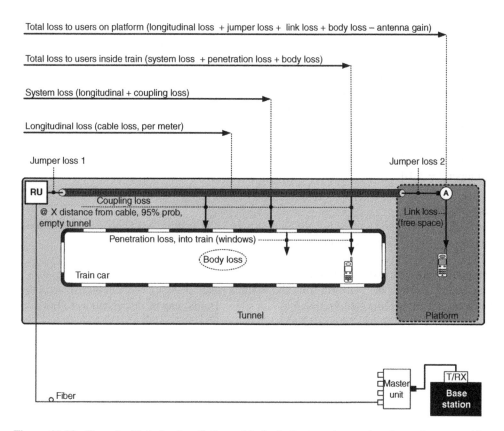

Figure 11.10   Tunnel with train. A radiating cable feeds the tunnel, covering the station area with a single antenna at the end of the cable. The system is based on a single remote unit (RU) in a DAS

Penetration loss

We have measured the penetration loss through the windows (with a section of radiating cable, placed exactly at the same location and alignment), and used the actual train, using the same frequency band (2100 MHz) that we plan to deploy in the tunnel.

The penetration loss is measured to be 12 dB through the windows of the train.

Output power from RU

Let us assume that we have a power output from the remote unit of +40 dBm (+30 dBm CPICH).

Tunnel length

This distance through the tunnel from the RU to the station platform is 300 m.

Body loss

Body loss can vary (3–15d B or even more) – there is a big difference between having a single user inside the train and the same metro fully loaded during rush hour. Let us assume 6 dB for this example, which is about 3dB more than the typical body loss for a standard indoor DAS design.

Jumper losses

We must remember to include every little detail in the calculations, all the components, cables, connectors, etc. In this example, we have a short 1 m jumper ('Jumper 1' in Figure 11.10) at the RU connecting it to the radiating cable that services the tunnel. We also have a longer jumper, 'Jumper 2', that connects the radiating cable at the station platform to the omni antenna, in order to provide service to users in the platform area.

We have measured the losses to be:

- Jumper 1 = 0.2 dB
- Jumper 2 = 3.4 dB

Antenna gain

The gain at the antenna servicing the platform is 2 dBi using a standard omni directional antenna and the mobiles are assumed to be 0 dBi antenna gain.

Calculating the link loss on platform

Link loss to users on the platform from the antenna (max. distance is about 80 m): we know the free space loss (FSL) (see Section 5.2.8 for more details) can be calculated as follows:

$$
\begin{aligned}
FSL &= 32.44 + 20\left(\log F[MHz]\right) + 20\left(\log D[km]\right) \\
FSL &= 32.44 + 20\log(2100) + 20\log(0.08) \\
FSL &= 32.44 + 66.44 + (-21.93) \\
FSL &= 76.94 \, dB
\end{aligned}
$$

We can then include the antenna gains and calculate the link loss from the omni antenna servicing the platform:

$$
\begin{aligned}
Link\ loss &= FSL - antenna\ gains \\
Link\ loss &= 76.94 - 2 - 0 = 74.94 \, dB
\end{aligned}
$$

Calculating the link losses to the users inside the train and signal levels

Using this information, we can now calculate the losses to the users inside the train. Look at Figure 11.10 to follow the parameters we are calculating. We will calculate the 'worst-case scenario' with the maximum distance from the RU of 300 m, just prior to arriving at the station.

$$
Longitudinal\ loss = loss / m \times number\ of\ meters
$$

$$
Longitudinal\ loss / m = Longitudinal\ loss\ per\ 100\,m / 100
$$

$$
Longitudinal\ loss / m = 8.2\,dB / 100\,m = 0.082\,dB / m
$$

$$
Longitudinal\ loss\ for\ 300\,meter\ is\ therefore = 0.082\,dB \times 300 = 24.6\,dB
$$

We need to add the 0.2 dB jumper loss from Jumper 1 to get the total loss = 24.6 + 0.2 = 24.8 dB

Now we can calculate the system loss:

$$\text{System loss} = \text{longitudinal loss} + \text{coupling loss}$$

$$\text{System loss} = 24.8\,\text{dB} + 71\,\text{dB} = 95.8\,\text{dB}$$

$$\text{Total loss to users inside the train} = \text{System loss} + \text{penetration loss} + \text{body loss}$$

$$\text{Total loss to users} = 95.8\,\text{dB} + 12\,\text{dB} + 6\,\text{dB} = 113.8\,\text{dB}$$

Now that we have established the total loss, it is quite simple to calculate the expected signal levels:

$$\text{CPICH signal level downlink}\,(\text{DL}) = \text{RU power} - \text{total loss} = 30\,\text{dBm} - 113.8\,\text{dB} = -83.8\,\text{dBm}$$

Conclusion on DL signal level inside the train

If we assume a fading margin of 10 dB, we can now assume the minimum signal to be $-83.8\,\text{dBm} - (-10\,\text{dB}) = -93.8\,\text{dBm}$. We are therefore below our target level of $-85\,\text{dBm}$, and even maintaining a 3G voice call will be a problem (see Chapter 8 for details of the LB planning). We also need to verify the UL. The fact is that even when we include the fading margin the DL is below a reasonable signal level. We should also include some 'degradation margin' due to dust build-up on the cable, antennas etc.

We might consider adding more RUs to compensate, or we could perhaps reconsider the location of the RU. Perhaps we should place the RU in the center of the tunnel and feed the cable into the middle – that will mean a 1:2 split – but the distance will now be over two individual 150 m sections of cable, with only 150 m in each direction. A distance of 150 m will result in a loss of 12.3 dB; then, if we add a 3 dB splitter loss to feed the cable, we will have a total longitudinal loss of 15.3 dB instead of 24.8 dB. This will give us about 9.5 dB more signal level inside the train so that we have $-84.3\,\text{dBm}$ instead of $-93.8\,\text{dBm}$ CPICH, which means we would be okay in terms of signal level.

Well, it might work – to quote my friend Jack Culbertson: 'It all depends.' Tunnel planning is not only a very detailed and thorough desktop exercise, but it is very important to perform all the calculations to the last detail. Can we actually place an RU in the center of a tunnel in a real deployment? Or will it compromise the 'kinetic envelope' (Figure 11.5) in the tunnel and therefore compromise security? That would definitely be a non-starter in a real-life situation. We would never be allowed to install any equipment that might compromise the safety and operation of the traffic in the tunnel.

Therefore it is always very important to know the exact possibilities and limitations in the actual tunnel in which you plan to deploy your DAS, especially when it comes to practical installation and access concerns. In the example here, say we decided to move the RU to the centre of the tunnel; it might even be possible to install it at that perfect location. However, what we hadn't realised is that post-installation access to the tunnel is extremely limited – we can only get access during the planned 'service slots', limited time slots during the month when the trains aren't running. Quite often, this might be a 'service window' of three hours every fourth week, so if the RU were to break down, it could be weeks before it can be serviced.

Calculating the signal level on the platform

The users on the platform will be serviced by the omni antenna that terminates the radiating cable as in Figure 11.10, and will in this following example be at maximum of 80 m from the servicing antenna, with a line of sight to it.

The previous information enables us to calculate the signal levels:

$$\text{Feeder loss to platform antenna} = \text{longitudinal loss in tunnel} @ 300\text{m}$$

$$\text{Longitudinal loss in the tunnel} @ 300\text{m} = 24.8\,\text{dB (as per previous calculations)}$$

$$\text{Feeder loss to platform antenna} = \text{longitudinal feeder loss} + \text{Jumper 2 loss}$$

$$\text{Feeder loss to platform antenna} = 24.8\,\text{dB} + 3.4\,\text{dB} = 28.2\,\text{dB}$$

$$\text{Total loss to users on platform} = \text{feeder loss to platform antenna} + \text{link loss}$$

$$\text{Total loss to users on platform} = 28.2\,\text{dB} + 74.94$$

$$\text{Total loss to users on platform} = 103.14\,\text{dB}$$

Now we can calculate the expected DL CPICH level as RU CPICH power – total loss:

$$\text{DL level CPICH} = +30\,\text{dBm} - 103.14 = -73.14\,\text{dBm}$$

With 10 dB fading margin we will be at –83.14 dBm. This is okay, as we are aiming for –85 dBm (we still need to confirm UL calculations; see Chapter 8).

Earlier, we concluded that moving the RU to the center of the tunnel was a good idea, so that instead of a distance of 300 m over the radiating cable there would only be 150 m and we would reduce the loss to 12.3 dB. With this change, the signal at the platform will 12.3 dB higher, at –70.64 dBm (including 10 dB fading margin). This will be more than enough to service the users at the platform.

Remember that the train will affect the coupling loss, so once again, always perform verification measurements, in the real tunnel, with the real train. (In the past, I have made the mistake of not doing this and it was quite an expensive lesson!)

We also need to make sure we provide for enough HO margin in a multi-cell solution, to allow for the needed overlap for HO retries and still maintain the signal level during HO retries. For more details, refer to Section 11.7

11.4.3   *Installation Challenges Using Radiating Cable*

In tunnel projects, installation challenges, limited space and lack of access play a major role in the project, especially with regard to time consumption and installation cost. Where there is less flexibility in terms of installation options, you simply have to 'make it work' given the circumstances; this often dictates the final design option. With regard to radiating cable installation pay close attention to:

- *Grounding*: installation of radiating cable inside tunnels, especially metro tunnels with operational train services and high-power DC lines to feed power to the trains, is a major challenge. The power supply to the trains via the typical third rail or overhead power line is so powerful that it is likely to induct power on the radiating cable in the tunnel, therefore grounding is very important.

- *Follow the instructions*: it is very important to install the radiating cable exactly as specified by the specific vendor of the cable. It is equally important to follow the local on-site guidelines 100%. Terminate all points in the system, do not leave the cable 'open' at one end, terminate with an antenna or use a terminator (dummy load).

- *Use cable clamps*: special clamps that stand off the wall have to be used to install the cable at a distance from the wall; if you do not do this, then the coupling loss will increase drastically. Normally you will need to use metallic fire protection clamps every 3 meters or so, to ensure that the cable does not fall off the wall in case of a fire and block rescue access. In tunnels with a high speed train service, special attention must be paid to the mechanical stress on the cable installation. Often a special type of heavy duty cable clamp has to be used to make sure the cable is secured to the wall.

- *Align the cable*: normally, especially on higher frequencies, such as 1800 or 2100 MHz, the cable has to be aligned perfectly so that the slots can leak to the environment with minimum coupling loss; a guide is typically printed on the jacket of the cable to assist with the alignment. Remember, if at all possible, to install the radiating cable level with the center of the windows (Figure 11.6), to optimize the coupling loss to the users inside the train, and maximize the RF level and performance. The difference between aligning the cable and having it offset 100 cm can easily be a degradation of the RF signal of more than 10 dB, thus impacting the quality of service (QOS), resulting in reduced data service, or possibly no data service and dropped calls.

- Installation distance to the wall: take careful note of the datasheet and follow the exact guidelines from the manufacturer. The typical radiating cable cannot be installed flush to the wall. A certain minimum distance is needed in order to allow the RF field to build up the RF signal and create the coverage in the tunnel.

- *Mechanical stress*: do not underestimate the mechanical stress on a tunnel installation; this applies to all equipment installed, and therefore also the cables. Normally you could get by with letting a coaxial cable lean against the corner of a concrete wall, but vibration on the coaxial cable caused by the vehicles or trains will cause the concrete to wear off the plastic jacket and into the copper of the cable. Over time, moisture and corrosion will degrade the performance. It is very important to observe that, if you have to pass the radiating cable through a heavy wall, you need to terminate the radiating cable, using a normal coaxial jumper through the wall, and continue with a new section of radiating cable on the other side of the wall. If this is not done correctly, the system loss will be increased.

- *Clean the cable frequently*: another practical issue with a major degrading impact on performance is dirt on the radiating cable. Dust from diesel exhaust residue, brake dust from trains, etc., will dramatically increase the system loss. Therefore routine cleaning of the cables will need to be scheduled to preserve performance.

- *Use DC blockers/surge protection* (see Figure 11.6): this is an inline device you mount on the RF connector of your remote units or base stations. These components will make sure that any unintentional DC on the cable will be grounded and protect your equipment and the surroundings, in case your feeder system connects to the DC feed that powers the train in case of an accident. You will need to follow the local installation code, and often have to mount these devices close to your equipment/equipment room. These devices are often (always) also used when deploying macro antennas, where the antenna feeders enter the equipment room, to protect the equipment room against lightning strikes on the antenna system.

The biggest challenge in using radiating cables is the installation. It is often very expensive and time-consuming. This is especially true if the system is to be installed in an operational tunnel. In this case you might only get access to the tunnel for a few hours every night, while it is closed for scheduled maintenance. Only rarely will you be allowed to close the tunnel, solely for the purpose of installing the mobile radio system. Installing radiating cable is a logistical challenge. The cables have to be transported on a special trolley, cable stands pre-mounted and connectors fitted. Everything needs to fall into place in order to be able to install just a few hundred meters of cable during a night shift. Remember, you need to evacuate and clean the site in good time prior to the traffic opening in the tunnel. Tracks need to be inspected in order to approve the tunnel for rail traffic again; it takes just a small cable clamp on the track to derail a fast-moving train – with potentially lethal effect.

11.5   Tunnel Solutions, Cascaded BDAs

One way to design a tunnel system with radiating cable is to use daisy-chained BDAs to 'ramp up' the signal at certain intervals (Figure 11.11) in order to compensate for the longitudinal attenuation and to keep the coverage level above the required design level. In this type of application the gain of the BDAs will typically be adjusted in order to compensate for the longitudinal loss of the preceding cable, but not higher – that would just cause noise build-up and degraded dynamic range (see Chapter 7 for noise calculations).

This design methodology is sometime referred to as 'zero loss systems'.

11.5.1   Cascaded Noise Build-up

Concerns with these types of system with cascaded amplifiers are noise increase and limitations in dynamic range. You will need to be careful when configuring the gain of the amplifiers in the system. This is because each amplifier will amplify the signal and the noise of the preceding link and also add its own noise figure to the performance. If you are not careful there

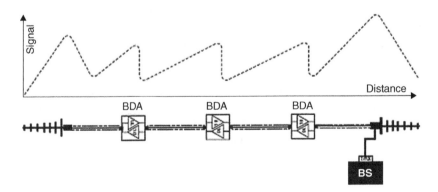

Figure 11.11   Radiating cable fed using in-lineBDAs (cascaded system), and the corresponding RF signal

will be a severe cascaded noise build-up in the system. Noise calculations and guidelines on how to design amplifier systems and control the noise are covered in Chapter 7. In practice, you should, at a maximum, cascade three or four BDAs, but this has to be verified in the link budget of the actual project.

Cascaded System, Principle

The principle of the 'zero loss system' is illustrated in Figure 11.11, where a base station drives the first section of radiating cable and the handover antenna, before the longitudinal loss attenuates the service level too much. The first BDA amplifies the signal and feeds the next section of radiating cable. This section then drives another section of cable, and so on. The two handover antennas will to some extent overpower the signal from the radiating cable when the mobile is close to these antennas. Note that the BDA works only in one direction (one direction for the downlink, the opposite direction for the uplink), unlike the 'T-feed system' (see Section 4.9), that radiates out in both directions of the radiating cable.

Power Supply to the Cascaded BDAs

To save on project cost, limit failure points and ease maintenance, centralized DC power is injected into the radiating cable near the base station. This DC will power all the BDAs on the line. Make sure that all the components on the line are able to handle the DC power, and are able to pass the DC when needed.

Supervision and Stability

One major weak point in this type of system is that, in the case of an amplifier malfunction or if a cable is cut, all signals could be lost in the remaining part of the tunnel system.

Special BDAs for tunnels will often have a 'bypass' function, to make sure that signals are passed on 1:1 with no amplification in case of an error. In most cases this will make only a small difference; the signal to the next BDA will be too low for the system to perform due to the lack of signal.

Specialized inline BDA systems are available that are able to communicate internally via the coaxial cable. Each active element will be assigned its own address, and the system will be able to communicate internally, perform auto calibration and have a central point where the status of the system, performance and alarming can be monitored and controlled locally or remotely via an RF modem or IP connection.

11.5.2   Example of a Real-life Cascaded BDA System

The measurement shown in Figure 11.12 is a real-life measurement of a cascaded BDA or leaky feeder system in a tunnel, showing a section of the tunnel with three BDAs.

You can clearly see the HO antenna that is placed just before the rightmost BDA, the 'peak' on the signal level, and on the left side of the graph is a similar peak from an antenna.

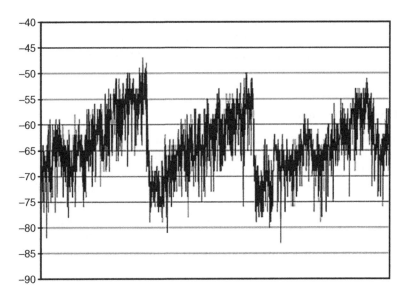

Figure 11.12 Real-life measurement of a cascaded BDA system in a metro tunnel

The measurement shown is over an 800 meter section of the tunnel, from one metro station to the next station. The radiating cable was placed in a 5 meter wide circular concrete tunnel, carefully aligned with the placement of the windows of the train. Measurements were made at 2 meters from the radiating cable, in the centre of the track during a walk test by foot in the empty tunnel. The distance between the BDAs was about 260 meters, and the radiating cable was a heavy 1.25 inch.

The design level was −75 dBm at 98% 2G-1800 in the empty tunnel (penetration loss into the train was verified by measurements to be only 5 dB). The longitudinal loss per 100 meters was 6 dB according to the specification of the cable. From the measurement, the longitudinal loss can be estimated to be about 15 dB. This confirms the information on the datasheet with regards to the longitudinal loss performance.

Note that the measurement was conducted during a walk test, at quite a slow speed compared with the speed of the final users (inside a metro train). This slow-moving measurement tends to emphasize the fading, as can be seen in Figure 11.12.

11.6 Tunnel Solutions, T-Systems

One of the most commonly used designs is to use a 'T-feed' system. This system has several advantages over the cascaded system in Section 11.5. Signal distribution to the RUs (Remote Units) is done in parallel with the radiating cable system. Each RU (Remote Unit) is fed an input, and provides output in both directions of the radiating cable. In the case of RU malfunction, or a broken radiating cable, most of the system will actually still work. Centralized control and monitoring are done at the 'optical fiber unit' co-located with the base station. This main unit of the tunnel system also feeds the base station with external alarms in case of malfunction. Remote configuration can be done via an RF modem or IP for easy trouble-shooting and maintenance.

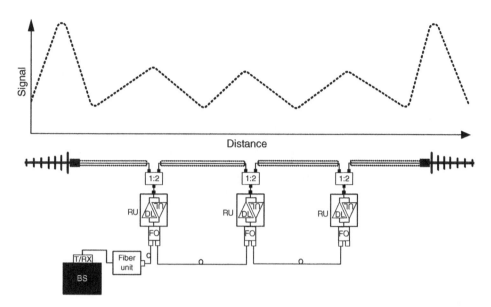

Figure 11.13 Radiating cable fed by a 'T-system'; the signal is distributed via optical fiber and fed via a 'daisy-chained' optical fiber system to each Remote Unit

11.6.1 T-systems, Principle

The principle for the 'T-feed' system (as illustrated in Figure 11.13) is that the base station connects to an optical converter that converts the RF into optical signals. Using optical transmission the signal can be transmitted over relatively long distances with only limited degradation compared with RF transmission over coaxial cable. Each RU (Remote Unit) has an optical interface and converts the signal back to RF, and then a power amplifier that transmits the radio power in both directions on the radiating cable. Naturally the RU has a reverse amplifier and RF-to-optical converter for the uplink signal.

The example in Figure 11.13 shows the principle of a commonly used 'daisy-chain' distribution of the optical signal, from RU to the next RU, thereby reducing the need for optical fiber to only one link. This is a big cost saver if you need to rely on renting pre-installed 'dark optical fiber' from a third party.

Better Noise Performance than Cascaded BDA Systems

Even with the optical links, converters and amplifiers, the 'T-feed' system is often to be preferred to the cascaded BDA solution (Section 11.5). The reason is better noise control, mainly due to the fact that the cascaded noise build-up is avoided because the BDAs do not feed each other, thus ramping up the noise figure.

11.6.2 Example of a Real-life T-system with BDAs

An example of the signal in a tunnel using a T-system for signal distribution can be seen in Figure 11.14. The measurement shows a section of about 2.6 km of a 4 km-long tunnel. As you can see there are four BDAs servicing this section of the tunnel, each supporting 660 m of

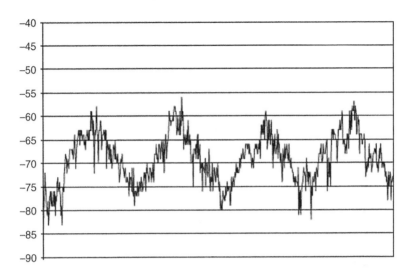

Figure 11.14 Measurement of the 2G-900 signal on radiating cable fed by a 'T-system', optically distributed BDAs/Remote Units

tunnel, 330 m in each direction. The system was designed using a 1 and 5/8 inch radiating cable installed in this large square-cut (about 4–6 meters) concrete rail tunnel. The measurement in Figure 11.14 was done sitting on top of an open diesel-operated rail trolley at 80 km/h (while Henrik Fredskild and I tried to cling on to our PC's, measuring receiver and a faint hope of continuing our lives!).

The design level for this tunnel was –75 dBm at 98% area on 2G-900 for the downlink, measured in the open tunnel. The measurement clearly shows that there are areas that are about 5 dB lower than the design level. In particular, the handover zone, illustrated in the leftmost part of the measurement, caused problems. The main problem turned out to be caused by the installer having used 1/2 inch coaxial cable instead of the specified 7/8 inch coaxial cables from the BDA equipment rooms to the actual tunnel (no equipment was allowed in the tunnel itself; this was restricted to emergency rooms placed along the tunnel), and the cable route was longer than expected. According to the installer it was much easier to install the thin cable! It probably was, but this resulted in about 10 dB extra loss, not accounted for in the link budget.

This is a perfect example on how the practicalities of tunnel radio planning can hit you like a ton of bricks!

Do include a level of safety margin, but on the other hand, over designing will be expensive. However, the reality is that in a tunnel most of the cost is related to installation work. Therefore, 10% extra on equipment expenses will have close to no impact on the final installed project price.

By the way, it also turned out that the penetration loss into the train was about 8 dB more than expected! So do remember to measure the penetration loss on the actual trains that are supposed to operate in the tunnel, at the serving frequency, and using a radiating cable outside the train, preferably in the actual tunnel, prior to committing to any design. The radiating cable part of the system in the tunnel was actually performing as specified.

As can be estimated from the measurement, the longitudinal loss is about 15 dB per 330 m. This is spot on with the data from the manufacturer of the cable. However, unfortunately the extra loss from the coaxial cable caused some concern and reinstallation of new 7/8 inch cable (at the expense of the installer).

11.6.3 T-systems with Antenna Distribution

Owing to the installation challenges using radiating cable and time to deployment, radio distribution using yagi antennas can be a viable option. If the distance between the tunnel wall and the vehicles allows, an alternative to the radiating cable could be to use distribution via yagi antennas. This still uses the 'T-feed' principle, but instead of a leaky coaxial cable you connect the BDAs to yagi antennas installed 'back-to-back', as illustrated in Figure 11.15.

You need to be careful though; vehicles, especially trains, will to some extent act as a moving 'RF-blocker' for the signal, but this approach is normally ok for car tunnels – and can save a lot of investment when compared to deploying a radiating cable. Make sure that there is only one train in each section of the tunnel at the same point in time.

For the link budget calculation, do realize that the longitudinal loss of a train can be very high. Using yagi antennas, clearance between the train and tunnel is important in order to get the signal into the users inside the train, by reflections through the windows. You need to make sure that the clearance between the walls of the tunnel is minimum 4λ in order to successfully service longer trains. Luckily many metro trains are quite 'open' from an RF perspective.

If the clearance is very tight and you have to rely on the radio signals to penetrate along the longitudinal axis of the train, the penetration loss can easily exceed 35 dB to the far end of

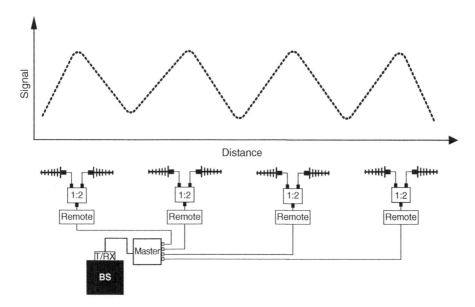

Figure 11.15   Yagi antenna distributed tunnel system

the train opposite the yagi antenna! This will depend on the RF frequency and how many passengers there are, i.e. how much 'body loss' there is, inside the train. In these cases you need to make sure that the train constantly is service by two yagi antennas, one from either direction.

Installation Challenges With Yagi Antennas

Special yagi antennas are available, designed mechanically for installation inside tunnels. These antennas will often be totally enclosed in a plastic tube (radome) so dust cannot degrade the performance over time.

Mechanical stress and reliability is a real concern, especially in rail tunnels, where there is mechanical stress due to air pressure and vibration caused by the moving train inside the tunnel. Special 'tunnel yagi antennas' will often have mechanical fixtures in both ends, making installation more stable.

Be careful with any hardware that you install inside a tunnel. If the equipment detaches from the wall, it can be a hazard for traffic and the users of the tunnel. There is a specific safety clearance profile around the profile of the train coach, where you are not allowed to exceed when installing the DAS antennas and equipment this is also referred to as the Kinetic Envelope in rail documentation.

Handover Design with Yagi Antennas in Tunnels

One major challenge with yagi-distributed systems is the handover zone design, if the system uses multiple cells. Refer to the following section for more detail on this issue.

11.7   Handover Design inside Tunnels

Large tunnel systems often need to be divided into several cells, especially if it is a rail tunnel system over a large area. Therefore you need to plan an adequate handover zone between the overlap of the adjacent cells that will ensure a successful handover of the traffic.

There are some basic parameters and tricks that are essential for a successful design handover design.

11.7.1   General Considerations

HO Zone Placement

In a metro rail application it is worthwhile considering having the HO zone planned at the station area, preferably in a well-defined zone, with clear definition of the difference in level of the two cells at the centre of the platform. This approach will have several benefits:

- *Slow moving traffic*: a train fully loaded with mobile users will load severely the signaling for the two cells handing over all the calls. The network will have to hand over all the users in the traffic within a limited time window if the train is moving at high speed (relative of course to the planned handover zone). Therefore it is preferred not to do the handover at high speed; you need a margin for 'retries' if the first handover fails.

- *Capacity overlap*: having the HO zone at the station means that there will be capacity overlaps from both cells at platform level where most passengers are placed. This helps in peak load situations as both cells will be able to carry the traffic (not for 3G designs).
- *Redundancy of the radio service*: if one cell is out of service, there will to some extent still be coverage from one cell at least at platform level, even if the preceding section of tunnel coverage is missing.
- *3G/4G*: for 3G/4G you need to be careful not to create extensive HO zones; ideally only the platform should be served by both cells, not the whole metro station.
- *HO zone size*: remember that HO zone size will depend on the speed of the train, the HO evaluation and the processing time, and an additional safety margin should be added to make further HO attempts, if the first handover fails.

11.7.2   Using Antennas for the HO Zone in Tunnels

As tempting as it might be, it is not recommended to use antennas as the primary source of signal in the HO zone in the tunnel. Antennas should clearly be used in the station hall area and pedestrian tunnels, but in the HO zone for a train tunnel the signal from two antennas is much too unpredictable and too easily affected by the physical presence of the trains or cars occupying the tunnel. It is very difficult to install antennas in the tunnel, in such a way that they cover a well-defined handover zone with clear signal margin difference between the cells in order to create a well-defined handover zone. The example in Figure 11.16 shows the potential problem: there is no clear definition of the HO zone due to the fading pattern of each antenna. There is a potential 'ping-pong' HO, where the mobile will HO carry out five HOs in each direction, before settling on the new cell. This will increase the signal loading by a factor of five, and increase the potential for dropped calls and quality degradation ('bit stealing' on 2G for HO signaling can be heard on the audio of the call during HO).

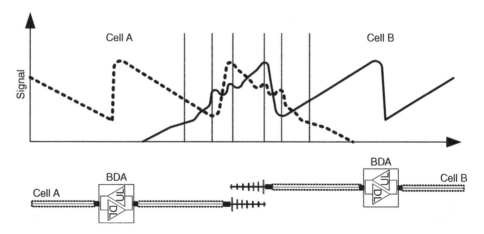

Figure 11.16   HO zone designed with termination of the radiating cables with yagi antennas

Example: Problems in an Antenna-controlled Handover Zone in a Tunnel

A real-life 2G example illustrates an extreme case of the problem (Figure 11.16), where there was an average of seven HOs in one direction and nine in the other; this is called 'ping pong' handovers and is described in Section 11.2.1. This multiplied the signaling load of the two cells when a train full of mobile users passed through the handover zone. After many nights of failed optimization sessions of the handover parameters, the problem was finally clarified and resolved by resorting to the design that can be seen in Figure 11.19.

The reality is, however, that some tunnels are indeed done with antenna distribution, and it can be an excellent solution if you avoid repeating my mistake, as just described. If you are using antenna distribution for the signal, the solution for the HO zone design, as illustrated in Figure 11.17, is described in the next section.

Controlling the HO Zone with Antennas

In some applications, like a car tunnel, it is not desirable to install radiating cable in the tunnel, but instead to implement a DAS based on yagi antenna distribution. It is a considerable challenge to control the handover zone design using antennas, but you simply must make the HO zone work, even with antennas, and you will have to provide a design that avoids 'ping pong' handovers. See the example in Section 11.7.2.

You might be tempted to use separate antennas with offset power levels in each direction, but even the exact same type of antenna will most likely have a non symmetrical propagation

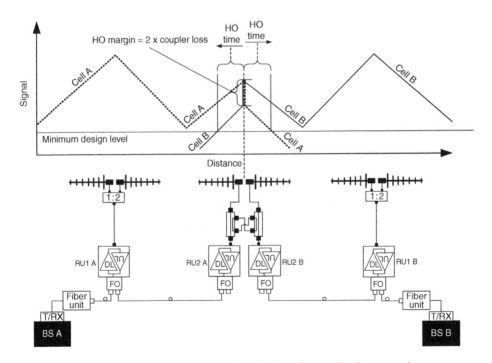

Figure 11.17 Yagi HO zone designed with a clear-cut handover margin

and fading performance when installed in close proximity in the confined space of the tunnel. It is most likely that it is not possible to install a total of four antennas in each HO Zone. The trick is to use the exact same antenna for both cells and directions ensuring 100% equal propagation and fading performance in order to obtain a clear-cut HO margin at a specific point, perfectly symmetrical in both directions. One solution is illustrated in Figure 11.17. The clear-cut handover is ensured if both signals shift their relative signal strengths at a certain point, and if both signals use the same signal path, in this case the same antennas.

The HO zone in Figure 11.17 works by offsetting the two transmit signals with a specific margin (thanks for the tip Henrik). By choosing other values for the directional couplers, you can design the HO zone to the exact margin you want; normally, always use the same coupler value to ensure the symmetrical design of the handover zone.

The signal will be coupled with twice the coupling value. Please note the directivity and verify the required isolation requirements between the two Remote Units RU2 A and RU2 B.

11.7.3   Using Parallel Radiating Cable for the HO Zone

A commonly used method for designing the HO-zone on a radiating cable system is to overlap the cable from the two cells (as illustrated in Figure 11.18). This is a valid design strategy, but it will take some distance (time) before the difference between the two signals is large enough to trigger the HO margin. In addition, you need to apply the time for the HO evaluation and execution, and time to retry the handover if the first attempt fails.

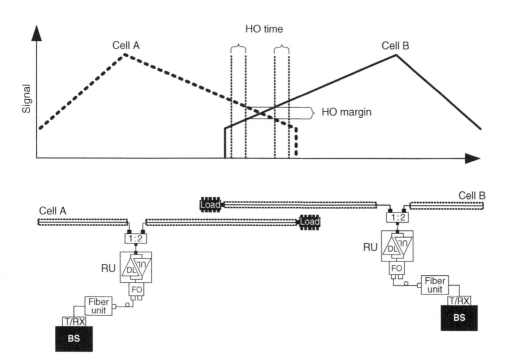

Figure 11.18   HO zone designed with overlapping radiating cable from each cell

Remember that the serving signal level still needs to fulfill the design criteria, so in practice you do not want to get below this level before, during or after the HO, or the following attempts should the first HO fail. Note that the signal from the one cell will be gone once the terminator has been passed.

In reality this means that you will need to start the HO procedure 6–10 dB before you reach the minimum design level, so the two adjacent cells in the handover zone will not be able to service the same distance of radiating cable as the other cells in the system.

In particular, if you need to place the HO zone in between stations in a metro, where the train is moving at full speed, you will need a long (expensive) overlap to the two cells per cable in the tunnel, in reality doubling the installation cost in this part of the system.

The overlapping cable handover zone is a possible solution, but it can be tricky to design and optimize due to the long overlap of the two cells. The two end amplifiers will be located relatively close to each other in order to be able to power the long handover zone. One big advantage, though, is the galvanic isolation between the two systems.

11.7.4 Using a Coupler for the HO Zone

The design that will give the best defined HO zone is to use a RF coupler to connect the two cells on the same leaky coaxial cable (as illustrated in Figure 11.19). Thus the two cells will have the exact same fading pattern and coupling loss to the mobile. However, the system loss will be different; this is the result of the coupler 'offsetting' the longitudinal loss at the point where the two cells are joined.

Normally a 3–6 dB coupler is used, giving a 6–12 dB signal offset between the cells. The HO offset is applied instantly to the serving and adjacent cell once the location of the coupler

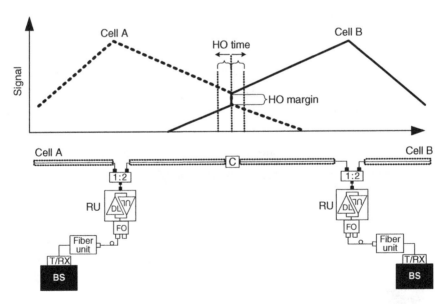

Figure 11.19 HO zone in a tunnel implemented with radiating cable distribution, connecting two cells A and B using a coupler/attenuator

is passed in the tunnel. The result is a well-defined HO point. You still need to make sure that the HO is performed before reaching the minimum required signal level.

The roll-off of the signal is not as steep as for the overlapping cable example in Section 11.7.3, but you need to incorporate a safety margin that allows for handover to be retried without the serving cell falling below the design level. There is only one problem worse than a handover problem in a tunnel, which is a handover problem in a tunnel at low signal levels!

This approach has been proven during many practical installations, and is highly recommended. But you have to bear in mind the DC/grounding issue when connecting the two cells, and you might be required to install a 'DC separator' that isolates the DC but passes the RF signals.

11.7.5 Avoid Common HO Zone Mistakes

Before we look at some examples of different types of sector plan there are some common issues that might be obvious, but never the less – remember:

- To provide sufficient HO zone size to accommodate multiple retries if the initial HO fails.
- That the HO-zone must be able to function in both traffic directions; sometimes trains will have to travel in the opposite direction of what was originally expected, during maintenance work in the original track.
- That when having the HO zone at the station area with slow moving traffic, make sure that trains skipping the station will still be able to make handovers; this is important to remember even for metro train solutions where it was originally expected that all trains will stop at all stations. However, a station might be closed for shorter or longer periods of time, so HO must be able to accommodate fast moving traffic, even if the HO zone was originally intended to service slow moving trains stopping and starting at the station.
- To be careful with placing the HO zone in the high speed area in the tunnel between the stations; remember that all passengers are in HO at the same time, this will place a high signaling load on the system.
- To be careful when using standard capacity calculations like Erlang in a tunnel environment; the traffic inside the train coaches might be Erlang distributed, however the trains themselves are not, resulting in a 'trainload' of traffic arriving at and leaving from the cell at the same time, not distributed as normal traffic. The traffic monitoring tools used by most mobile operators might not display any blocking problems, due to averaging algorithms, but if you look closer at the exact statistical counters in the network you might actually have 'hard blocking' anyway.
- That HO Zone requirements and preferences will be different in 2G and 3G/4G.

11.8 Redundancy in Tunnel Coverage Solutions

It is often required to design a tunnel coverage solution that has to meet some level of redundancy so that in case parts of the system fail there will still be some service level for basic services, voice SMS, etc. Often, it is not required to have the same high quality of service during the fault scenario. The main concern is normally the active elements of DAS, but depending on the application it can sometimes be necessary to plan the tunnel coverage system so that the service will be on air, even if the radiating cable is cut.

The precise solution depends on many factors, ultimately the actual specification for the specific service for the project at hand. It will add cost to the project, especially if 100% redundancy is needed and a 1 + 1 redundancy is required; this will mean that no matter what part of the system fails, the service will still be at 100%. In practice this can take a parallel system, and so a 100% increase, doubling the required hardware and installation – and cost.

On the DAS part of the solution, the passive portion, one easy way of providing redundancy could be to feed the system, in a 'ring', as illustrated in Figure 11.20.

In this solution the same cell will be distributed in a ring type scenario in both directions. One must make sure, however, that there will be enough signal to the most distant antenna when the feeder cable is cut. Also, in the normal scenario when the ring is intact, there can be some inter symbol degradation as a result of the same RF signal arriving at the antenna from two branches of cable with different delays.

This is applicable to both antenna solutions and radiating cable solutions, but the weak point is that you must be able to feed the same cell to both ends of the radiating cable/DAS. This makes this solution less attractive for tunnels, but useful for buildings, or the station area of a tunnel solution.

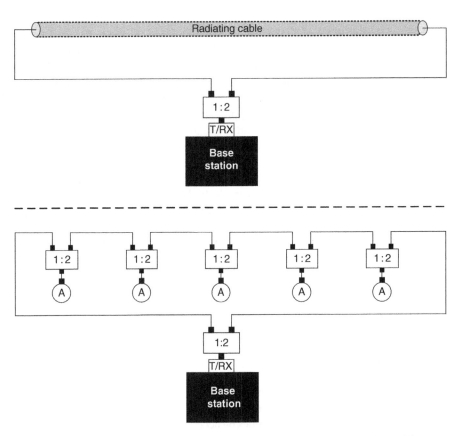

Figure 11.20   Examples of 'Ring feed' types of redundancy on a passive DAS

11.8.1 Multiple Cell Redundancy in Tunnels

For larger tunnel coverage solutions with multiple cells and multiple services – typically mobile coverage and emergency system coverage such as Tetra – redundancy is often a requirement, especially for the emergency system, which often operates on a lower frequency as compared with mobile services.

The fact that the emergency system is often on a lower frequency can actually assist with the overlap needed for redundancy of the emergency system service. The lower frequency will typically mean lower loss in the coaxial part of the system, both feeders and radiating cable. So, redundancy on the mobile telephony part is often 'nice to have' but for the emergency system it is a must.

A simple way of ensuring this between two cells can be seen in Figure 11.21; a typical tunnel coverage system designed for Tetra in a rather long tunnel with two cells and two parallel tunnels, one for each direction of traffic.

Two separate base stations/cells feed each end of the system and at the midway point in the centre of the tunnel; the two cells are coupled with a 3 dB Hybrid Coupler (see Section 4.2.8 for more detail of the Coupler). As you can see on the signal level graph in Figure 11.21, the

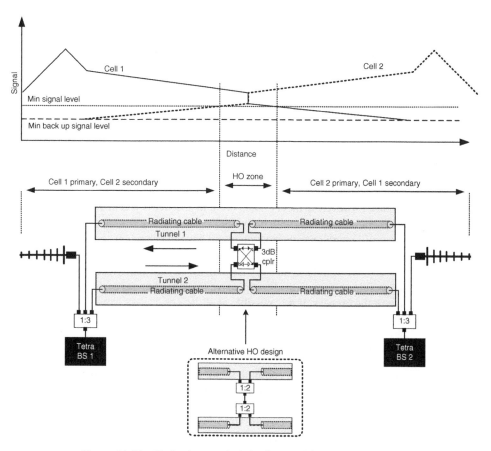

Figure 11.21 Redundancy principles for a multi cell tunnel solution

two antennas in the handover zone will give a peak in the level, ensuring the handover, and halfway through the tunnel the Coupler ensures a nice clean offset of the signal level from Cell 1 and Cell 2 to provide a well-defined handover point and a nice overlap of the two cells. As illustrated in Figure 11.21, the coverage will stay above the minimum signal level required in the design specification throughout the tunnel. Please also note that in Figure 11.21 the redundancy signal level is 'minimum back up signal level'. So you must design each part of the system, both Cell 1 and Cell 2, so as to ensure that the service will stay above the redundancy signal level.

In a case where one cell is not operational (or traffic capacity is used) then the other cell can act as a back up.

The 'redundancy' also applies in the case of malfunction or a cut cable; coverage above the 'minimum back up signal level is ensured. The only thing that is compromised in the case of a cut cable is the handover zone; in the area where the cable might be cut you will be able to make a call, but moving traffic might suffer with dropped calls due to the lack of overlapping cells for handover. This is not a 100% failsafe, but close to 99% and will be sufficient for most systems. It is important to note that in the case of a cut cable, both the coupler and the radiating cable is not terminated with 50Ω as expected, and this will affect the coupling values and behaviour of the components, making the precise service unpredictable.

11.9   Sector Strategy for Larger Metro Tunnel Projects

It is very important at an early stage to consider the sector plan, cell lay out and capacity for large metro tunnel projects. This will typically be a sub level railway application, and this application will be used as an example in the following chapter. You must make sure to design the cell structure and plan to accommodate current and future needs; typically you will face the challenge of providing a system that can handle 2G, 3G and future services such as 4G, etc. In addition you might have to provide coverage and redundancy for emergency systems.

These, often quite different, systems may not require the same type of strategy or even the same sector plan. At some point there will have to be a compromise where the business case plays a significant part; however, in theory – and often in practice – you can design a cost efficient multi cell/multi operator system with individual sector plans and strategies depending on the service – or in a multi operator scenario, you might have independent sector plans for each of the mobile operators.

So consider carefully the options at an early stage in the design specification of the project; it will often be impossible or unrealistically expensive to make changes to the design or sector plan at a later stage in the process, if it is possible at all.

11.9.1   Common Cell Plans for Large Metro Rail Systems

Let us have a look at some of the more common sector options in large rail tunnel projects.

One of the most common sector plans for large metro rail projects can be seen in Figure 11.22; each sector will service a metro station as well as the two tunnels to the next station. The advantage of this option is that the HO zone is just at the edge of the next station, preferably at the first 10–30 meters of the tunnel where you must make sure to design an overlapping HO zone. This will result in a limited HO zone (3G/4G) and relatively slow moving

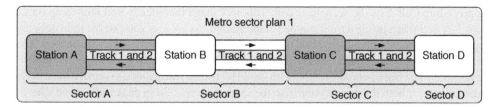

Figure 11.22   Metro tunnel solution sector plan with one cell per station and the handovers occurring right at the entrance/exit to/from the station

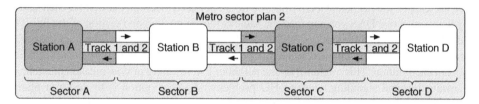

Figure 11.23   Metro tunnel solution sector plan with one cell per station and the handovers occurring between the stations in the tunnel

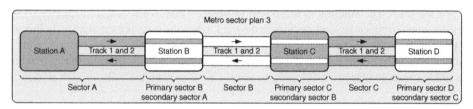

Figure 11.24   Metro tunnel solution sector plan with one cell per station, but at the stations the rail area and platform have secondary coverage from the cell from the preceding station

traffic, with good margin for HO retries in case of HO problems. This approach is applicable for 2G/3G/HSPA and 4G – but be careful with isolation between the cells at the platform area, as this can be a challenge.

Another widely used approach for sector plans can be seen in Figure 11.23; each cell will cover the metro station and halfway through the tunnel to/from the next station. It seems like a nice logical solution to the sector plan and HO zones. However, you must be careful and make sure that sufficient overlap is provided in the HO zone, and it is recommended only to use this design when designing a tunnel coverage system based on radiating cable. It is recommended to use the principle of coupling the two cells together using a coupler (as seen in Figure 11.19) to ensure a well-defined HO zone. Please remember that this approach will often result in having the HO zone in the area where the train is moving at maximum speed, so a good overlap must be provided. However, this solution is good for 2G/3G/HSPA and 4G.

For 2G systems (or other FFD systems with hard HO) the sector plan illustrated in Figure 11.24 can be considered. It consists of a cell structure with primary and secondary cells. So, for example (see Figure 11.24), the Station B platform area is primary served by Sector B. But along the rail track aside the platform on Station B the radiating cable will still be servicing the preceding Cell A as the secondary cell. This will mean that ideally you will camp on the secondary cell when inside a train stopped at the station. As soon as you exit the train door and step out on the platform you will hand over to the primary cell covering the station area. This is a tricky balance and requires a very uniform level from the primary cell along the edge of the platform. The big advantage for 2G with hard HO is that you are able to service most of the platform area by both the primary and also the secondary cell. So in peak traffic time where a platform will be packed with passengers you will be able to use the capacity of both cells, primary and secondary, to carry the traffic. You must be careful with 3G/4G systems, due to the fact that this creates a large HO zone, and will have the opposite effect on the 3G/4G system; it will drain capacity due to HO load and data bandwidth degradation on 4G.

For low capacity areas on the metro line you might not need to assign one cell for each station, so by using an active DAS you can feed the same cell to more than one section of the line. One cell could cover two to three stations and so on. However, you must be careful with neighbor list limitation; geographically this will mean that two station locations will require the set of neighbors from each station. This might fill up the neighbor list, resulting in slow measurement, handover evaluation and signaling, causing HO problems. Another option is to design a plan as illustrated in Figure 11.25 where tree cells will service four metro stations/sections.

However, it is not recommended to split the station into two cells for 3G/4G systems – this will result in a large HO zone in most of the station area.

There is a way of avoiding handovers in the tunnel, and only handing off to other cells when you exit the train. The solution is illustrated in Figure 11.26 where one cell (Sector E) will cover the tunnel from end to end, including the track when the train is a the station. In this way you have created a 'corridor' of one cell throughout the entire metro. This solution might be ideal for fast moving train lines, but it requires some careful consideration at the station area, so as not to create too large a soft handover zone. One uses separate cells on each station and

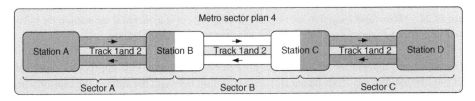

Figure 11.25 Metro tunnel solution sector plan with one cell covering more than one station, typically for low capacity solutions

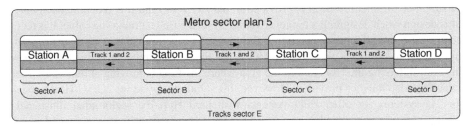

Figure 11.26 Metro tunnel solution sector plan with one cell covering the track end to end, each station has separate cells that are gateways to the outdoor macro network

at the end of the train line where the system needs to hand over traffic to the outside macro network (gateway cells). These platform-cells (Sector A–D) will have to function as gateway cells to the local macro network. It is not likely that the corridor cell will be able to handle the extensive neighbor list when covering a large geographic area; it will be a problem when interfacing the above ground macro network with the metro DAS system. You are likely to be limited by frequency planning or code planning.

11.9.2   Using Distributed Base Station in a Metro Tunnel Solution

Some tunnel solutions are relatively small, with relatively short tunnel sections. In some cases they do not even need an active DAS to distribute the signal, as the base stations themselves can feed the passive DAS sections if the base stations are distributed throughout the metro system, typically located at the metro stations, as illustrated in Figure 11.27. At first glance this seems like a cost efficient solution, and it is. However, by using this strategy you have made a large compromise and will have to stick to that static sector plan in the future. This might not be ideal for future services or capacity needs. But, on the other hand, it is a fast way in, with no additional active DAS hardware to worry about.

You will be stuck with a similar sector plan as that illustrated in in Figure 11.27. Also, for multi operator solutions you will have to co-locate the base stations with the other mobile operators and combine the base stations at a relatively high transmit power, likely increasing the risk of inter modulation problems (refer to Section 5.7 for more detail on these potential problems, which are not to be underestimated). In a case where one is operating with multi operators and multiple technologies the potential problem when combining at high power will become an even more likely issue (see Section 5.7 for the details).

11.9.3   Using Optical Fibre DAS in a Metro Tunnel Solution

One of the best methods of providing coverage and capacity for larger metro tunnel systems is to use one or more centralized Base Station Hotels and distribute the coverage and capacity from there. The principle is illustrated in Figure 11.28. This DAS comes in two main types,

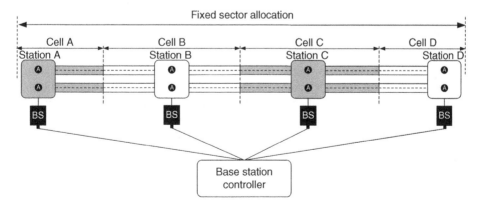

Figure 11.27   Metro tunnel coverage solution with distributed base stations

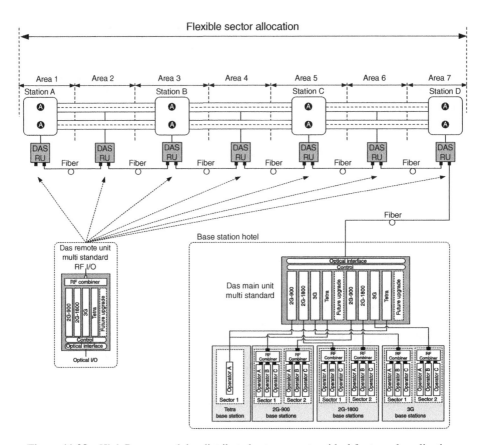

Figure 11.28   High Power, modular distributed antenna system ideal for tunnel applications

one with analogue transport over the optical fiber while some versions of high power DAS utilize digital transport of the signal between the Main Unit of the DAS located at the Base Station Hotel and the Remote units. The result has some advantages over traditional analogue transport over the optical fiber, downside is that it is realative complex to digitize analog RF signals. It thakes a high datarate (bandwidth) on the fiber, and in addition adds delay, that canibilize the reach over the fiber:

- *Sturdy Optical link budget*: owing to the digital nature of the transport of the signal, the system is not susceptible to degrading effects due to optical fiber loss and reflections, etc. on the optical interface. Typical analogue DAS systems can handle about 10 dB of loss on the optical before the link breaks down. The typical digital DAS can handle more than 20 dB of loss, and the link quality will remain perfect. See Chapter 12 for more detailed information on digital transport DAS.
- *Simulcast and flexible sector plans*: owing to an RF switch matrix system, the DAS will be able to transport any of the sectors that are connected to the Main Unit to any of the Remote Units (possible also for "analog DAS"). This makes it possible to divide the cell lay out into any plan one desires – one can have four Remote Units simulcasting the same sector in

unison, or divide them into any sector configuration that one might desire. Each service and/ or mobile operator might even have its own sector plan. So one can design ideal handover zones dedicated to 2G, 3G and 4G – ideal for future proofing and optimization.

- *Multi band support*: most types of high power DAS supports multiple standards and frequency bands in the same system. As seen in Figure 11.28, the Main Unit located at the Base Station Hotel as well as the Remote Unit located in the tunnels and stations are modular. Typically the upgrade to a new band is merely a matter of installing a new card in both units in the field.

These parameters make High Power DAS ideal for tunnel environments. In Figure 11.28 one DAS Main Unit at the central Base Station Hotel supports 2G-900 two sectors, 2G-1800 two sectors, 3G2100 two sectors, each band for three operators. On top of the mobile service the DAS also supports the emergency service on Tetra for the 'Blue Light Service' for one operator one sector throughout the tunnel system. All of the mobile operators can adopt any sector plan they desire, divided over the seven multi-band Remote Units.

The optical fiber interface on the DAS is well advanced and by the use of WDM (Wave length Division Multiplexing)/CWDM (Corse Wave Length Division Multiplexing) the optical fiber installation can be utilized to a maximum. Sometimes, the optical transport system or the DAS (analog or digital) can even use existing optical fiber with active traffic – by using a different wavelength/colour.

This addresses one of the main cost components in a typical deployment of DAS in a large scale tunnel system; optical fiber installation.

11.10   RF Test Specification of Tunnel Projects

Some of the principles of RF measurement and verification that are typically used for indoor coverage systems (see Section 5.2.2) can also be applied to tunnels. However, it is crucial to realize that tunnels are a special environment and specifications on how to perform and evaluate RF measurements of the implemented system are very important.

Due to the great complexity of a tunnel project, mobile operators will typically engage with a third party supplier/installer to design, implement and verify the system.

Therefore the RF test procedure and specification has to be set in stone at an early stage to make sure that all parties uses the same metrics and procedures for the design verification measurements.

You must as a minimum specify these points in the design and test specification:

- Type of calibrated measurement tool to be used (not any test mobile from a random drawer somewhereâ€¦);
- Calibrated antenna type/gain;
- Test set-up, antenna on tripod, location on the test vehicle;
- Test frequencies;
- RF measurement data sample speed/distance, it is recommended to use *William Lee's criteria*; Lee calculated that for a valid RF measurement you need to sample an average thirty-six samples over a distance of 40λ (wavelengths);
- Speed of measurement vehicle;
- Where the test is done (inside train coach, open track, location on the track);

- Success of KPIs; level, distribution, min. level;
- Typically the KPI would state that you need to confirm –XX dBm in 95% of the samples. However, one should not conclude that 5% of 'bad' samples is not a problem – it also depends on the distribution of the 'bad' samples. For example, it would not be a problem that a few samples every 2 metres falls below the –XX dBm level if in the Handover Zone (even exceeding 5%) – whereas 4% of concurring bad samples for a solid distance of 25 meters could lead to a dropped call – even when fulfilling the '95%' coverage;
- Measurement route, tracks, station areas, escalators, elevatorsâ€¦;
- HO zone length;
- Primary cell level, secondary cell level;
- Perform RF measurement on all existing radio systems;
- Coupling loss measurement on the radiating cable along the total length of the tunnel;
- Post processing procedures of the results;
- RF-test report format and required KPIs.

11.11   Timing Issues in DAS for Tunnels

For long optical fiber deployments in a DAS system one must be careful with the delay of the DAS and timing in the system. This is something that is often forgotten but in an extreme case this could have a fatal impact on the service if one fails to realize this. In particular when using DAS that lies on digital transport or digital signal processing, this will introduce relative long delays to the signals.

It is a fact that all cables, coaxial or optical fiber, will have a certain delay; the velocity of the electric signal on coaxial cable and the light in the optical fiber will simply add delay depending on the specification of the cable or optical fiber. In addition to this the DAS hardware, Main and Remote Unit will also add delay to the signal due to internal digital signal processing (if used), amplifiers, etc. Aanalog DAS will have less delay than digital, colse to zero of DAS HW itself.

The effect of this delay will be to cause the mobile network to react as if the mobiles served by the DAS are more distant than the actual physical distance from the cell/ DAS antenna. In DAS systems with active elements, Remote Units, Repeaters, antennas, etc., radiating from the same cell as the adjacent system, covering the area ordering the antenna, you might cause some unwanted effects such as inter symbol interference, co-channel interference, etc. depending on delay offset of the simulcasted signals the type of mobile service the system supports.

For normal indoor DAS solutions this timing offset effect will rarely be noticed, but for installations with several kilometers of optical fiber, and sometimes daisy chaining of active DAS equipment, it is very important to include it in the design considerations.

The exact timing/delay calculation will obviously depend on the precise installation, cable routes and data, the data on the active DAS hardware used, etc., and an example that demonstrates the potential problem can be seen in Figure 11.29. This application is a typical tunnel solution for a tunnel covered from a base station shelter outside the tunnel due to lack of space in the tunnel itself, at about 1000 meters from the tunnel entrance. The first part of the rail track outside the tunnel on the side where the equipment shelter is located is down in a cutting, so there is a lack of signal level from the macro network. To resolve this issue it is decided to deploy a low 5 meter mast at the equipment shelter feeding two antennas 'back to back' in order to extend the coverage from the tunnel system to the handover zone in the macro network.

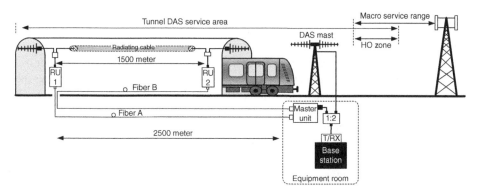

Figure 11.29   A typical scenario of a tunnel DAS and extended coverage just outside the tunnel interfacing to the nearby macro. This scenario can create a timing problem

The power to feed the antennas located at the equipment shelter is taken directly from the output at the base station with a splitter. The same cell is split off to the Main Unit feeding the active DAS with the same cell. From the equipment shelter the Main Hub in the Active DAS feeds optical fiber to the only cable access point about 2500 meters away from the DAS Master Unit. From the cable access point the optical fiber feeds two Remote Units that provide coverage inside and just outside the tunnel. One Remote Unit (RU1) is located near the cable entrance point, the other in the far end of the tunnel at about 1500 meters distance (RU2).

In this typical scenario (based on an actual deployment I was involved with) there are some potential pitfalls that we must be sure to avoid.

First, there are all the normal coverage and HO cell overlap concerns addressed in the first part of this chapter. In addition, there is the issue with delays in the system, offsetting cell size and finally the potential problem one might face when signals from the same cell arrive at the mobile user by two or more different signal paths in the DAS with offset delay. Obviously the mobile communication standards in both 2G (the Equalizer in the receiver) and 3G (the fingers in the Rake Receiver) will be able to handle this to some extent; one have the same effect in any typical multipath RF environment with reflections. However, the relatively long delay offsets caused by the DAS HW and the cables might be too long for the systems to handle; let us have a look and calculate the example in Figure 11.29.

Timing Calculation Example

First of all we need to know the timing data, the delay of the DAS system components, in this example:

DAS System delay (Master Unit and Remote Unit in total): 13μ Seconds Optical Fibre Velocity

So what does this mean? Let us start by considering the 13 μS delay of the Active DAS HW. RF travels with the speed of light 'C' at about 300.000 km/Second so in 13 μSeconds; this corresponds to 0.000013 * 300.000 = 3.9 km. So the total delay of the DAS hardware, the Main Unit and Remote Unit will increase the cell size 3.9 km. So if a mobile is in traffic 100 meters from the Remote Unit, that delay will make it appear to the mobile network as if the mobile is at 4 km distance.

In addition to the delay and thereby timing offset we have the delay in other parts of the system. For tunnel systems there is another important component adding to the delay and timing offset; the optical fiber.

The propagation speed; the velocity of the optical communication on the optical fiber needs also has to be considered. This effect is often forgotten, optics are propagating with the speed of light, right? – Not! Let us take a closer look.

You will have to look up the specific data of the optical fiber cable used in the specific system, but a typical propagation speed (velocity) is not C at 300.000 km/Second but a factor of 1.468 (0.681) slower, so in reality the speed of the optical signals on the optical fiber is abut about 204.500 km/Second.

This will in reality offset the cell size with 1.468 km per kilometer of optical cable.

11.11.1   Calculating the Total Delay of a Tunnel Solution

Considering the the example in Figure 11.29 we now can establish that the DAS HW will add 3.9 km to the cell size, for Remote Unit 1 the 2500 meters of optical fiber will add another (2500*1,468) 3.67 km to the cell size. So the signal to/from RU1 is offset by 3.9 km of the DAS HW and 3.67 km of the optical cable, a total 7.57 km has been added to the cell size. For RU2 we have 1500 meters extra optical fiber, so the timing to RU2 is offset by 3.9 from the DAS HW and (4000*1.468) 5.872 km, a total of 9.772 km.

However, there is more adding to the delay of the signal; the radiating cable! Yes – again this is something we do not normally consider; the velocity factor on the coaxial cables typically a factor of 0.88 * C. So each km of radiating cable will offset the cell size by 1.1364 km; this effect needs to be considered when designing the HO zones of the cable, or if you like, in the example, simulcast the same cell from two Remote Units from each end of the cable.

The Impact of the Delay in the DAS

There are several effects and impacts on the service due to this delay that one must consider.

Offset of the Cell Size

First of all there is the sheer offset in cell size. Many mobile systems have a parameter setting in the network for maximum cell size. This to prevent distant mobiles picking up unintended spillage from macro cells which are not meant to service that area. Sometimes this parameter is set to a default value based on the type/class of the cell – urban cells max. 3 km, rural cells max. 6 km, country side cells max. 25 km –indoor cells will often be defined with a default of max. 1 km.

The effect of this setting is that the cells will simply not be allowed to set up calls when the timing of the mobiles accessing the cell is greater than the predefined maximum cell range. From the example in Figure 11.29 and the calculations we have just done, one needs to ensure that the maximum cell size is set with the offset of RU2 in mind, so in reality it is not shorter than about 10 km due to the longest delay; the DAS HW plus optical fiber to RU2 plus the radiating cable towards RU1 in the tunnel.

Multi Signal Path Issues

Tunnel systems will often be designed with more than one active element broadcasting from the same cell: simulcasting. This is also the case with the example in Figure 11.29, where we are feeding the same area, the tunnel, with two Remote Units within the same cell, covering the same area – the tunnel – via the radiating cable. We need to take note of the relative delay offset introduced by the different lengths of the optical fiber feeding RU1 and RU2. The relative difference is 1500 meters, and at the centre of the radiating cable, where both Remote Units will have same signal level, we can calculate the timing offset between the two signal paths. Given the offset caused by the velocity factor of the optical cable, this 1500 meters of optical fiber will offset one signal (RU2) of $1500*1.468 = 2.202$ km. So in reality this is like a multipath environment in an urban environment with one reflection offset by 2.202km – the mobile and base station will easily be able to handle this offset of 7.34 μSecond – the equalizer in the 2G mobile phone will handle up to 16 μS and the Rake Receiver in the will also handle the two signals from the same cell.

So at first glance the system seems to work, but in order to be sure we must analyse the area in the DAS where there is a relative difference between multiple signal paths.

In the example in Figure 11.29 we have decided to split the signal from the base station to feed the 'back to back' antennas on the local mast near the equipment room, and the same cell is feeding the signal to the Master Unit of the DAS.

By doing so we have created a potential problem; Remote Unit 2 has a long signal path. Feeding RU2 is 4000 meters of optical fiber having a velocity factor of 1,468, this corresponds to 5.872 km, a delay of 19.57 μSeconds, we also need to add the delay of the 13 μSeconds – resulting in 32.57 μSeconds in relative difference between the signal from RU2 and the base station signal in the area between the tunnel entrance and the antenna located at the equipment shelter.

This delay exceeds the capabilities of the equalizer for the 2G receiver, resulting in co-channel interference and dropped calls. This is also a potential problem for 3G services – the network is simply not designed to handle multipath environments with the range of offset between signal paths, due to the fact that this is not likely in a typical multipath environment.

Maximum Service Range in a Tunnel DAS

Knowing the delay data and the optical fiber and radiating cables we can calculate the maximum possible service distance. This obviously depends on the type of service, but one example is that the maximum cell size for 2G is about 32 km; however, remember to include the delay of optical fiber and active DAS elements – in practice this will limit the 2G tunnel DAS to no more than 20 km, depending on the actual data of the DAS equipment and cable. So, consider carefully the limitation on the actual mobile service that the system is designed for and the system data.

Delays in the Handover Zone

We must also take the delay into account in the end of the tunnel (Figure 11.29) opposite the equipment room on the left side. Here we must make sure that the mobiles handing over to and from the macro network will be able to detect the neighbor cell and perform signaling – for a

WCDMA (3G) system it is important to ensure that the 'Search Window' will be able to handle the relative difference in delay between the tunnel system and macro network – if not, left unadjusted, the system might not be able to perform handovers

11.11.2 Solving the Delay Problem in the Tunnel DAS

Delay is a physical fact related to the DAS equipment and cables – so there will be an absolute maximum service range/size of the DAS system, in order to maintain the needed quality of service for the specific service/services accommodated in the Tunnel DAS.

Calculating this is as crucial a part of the design process as the coverage and capacity calculation.

So how do we cover tunnel systems exceeding the maximum service distance on the DAS? Often we will not have to deal with the problem in practice due to the fact that long tunnel systems will typically be designed with multiple cells. In reality a small large tunnel DAS system is made up of sections of individual tunnel DAS systems owing to capacity concerns and limitations on the number of possible neighbor cells.

Internal delay problems can be solved by delay compensation in the DAS system. The newest type of digital distributed DAS will allow one to add additional delays to each of the Remote Units in order to solve internal delay issues in sections of the tunnel DAS, introduced mainly by the digital signal processing.

In theory one could also add delay in the traditional analogue DAS, by simply adding more optical fiber – a spool of optical fiber at the Master Unit or at the Remote Unit. One must, however, be careful with the extra loss that one is adding to the analogue DAS – typically 0.4 dB per km, and not exceed the maximum optical loss in the optical link budget – typically a maximum of 3–8 dB.

11.11.3 High Speed Rail Tunnels

Obviously high speed rail lines will pose a specific challenge, in particular in the handover zones – simply because of the high speed one will need to provide sufficient overlap of the handover cells (see Section 11.2.3 for more detail).

Designing DAS for high speed tunnels is also reflected in the sector plan; for this application DAS systems such as that illustrated in Figure 11.28 are often to be preferred, and are often the only solution capable of extending the sector over the long distances required to cater for long cells. The advantage is optical fiber distribution and the option of simulcast (see also Section 12.4.2 on how to cover the rail tracks outside the tunnel). DAS would allow one to extend the same cells along the rail track, thereby limiting the number of handovers in the tunnel, for example the sector/cell lay out illustrated in Figure 11.26.

Increased Mechanical Stress

It is also very important to realize that the sheer speed of the train imposes mechanical stress on antennas and equipment, radiating cable, antennas, etc. Do not underestimate this fact. It is highly recommended that one only uses equipment certified to handle this mechanical stress.

Another thing to remember is that equipment boxes, such as antenna enclosures, base stations or DAS remote units mounted in the tunnel, will be loaded with high pressure from an approaching train, to be followed by a strong vacuum once the train has passed by the equipment location.

11.11.4   Road Tunnels

Even though most of this chapter gives 'rail' as the application, all of the planning and implementation techniques also apply to car tunnel applications. One of the big advantages – especially when it comes to cost – is that road tunnels can normally be covered using antenna distribution, saving on that expensive radiating cable (see Section 11.3).

Many smaller road tunnels are quite short, a few hundred meters, and can often be covered by a solution such as that illustrated in Figure 11.30. In this example a nearby macro cell is capable of providing a good signal level at both ends of the tunnel, yet struggles to reach through the tunnel and service the users inside the tunnel itself. So a repeater is deployed at one entrance to the tunnel, which picks up the macro cell already servicing the road leading up to the tunnel. The repeater amplifies the signal and re-radiates the same cell inside the tunnel.

To provide coverage inside the tunnel a Yagi antenna is installed some 20–50 meters into the tunnel, in order to provide sufficient isolation (for more detail on repeater solutions see Section 4.8). This Yagi antenna beams down the tunnel and services all the way to the end of the tunnel, with a few meters of overlap at the far end of the tunnel.

We need to pay close attention to this set-up (Figure 11.30), on top of the challenges we normally face when deploying repeaters (Section 4.8). We also have a potential issue with the overlap of coverage at each end of the tunnel. Owing to the delay of the repeater there will two strong signals from the same cell present at these locations, one the direct signal from the macro cell; the same macro cell, however, is also used as the donor cell to the repeater, amplified and re-radiated in the overlapping area – delayed in time due to the delay of the repeater. This could lead potentially to a problem depending on the type of system one is deploying. For example, for 2G one will need to ensure that the delay offset is within the capabilities of the equalizer of the mobile($16\,\mu s$) – for 3G systems one needs to make sure that the delay is longer than one chip for the rake receiver to work ($0.25\,\mu s$).

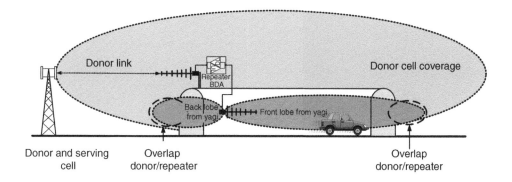

Figure 11.30   A short road tunnel covered with a single repeater

For longer road tunnels a dedicated base station and cells might be deployed to cover the road tunnel; if this is the case one must pay the same careful attention to the handover zones at the entrances of the tunnel (see Section 11.2) but because of the more uniform penetration into a car as compared with a rail coach (Figure 11.2) the design of handover zones for road tunnels is more 'easy'.

Final Notes on Tunnel DAS Design

I hope that this chapter has helped you to understand some of the key issues and concerns when designing tunnel coverage and capacity systems. Most of the potential pitfalls that I have indicated are actually based on hard (expensive) personal experience; I hope that this can help one to avoid repeating the same mistakes that I have made.

However, please realize that 'it is just a tunnel', that no tunnel is perfect when it comes to RF implementation, and at the end of the day it will always be a compromise to some extent.

Please be sure to use only good quality components for tunnel applications; this goes not only for the passive components, mechanical brackets, etc., but also for the repeaters and active DAS units – both from a mechanical viewpoint and for the electrical and RF specifications. One very important parameter is the IP3 and dynamic range (see Section 4.12.10). One could easily imagine a user standing near the servicing Yagi antenna, and another at the far end. The system must be capable of handling this, by having a sufficient dynamic range and high IP3 (see Section 4.12.10).

The clever RF planner will know when and what parameters to compromise on, and when not to.

To quote my good friend Henrik Fredskild: '*A poorly designed or implemented tunnel solution with a drop/problem will haunt you like a ghost – long after the system is set into operation*'.

Most of my tunnel experience has been on mutual projects with Henrik, and we have the scars to prove it!

12

Covering Indoor Users From the Outdoor Network

It is a fact that most mobile users reside inside buildings. Depending on where they are located in the world mobile operators are facing the challenge of having more then 80% of mobile traffic originating from users inside buildings.

12.1 The Challenges of Reaching Indoor Users From the Macro Network

The main focus of this book is to provide knowledge on how to design and implement coverage solutions deployed inside buildings so as to overcome this challenge.

It would be nice if we at some point could have indoor coverage solutions inside each and every building, and we do envisage a solution that addresses even small buildings, such as the femtocell deployed in residential buildings and houses (see Section 4.11 for more detail about femtocells). For shopping and commercial buildings dedicated indoor DAS would often be the best choice from a technical point of view. Even though it would be a theoretical possibility to deploy a dedicated indoor DAS inside all medium to large buildings the challenge remains to make it an economically viable and attractive solution.

We know for a fact that the macro network will struggle to achieve deep indoor penetration of the RF service a few kilometers from the site – so we will have to consider other options. As we see in Figure 12.1 the macro site deployed above the clutter (height of the buildings) will struggle to reach indoor users. Most signal paths will have to rely on reflections, and this gives rise to the challenge of multi path and high link losses between the serving base station and the mobile. The high losses will cause the base station to use high DL transmit power, potentially causing unwanted DL interference in the surrounding network, degrading service and capacity in the overall network. The same effect can be seen in the UL where the mobile

Indoor Radio Planning: A Practical Guide for 2G, 3G and 4G, Third Edition. Morten Tolstrup.
© 2015 John Wiley & Sons, Ltd. Published 2015 by John Wiley & Sons, Ltd.

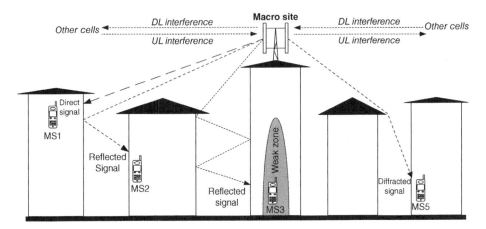

Figure 12.1 A macro base station will rely on reflections and high losses to reach users inside buildings

is likely to create unintended UL interference in distant macro cells operating on the same frequency. This excessive UL power will degrade UL service and capacity in the network. This problem (Figure 12.1) can present a challenge especially if we try to use macro sites to cover 'hotspots' in a city area. A 'hotspot' is an area with a large concentration of mobile users that virtually soaks up whatever capacity you can service from the macro site.

As calculated in this book, we need to make the distance between the serving base station and the mobile as short as possible; this will limit path loss, minimize fading, minimize UL and DL power and minimize interference, thus maximizing network capacity and the business case.

The problems and challenges we face when deploying macro cells in an urban hotspot area (Figure 12.1) can be summarized as:

- High path loss, lack of deep indoor RF penetration.
- High power load per user.
- Drains capacity in the RF network.
- Lack of isolation between cells.
- DL interference in distant cells.
- UL interference from distant mobiles.
- Lack of single cell dominance.
- Large handover zones (3G/4G).
- Limited HSDPA performance, due mainly to lack of isolation between serving cells.
- Degradation of capacity and impact upon the business case of the network.

12.1.1 Micro Cell (Small Cell) Deployment for IB Coverage

The alternative to deploying macro sites, and concentrating the coverage and capacity at street level, is widely known as Micro Cells/Small Cells. Micro cells will typically be deployed as outdoor base stations below the local clutter height (minimum 2–3 meters below the building heights), at a minimum of 3–10 meters above street level. Micro cells are often used to address the problems described above with macro sites in hotspots in an urban network. These micro cell base

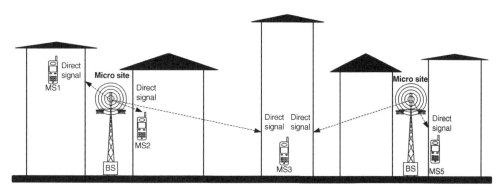

Figure 12.2   Example of micro cellular deployment utilizing street furniture, in this case lamp-posts for antenna and base station locations

stations typically operate at a power level about 10 dB below the power level from the more traditional macro network, but have the same functionality as the normal macro network.

An example of a typical micro cell deployment can be seen in Figure 12.2. The two micro cells could be deployed using lamp-posts already installed in the street for antenna locations, each broadcasting the signal from the base station co-located with the lamp-post/antenna mast. When comparing the traditional macro deployment in Figure 12.1 with the Micro cell deployment as illustrated in Figure 12.2 we will benefit in several parameters due mainly to the fact that each of the micro cells has a relatively short distance between the serving antenna and the mobiles being served by the cell.

This, more direct signal path between the base station and mobile gives less path loss and less fading, in many cases 20–40 dB lower link loss, simply by means of moving the antennas down into the street, even when compared with a macro base station deployed at roof top level on a neighboring building.

Micro Cell Advantages

Compared with the traditional macro deployment the micro cell provides these benefits:

- Lower path loss, better deep indoor RF penetration.
- Less fading margin.
- Lower power load per user.
- Drains less capacity in the RF network.
- Better isolation between cells.
- No DL interference to distant mobiles.
- No UL interference from distant mobiles.
- Better single cell dominance.
- Better defined handover zones.
- Better data performance, due to better dominance and isolation between serving cells.
- Tighter re-use of frequencies/codes, more network capacity and improved business case for the network.
- Easy zoning and fast deployment.

- Less costly deployment.
- Lower roll out cost compared with macro sites.
- Better business case.

Micro Cell Drawbacks

Micro cells appear to be an ideal solution and can often be implemented for a fraction of the construction cost when compared with a traditional macro site. In many cases there is a long zoning procedure with regard to acceptance for roof top Macro deployment but there are other parameters to consider:

- Micro cells will cover a radius of 200–300 meters maximum.
- You must make sure to deploy the micro cells near or on top of the capacity/data 'hotspot'.
- Often a cluster of micro cells will be needed to cover an area.
- Discreet HW and antenna deployment will be required, antennas often need to be integrated into street furniture.
- Obtaining sufficient transmission to the micro cells can present a challenge.
- Space for base station and supporting HW.

12.1.2 Antenna Locations for Micro Cells

Selecting antenna types and locations for micro cells is a very important part of the overall micro cell solution design. The exact location of the antenna will play a significant role in RF performance and business case. Several parameters play a role when selecting the antenna location; obviously one must ensure that one can achieve the intended coverage area and at the same time come up with a practical deployment that the building owner/local authority will accept for implementation implement in the building/area.

Be sure to fully document the desired antenna location with photographs or drawings; for indoor antenna deployments, an offset by a few meters of antenna location can have negative effect on service level and the business. The main issue is that the antennas must be deployed below the height of the local clutter/buildings in order to maintain the advantage of micro cell deployment: confined coverage in a target hotspot. Examples of antenna deployment can be seen in Figure 12.3. The final location will to a great degree depend on the actual limitations of the building, or in the area.

Examples of Antenna Placement

Let us look at the antenna locations in Figure 12.3 one by one.

Location A (Figure 12.3) is a directional antenna or antennas beaming down the canyon of the street (as in Figure 12.4C). Alternatively, this antenna could be intended to target the building opposite the street and be deployed perpendicular to the building. This antenna location (Figure 12.3A) should only be considered for relatively short buildings, i.e. a maximum of eight floors, otherwise one might lose some coverage on the lower floors of the buildings in the area, especially near the building where the antenna is deployed. Typically this antenna location will be selected as the last resort; be careful to keep the radiation going down the street – so no closer.

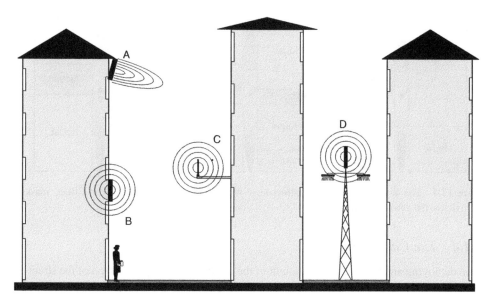

Figure 12.3   Examples of antenna locations applicable for micro cell deployment

Location B (Figure 12.3) is used typically when deploying 'back to back' antennas providing coverage up and down the street canyon, as illustrated in Figure 12.4C. The two antennas mounted back to back can broadcast the same cell, fed by a splitter to the two antennas, thus utilizing the directivity of the antennas on both directions. This antenna location (Figure 12.3B) is ideal for discreet deployment and is relatively easy to implement.

Locations C and D (Figure 12.3) are widely used for micro cells; the micro cell is feeding an Omni antenna mounted ideally on a lamp-post at some distance from the walls of the building (Figure 12.2D). This will maximize coverage in the canyon of the street. One other advantage of using a lamp-post (Figure 12.2D) is that the base station can be integrated in the pole design. Sometimes mobile operators will even custom design a lamp-post that integrates both the base station and the antennas (inside a plastic enclosure) as a totally invisible discreet solution.

It is often not possible to use lamp-posts and one will have to find a location on the facade of the building where one can deploy a bracket that ensures maintaining some distance between the Omni antenna and the wall of the building.

12.1.3   Antenna Clearance for Micro Cells

One must be careful with clearance from other objects close to the antennas that might impact upon the radiation from the antenna. Obviously antennas must not point directly at any metallic objects, and one must make sure to have a clearance of minimum 10λ from small to medium obstructions, and be sure to stay clear of major obstructions, such as walls, etc. When deploying Omni directional antennas on the facade of a building, as illustrated in Figure 12.3C, one must ensure that the Omni antenna has a minimum distance from the surface of the building of a minimum of 2λ preferably more.

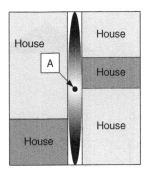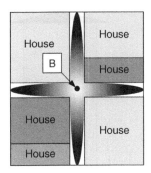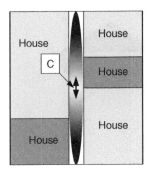

Figure 12.4 The antennas installed at street level will be affected by the 'Canyon Effect' masking directivity to the antenna footprint

12.1.4 The Canyon Effect

When deploying antennas below the height of the building in the open canyon of the street the directivity of the antennas will be dictated by the shape and size of the street. This is known as the 'Canyon Effect'.

The exact footprint of the antennas can be very hard to predict, but in general the open street will add directivity to the footprint and guide the RF signal in the open area of the street.

The effect can be seen in Figure 12.4 and applies to both directional antennas (Figure 12.4C) and Omni antennas (Figure 12.4A and B). Omni antennas will to some degree behave as a directional antenna. A clever RF planner will use this effect to direct the signal to the planned coverage area and at the same time maximize isolation and avoid leakage to other unintended areas.

Special attention must be given to antennas installed in or near the intersection of two streets, as shown in Figure 12.4B, where one Omni antenna will provide service in all four directions.

12.2 Micro Cell Capacity

Even though micro cells are a strong tool for reaching indoor users in dense areas, typically capacity hotspots, there remains one major challenge: capacity design and the lack of resource sharing. Ideally, one should be sure to implement the individual micro cells precisely in the traffic hotspot in order to maximize capacity and performance, as discussed in the first part of this chapter. That however, can be a challenge; you might not get permission to deploy the micro cell in the exact desired location but perhaps 100 meters from an ideal location and that can easily represent an offset of half the size of a cell. This results in having the HO zone directly on top of the hotspot, which is not ideal – certainly not for 3G and 4G. In Figure 12.5 we can see this principle in a micro cell network structure; a centralized controller will connect to a number of micro cells – in this case four, cells 1 to 4, deployed in the targeted hotspot area. This structure is static in terms of capacity allocation once deployed; the capacity offering will be assigned to the individual locations of the individual cells, each serving a fixed area – greatly dependent on the actual installation challenge of each of the cells and servicing antennas.

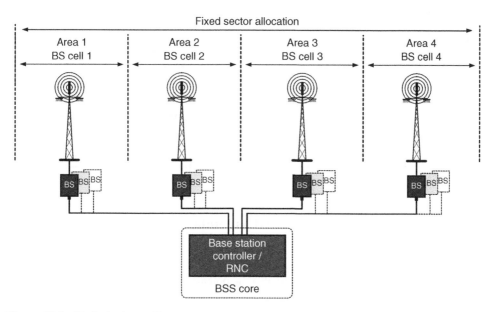

Figure 12.5 Typical micro cell structure, one central controller connected to a number of micro cells in a fixed sector allocation

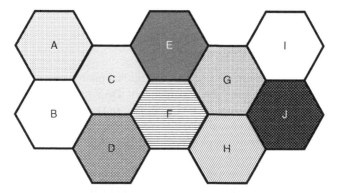

Figure 12.6 Example of Micro cell cluster, 10 micro cells in a fixed cluster servicing an area, each micro cells capacity is fixed to the individual service area

Static Capacity Allocation

Static micro cell capacity can also be depicted as in Figure 12.6; this shows a micro cell cluster of 10 cells A to J, with capacity fixed to the actual location of each of the micro cells A to J. Each of the micro base stations is operating as an individual cell on a dedicated frequency/code.

This traditional micro cell solution is sufficient if you have a relatively uniform distribution of traffic load throughout the micro cell service area. In practice, however, a cluster of micro cells will service an area with uneven distribution of the capacity/data need; there might be some extreme hotspots where capacity is greatly needed and other areas where only a few calls are made, but where you need to provide sufficient coverage level for occasional traffic. The traffic pattern in the area may change over time and that can lead to an expensive re-design of the capacity configuration and physical de-installation of equipment, re-locating it to more loaded cells over time. We then add the complexity of a mobile network that must tender for several layers of different technologies such as 2G, 3G and 4G in the same micro cell cluster. In this multi layer deployment one might wish to design for one cell size for 2G, another for 3G and a third for 4G.

The inherently static nature of micro cell deployment is its Achilles heel, and has in many networks led to less use of this strong tool for reaching deep inside buildings. This is a shame because the street level deployment of cells is a strong tool; and recently a new addition to the solution portfolio has been added that resolves the issue with static capacity and multi layer/ multi operator deployment: Street Distributed Antenna Systems – in short; Street DAS – or Outdoor Distributed Antenna Systems – ODAS.

12.3   ODAS – Outdoor Distributed Antenna Systems

As described above, micro cells can be a strong tool for reaching deep inside buildings with high capacity and data speed. The main challenge is the static capacity allocation that is attached to the individual micro cells and also the potential issue of having a compatible cell deployment of sector plans for mixed operations of 2G–3G–4G. Another drawback with traditional micro cells is the lack of multi operator support; each mobile operator will have to deploy their own individual base station for each technology in each cell. This can sometimes make it close to impossible to share the infrastructure for multi operator deployment of micro cells, due to lack of installation space for the equipment needed at the antenna site.

Unlock the Capacity from the Coverage

There is an answer to the problem of deploying traditional micro cells; 'Outdoor Distributed Antennas Systems' – ODAS. Like the active indoor DAS covered in Chapter 5, ODAS employs the same principle of a single centralized base station supporting multiple antennas allocated inside a building – with one difference we can use the same principle in an outdoor network for a micro cellular type of deployment utilizing all the advantages of the micro cellular type of deployment.

The principle of ODAS can be seen in Figure 12.7; the actual antenna deployment is similar to the traditional micro cell implementation, but instead of having distributed base stations at street level, we now service the capacity from a centralized Base Station Hotel to outdoor Remote Units located at the antenna points.

Base Station Hotel

The Base Station Hotel (Figure 12.7) can be a central location, in an easily accessible room, typically an existing equipment room that is already used by the mobile operator.

In the centralized Base Station Hotel, multiple services can be fed to a central Multiple Standard Main DAS Unit. From the DAS Main Unit optical fiber is fed to the individual antenna locations connecting Remote Units to the DAS. The DAS Remote Units have RF power comparable to a Micro base station but will be able to support more than one technology, depending on the configuration.

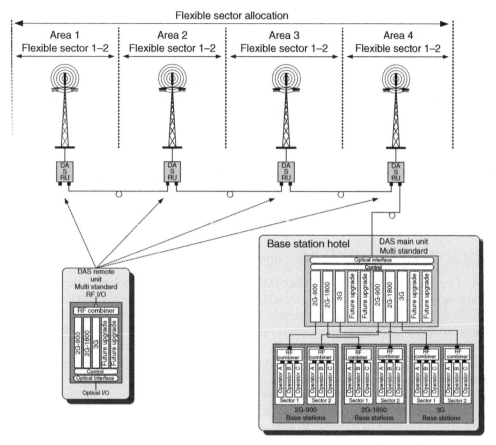

Figure 12.7 A simplified example of a micro DAS cluster served by a central base station where all the base station resources are located. This is ideal for flexible sector plans and benefits from simulcast options where several remote units might service the same sector, increasing the footprint of that cell. The DAS can easily be upgraded to support several services at the same time

As we can see in Figure 12.7, both the DAS Master Unit and the DAS Remote Unit are based on a modular concept, where support cards/amplifiers can be added in the field to tender for new services in the future.

For HSPA+ and 4G networks we could consider implementing MIMO in order to provide the highest possible data speeds. One needs to ensure that one selects a DAS system that will allow one to separate the different MIMO paths throughout the DAS system.

12.3.1 The Base Station Hotel and Remote Units

The central Base Station Hotel (Figure 12.7) could be located in an existing equipment room at the mobile operator, taking advantage of easy access for service and upgrades, battery backup, transmission, air-conditioning, etc. This could resolve a major issue with lack of space in remote locations when deploying micro cellular type solutions that are designed to support multiple mobile standards and sometimes multiple operators. This topology is possible because of optical fiber distribution, that for the DAS will be able to reach more than 20 km over the optical fiber between the central Base Station Hotel and the remote units (less distance on digital DAS due to the addede delay).

You will not need space at each remote location to accommodate all the base station and site support equipment. The same goes for the remote antenna location where the Remote Unit is installed. The DAS Remote Unit is an outdoor sturdy type; it needs no equipment room and new modules can be added internally for future expansion and upgrades so as to accommodate new services and wireless standards supported by DAS.

This will have a major impact on the business case of the implemented system; you simply have the structure in place with the Base Station Hotel and remote units. Come future services and standards it is relatively easy to upgrade, compared with the traditional option of replacing all micro cell base station hardware.

12.3.2 Simulcast and Flexible Capacity

One of the greatest advantages of outdoor DAS is flexible simulcast. Simulcast is the same configuration we use in a normal active indoor DAS, where all antennas broadcast the same sector/cell. So for outdoor DAS all remote units in a selected area will be broadcasting the same cell that is connected to the Main Unit. In other words, the remote units are simulcasting the same cell. This gives one the opportunity to apply a configuration as illustrated in Figure 12.8 where one has four cells connected to the central Main Unit, Cells A, B C and D feeding signals to 10 remote locations, as illustrated in Figure 12.8. Cell A is simulcasted by five remote units, Cell B by three and then Cell C and D are each broadcast by one remote location. This demonstrates the ability of the ODAS system to provide the required coverage footprint in the target area and at the same time adapt the capacity to what is needed – the total area will have improved coverage, and then extra hotspot capacity in more dense areas, in this case (Figure 12.8) covered by Cells C and D. This is a much more efficient utilization of the network's resources, and unlike traditional micro cells one can implement coverage and capacity individually and

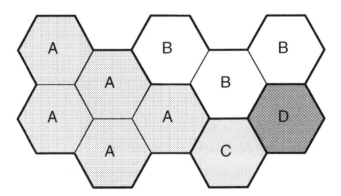

Figure 12.8 Several DAS remote units simulcasting the same sector, resulting in a high utilization of the network recourses; and hotspots can easily be served with high capacity

optimize the business case. If we compare the ODAS coverage and capacity area in Figure 12.8 with the traditional micro cell application in Figure 12.6 the difference is quite evident.

In both cases the RF signal level is raised to the desired signal strength, but for the traditional micro cell deployment in Figure 12.6 the coverage and capacity needs to follow 1:1; so in order to support the required coverage area nine micro cells antennas are needed to provide the signal level, thus ten micro base stations will be deployed (Figure 12.6 Cells A–J). However, in reality the capacity is not uniformly distributed in the area, so if we use the ODAS implementation, as illustrated in Figure 12.8, we will provide the required RF coverage level, and we can adapt the capacity – having clusters of remote locations simulcasting the same cell, while other remote locations only broadcast one cell – concentrating hotspot capacity to where it is needed. And in this case (Figure 12.8) we can cover the same area with the required signal level and tender for the actual capacity need by using only four cells not ten as in traditional micro cell deployment (Figure 12.6). Another strong application for these ODAS solutions is for high speed rail deployment, boosting both performance and the business case, as described in Section 12.5.

12.3.3   Different Sector Plans for Different Services

The ability to provide flexible and adaptive sector plans, using different simulcast schemes, improves the business case and lowers production costs. Depending on the type of ODAS that one is deploying the system could even be able to provide one sector plan on 2G, a separate sector plan for 3G and a third for 4G. Some configurations of ODAS will even have different sector plans for different mobile operators using the same frequency band and sharing the DAS infrastructure.

12.4   Digital Distribution on DAS

The some versions of DAS system utilize digital processing and transport which gives rise to some challenges and many advantages with regard to performance and application.

There are several ways of utilizing digital technology, and the following is a general description.

The latest versions of DAS will interface directly with the base station at the digital baseband level interfacing to the OBSAI (Open Base Station Architecture Initiative) and CIPRI (Common Public Radio Interface). This can be done to an analogue or digital DAS. This has several advantages and saves cost by having the DAS as a 'Remote Radio Head' saving the cost of the PA and power supply, etc. at the base station location.

In principle, digital DAS works like this (Figure 12.9):

1. If the signal from the base station is not already at the digital level, using CPRI or OBSAI, the Central Master Unit (MH), the MU will convert/digitize the RF from the base station.
2. Then via signal processing the Digital DAS will divide the signal into a timeslot/frame structure.
3. Then internal timeslots/frames are then converted into optical signals and distributed over the fiber. This can be to/from a single remote unit (RU) or the same signal/frame could be fed to/from several RUs in a simulcast scheme, as shown in Figure 12.8.

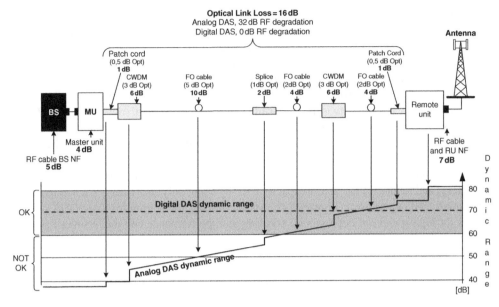

Figure 12.9   Comparing the dynamic range of digital DAS (stippled line) with the dynamic range of analogue DAS (solid line)

The same procedure works in reverse; where the analog RF on the UL from the mobiles is converted into digital and sends back from the RU to the MU.

The advantages of digital DAS for RF performance (Figure 12.9) is compared to the traditional analog DAS in Figure 12.9. The main issue with analog DAS is degradation of the dynamic range due to losses, for exactly 1dB of optical degradation RF performance on the analog DAS will be degraded by 2 dB. This is unlike digital DAS where RF performance is maintained independent of the passive losses on the optical fiber (up until the link eventually breaks down, typically after more than 25 dB of optical losses). The standard analog DAS will typically only be able to handle a maximum of 10 dB of optical loss. The drawback of digital transport is added delay to the DAS chain, limiting the reach. Digitizing the RF signals also results in a very high digital bandwidth (bitrate) requirement on the fiber transport. Analog transport will have a less sturdy optical link budget, but reach longer and typically be wideband on the fiber transport.

12.4.1   Advantages of ODDAS

We can summarize some of the main advantages of outdoor DAS from Figure 12.7 and Figure 12.9:

- *Sturdy optical link budget*: due to the digital nature of the transportation of the signal, the system is not susceptible to degrading effects due to optical fibre loss and reflections, etc. on the optical interface. Typical analogue DAS systems can handle about 10 dB of loss on the optical before the link breaks down. The typical digital DAS can handle more than 20 dB of loss, and the link quality is still perfect.

- *Simulcast and flexible sector plans*: both analog and digital DAS will be able to transport any of the sectors that are connected to the Main Unit to any of the remote units. This makes it possible to divide the cell lay out into any plan that one desires – one can have four Remote Units simulcasting the same sector in unison, or divide them into any sector configuration of one's choosing. Each service and/or mobile operator might even have its own sector plan. So one can design ideal handover zones dedicated to 2G and 3G – ideal for future proofing and optimization.
- *Multi band support/multi operator*: both analog and digital types of high power DAS typically supports multiple standards and frequency bands in the same system. As seen in Figure 12.7, both the Main Unit located at the Base Station Hotel as well as the Remote Unit located in the tunnels and stations are modular. Typically the upgrade to a new band is merely a matter of installing a new card in both units in the field.
- *The Outdoor Distributed Antenna System*: ODDAS will provide an upgradeable RF back bone in a network; once the BS hotel and remote locations are in place, future upgrades are relatively easy, making a network more future proof.
- One can utilize existing optical fiber that is already carrying traffic, by using a different colour CDWM (coarse wavelength division multiplexing) on both analog and digital DAS.

These parameters make High Power ODAS ideal for microcellular type deployments. In Figure 12.7 one DAS Main Unit at the central Base Station Hotel supports 2G-900 two sectors, 2G-1800 two sectors, 3G2100 two sectors, each band for three operators.

The optical fiber interface on the DAS is quite advanced and by the use of WDM (Wave length Division Multiplexing)/CWDM (Coarse Wave Length Division Multiplexing) the optical fiber installation can be utilized to a maximum. Sometimes the optical transport system or the DAS can even using existing optical fiber with active traffic – by using a different wavelength/colour.

This addresses one of the main cost components in a typical deployment of DAS in a large system: optical fiber installation.

12.4.2   Remote Radio Heads

Base station manufacturers make increasing use of optical fiber to feed Remote Radio Heads (RRH), using an optical fiber connection between the base station processing/baseband unit and the remote amplifier.

This application saves the passive coaxial cable from the base station to the antenna – so for a typical macro base station deployment, your base station will consist of two elements, the processing/baseband unit in the equipment and the Remote Radio Head located on top of the macro tower. This topology could obviously also be used to feed RRH distributed in a tunnel, an indoor or outdoor DAS solution.

RRH is a strong tool and a very good alternative compared to traditional base stations, you have a central baseband unit, and a fiber fed remote unit that can be located several hundred meters apart.

At first glance this seems like an attractive way of designing a DAS, where a RRH will drive a local network of passive DAS or for our application a micro cell deployed at street level.

Even though this will provide some advantages when compared with traditional passive coaxial distribution there are some challenges and limitations that we must consider:

- RRH will typically be limited to single band support; the current RRH solutions only support a single band, so one RRH will be needed for each service. This can be a considerable challenge for tunnels where limited space is available at the far end of the tunnel.
- For single operator support, one set of RRH will be needed per mobile operator, so for larger deployments of multi band, multi operator solutions one will need one RRH per mobile operator, per band. So imagine a large tunnel deployment with three mobile operators each deploying 2G-1800, 3G2100 and 4G900 – one will need nine RRH at each remote location, and one will have to combine the signals from these nine RRH in a high quality passive combiner to avoid PIM (see Section 5.7 for more detail) – requiring a 19' rack for the combiner alone. This will often not be possible to install in the confined space of the tunnel.
- RRH will normally not apply simulcast like the intelligent simulcast schemes and sector plans of DAS. The ability to have flexible capacity and several DAS Remote Units servicing the same sector, compared with a single sector, locked to each RRH is very important for capacity efficiency and the overall business case. This will maximize the utilization of network resources by the simulcasting of single sectors by several DAS RU's.
- Using DAS it will be easier to upgrade capacity offering; this capacity offering could also be controlled dynamically via traffic counters in the network controlling the simulcast schemes in the DAS so as to adapt and control the offered capacity to current hotspots.
- Capacity and coverage can be implemented independently using DAS; the optical transmission interface is sturdy and can handle losses of more than 25. Traditional RRH is deployed as one RRH per sector, with no simulcast.

12.4.3   Integrating the ODAS with the Macro Network

When implementing an outdoor DAS solution it is apparent that one has to take into account the nearby macro network. The ODAS will be implemented in an area where it can be hard to reach indoor users from the traditional macro network (Figure 12.10) and the ODAS is considered typically when:

- It is difficult to get permission to deploy traditional roof top macro sites.
- The capacity and data speeds required cannot be implemented by macro sites.
- The buildings are simply too high for roof top deployment to provide deep indoor coverage.
- The power load per user will be too high for the macro layer to make an attractive business case due to high link losses from the macro layer to the users.

Buildings (Figure 12.10) 1 and 5 near the existing macro antennas might have sufficient data service to some degree, especially the side of buildings 1 and 5 that are facing the macro antennas. But the other side of these buildings that face away from the macro site and the other buildings 2, 3 and 4 might have large areas indoors with an insufficient service level.

In order to overcome the problems described above there are several options to consider. One could implement individual micro cells at street level but, as described in Section 12.3, street DAS will have several benefits. This problem can be seen in Figure 12.10, where an area

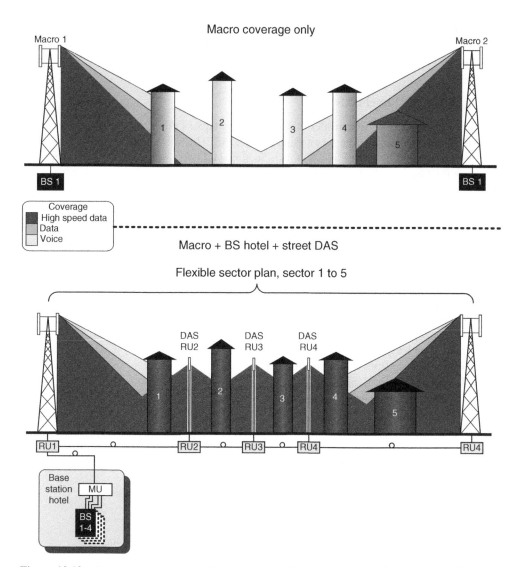

Figure 12.10 An area of a city where the macro network struggles to reach indoor users and the data service offered is insufficient. Problems can be solved by ODAS, adding both the needed coverage level and capacity. The ODAS simulcast configuration could even include some of the macro sites

of an urban city has insufficient coverage and data capacity. If this is a 3G network one of the main issues could be the high penetration loss to the users, cannibalizing capacity due to high power load per user (see Section 3.3 for more detail).

In this case (Figure 12.10) one can use the capacity of the macro net to drive the street DAS and include the macro network in the DAS, thus utilizing the existing base station resources and maximizing the business case.

The solution, as can be seen in Figure 12.10, is ODAS raising the coverage level, minimizing the load per user and maximizing the capacity. This, combined with the ability to tender for flexible and adaptive sector plans, can often be an ideal solution for both 2G, 3G and 4G deployment, not only from a performance point of view but also for the business case of the network, maximizing the use of resources and investment.

Timing Issues with Street DAS

One very important issue to consider when deploying street DAS is the timing delay introduced in the system, in particular when using digital DAS. This will impact upon the network – offsetting timing and cell sizes, etc. Refer to Section 11.11 for more detail. This is depends greatly on the fiber installation and distances of the fiber.

In Conclusion, Street DAS/ODAS

If we add all the advantages of street DAS to those of the traditional micro cell we have a very strong tool when using street DAS, especially when using the new generation of distributed DAS:

- Low path losses, better deep indoor RF penetration.
- Lower power load per user.
- Drains less capacity in the RF network.
- Better isolation between cells.
- No DL interference from distant macro areas.
- No UL interference from distant mobiles.
- Better single cell dominance.
- Can limit and optimize the number of handovers along a street by utilizing simulcast, improving overall performance and decreasing dropped calls (see Section 12.5.3).
- Better defined handover zones (3G/4G).
- Better 3G/4G performance, due to better isolation between serving cells.
- More network capacity and improved business case of the network.
- One shared infrastructure for multiple technologies and multiple operators.
- Flexible capacity allocation.
- More future proof and easy to upgrade to future services.
- Easier and faster to implement, less equipment is required at the remote location.
- Simulcasts and integrates easily with the existing macro network.
- Low visual impact roll out.
- No need for base station equipment room.
- Fast deployment.
- Saves on OPEX.
- Can support multiple operators.
- Independent 2G, 3G, 4G sector plans.
- Can be combined with the macro network.
- Better business case, due to better trunking efficiency than micro cells – so better utilization of the capacity deployed.

These advantages of ODAS, provides a strong tool for operators. Street DAS will in some cases be more attractive for deployment of deep indoor penetration and coverage over a larger area.

Even though a dedicated indoor solution will often be superior in terms of RF performance, one must realize that the business case of the network plays an important role, and it would be difficult to implement indoor systems in each and every building in the area. ODAS can often be implemented at a fraction of the cost, and with a considerably shorter implementation process.

12.5 High Speed Rail Solutions

Outdoor Distributed Antenna Systems (ODAS) are also a strong tool for deployment along a rail track. The big advantage is the long reach of the optical fiber, ODAS can handle an optical link budget in excess of 8 dB (digital) without any degradation of RF performance, but mind the added delay if applying digital DAS, for more detail see Section 12.4. This is a practical, yet important parameter to consider when implementing DAS in the harsh rail environment.

Let us take a closer look on some of the unique challenges related to high speed rail, on top of the general aspects of rail tunnels discussed in Chapter 11, and outdoor DAS in this chapter.

12.5.1 Calculating the Required Handover Zone Size for High Speed Rail

As was discussed in Section 11.2.3 one needs to pay a great deal of attention to the HO zone, how we configure the antennas of the DAS in order to ensure that we have a usable and successful HO zone.

Refer to Section 11.2.3 for all of the detail, but one thing one needs to realize is the sheer size of the HO zone required for high speed rail.

Calculating the Handover Zone Size Required for High Speed Rail

One can easily calculate the required HO zone overlap if one knows the speed of the vehicle and the time required to perform the handover, remembering to include the time for handover retry if the initial handover is not possible.

Let us look at one example for a high speed rail solution:

We have a vehicle with a speed of 350 km/h, and in this example we need 6 seconds for Handover (once the HO trigger value is fulfilled), and then we will add extra margin in the HO zone size to accommodate, measurement evaluation, HO execution, signaling and retries – a total HO zone that last 18 seconds.

First we will calculate how many meters per second the train moves:

The relation between km/h and meters per second [m/s] is: m/s = (km/h)/3.6

At a speed of 350 km/h this would be 350/3.6 = 97.2 m/s.

So if we need to provide 18 seconds of valid HO zone we would then need $18 \times 97.2 = 1750$ metres of valid overlapping HO zone; and remember that throughout the zone both cells (serving and candidate) would need to have sufficient signal level to fulfil the HO trigger and provide signaling, etc.

This is a challenge to design for, and especially for high speed rail. But it is a vital requirement – even under ideal conditions, when all of the antennas are planned and installed correctly.

Optimized Handover Zone

The calculation of the required handover zone needed for high speed rail clearly indicates the huge – and expensive – challenge of tendering for handover zones, in particular for high speed rail. One needs to ensure that one pays close attention to all of the issues described in Section 11.1: optimized antenna locations, etc.

In addition to this it is often required to optimize the handover evaluation and trigger parameters between the two handover candidates, in order to make sure the handover evaluation is optimized. One does not want to end up with a situation where one has a 'ping pong' handover zone. 'Ping pong' handover zones are where one has multiple handovers forward and back again until one finally settles on the new cell. This will, in practice, entail that one handover will result in a series of two to five handovers, multiplying the signaling load on the two cells.

Keep in mind that one will literally have a train load of calls that all need to perform handover within a few seconds; this is a heavy challenge for the network and base station, and so if one multiplies this load with an unintentional 'ping pong' handover zone, one will most likely be facing a trainload of dropped calls. Keep in mind that the signaling channels in the network per default are configured to handle normal traffic and signaling. Therefore it is recommended to revisit the configuration of the signaling channels and verify that the network can handle the entire load. Some networks allow one to 'chain' the cells – so that one has a sort of pre-synchronization between the handover candidate cells; this optimizes the success rate and provides swift handovers – it is highly recommended to investigate all possible parameter options.

The Capacity Challenge

Another concern is traffic dimensioning. In Chapter 6 we looked at normal traffic behaviour – Erlang distributed traffic, which is the basis for calculating the number of traffic channels for a specific traffic load and blocking rate. However, Erlang's basis for all of the formulas is a normal distribution of the arrival of calls, etc. However, even though the traffic inside the train might be Erlang distributed, the trains are not – they follow the train schedule. So one will have a trainload of calls hitting one's cell all at once, and a few seconds later they will be handed over to the next cell – not Erlang distributed.

This can also fool some of the standard statistical evaluation tools that one has in the network that evaluate traffic blocking, etc. I have seen cases where we did in fact have lots of blocking in a metro tunnel system that did not appear on the overall dashboard of the statistical tool – but when carefully checking the counters in the network we could see clear hard blocking!

12.5.2 Distributed Base Stations for High Speed Rail

In Figure 12.11 we can see a typical method for providing coverage along a rail track.

Multiple base stations are deployed along the track with an inter distance based on our link budget calculations. The challenge, however, is the penetration loss into the train coaches. As described in Section 11.1.1, one will have a quite high penetration loss into the train when one covers it from a longitudinal direction, sometimes up to 40 dB, unless one has an on-board

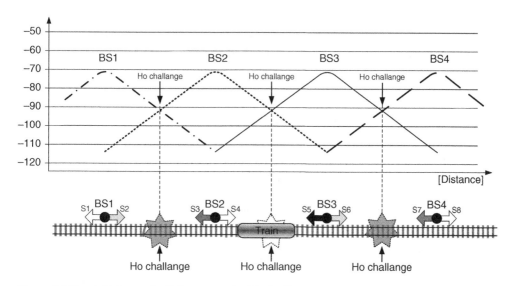

Figure 12.11   A high speed rail line covered with distributed base stations alongside the track, with potential HO problems in the overlap and potential issues with the Doppler effect

repeater system such as that described in Section 4.9. If one has no on-board repeaters deployed in the train and one indeed faces longitudinal loss exceeding 30–40 dB the result will be a very short distance between the base station deployments due simply to the need for signal strength along the track to compensate for the high penetration loss.

Calculating the Number of Required Base Stations

Typically one will use low masts that are already present on the track side, with antenna installation of perhaps 15 meters in height. Depending on one's link calculations and design levels one might likely end up with a base station for every 4 km along the track, or even closer. If one then considers the high speed train scenario in Section 12.5.1 where the train actually moves at 97.2 meters per second, it will in effect mean that one will have a handover every 41 seconds (4000 meters/97.2 = 41.1seconds). Keeping in mind that we calculated a required handover zone of 1750 meters (Section 12.5.1) we would in fact have taken this into account and made sure that both handover candidates fulfilled the design requirements in all of the handover zone – so there an overlap of 1750 meters!

Even if one calculates the signal level to reach 4000 meters one will have to space the base stations according to the overlap with an inter distance of the base stations of 4000–1750 = 2250 meters, almost doubling the number of required base stations, increasing the cost heavily.

This is the scenario illustrated in Figure 12.11 where four base stations (BS1–4) service a rail track using eight sectors (S1–8), with a potential handover problem between the cells.

If we take an example of 100 km of track under the conditions above for a high speed rail track we would have a need to deploy 45 base stations, 90 sectors and 90 handover zones to tune and optimize and with a high risk of potentially dropped calls in each handover zone.

If one is designing for multiple technologies, which could be 3G and 2G in the same deployment, then one needs to realize that one has to design for the weakest link, so if one system could make do with a base station spacing of 5000 meters, and the other 4000 meters then we would have to space the base stations 4000 meters apart, slightly over designing one system.

12.5.3   Covering High Speed Rail with Outdoor Distributed Antenna Systems

Compared with distributed base stations one might have a better option using an outdoor distributed antenna system – ODAS, as described in Section 12.3, for several reasons.

The big advantage in using ODAS for high speed rail is the simulcast option, as described in Section 12.3 and illustrated in Figure 12.7 and Figure 12.8. This is a great advantage when providing solutions for high speed rail, as the example in Figure 12.12 illustrates.

Utilizing Simulcast with ODAS for High Speed Rail

By simulcasting the same sector over multiple remote units one limits the number of needed handovers significantly. In this example (Figure 12.12) we have deployed four remote units (RU1–4); each of these remote units is equipped with two amplifiers/PAs that broadcast in two

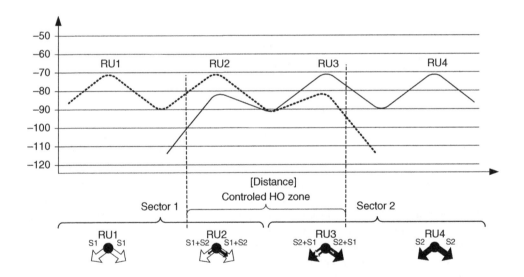

Figure 12.12   A high speed rail line covered with an outdoor DAS some few hundred meters from the track. The cells are extended over more remote units, limiting the number of cells – thus limiting the number of handovers and the possibility for HO failure. The remote units RU2 and RU3 will simulcast both cells offset in level, extending the HO zone to the required size with a well defined HO margin

directions. This example shows the four remote units simulcasting the two sectors (S1 and S2) over an extended area compared with the relatively limited area covered by the base station deployment illustrated in Figure 12.11. At first glance we can see that we have only two sectors and one handover using the ODAS, compared to the base station distribution that has eight sectors and multiple handovers.

It is evident that using ODAS limits the number of handovers, thus limiting the number of potential handover failures and problems. But looking carefully at Figure 12.12 there is actually more to it; if we study the handover zone more carefully we can see a special configuration of remote units 2 and 3. R2 and RU3 are actually simulcasting both sector 1 and sector 2 (S1 and S2) but offset in level. Using this approach one can configure the ODAS with a specific offset of the power level broadcaste on each antenna for the two different sectors, thereby controlling the handover margin and zone 100%. The propagation loss from the antennas would be the same (more or less), one is in control of the difference in signal level over all four coverage areas of the two remote units (RU2 and 3) this gives one a well designed and controlled handover zone that will support high speed rail easily.

Big Savings and Better Performance with ODAS for High Speed Rail

If we assume the same location and power level from the ODAS as for the distributed base stations the difference is quite apparent. In Section 12.5.2 we calculated a need for 45 base station locations in order to cover 100 km of rail track, dictated by the need for the 1750 meters overlap between the sectors in order to ensure a sufficient handover zone.

If we compare the layout of the ODAS in Figure 12.12 with the distributed base stations in Figure 12.11, we do not need to space the remote units 2250 meters apart owing to the requirements of the handover zone, but can actually place them 4000 meters apart as we assumed in the link budget. In effect one will then need only 25 remote units and not 45 base station locations – this is a significant saving for the implementation budget.

However, there is more to it; assuming that we simulcast the same sectors over four remote units we would only need about 16 sectors, not 90 as in the case of the base station. This limits the number of handovers and possible dropped calls significantly and, with the controlled handover zone, by utilizing simulcast of the ODAS these handovers would be performed much better.

The ODAS has yet more advantages. As described in Section 12.3, we can deploy multiple technologies using the same ODAS system, thus combining 2G, 3G GSMR, etc. and giving us even more incentive for using ODAS when compared with distributed base stations, from both a performance, implementation and cost perspective.

12.5.4 Optimize the Location of the ODAS and Base Station Antennas for High Speed Rail

If we look carefully at Figure 12.11 and Figure 12.12 there is a significant difference in the location of the antennas. In Figure 12.12 the antennas are not placed alongside the train line as in Figure 12.11, but are placed at some distance from the rail track. Using this approach for both distributed base stations and ODAS is worth considering for several reasons.

Plan for Less Penetration Loss

As described in Section 11.1.1 the penetration loss in the longitudinal direction of the train coaches is considerable, sometimes more than 40 dB, whereas the penetration loss through the windows could be significantly lower, quite easily in the region of 20 dB or less.

This difference in penetration loss will have a significant impact on our link budget calculation for the system. It will reduce the number of required base stations/remote units significantly. However, the challenge lies in acquiring usable antenna mast locations for this type of deployment. This type of deployment will be a challenge, but is definitely worth pursuing considering the savings in the total number of required antenna locations to cover the same section of rail line. This is all an effect of the improved link budget by reaching the users inside the train through the windows with relatively low penetration loss. There is actually more to it; this approach will also reduce the Doppler Effect, decreasing the relative velocity of the high speed train to the antenna locations.

12.5.5 The Doppler Effect

One of the main challenges for high speed rail and radio systems is the impact of the so called 'Doppler Effect'. Owing to the velocity of the train the frequency from the base station relative to the users onboard and vice versa will be slightly offset. The frequency will simply be skewed slightly in the spectrum depending on the speed of the vehicle relative to the frequency.

This Doppler Effect, together with something called 'Coherence Time' is what limits the maximum speed a mobile system can support. In the early days of 2G systems we did not expect to see high speed trains moving close to 500 km/h – this is the reality today.

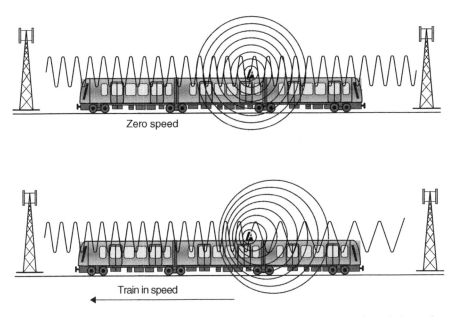

Figure 12.13 Owing to the velocity of the train the frequency from the base station relative to the users onboard and vice versa will be slightly offset

The principle behind the Doppler Effect is illustrated in Figure 12.13. I am sure that you have already experienced the Doppler Effect, when onboard a train. When the train approaches a rail crossing with a signal bell chiming the pitch of the tone of the bell is high, but as soon as you pass the crossing – and the bell – the pitch of the tone of the bell changes to a lower frequency. The same effect applies to radio waves and ultimately limits the highest possible speed that we can service with a mobile system.

Limit the Doppler Effect

We can– to some extent – limit the Doppler Effect; at first it may sound like an impossible idea – bit it is actually possible. Remember that the Doppler Effect is caused by the speed of the transmitter relative to the receiver and vice versa – herein lies the solution; '*relative*'.

Obviously we cannot change the speed of the train, but we can change the relative speed to the serving ODAS/base station antennas. Look at the difference between Figure 12.11 and Figure 12.12 in terms of the location of the antennas from the ODAS/base stations. In Figure 12.11 the antenna location is a few meters fom the track; thus having a 1:1 relationship in relative velocity to the train. In Figure 12.12 the antennas are placed off the track; depending on the distance from the track the relative velocity will be lower and lower the further away from the track you place the antennas.

This is why it is sometimes possible for people (in extreme emergencies) to use their mobile phones onboard airplanes while moving at speeds of close to 950 km/h. And this despite the potential timing issues that one might expect moving at such a high speed. One must remember that when the airplane is 10 km above the earth, the relative velocity of the mobile to the very distant base station on the ground is very low compared to the velocity of the airplane. Whereas if the plane was only 10 meters off the ground the relative velocity would be close to the speed of the airplane.

So not only can one solve the problem with penetration loss described in Section 12.5.4 by offsetting the antenna locations to the side of the track, thus saving on infrastructure cost, but one will also be able to deal with the limitations caused by the Doppler Effect.

Final Note on ODAS

I hope that this chapter has shown just how efficient ODAS solutions are. ODAS is a very good alternative to micro cells. ODAS is a very comprehensive and flexible tool that provides one with the ability to unlock the tie between the base station coverage and capacity by offering the flexibility of simulcasting several cells over multiple remote units.

The ODAS application for rail deployment covered in this chapter shows how versatile a tool ODAS can be, solving the problem of the handover zone, limiting the number of handovers and the required number of antenna locations alongside the track. It provides better performance at a lower cost, and sometimes a good alternative to dedicated indoor DAS in each building.

13

Small Cells Indoors

What is a small cell? Well, there is no clear definition – I suppose it must be smaller than a macro cell, for sure! I recall working with the 1G system in the 1980s, when I heard that this new fancy mobile system called '2G' with a limited coverage range would be deployed in the early 1990s. This new 'all digital' mobile system had a cell range limited to 32 km. Relative to the old 1G analog system, the 2G system was definitely a small cell. And then along came 3G, which typically uses smaller cells than 2G, and this was followed by the current 4G deployments with even smaller cells.

For the purpose of this book, I will use 'small cell' as a generic term, typically implemented to solve both coverage challenges inside buildings via femto cells and in high-capacity areas in outdoor network hotspots (as described in Chapter 12) using 'micro cells'.

In this chapter we will focus mainly on femto/pico cells, which are typically applicable to indoor deployments.

Small Cells Are a Strong Tool

Small cells are a strong tool for use by mobile operators to solve some of the major issues they are facing with regard to keeping up with the requirements of 'mobile broadband everywhere' in today's connected world. At some time in 2015–16 it is predicted that the wireless data volume will exceed the data volume transported on the wired network (of course, the wireless devices will still need a lot of wires to support that volume of data). The fact remains that we need to implement solutions that can support this ever-increasing demand for mobile data at a cost-efficient level. Small cells inside buildings could, in many cases, be the answer, sometimes as 'stand-alone' deployments, sometimes in combination with a DAS deployment to extend the footprint.

Indoor Radio Planning: A Practical Guide for 2G, 3G and 4G, Third Edition. Morten Tolstrup.
© 2015 John Wiley & Sons, Ltd. Published 2015 by John Wiley & Sons, Ltd.

DAS Is Actually a Type of Small Cell

Some might say that small cells and DAS are very different, but a DAS is actually a very versatile type of a small cell concept. Obviously a 'donor' cell (or cells) needs to be connected to feed the capacity and functionality, and then the DAS will transport and distribute the coverage and capacity in a flexible and versatile way for various application types. We will also see small cell/DAS products of the future that are more integrated and, over time, they will perhaps become one product, a small cell with multiple radiation points (a DAS), and perhaps a type of intelligent DAS where the RF energy and capacity is diverted and switched to the exact antenna for the specific user or users. These types of solution will have a huge impact on the capacity and network performance in terms of making sure that the signal only 'hits' the needed user/users, thus limiting the overall interference in the network, increasing reuse and increasing capacity.

The following chapter is merely a short introduction to indoor small cells for indoor use (pico and femtocells). For more detail about femto cells refer to [9], from which much of the inspiration for this description of femto cells comes.

What Is a Small Cell?

Although this book is focused on indoor coverage and deployments, the term small cells is used to describe any cell smaller than a 'macro cell', as we can see in Figure 13.1 Small cells can be applied in a number of situations, ranging from single rooms or houses in a residential scenario to urban and rural deployments. The enterprise and urban areas are driven by both coverage and capacity, while rural and residential areas are driven mostly by coverage. DAS combined with pico

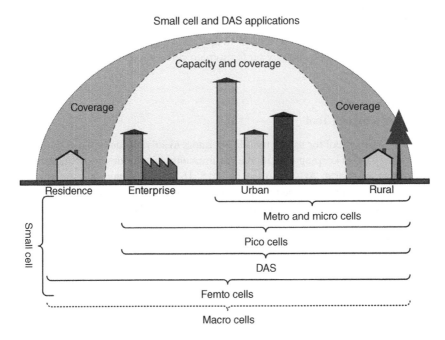

Figure 13.1   Small cells, femto, pico, and micro cells

cells is applicable in both urban and rural/suburban deployments (see Section 12.3 for outdoor DAS deployment). For more details on types of micro cell deployment, refer to Section 12.1.

Small cells will typically be deployed to solve coverage issues in residential and rural areas, as we can see in Figure 13.1, and in enterprise and urban applications will typically be implemented to increase capacity, and often close coverage gaps in the network as well.

13.1 Femtocells

A femtocell is a kind of wireless access point radiating only a relatively low power that is just enough to cover a limited area, typical residential home or small office space (femtocell class 1). Unlike a Wi-Fi access point, the femtocell operates in a licensed spectrum that belongs to one of the licensed mobile operators and has to be connected to that particular network, via a typical backhaul connection provided to the location – this could typically be standard internet DSL transmission – in order to be operational.

Femtocells are normally quite limited in capacity and support only a limited number of voice users, and will usually have only a relatively limited data bandwidth. However, femtocells are evolving quickly and will soon be available with more data support, and will quite possibly combine multiple mobile broadband access technologies within the same unit, for example Wi-Fi with 3G, 4G etc.

Bringing a low-cost base station, with somewhat limited RF performance, into a household with multiple transmitting devices, such as Wi-Fi access points and DECT base stations, could cause some concern in terms of receiver blocking these other transmitting devices which are in close proximity to the femtocell and are quite possibly transmitting within a nearby RF spectrum. Therefore it is important for the mobile operator to test and verify the performance of the femtocells thoroughly when qualifying and selecting vendors and suppliers of femtocells.

Femtocells (class 1) should not be considered as a replacement for high-capacity indoor DAS solutions, but rather to provide indoor coverage and capacity in private houses and small offices – typically less than 200 m² and with a handful or fewer numbers of users. However, the future will bring femtocells with more capability and functionality, and power and capacity and will be a very important and strategic tool. The network operator will have to invest in more than the small cell hardware; a femto gateway and upgrades in the central core network are needed too, as shown later on in Figure 13.10. Therefore the adoption of femto cells will be a wider strategic decision, and not merely a few small cells to solve a number of limited issues.

The key attributes of a femtocell are as follows:

- **Dedicated capacity, limited coverage**
  A femtocell will provide dedicated capacity for the residential house or small office where it is being deployed; thus the femtocell will offload the surrounding macro network – those resources in freed-up capacity can then be assigned to support other users in the network. Due to the low power radiated by the femtocell, the coverage range would be limited, comparable to a Wi-Fi access point.
- **Good dominance**
  Femtocells (if installed correctly) will be relatively well isolated from the surrounding network, by the attenuation of the outer walls in the building. There are some application examples later in this chapter (see Figures 13.11–13.13).

- **Operates in a licensed spectrum**
  Femtocells will operate within the licensed spectrum of the mobile operators; they could support 2G, CDMA, 2G (3G) or 4G, etc. In the future it is likely that femtocells will have combined wireless access and also support Wi-Fi. This makes the femtocell a strong tool for supporting mobile broadband for multiple access standards and user devices.
- **Self-installed by the user**
  The aim is to support installation by the consumers themselves. The femtocell, once connected to the internet, will self-configure and optimize its setting adapted to the surrounding mobile network in order to maximize its performance, and not degrade the performance of the surrounding network. However, full control over the femtocell remains with the mobile operator that owns the licensed spectrum supported by the femtocell.

  Typically the user of the femtocell could decide to make it 'closed' to only support specific users, or 'open' to support all users (on the particular mobile network) when in range. Allowing other users access to your femto will then often also give you access to other open femtos when you are outside your own coverage area.
- **Automatic control and configuration**
  The femtocell will, within certain limits (determined by the mobile operator), be self-configured, self-controlled and self-managed. Over the first few minutes of operation when it is connected and powered up for the first time, it will perform some automatic configuring routines, measurements of the surrounding network and RF environments and perform self-optimization accordingly. Once the femtocell enters operational mode it will continuously self-optimize according to the local RF environment and macro network so as not to degrade performance in the coverage service area of the femtocell or degrade the surrounding macro cellular network.
- **Low production cost for the mobile operator and good quality for the users**
  Depending on the mobile operator's business model for offering and implementing femtocells, this could be quite an interesting approach and strategy – not only from a pure technical performance point of view but also for the overall network business case. The scenario of deploying femtocells to support residential users could be attractive for both the users and the mobile operator. One could imagine that the users will co-finance the femtocell; in return they get good signal quality and a home zone concept with attractive tariffs and increased functionality. One example of this functionality could be a virtual home phone number, with all mobiles (having different individual numbers) being paged in unison provided that they are within range of the femtocell in the home zone. These added services and functionality may also lead to more loyal users – saving the mobile operators from expensive churn of customers (when users go to another mobile service provider/operator).
- **Backhaul over the internet**
  The typical femtocell will rely on transmission backhaul over a standard internet DSL line, already present in the home or in the corporate building. This DSL line would need to meet specific demands with regard to quality of service (QOS). Obviously the data offering in the footprint of the femtocell is tied to the capabilities, data speed and latency of the DSL line.
- **Location tracking**
  It is important for the optimization and functionality of the implemented femtocell and the surrounding network that the location of the femtocell is known and registered in the network. This location detection can be based on GPS, the femtocell reporting of the surrounding macro network cells, signal strength, self-reporting by the user, and so on.

13.1.1   Types of Femtocells

Even though we at this point in time consider femtocells to be applicable only for households and small offices (femtocell class 1), in the near future new types of femtocells will emerge, with more capabilities, more coverage, capacity and functionality.

* **Femtocell class 1**
  This is the first type of femtocell currently available on the market; the purpose of this type of femtocell – class 1 – is to cover a typical residential household or a small office building with relatively limited capacity. This type of femtocell would be installed by the users of the building themselves. The transmitting power of class 1 femtocells is typically about 20 dBm, and thus the serviced coverage area is comparable to a Wi-Fi access point. The voice call capacity is limited to a handful of simultaneous calls (typically about five to seven) and femtocells of class 1 will have limited functionality beyond the basics of simply passing the calls and data traffic to the core network.
* **Femtocell class 2**
  This type of femtocell – class 2 – has more capacity than class 1 femtocells and would typically support up to 10–15 simultaneous voice calls. This type of femtocell has more transmitting power, typically in the range of about 23–26 dBm, thus providing coverage over larger areas as compared with class 1 femtocells, making it more suited to small to medium-sized office building deployments. Due to the radiated power from this class 2 'femtocell', one could/should really consider this as an advanced version of a pico cell; it could have some internal functionality in terms of local call switching and other more intelligent service-oriented functions. This type of femtocell will typically be installed by the mobile operators or alternatively by the users in the building themselves, if they have IT capabilities. This class of femtocell could be ideal for the application illustrated in Figure 13.7, and described in Sections 13.1.4 and 13.1.5 where the small active DAS extends the coverage and capacity area beyond the coverage capabilities of the femtocell.
* **Femtocell class 3**
  This femtocell type of base station has yet more capacity and could support more than 20 simultaneous users, and, with a radiated power of 30 dBm or more, could cover a relatively large area as compared with class1 and 2 femtocells.

  This femtocell will typically only be installed and managed by the mobile operator, and could be considered to be a micro cell due to its high power, but with added functionality based on femtocell technology. One could even consider this type of 'femtocell' for outdoor coverage areas. Another interesting application for a class 3 femtocell is illustrated later on in the chapter in Figures 13.7 and 13.8 and is described in Sections 13.1.4 and 13.1.5 However, do not combine several femtocells within the same DAS; this will cause 100% overlap of the cells within the coverage area, with degraded performance of handover (HO) and quality as a result.

13.1.2   The Pico/Femtocell Principle

The typical pico/femtocell base station (as illustrated in Figure 13.2) is a very effective tool for deploying in-building coverage capacity and increased network functionality at a relative low cost and on a stand-alone basis. The femto cell hardware is a small device with an integrated antenna, typically designed for visible wall mounting in an office or residential

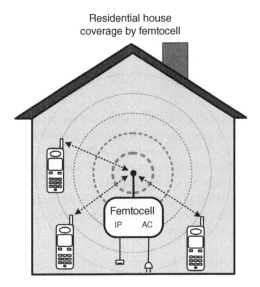

Figure 13.2   Pico/femto, low-power base station with internal antenna

environment, the same type of design that you probably already know from Wi-Fi access points. Most pico/femtocells will use IP (DSL)for transmission backhaul, making the pico/femtocell a very fast solution to implement, in many cases installed by the users of the buildings themselves – for femtocell classes 1 and 2.

The output power from the pico/femtocell is sufficient to cover about 20–40 m depending on the service requirements and the class of femtocell, radio design level and the environment in which the pico/femtocell is installed. In most cases the pico/femtocell will need a local power supply, although some types can be powered over an ethernet cable that also supplies the IP connection. A local power supply can be expensive, and often you are not allowed to plug it into the existing power group in the building if you deploy multiple femto/pico cells in a larger corporate building, but will have to install a dedicated AC group for all of the pico cells in the building.

Advantages for the users by Added Functionality in the Femtozone

It is apparent that femtocells will solve the coverage challenge and add capacity to the area they cover, but there is additional potential if you have a dedicated small cell for a specific area, regardless of whether the small cell services the area directly (Figure 13.2) or via a DAS (Figure 13.7). This is mainly because you know the mobile is registered at that specific cell within a limited area, and this 'femtozone' location enables new services.

Femtocells can add various new services and functionality when users are in the femtozone. For a private house this could be:

- Low or free tariffs when calling in your femtozone.
- Virtual home number, such as those we had (those of us who are old enough) with a fixed phone line with one number that people could use to call the 'household'. This function

could be enabled in the 'femtozone' where all mobiles ring simultaneously when dialing the specific number.

- Automatic 'alerts' via text or mail massages when people are arriving at the house, for example when your children arrive home safely after school.
- Interaction with databases on your mobile, such as podcasts, or photos being automatically synchronized with 'the cloud' when you are in the 'free' femtozone.
- Interaction with or control of other equipment in the house, such as TV, entertainment and alarm systems.

13.1.3   Typical Pico Cell Design

The typical pico/femtocell application is a small to medium-sized office building (as illustrated in Figure 13.3), where two pico cells are covering an office floor. This could be repeated throughout several floors. In the case of a building with several floors, be sure to place the pico cells 'symmetrically', that is, in the same place on each floor – so that the one pico cell will not leak coverage to adjacent floors, creating soft HO zones or degraded 3G/4G capacity. Remember that pico cells are individual cells, and therefore the 'interleave' concept (see Section 5.3.9) cannot be applied. Femtocells will normally be deployed so that a single femtocell will cover the entire area of that relatively small building (private house or small office). To support a larger office building with several floors with femtocells, as illustrated in Figures 13.3–13.6, it is recommended that you do not take the approach illustrated in Figure 13.8, avoiding coverage overlaps and instead that you use a femtocell with sufficient capacity (typically class 2 or 3).

Coverage Overlap

As for any other multi-cell environment, you need to design for some overlap of the cells so as to ensure coherent coverage and a sufficient HO zone; however, try to minimize the HO zone by using heavy structures, firewalls, elevator shafts, and the like, in order to separate the cells.

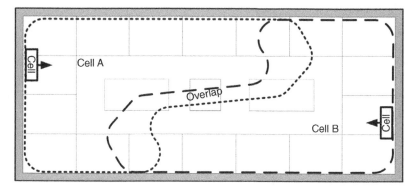

Figure 13.3   Two pico cells covering an office floor

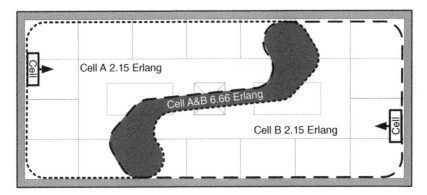

Figure 13.4 Two 2G pico cells with overlapping coverage and capacity trunking

Capacity Overlap (2G)

The coverage overlaps of the pico cells will, for 2G, give capacity overlap and trunking gain in the overlapping area (as illustrated in Figure 13.4). In the area where only one pico is covering, the capacity will be 2.15 Erlang (2G, 7TCH 0.5% GOS; see Chapter 6 for more detail), and in the area where both cells are overlapping, the users will have access to both cells, by means of traffic-controlled HOs; if one cell is loaded, they can set up calls on the other cell.

This capacity overlap of the cells yields 6.66 Erlang in trunked capacity in the overlapping area (refer to Chapter 6 for traffic calculations).

Hotspots

The capacity overlap makes it possible to cater for 2G traffic hotspots in the building by deploying overlapping pico cells in these hotspot areas. However, in reality, these hotspots tend to be dynamic and do not always occur in the same place in the building. This makes 2G hotspot design with pico cells a challenge.

Coverage Overlap, 3G

In a 3G pico cell/femtocell application (as illustrated in Figure 13.5) with the same two pico cells/femtocells overlapping with a margin of 3–6 dB, there will be a large soft HO zone. If you are not careful and do not limit the overlap of the pico/femtocells using internal heavy structures in the building when planning the deployment, the overlap might cannibalize the capacity of the system; remember that these are two different cells so you need both radio processing and transmission capacity on both cells for one call in soft HO.

Coverage Overlap HSPA

The two indoor pico cells/femtocells (as illustrated in Figure 13.6) will typically operate on the same frequency; unlike one carrier assigned to 3G R99 traffic, HSDPA does not use soft HO.

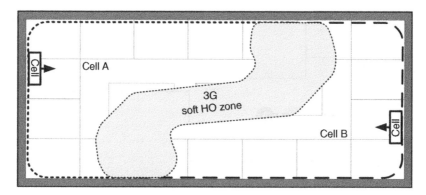

Figure 13.5 Pico cell overlap on 3G creates a soft handover (HO) zone

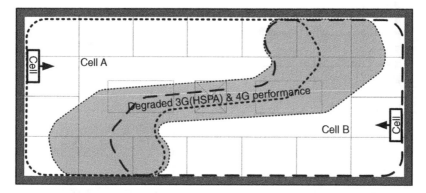

Figure 13.6 Pico cells causing 'inter-cell interference', degrading the HSPA performance in the building

The problem is that the cells will cause co-channel interference with each other ('inter-cell interference'). This will severely degrade HSPA performance in the building. There will be HSPA throughput in the overlapping area, but the speed will be limited as the two signals become more and more equal. When relying on pico cells/femtocells for indoor HSPA applications, it is of the utmost importance to use internal structures and firewalls to separate the cells and to avoid coverage overlap in areas of high traffic. With careful planning, pico cells/femtocells could be a very strong and viable tool for indoor applications, but try to avoid using more than one pico/femtocell in an 'open' environment, with high traffic density.

13.1.4 Extending Pico Cell Coverage with Active DAS

The pico base station is a very effective tool for deploying in-building coverage at low cost on a stand-alone basis. However, there are circumstances in which the coverage area of the pico base station/femtocell needs to be extended without increasing the capacity of the system, e.g. in warehouses (as illustrated in Figure 13.7) and residential building towers.

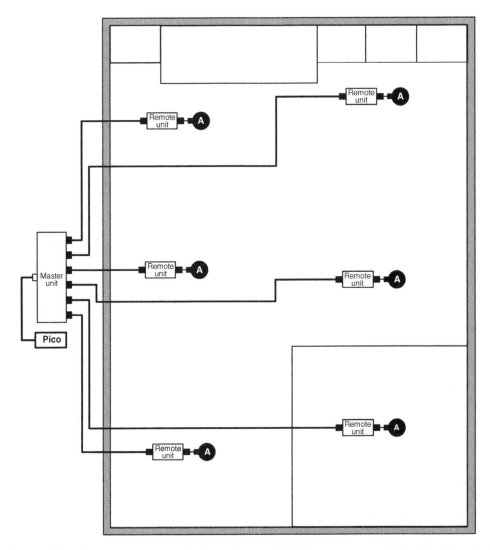

Figure 13.7   The coverage of a single pico cell can be extended, using active DAS to distribute the signal over a wider area

If you combine the small active DAS system from Section 4.4.3, you can extend the coverage footprint of a single pico cell/femtocell. The reason for selecting the active DAS is the 'no-loss concept' of the active system. The attenuation of passive distribution will degrade the performance of both the downlink and uplink.

In 2G applications, you can also increase the capacity by connecting two pico cells or more to the same active DAS, and letting traffic-controlled HOs take care of the traffic flow between the two cells. You do not want to combine two or more pico cells/femtocells for 3G or HSPA into the same DAS; this will turn the whole cell into a soft HO zone for 3G or give co-channel interference in 100% of the area and thereby degraded HSPA performance.

13.1.5   Combining Pico Cells into the Same DAS (only 2G)

In some pico cell applications, there are scenarios where a highly mobile user community within a building that has several pico base stations can overwhelm the capacity of a single pico base station system (hotspot blocking). An example of this is when employees congregate in the cafeteria or similar lunch area. This situation can be improved by taking the same number of 2G pico base station systems and feeding them into an active DAS system to cover the same area.

The results is all the radio channels being available in all of the building, leading to higher trunking efficiency, more capacity available over a larger area (as illustrated in Figure 13.7) and hotspot capacity throughout the building.

Some manufacturers of 2G pico cells let you combine the pico cells into one logical cell with only one broadcast control channel (BCCH), thus freeing up capacity on the individual pico cells and boosting the trunking efficiency even more.

This approach can also be considered for femtocells of class 3; however, with the exception of combining several femtocells within the DAS; one should only feed one 3G/ 4G into the DAS. If you combine several femtocells into the same DAS, you will create a HO mess (100% overlap) with degraded performance and interference between the cells. This will lead to poor quality and possibly dropped calls.

Only for 2G!

Do not combine two or more pico cells/femtocells for 3G or HSPA into the same DAS; this will turn the whole cell into a soft HO zone for 3G or give co-channel interference in 100% of the area, degrading HSPA performance over the whole coverage area. For extending coverage for femtocells via an active DAS (Figures 13.7 and 13.8), it is not recommended that one uses femtocells with higher capacity (class 2 and 3) to accommodate the demand for capacity in the building.

13.1.6   Cost Savings When Combining Capacity of 2G Pico Cells

There is great potential for cost saving by combining the capacity for two or three pico cells rather than deploying several more pico cells to cover the same area. This saves on the number of pico base stations needed to cover a building and provides cost savings on backhaul costs due to:

- better trunking efficiency of the pico base station (more efficient use of the deployed radios; see Section 6.1.9);
- saving on base station controller (BSC) cost (need to support fewer base stations in the same building) – each base transceiver station is an 'element' on the BSC, and subject to annual license fee, saving on transcoder and interface costs;
- saving on mobile switching center costs;
- saving on ADSL costs.

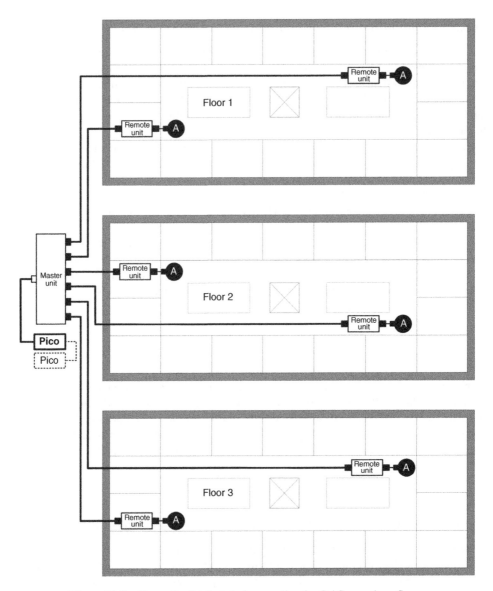

Figure 13.8   Pico cells distributed via a small active DAS over three floors

Fast deployment will provide savings because of:

- the concept of distribution of pico cells via a small active DAS providing the mobile operator with the ability to react quickly to coverage requests from customers;
- less churn;
- more revenue;
- easy and fast extension of the coverage when new areas in the building need to be added to the system.

Operational savings will come from:

- uniform coverage from the DAS system, which is easy to plan, implement and optimize;
- fewer 2G RF channels used in the building, with less interference leaking out;
- detailed alarming on the DAS, which will limit down-time.

13.2   Heterogeneous Networks (HetNets)

Today's wireless service offerings are based on multiple access technologies: 2G services for voice and data in areas where there is no other option; 3G for voice and high-speed data; and then 4G cells for even faster data speeds. In some places you will even get access to Wi-Fi areas to offload the data traffic load from the traditional 2G, 3G and 4G networks. A simplified version of the wireless offerings can be seen in Figure 13.9 where multiple layers of mobile technologies, macro and various types of small cells and Wi-Fi make up a seamless wireless network.

One of the main challenges with a HetNet like the one in Figure 13.9 is the planning and management of such complex network architecture, between several layers of radio access layers and radio interfaces. Luckily the standardization bodies are taking care of this complex task by developing the appropriate functions and features to support mobility, HOs, etc. The benefits for the mobile data users and network operators is that a very complex network with this many technologies enables the highest possible network performance and user experience at the lowest possible network investment and cost for the users.

The HetNet approach will also ensure a swift implementation of new small cells and new radio access technologies in the future, providing mobility and compatibility with existing network layers. It means a time will come when you will be able to roam from a rural area on 2G and 3G, to an urban area where you will hand over to 4G, eventually ending up on a high-speed Wi-Fi connection at home with no data loss and full mobility throughout the connection, and all this done automatically. You could even divide the data stream over several access technologies, so you could, for example, get the downlink data stream from the servicing macro cell, while the uplink from the mobile could be serviced by a local small cell.

13.3   Implementing Small Cells Indoors

Small cells are a very strong and efficient tool to solve certain challenges in the mobile network. In the following section, we will look at the femtocell network elements and some of the basics you need to consider when planning femtocells inside a building when deploying multiple small cells.

The Femtocell Network Elements

If we look at the generic femtocell network layout in Figure 13.10, we can appreciate the principle behind the network architecture The 3G femtocell network interfaces via a femtocell gateway to the radio network controller via the normal core network, ensuring 100% compatibility with mobility management and other features/functionality of the network.

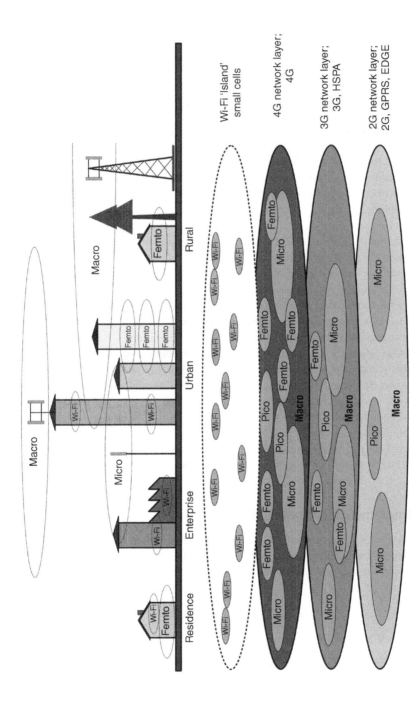

Figure 13.9 where multiple layers of mobile technologies, macro and various types of small cells and Wi-Fi makes up a seamless wireless network

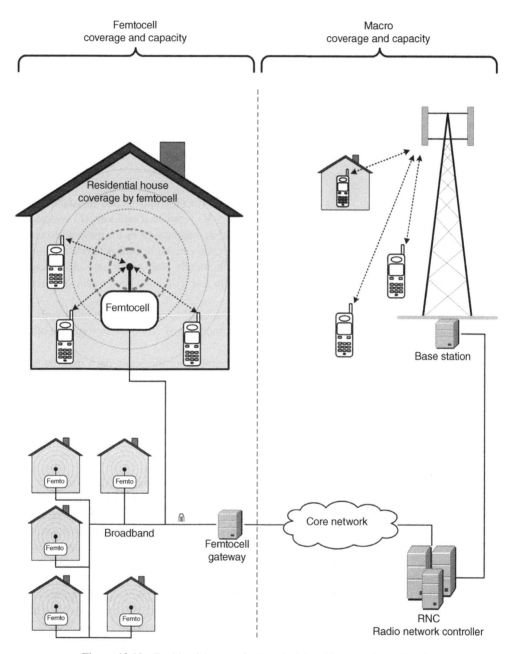

Figure 13.10   Residential area – femto principle with supporting network

The femtocells connect via standard 'carrier grade' asymmetric digital subscriber line (ADSL) or fiber to the home (FTTH) network to ensure enough bandwidth and QOS in the backhaul to maintain the service levels and functionality in the designated femto area/zone.

13.3.1 Planning Considerations with Indoor Small Cells

Let us have a look at some of the basic design considerations you face when planning with femtocells inside buildings. GPS and other location-detecting algorithms will aid the implementation of femtocells, and remember that the majority of the femtocells are installed by the users themselves in an uncontrolled environment and that poses specific challenges, as follows:

- Location of the femtocell – how do we detect if the femtocell is moved and reinstalled in a controlled way? In theory you could have a 'mobile' 3G femto that relies on 4G backhaul. That could be a challenge in any network as it could degrade the QOS for users who camp on the femto, and also impact performance of other cells and the overall network.
- Power stability – the femtocell must 'wake up' automatically after power cuts and outages. It is absolutely essential that the femtocells are stable and will come back online automatically if anything happens.
- DSL transmission instability must not cause the femtocell to interfere with other cells or network elements, and the femto must go 'off air' if there is any risk of compromising other network cells.
- Fraud – we need to ensure a secure network connection from the femto back into the network to prevent hackers from accessing the vital network infrastructure if they try to hack into the network via the DSL/fiber connection from the small cell.

Other Network Planning Challenges with Femtocells

- **Frequency and code coordination**
  Femtocells operate within the licensed band of the mobile operators and therefore have to be considered in the overall network deployment planning. Frequency planning and code coordination are obviously parameters that are important to ensure performance in the femtozone but also to avoid that the femtocell potentially generating interference in surrounding macro or other small calls in the vicinity. Both co-channel and adjacent channel interference are a concern, in particular when deployed close to existing cells in the mobile network.
- **Link budget for femto cells**
  Like any other RF system, you will need to do a link budget when planning with femtocells. Often you will plan with a few selected types with the same output power, noise figure and capacity, and will be able to generalize their coverage range in certain area types. With these general guidelines, you can create a set of 'rules of the thumb', e.g. that femto type 1 will have x meters of coverage in a typical residential building, y meters range in an open office environment, and z meters in a dense office building. Refer to Chapter 8 for more detail on the link budget and see Section 9.7.5 for an easy-to-use mobile planning tool for small cell deployments indoors.

I believe that users relying on the service via self-deployed femtocells will be more for-giving in terms of coverage issues. For them, the femto looks and feels pretty much like a Wi-Fi access point, and most know from experience that there is a limited range of 20–30 m in their house, and that they will need to move the access point to a certain location to get service throughout the house, similarly to the femtocell they installed themselves.

Bear in mind that users might implement femtocells even in areas with high levels of nearby macro sites. This is obviously not to improve the coverage level, but to provide more dedicated capacity and functionality in the 'femtozone' as described in Section 13.1.2. This will pose a special challenge when the femtocell needs to overpower a high signal from a nearby macro cell.

- **Capacity planning for femtocells**
  When designing solutions based on femtocells, it is obviously important to consider the capacity planning as well as the coverage design. We can use the exact same approach as described in Chapter 6. We do, however, need to be very careful with 'hotspots', i.e. areas with a high concentration of users. Femtos' biggest weakness is the lack of ability to share capacity and trunking the resources, as we can in a DAS. With femtocells you need to know the exact physical location of the 'hotspot' and that is not easy in, for example, a conference or exhibi-tion center, sports arena, or airport (see Section 5.4.1 for more details about 'hotspots').

- **Practical deployment concerns using femtocells**
  Like any RF system, the location of the antennas, in this case the femtocells, plays a big role in the performance. Just like normal radio planning we need to make the antennas, in this case the femtocell, 'visible' for the users (see Section 5.3.4 for details).

  Other practical concerns also need to be considered, e.g. the power supply – should we use a local AC outlet , in which case we need to accept that if a fuse blows there is no cov-erage? Do the users inside the facility expect power back-up for such occasions? It is expen-sive to distribute a new AC power group/grid in the building solely to power a cluster of femtocells. Some femtocells use power over ethernet (POE) or power over fiber (POF). This approach makes it possible to centralize the power supply and any battery back-up needed and cuts down on small cell solutions where you deploy clusters of femtos in a facility.

  As these femtocells are, to a large extent, meant to be 'install yourself' items, challenges with configuration and deployment are apparent. But in the end they are an extension of the mobile network, and the mobile operator will ultimately be held responsible for the key performance indicators of the network.

- **Distance to other RF-sensitive equipment**
  Do follow the guidelines from the specific vendor, and remember to maintain a 'safety dis-tance' to other RF equipment, such as other femtocells from other networks, wireless access points, wireless passive infrared detectors (PIRs) and other 'alarm' equipment. Normally I would recommend a minimum of 100 cm, but consult the guidelines of the specific vendor.

13.4   Planning Examples with Femtocells

Let's have a look at a few typical deployments of femtocells on actual floor plans and appli-cations. This is a good way of exemplifying some of the concerns and issues from a radio planning perspective

13.4.1   Small Office Space

The first example is shown in Figure 13.11, a small office space of about 30 × 30 meters in a fairly open environment. The center section consists of 'office cubes' with no actual walls, and along the perimeter of the office space are management offices with light walls and glass walls facing into the center of the building.

The obvious thing would be to place the femtocell in, or near, the center of the floor space as shown in Figure 13.11. This might work fine but you should be aware of the following potential pitfalls:

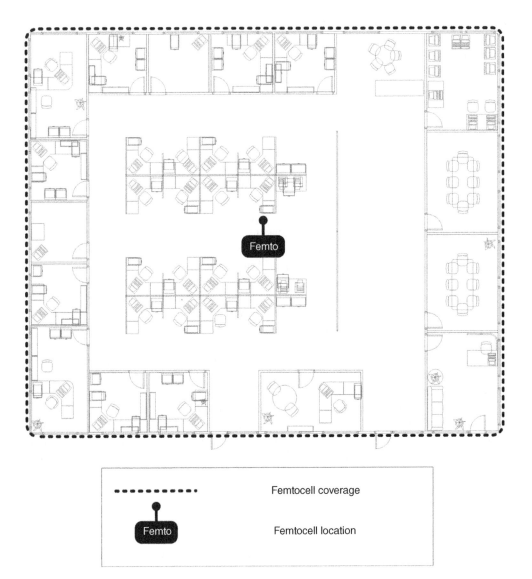

Figure 13.11   Small office single femtocell deployment

- Make sure that the femtocell can dominate all of the floor space, so that you do not end up with soft HO issues or macro interference along the perimeter, or with a 'corner office problem' (see Chapter 3.5.3 for details).
- How about the capacity? Will one femtocell be sufficient to cope with all the users in the office? Verify the traffic calculations and ensure that the femto and backhaul can support the requirements (see Chapter 6 for details).
- Provision for future upgrades – if you need to add more capacity what is the plan? Deploying multiple femtocells to add capacity could result in extensive soft HO areas in this type of open environment.
- Make sure you have a symmetrical design to adjacent floors, as described in Section 5.3.9, to avoid leakage to those floors.

13.4.2   Medium-sized Office Space

Another example is shown in Figure 13.12, a medium-sized office space of about 30 × 40 m in a medium-dense environment with light plasterboard-type medium walls and relatively limited loss.

Although the capacity load in this office space might not exceed the capacity resources of a single femtocell, we need to deploy two small cells to ensure sufficient coverage throughout the floor. Therefore the motivation for deploying two femtocells is coverage rather than capacity.

We need to aware of the following potential pitfalls in the design shown in Figure 13.12, in order to get the best performance experience for users:

- Be sure to utilize the structure and walls of the building to minimize the HO zones, as shown in Figure 13.12.
- Make sure to use the hallways to distribute the service from the two femtocells to maximize the performance, and make the RF signal from femtocells 'visible' to the users to ensure the best possible coverage.
- Make sure that the two femtocells dominate the whole floor space and that you do not end up with soft HO issues(3G) or macro interference along the perimeter, or with a 'corner office problem' where you do not dominate the full floor space, or interference along the windows (see Section 3.5.3 for details).
- In the case of a multi-story building, make sure your design and implementation are symmetrical to those on adjacent floors, as described in Section 5.3.9.

13.4.3   Large Office/Meeting Space

The last example of using small cells inside a building is given in Figure 13.13, a large office space (about 40 × 60 m) with meeting and conference rooms in a medium-dense environment with plasterboard-type medium walls. As shown in Figure 13.13, we initially tried to deploy four small cells (Femto 1–4, black rectangles). At the point of planning we were unaware that there were several important conference areas and VIP meeting rooms that we needed to take into account (this is one reason why a site survey is important). As you can see on the floor plan in Figure 13.13, this initial design led to a soft HO zone right through one of the VIP meeting rooms – the boardroom actually! Having realized this, we were forced to relocate femto 3 and 4.

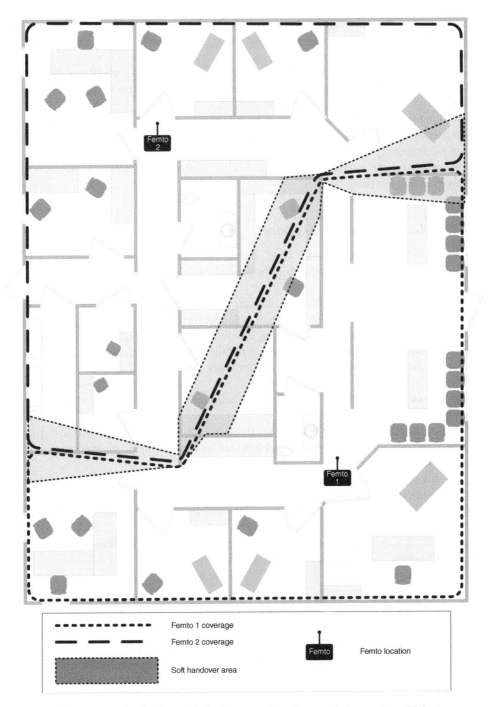

Figure 13.12 Medium-sized office, with deployment of two femtocells to provide sufficient coverage and capacity

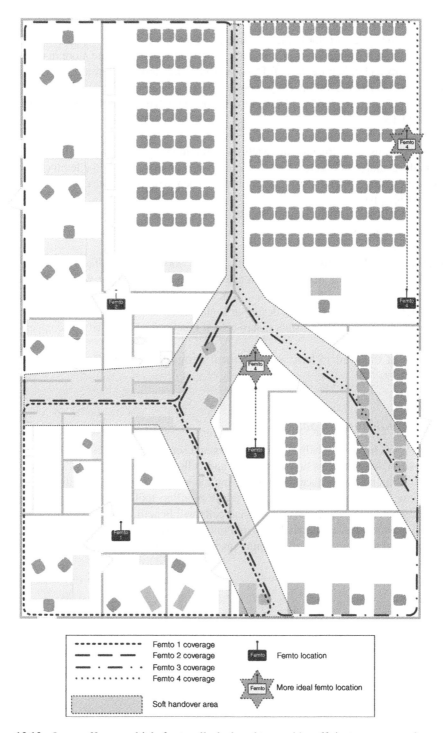

Femto 1 coverage
Femto 2 coverage
Femto 3 coverage
Femto 4 coverage

Soft handover area

Femto location

More ideal femto location

Figure 13.13 Large office – multiple femtocells deployed to provide sufficient coverage and capacity

Femto 3 was moved up the corridor to a new location (marked with the star) to make sure it could reach down the corridor to the boardroom. At the same time we moved femto 4 further into the center of the large coverage area, allowing femto 3 to dominate more in the two VIP meeting rooms and offsetting the soft HO zone to the area around the wall between the large conference room and the VIP meeting rooms.

As radio planners, we need to aware of the following potential pitfalls regarding the design in Figure 13.13:

- Make sure to use the structure and walls of the building to minimize the HO zones. As we saw in the example in Figure 13.13, we actually had to move two of the femtocells to solve a problem with a soft HO zone. This could have been avoided had we considered this issue in the design phase.
- Use the hallways to distribute the service from the two small cells to maximize the performance, and make the femtocells 'visible' to the users, ensuring a uniform distribution of the coverage.
- Make sure that the femtocells dominate the whole floor space and that you do not end up with soft HO issues or macro interference along the perimeter, or with a 'corner office problem' (see Section 3.5.3 for details). It is apparent that the topmost left corner and the lower right corner of the revised design in Figure 13.13 are prone to isolation issues with nearby macro sites, or even distant macro cells if this floor was located in the upper reaches of a high-rise building.
- Be aware of the need to have a symmetrical design to adjacent floors, as described in Section 5.3.9.

13.4.4   Final Word on Small Cells

I hope that this chapter has helped you to understand some of the basics of small cell deployments and HetNets. There is no doubt that small cells are a very important tool for solving many of the challenges we face when designing indoor solutions to cope with both coverage and capacity. They also have a role as a future-proofing tool for many applications, both indoors and outdoors. Sometimes these small cells will benefit by being extended with DAS solutions and sometimes as stand-alone deployments. In Section 9.7.5 you can see an example of a very useful tool for small call planning inside buildings.

14

Application Examples

The intention of this chapter is to bring all the preceding information and knowledge from the book into a few typical examples, and to try to address some of the main considerations and concerns for specific applications.

For more details regarding detailed link budgets, noise and capacity calculations, etc., please refer to the specific chapters covering these details.

The most typical applications will be covered with the highlights, main points, and a checklist to help you remember the key things you should consider. Obviously you can 'stitch' the knowledge together to do some indoor radio design by using the various chapters in the book, but I thought it would be good to apply some of my real-life experience into some examples, in various sizes.

We will not cover all details of the individual designs, but I will highlight some of the features that are specific, and sometimes unique, to the application. Take the following application examples as general ones, and then apply all the knowledge from the previous chapters as well as your own experience.

14.1 Office Building Design

Office buildings represent the bulk of the indoor planning designs we implement. These buildings can frequently be considered as 'standard' designs, and they are often of similar structures, wall types, and, to some extent, layouts. There will obviously be differences between offices in old traditional type buildings, and modern new constructions.

Indoor Radio Planning: A Practical Guide for 2G, 3G and 4G, Third Edition. Morten Tolstrup.
© 2015 John Wiley & Sons, Ltd. Published 2015 by John Wiley & Sons, Ltd.

14.1.1   Typical Features and Checklist for Office Buildings

In addition to the standard considerations, there are a few important things to remember:

- They are often heavy users of voice and data.
- Remember to 'profile' the users in terms of data and voice load in your actual design.
- Do you need Wi-Fi offload?
- For large office buildings, remember to provision for more sectors in the future, to be able to add capacity.
- Is there metallic coating on the windows? This could help with isolation between macro and indoor.
- Should elevators and stairwells be included?
- Any meeting and conference rooms that needs special attention?
- Any management offices and areas with high load?
- How about wireless services in IT and technical rooms?
- Any sub-level service and parking areas that need to be included?
- Do we need a multioperator solution?
- Any other radio services we need to plan for or take into account, like 'Tetra' and other private mobile radio (PMR) services?
- Do we need to consider future upgrades to the system?

14.1.2   Small to Medium-Sized Office Building

A typical floor plan for a small to medium-sized (single floor) office building can be seen in Figure 14.1, in this case showing the typical antenna locations you would normally consider. You would typically select one or more of these locations depending on the requirements and the present signals in and close to the building. You would usually avoid a single antenna solution (location 1), unless you were designing for a building with really good isolation i.e. a high attenuation los to/from the outside macro environment.

One concern is the fact that the single antenna (location 1) would probably need relatively high power to reach the corners of the structure, and remember that the mobiles in those areas are typically the heavy users (managers) (see Section 5.3.8). There would also be potential electromagnetic radiation (EMR) concerns close to the antenna due to the high power required.

If you have a typical modern office building with 'standard' tinted windows (metallic coating for sunscreen), there will normally be a relatively good degree of isolation (attenuation through the windows) of 20–30 dB, sometimes even higher, and you can probably use antenna locations marked as '2' in Figure 14.1. These antenna locations will ensure a uniform coverage level and good performance across the floor space. Location 2 is probably also the best installation option, as it is in the walkway, which is unlikely to change in the future and be affected by new walls and so on. Typically there will also be installation trays and ducts over the suspended ceiling for 'easy' installation of the required DAS cables.

In cases where you have strong macro signals from the outdoor network and need to dominate the inside of the building, antenna locations 3 in the corners of the building (Figure 14.1), using corner-mounted directional antennas on one or more sides, could help you address this challenge.

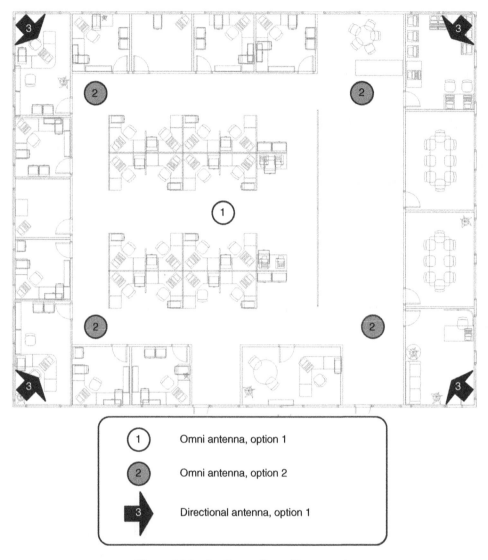

Figure 14.1   Small – medium office example

The advantages with these corner antennas are that you will dominate with strong signals inside the building and still, to some extent, control the leakage from the indoor DAS. The challenge is often the installation of the antennas and the cable needed. In extreme cases, with really high signals from a nearby macro, you might need to supplement the corner antennas with directional antennas at the perimeter walls between the two corner antennas, facing inwards. For larger floor spaces, you will probably combine the antenna locations to achieve good isolation at the perimeter with the location 3 antennas, and then cover the center of the building with the locations 1 and 2.

Small Cell Alternative

For these small to medium-sized office buildings, with single levels and relatively few users, a small cell solution could be a good alternative by deploying one or more femto-type cells. Please refer to Chapter 13 for more details on designing with small cells.

14.1.3   Large Office Buildings

Besides the considerations described in the previous section, large office buildings (high-rises) will have specific challenges and pitfalls to watch out for. One of the main considerations for large office buildings, besides the coverage challenges described in the previous section, would be to ensure enough capacity and also to prepare options for future upgrades of capacity.

In Figure 14.2 you can see a versatile strategy when, from day one of the initial design, the DAS design is divided into sections that can be combined in any way you want to create areas of logical sectors. With this approach you can initially decide to configure the DAS as one sector for the whole building, and then over time, as the capacity load increases, divide it in to more and more sectors. In Figure 14.2 we have a 32-floor high-rise; the DAS is designed to support sections of two floors. We can then combine these sections of DAS into the sector/cell configuration that will support the given capacity load. In the example, we can start with configuration 1 and simulcast one cell across the 16 sections of DAS covering all 32 floors. This is a good starting configuration if the capacity load is low in the building. Over time you can upgrade the capacity by dividing the DAS in the building into more sectors; DAS configuration

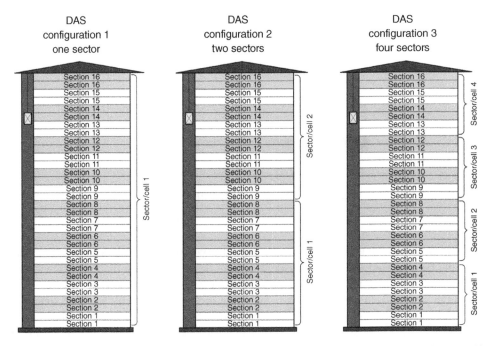

Figure 14.2   Large office building designed in coverage areas that can be combined into various combinations of sectors. This is a good strategy because you can divide the DAS into more sectors in the future

2 will give you double the capacity with two sectors, or you could upgrade to DAS configuration 3, which will give you four sectors, representing a factor of four in capacity compared with configuration 1. In this example you have provisioned for 16 sections, so you could have up to 16 sectors in the maximum configuration.

If it makes better sense in your design, you do not have to lock the sector configurations to be the same for 2G, 3G, and 4G. You can have one sector plan for 3G and another for 4G if you want in the same DAS. In other words, it is very easy to upgrade in the future, if you consider this approach from the outset of the initial design. If you are designing for a multioperator DAS, the different operators could have different configurations, so it is indeed a flexible approach.

Please refer to Section 5.5 for more details on strategies you might use for cells inside the building.

The High-rise Problem

As described in Section 3.5.3, the 'high-rise problem' in the uppermost part of a high-rise building, which is a result of interference from the outdoor network, can be a big concern. Even distant base stations from the outdoor network can 'hit' the users in the upper reaches of a high-rise, degrading the performance and causing dropped calls. So be careful and establish the correct design levels, and please do refer to Section 3.5 for more details on solving this problem.

Elevators

High-rise buildings will obviously have elevators, and the concomitant challenges of signal levels inside the elevator cars. You should aim to have the same sector, even a dedicated elevator sector like the one shown in Figure 14.2, servicing the elevator throughout the high-rise building with no handover (HO) challenges. Achieving sufficient signal inside the elevator car is indeed a challenge, and several options can be considered. Please refer to Section 5.6 which addresses the topic of elevator coverage in detail. It will be a good idea to use the same sector to cover the vertical elevator in a high-rise with multiple sectors in order to avoid HO issues. However, it depends on how you can 'isolate' this vertical sector in the building (as discussed in Section 5.6).

14.1.4   High-rises with Open Vertical Cavities

We have already covered the main challenges of high-rise buildings in Section 3.5.3 and earlier in this chapter (Figure 14.1, corner mounted antennas). In addition, we often see high-rise buildings with vertical open cavities from ground floor and up to several, sometimes all, floors above (see Figure 14.3).

As we addressed in Section 5.3.9, it is not recommended to 'interleave' the coverage footprint between antennas on adjacent floors to achieve the required coverage levels in the building. In high-rise buildings with open vertical cavities between multiple/all floors, in particular, we need to be careful. It is very likely that we need to deploy multiple sectors, vertically separated in the building, and over time divide the building into even more sectors to keep up with capacity demands, as described in Section 14.1.3.

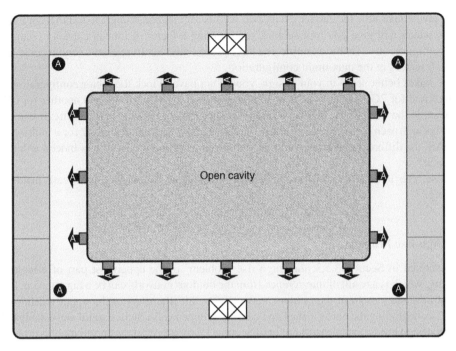

Figure 14.3 Large high-rise building – open cavities and antenna placements using many low power and symmetrically placed antennas to limit 'spill-over' between the floors

The main challenge with these buildings with an open cavity is to avoid (limit) the leakage/spillover between adjacent floors. You need to use more antennas with less uniform power and have a symmetrically distributed cluster of antennas between the adjacent floors. Another option is to supplement these antennas with directional antennas facing away from the cavity, to ensure strong dominance on the actual floor with very limited spillage to other floors.

It is recommended to use all possible physical elements of the building structure to minimize spillage of signal to other floors through the open cavity. Hence, use the concrete beams and pillars as antenna locations for the directional antennas to shield the back lobe from these antennas even more, as shown in Figure 14.3, where the directional antennas are mounted on pillars. It is also recommended to make sure the building is using metalized windows with high RF attenuation, in order to ensure that the leakage from the DAS does not create any problems in the outdoor area. Otherwise, antennas facing inwards along the fringe toward the center of the building should be considered as an alternative – perhaps even a combination.

14.2 Malls, Warehouses, and Large Structure Design

Large shopping centers and malls (Figure 14.4) are typically also on the top priority list of buildings where mobile operators would like to have excellent mobile service for both voice and data. The owners of the shopping center will also want to make sure that shoppers are experiencing good wireless services, otherwise they might go elsewhere to shop. Often there is a need for a multioperator, multi-technology (2G, 3G, and 4G) solution, and you will frequently find there is an existing Wi-Fi system or systems already in place.

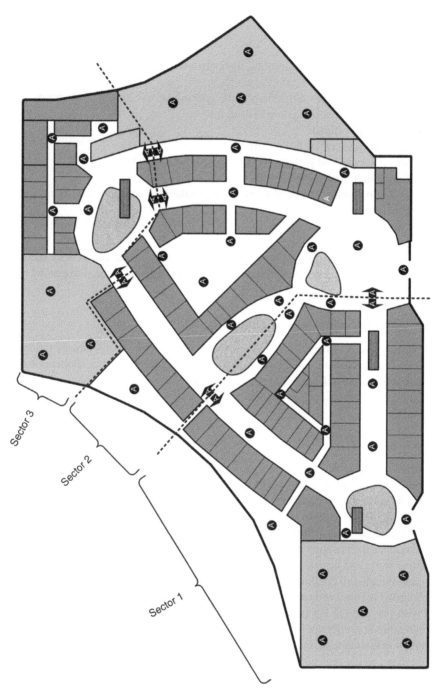

Figure 14.4 Large shopping mall example showing a single level with open cavities, internal streets and a few large prime shops divided into three sectors on the level shown. Other levels might have different layout and sector plans

14.2.1   Typical Features and Checklist for Malls, Warehouses and Large Structures

In addition to the standard considerations, there are a few important things to remember:

- Do plan for future upgrades, extensions to the system, more sectors and new bands, space in equipment rooms and racks – you will need it.
- Do you need Wi-Fi offload?
- Should elevators and stairwells be included?
- Any sub-level service and parking areas that need to be included?
- Do we need a multioperator solution?
- Any other radio services we need to plan for or take into account, like 'Tetra' and other PMR services?
- Include 'back office' areas – there is often a big portion of the potential revenue in those staff areas.
- Don't be tempted to service outdoor areas from the same sectors that service the indoors. You are likely to pick up noise and interference (uplink, UL) from the surrounding network

14.2.2   The Different Areas of Shopping Malls

Large shopping malls will typically have some common features and considerations. With reference to Figure 14.4 we can take a brief look at some of the typical challenges and features, such as food courts, large prime shops, parking areas, open areas, cinemas, open vertical cavies between floors and open skylights.

Antenna Locations

Quite often you will be restricted mainly to installing the antennas in the 'streets' in the mall, as in Figure 14.4. Omni antennas would typically be a default choice, especially installed in the intersections of the streets to maximize the footprint. Sometimes you would consider sector antennas mounted 'back-to-back' as a viable solution when you need multiple sectors on the same floor in an open area like a shopping mall. Try to use the physical structures in the mall to maximize the isolation between the different sectors. This could, for example, be using the concrete walls and beams in the border areas between sectors to maximize isolation and help improve single cell dominance. In areas where it is not possible to use directional antennas 'back to back', it is recommended to use symmetrically placed omni antennas (one in each sector) with just enough RF power to stay within design requirements. Minimizing the RF power is required in order not to 'spill over' to the adjacent sector/area, but it still needs to be high enough to maintain the coverage key performance indicator (KPI) in the area.

Although we would typically try to limit the antenna installation to the 'streets' of the mall for easy access and permission to deploying antennas, we sometimes need to have antennas inside the large prime shops to achieve the required service inside these large shops.

We need to place antennas at the entrance of the mall to make sure we dominate any macro signals, ensuring the users will camp on the indoor DAS as soon as they are inside the mall, or even just before they enter it (but not overspill to outdoor area). If we make sure that the soft

HO zone (3G) is not extended into the mall, we can limit the noise load on the UL. The potential of unwanted increase of the UL noise is due to the fact that if you 'stick' on the macro inside the mall, the mobile will have to power up the transmit power to reach the outdoor macro and potentially hit your high-capacity indoor system with noise load on the UL, thus degrading the performance and capacity of the indoor DAS.

Skylights

Many malls often have a large skylight to let in daylight. These large open windows in the roof also gives the nearby macro network base stations a chance of reaching users inside the mall. It can be tempting to place a nearby macro and then 'blast' a signal into the mall, thus saving an indoor system – this is a classic mistake. Over time you will realize that the traffic load from the mall will drain the outdoor macro, and that the service inside the mall needs to be better than the macro can provide. You will then end up installing a DAS inside the mall anyway. The problem is that now you need to overcome a high signal in some areas of the mall from the nearby outdoor macro. Ideally you should shut down the outdoor macro base station once you have implemented the indoor DAS in the mall; however, it is likely the outdoor base station also serves other areas outside the mall so you cannot switch it off or realign the antennas without impacting the users outside, perhaps even in other buildings. The only way out is often to raise the design levels inside the mall, installing many antennas in a costly solution to patch the dominance problem and ensure 100% indoor dominance from the DAS to avoid soft HO and quality issues. Sometimes it is possible to install directional antennas at the edge of the large skylight facing into the mall, to overcome a high signal level from a nearby macro. Make sure to use all possible structural elements to block the signal from the indoor DAS from leaking to the outdoor area and creating interference. This will also prevent these antennas from picking up too much UL interference from the outdoor network and impacting the capacity of the indoor DAS.

Do Not!

The previous section clearly shows that you should not be tempted to use a nearby macro to cover a large mall. You could perhaps us a temporary 'cell on wheels' (COW) during the construction phase, but you should plan to install a DAS to be on-air from day one of the opening of the mall. This is the best way to ensure coverage, capacity, and the ability to upgrade in the future.

HotSpots Inside the Mall

Like other large buildings there will be areas of expected high usage of wireless services and data load. These areas need special attention in terms of ensuring the best possible mobile signal service, high quality and best data speeds. The obvious hotspots for data usage would be the food court, and café and restaurant areas. Sometimes there will be administration offices, meeting rooms and the like which might need extra attention and this should be included in the design of the DAS.

Open Vertical Cavities

Large shopping malls will often be multi-level buildings, and frequently with open vertical cavities between multiple floors. Here you will typically also find escalators and stairs. You need to pay attention to these areas when dividing the building into multiple sectors vertically. Often you will have different sectors on different floors. Carefully evaluate how this will be practically applicable and how to limit the leakage between the floors. Make sure to limit the HO areas – in extreme cases you can apply same strategy as described in Section 14.1.3.

Indoor Parking Areas

If indoor parking is part of the project, you would typically get by with relaxing the design requirements, especially in sub-level areas. Perhaps only voice and slow-speed data will be needed (3G). You might save considerable costs if you don't have to plan for 4G and MIMO in these typically low load areas. However, you need to check this in the specification phase and ensure you are aligning the design to the exact requirement and expectations of the users.

Administration and 'Back Office' Areas

Pay attention also to the areas to which the typical shopper will not have access, e.g. the administration offices, and the storage and service areas. These areas might only require a few extra antennas and DAS equipment to be covered, but they will be very important for the staff, and perhaps help you ensure an extra boost in traffic and revenue for the implemented solution.

Entertainment Areas

Some malls feature indoor water parks, ice rinks, amusement parks, cinemas or other entertainment areas. You will need to tailor the solution to the specific needs, requirements, and expected data loads.

14.3   Warehouses and Convention Centers

Large open halls such as convention centers and warehouses would often be very heavily loaded with traffic during events or have no load at all. Another challenge is the open nature of the halls: how do we solve the problem of multiple cells in this open environment and provide sufficient isolation between the cells at the same time.

Let us have a look at the example in Figure 14.5 and see if we can address some of the main challenges. In this example we have four open halls: A, B, C and D. In fact, there is the possibility of combining these halls individually, or even opening all the halls into one big open area using a 'sliding wall section' between the individual halls.

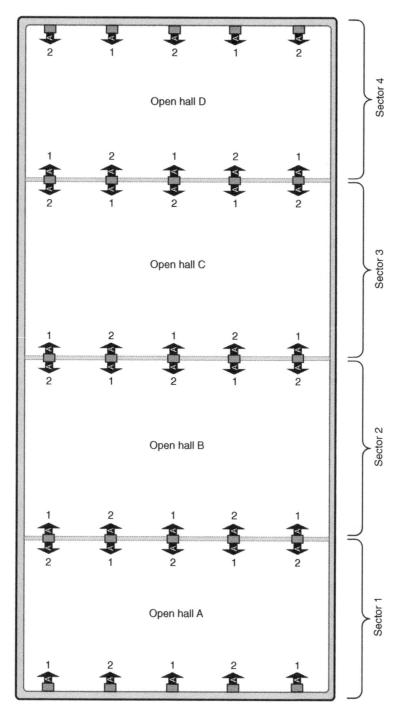

Figure 14.5 A large open convention center with open halls – back-to-back antennas are deployed to maximize isolation between sectors

14.3.1   Typical Features and Checklist for Warehouses and Convention Center DAS Deployments

In addition to the standard considerations, there are a few important things to remember:

- It is advantageous to use a fiber DAS, so you can distribute coverage and capacity from one central base station hotel in a large convention center.
- Be careful when feeding signals between multiple buildings – you will need to use fiber between the buildings to ensure galvanic isolation, and prevent severe grounding issues.
- Do you need Wi-Fi offload?
- Any sub-level service and parking areas that need to be included?
- Do we need a multioperator solution?
- Any other radio services we need to plan for or take into account, like 'Tetra' and other PMR services?
- Any 'hotspot' buildings such as conference/meeting areas that need extra capacity, such as press areas administration offices?
- Do we need to consider future upgrades to the system?

High Capacity and Performance in Open Areas

The challenge is that we have four open halls, as shown in Figure 14.5. These can be joined by sliding walls in-between Halls A and B, Halls B and C and Halls C and D. We need to provide high signal levels, high data capacity, and multiple sectors/cells to support the capacity.

If we utilize the 'divider beam' (typically a large metal or concrete beam supporting the structure between the hall sections, supporting the roof structure) and deploy back-to-back directional antennas, we can isolate these halls and open areas with individual sectors/cells. We could, if we wanted to, simulcast all four halls as one large sector/cell if the capacity load was relative low and easily supported by a single sector. Alternatively, we could, as in this case, divide the halls into four individual sectors/cells for more capacity, as shown in Figure 14.5.

You can apply the same strategy as used in stadium design when designing for multiple sectors in an open area, maximizing the isolation in open areas (see Section 14.6.8 for more details and guidance).

MIMO in the Open Halls

In case we need MIMO performance, we could, in this case (due to the line of sight to all antennas in this open area), use every other antenna for MIMO 1, and the other antennas for MIMO 2 – thus providing MIMO performance in a majority of the area, for the deployment cost of a SISO system. You will need to ensure that each of the MIMO antennas has a full coverage footprint, as this approach will only support full MIMO in the areas where both antennas fulfill the design requirements.

It is very likely that there will also be a Wi-Fi system implemented to support the heavy data load during conventions and similar events. For large exhibition centers, with multiple halls spread across a large campus-type area, base station hotels should be considered (see Chapter 12).

14.4   Campus Area Design

Campus areas refer to clusters of various buildings spread across a larger geographical area. Typically this could be a factory complex with various building types of various functions, such as administration (office) buildings and production buildings (large open halls). Campus areas could also be large hospitals or universities. You could therefore consider each of these building types individually, as per the applications in this chapter.

Therefore I will not spend too much time describing campus deployments, as they are actually just multiple buildings scattered over a specific area. Most of the individual building types and challenges are already described in this chapter. However, there are some unique considerations to take into account when deploying campus-wide DAS solutions.

14.4.1   Typical Features and Checklist for Campus DAS Deployments

In addition to the standard considerations, there are a few important things to remember:

- It is advantageous to use a fiber DAS, so you can distribute coverage and capacity from one central base station hotel.
- Be careful when feeding signals between multiple buildings – you will need to use fiber between the buildings to ensure galvanic isolation, and prevent severe grounding issues.
- Do you need Wi-Fi offload?
- How about wireless services in IT and technical rooms?
- Any sub-level service and parking areas that need to be included?
- Do we need a multioperator solution?
- Any sub-level tunnels between the multiple buildings that need special attention?
- Any other radio services we need to plan for or take into account, like 'Tetra' and other PMR services?
- Any 'hotspot' buildings like conference/meeting areas that need extra capacity?
- Do we need to consider future upgrades to the system?

14.4.2   Base Station Hotels Are Ideal for Campus DAS

Utilizing all the benefits of shared resources and centralized base stations, sharing a common DAS (as described in Chapter 12) is appropriate for campus deployments. For campus solutions you would often be able to provide a very efficient solution that is swift to install, and that will be value for money if you utilize both outdoor and indoor DAS. Please refer to Chapter 12 for more details on outdoor DAS. You would be able to cover a lot of the buildings and general areas from outdoor DAS nodes, and then implement indoor DAS in the more demanding buildings, utilizing simulcast of the same cells in some areas for both outdoor and indoor DAS, avoiding/minimizing HO areas and challenges between indoor and outdoor areas.

Utilizing a central base station hotel for all the services will also make the total solution very scalable, upgradable and future-proof. In many campus projects, you will find there is relative easy access to existing fiber or cable ducts between the various buildings, centralized technical rooms that can be used as the base of your DAS design.

14.5   Airport Design

Airports are very strategic structures for the mobile operator, and often the most important indoor DAS you will design in your network. International airports will require special attention owing to the number of potential roamers from abroad you can 'stick' to your network if you capture them upon arrival. These roamers are typically high-revenue users, both inside the airport and, in particular, when using mobile services 'landside', after arrival.

Airports are notorious for an ever-changing series of refurbishments and extensions. Therefore you need to ensure that your DAS design and strategy can cope with future demands as well as solve the current challenge. Although the main targets of deploying a DAS in the airport are the many passengers in the airport itself, you should keep in mind the staff areas. You might have 30 000 passengers per day, each with an average time in the airport of two hours, representing a total of 60,000 potential hours, but with a staff of, say, 3000 people, all of whom will be in the airport for eight hours per day, you are looking at a total of 24 000 potential hours. If we then take into account that the staff are using their mobile an average of 30 minutes/ day, and the travelers perhaps only four minutes, you have a potential revenue of 90 000 call minutes from the staff and 120 000 call minutes from the passengers. It is clear, then, that there are good reasons to include the 'back office' areas for the airport staff as well as the passenger areas.

Airports will typically be multi-operator solutions with all the associated challenges, mainly to do with agreeing on sector plans, design requirements now and in the future, commercial agreements etc., as well as political challenges between the operators and sometimes with the airport.

14.5.1   Typical Features and Checklist for Airports

In addition to the standard considerations, there a few important things to remember:

- Be sure to plan for future upgrades, extensions to the system, more sectors and new bands, space in equipment rooms and racks – you will definitely need it.
- Do you need Wi-Fi offload?
- Should elevators and stairwells be included?
- Any meeting and conference rooms that needs special attention?
- Any sub-level service and parking areas that need to be included?
- Do we need a multioperator solution?
- Any other radio services we need to plan for or take into account, like 'Tetra' and other PMR services?
- Remember special attention to frequent flyer lounges and other VIP areas.
- Include 'back office' areas – there is often a big portion of the potential revenue in those staff areas.
- Don't be tempted to service outdoor areas from the same sectors that service the indoors, you are likely to pick up noise and interference (UL) from the surrounding network.
- If possible most of the major active equipment should be placed 'landside' so you don't have to pass security for access.
- Any airport hotels or other accommodation in the airport campus that should be considered?
- Is it, or will it in the future be, a multi-terminal solution?
- For large airports, is there a 'shuttle train' between terminals, sub-level, that needs a 'tunnel DAS'

14.5.2   The Different Areas in the Airport

Let us have a look at the few highlights for the different areas and see how we can address the typical challenges and ensure best performance and value for money.

Sector Strategy

A sector plan for a typical airport can be seen in Figure 14.6; it is likely that you will not need to divide the system into this many sectors from day one, but you should provision for this type of sector strategy from the start of the design process, implementing your DAS in several sections/areas that can be split or combined. You may not start with six sectors as shown in Figure 14.6, but if you are already up-front in your design phase, make sure to plan for

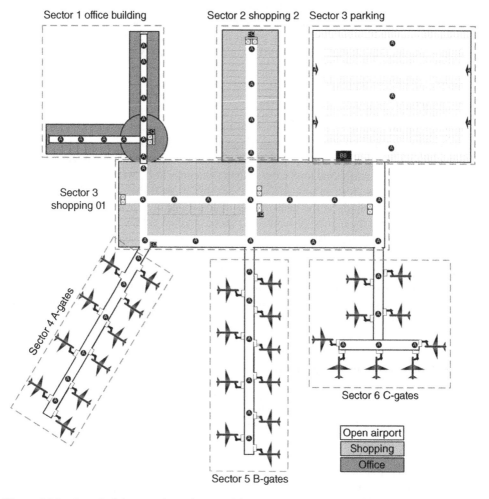

Figure 14.6   A typical layout of an airport, with gate areas, shopping, office building and nearby parking

sectorization in the future by planning different logical areas. You can then easily start by 'simulcasting' several areas using a few cells, but as the need for capacity arises in the future, you will be able to split these combined DAS areas into several sectors as necessary.

You should ensure that from day one this DAS area separation is utilizing the structural elements between the different areas of the airport to achieve the best possible isolation between the future cells, after you 'split' to more cells in the future. Normally you would have the same sector layout for all the services, so 2G, 3G and 4G would have the same sector plans. However, this does not need to be the case – you could, for example, have three sectors for 2G, three sectors for 3, but six sectors for 4G if you are designing for a 'data heavy' solution.

Different 4G Bandwidths in Different Areas

For 4G you could decide to use a narrow bandwidth in sector 3 (Figure 14.6), e.g. 5 MHz in the parking area where you are more coverage-driven than capacity-limited, and then design with a 20 MHz 4G channel in the sectors covering the capacity-heavy areas in the terminal. This will give you 6 dB more RS transmit power in the parking area using the 5 MHz sector, compared with the sectors supporting 20 MHz (refer to Section 8.3 for more details).

The Gate Areas

As shown in Figure 14.6 we try to place most of the antennas as close as possible to the gate waiting lounge. The rationale for this approach is standard 'hotspot planning' (see Section 5.4.1 for more details). You might be able to use the same number of antennas, achieving the same KPIs in terms of minimum signal levels, but with the antennas in-between the gates. However, having the antennas as close as possible to the seating areas at the gates will ensure good service for the users waiting at the gates, using their data-heavy devices. Thus the users in these areas are assured the highest possible data speed, and the DAS will therefore be able to carry much more data through the system, improving both user experience and revenue for the mobile operators.

Gate areas for international flights require special attention and focus. These gates will typically open one to two hours prior to departure, so passengers will often have more time to sit down and use data services, empty their email boxes, download the needed files and so on prior to their flight.

Outdoor Antennas on the Bridge to the Plane

Airlines are relaxing the restrictions on using mobiles on board, even sometimes allowing passengers to use them after boarding and before take-off. You might therefore be considering providing strong signal levels close to the planes when they are parked at the connecting bridge. The other important reason could be that you want to capture the roamers as soon as they switch on their mobiles, and ensure they have good service throughout the airport and stay on your network whilst they are visiting your country.

If you can get the needed approvals and permission, it could be very tempting to place small outdoor antennas at the connecting bridge, or close to the plane. But you must remember the possible impact on the overall performance: if you connect these antennas to same sectors as the high-capacity sectors inside the airport, then the risk is that this outdoor antenna will pick up UL interference/noise from the outside network/environment. On the downlink (DL) you should consider the potential interference leakage from the high-capacity indoor cells, and the impact on the outdoor network and users.

VIP and Business Lounges

In airports, the hottest of hotspots would be the VIP/frequent flyer lounges. Here you will have a high concentration of the heaviest users in the airport. These will be business travelers with high demands for data speeds, voice quality and overall capacity. Be 100% sure to include all corners and areas of these lounges with the best possible service. These will often be the most valuable users in your network, decision-makers who could have a big impact on your revenue.

The Main Terminal and Shops

You will face similar challenges to those in the example from the shopping mall, such as only installing antennas in the 'streets', multiple open areas and levels. Modern airports share a lot of features with shopping malls (refer to Section 14.2 for more details); the shops and restaurants would be fairly open and hotspots would be concentrated to the food and seating areas. If you need to divide the main terminal buildings into multiple sectors, you should consider the same approach and strategy as for the shopping mall. Often a vertical separation of sectors is preferable if you have solid concrete slabs between the floors. If you have large open cavities between the levels, you could use same sector separation with 'back to back' antennas as described for the shopping mall.

Office Buildings

Don't forget the airport administration buildings and back office areas, as described earlier a major portion of the revenue opportunity and airtime could be from the staff. Apply same strategy for these office buildings as described in Section 14.1.

Wi-Fi Offload

Of all places, airports will definitely have a Wi-Fi system. This could be part of your data offload strategy, or it could be a competing Wi-Fi operator. Ensure that you carefully coordinate the location of your DAS antennas and the Wi-Fi APs. A guideline minimum of 100 cm between DAS antennas and Wi-Fi APs is recommended to avoid inter-system degradation.

The Parking Area

In Figure 14.6 you have a typical example of a parking area connected to an airport. You would typically get by with relaxing the design requirements. Perhaps only voice and slow-speed data are needed (3G). You might save considerable costs if you don't have to plan for 4G and MIMO in these typically low-load areas. You could also get by using a narrow channel for 4G, ensuring sufficient coverage, and not overkilling the capacity. However, you need to check this in specification phase and ensure you are aligning the design to the exact requirement and expectations of the users.

As you can see in Figure 14.6, the distance between the antennas is greater; due to the open nature of the parking area with only minor 'clutter loss', typically 100% of the users will be in a straight line of sight to at least one antenna in the parking area. We might choose to use wall-mounted directional antennas as shown, and this will often make sense from an installation point. Typically the capacity need in the parking area does not justify its own sector, and in this design example we have included the parking area in sector 3 (cell 3), sharing the same capacity as the main terminal building, and also avoiding handovers when passengers move between the car park and the main terminal building.

DAS Equipment Locations

As shown in Figure 14.6, we often prefer to have the main equipment room on the land side of the airport, easing access for upgrades and service. Active DAS would be default selection for most major airports, so you need to clarify the required and appropriate equipment locations throughout the airport and buildings.

14.6 Sports Arena Design

Sports arenas are among the most challenging designs you will encounter on several fronts. The main challenge is the extreme concentration of users requiring very high peak capacity within an open area, where it is very hard to control the isolation between the cells, and the unpredictable load pattern. Adding to these challenges is the mobility of users entering and leaving the event, who might spend some time outside the arena waiting to get in.

Modern stadiums and arenas will usually have multiple functions, such as concert venues, conference areas, meeting areas, hotels and hospitality within or near the same structure. Therefore, the capacity demands will shift depending on the function. The coverage requirements will be different if you have a big event on the playing field, e.g. a concert where the users will need coverage and capacity during the event, than they will be during a sporting event. The athletes will not require any capacity and you might want to switch those cells off in this case to limit the interference to the other cells. Sometimes all the capacity load could be in the conference areas and not in the open seating area.

If the arena is located in a dense urban area or close to any macro area where you already have strong signal levels, the challenge gets even bigger. You need to overcome the existing signals to have dominance of the cells within the arena, but at the same time avoid leakage from the arena to the nearby macro area.

If you are lucky enough to be involved with a stadium project, you will learn a lot – and most of all you will need to be prepared to spend a lot of time before, during and after the design in the actual arena, performing measurements and optimization.

I believe that you could write a whole book on stadium and arena planning alone, so consider the following inputs to your planning process. Adapt to the actual structure and demands of the project and surroundings, and prepare to spend a lot of time outside the office during site surveys, measurement campaigns and so on. Oh, and good luck!

14.6.1   Typical Features and Checklist for Stadiums and Arenas

In addition to the standard considerations, there a few important things to remember:

- The biggest challenge is to ensure isolation between multiple sectors in the open area of the stadium.
- Very unpredictable behavior of the user load on data and voice; traffic is related to events, e.g. a goal during a soccer match.
- Wi-Fi offload is often needed to make sure you can get sufficient overall capacity.
- If possible, design for a 'flexible' sector plan, to cater to the shift in capacity need – most of the time the arena will need only limited capacity, but during events a very high capacity is required.
- Any meeting, press, VIP and conference areas that need special attention?
- Any sub-level service and parking areas that need to be included?
- Do we need a multi-operator solution?
- Any other radio services we need to plan for or take into account, like 'Tetra' and other PMR services?
- Do we need to consider future upgrades to the system?
- How does this solution fit into the existing macro layer?

14.6.2   Arenas Require 3D Coverage and Capacity Planning

Many of our indoor designs can be considered as 'flat', i.e. one floor plan per floor, typically isolated by a concrete slab giving us the benefit of high isolation between the floors. Arenas are 3D structures, and you need to take this into account both when performing the capacity design and when doing the coverage design, with special attention to the isolation between the cells.

The 'vertical mobility', when users move up and down the staircases while entering and leaving the stadium or seating areas, needs to be taken into account from the start of the planning process.

A generic layout of a typical 'bowl' shaped arena can be seen in Figure 14.7. We need to consider both the horizontal view and the vertical view when we plan for capacity and coverage. There are more details on this in the following chapter, but the principle it to 'slice' the arena into areas that can be joined into sectors (cells). For example, area Y1 and X1 in upper seating could be one logical sector (cell) or areas X1, X2 and X3 could be connected to one logical sector (cell) so you combine upper, mid-level and lower seating and the vertical staircase into one sector (cell).

14.6.3   Capacity Considerations in the Arena

Before we even start doing any coverage design, making antenna location decisions, and selecting the type of DAS and antennas, we need to take a close look at the main challenge: the capacity load. Achieving the capacity is the main challenge in a stadium. In principle, the coverage for the seating areas could be achieved with a few high-power antennas in the venue, but it is obvious that one to three sectors (cells) will not come even close to catering for the

Top view

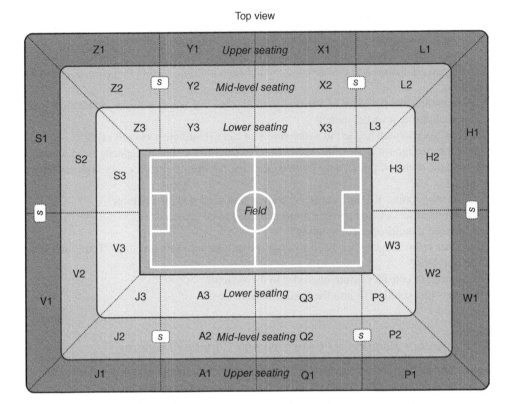

Side view

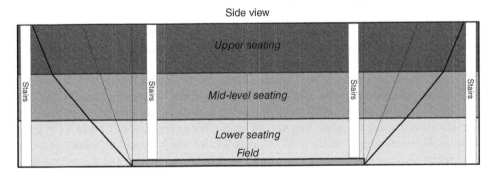

Figure 14.7 A stadium is a 3D complex, and this needs to be carefully considered in both the capacity and coverage planning

traffic in an arena with 50 000+ seats. It all comes down to the capacity, and how we best manage to implement the needed capacity, with a minimum of overlap between the sectors (cells) to limit soft HO zones. The main issue we have to deal with in terms of performance is the 'isolation' between the cells.

We know from Chapter 6 how to perform the basic voice and data capacity estimate – and it will only be an estimate!

Traffic in the Arena is Unpredictable

We must understand that the behavior of users, with regard to both voice and data, will not be Erlang or uniformly distributed during the period they are at the venue. Typically the voice load during a game will be concentrated before and after the game, and during the breaks. If there is a concert at the arena, the voice load might be highest when the band plays their top hits, and people want to 'live share audio' via a voice call. The data load will also be unpredictable; people will perhaps upload the latest goal on their social media platform, or post a picture just as 'The Wall' comes down in Roger Waters' show.

The nice thing about the data load is that it does not matter too much if the upload is 'buffered' and is three to five minutes late, whereas the voice services should be maintained at 1:1 in real time, so often you will prioritize the voice side of the capacity design.

Capacity Strategy, Now and in the Future

Flexibility and upgradability are key. As discussed in the preceding sections, the coverage areas of the arena should be divided up in sections, which can then be combined into various sector/cell configurations that can support the current and future capacity needs.

In Figure 14.8 we can see the principle of planning the stadium in areas and then feeding these areas in the sector (cell) layout that supports the required traffic. In this example (Figure 14.9), we have divided the arena into the following sectors (cells):

- **Sector 1 (cell 1): DAS remote 1, 2, 3, 4, 5, and 6**
  Area S1 and Z1 (upper seating)
  Area S2 and Z2 (mid-level seating)
  Area S3 and Z3 (lower seating)
- **Sector 2 (cell 2): DAS remote 7,8, 9, 10, 11, and 12**
  Areas Y1 and X1 (upper seating)
  Areas Y2 and X2 (mid-level seating)
  Areas Y3 and X3 (lower seating)
- **Sector 3 (cell 3): DAS remote 13, 14, 15, 16, 17, and 18**
  Areas L1 and H1 (upper seating)
  Areas L2 and H2 (mid-level seating)
  Areas L3 and H3 (lower seating)
- **Sector 4 (cell 4): DAS remote 19, 20, 21, 22, 23, and 24**
  Areas W1 and P1 (upper seating)
  Areas W2 and P2 (mid-level seating)
  Areas W3 and P3 (lower seating)
- **Sector 5 (cell 5): DAS remote 25, 26, 27, 28, 29, and 30**
  Areas A1 and Q1 (upper seating)
  Areas A2 and Q2 (mid-level seating)
  Areas A3 and Q3 (lower seating)
- **Sector 6 (cell 6): DAS remote 31, 32, 33, 34, 35, and 36**
  Areas V1 and J1 (upper seating)
  Areas V2 and J2 (mid-level seating)
  Areas V3 and J3 (lower seating)

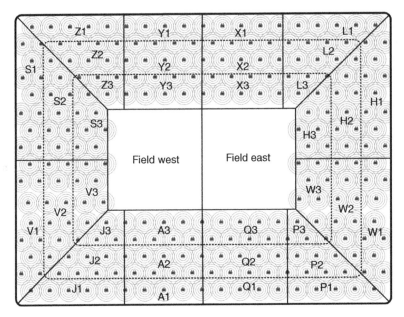

Figure 14.8 The different areas in this capacity layout are now divided into sectors (cells) by joining (simulcasting) the cell over multiple DAS areas. We are using Wi-Fi offload on a separate layer, and are observant of the HO zones, especially in the field

- **Sector 7 (cell 7): DAS remote 37 and 38**
  Field west
- **Sector 8 (cell 8): DAS remote 39 and 40**
  Field east

This configuration could support the current load, as well as what we can foresee in the upcoming years. In addition to the mobile services, we could use a dense cluster of Wi-Fi access points as indicated in Figure 14.8 to support data offload to Wi-Fi, in order to release resources for 3G and 4G.

In Figure 14.9 we can see the principle of how we feed the various sections of the DAS and the local DAS remote units (RUs) from one central base station location that supports multiple operators and multiple services (2G, 3G, and 4G in this example). In this example we have eight sectors/cell, and we may not switch on sectors S7 and S8 unless there is an event where mobile users on the field require the capacity. This will prevent extensive HO zones and isolation/noise challenges from S7and S8 during normal sports activities.

Minimize Cell Overlap

The trick to having many sectors in an open environment like a sports arena is to have a well-defined footprint of every section of the DAS, right down to the individual footprint of each antenna. We will take a closer look at this in Section 14.5.4, where we will strive to get only a few overlapping HO areas to maximize the capacity in the area.

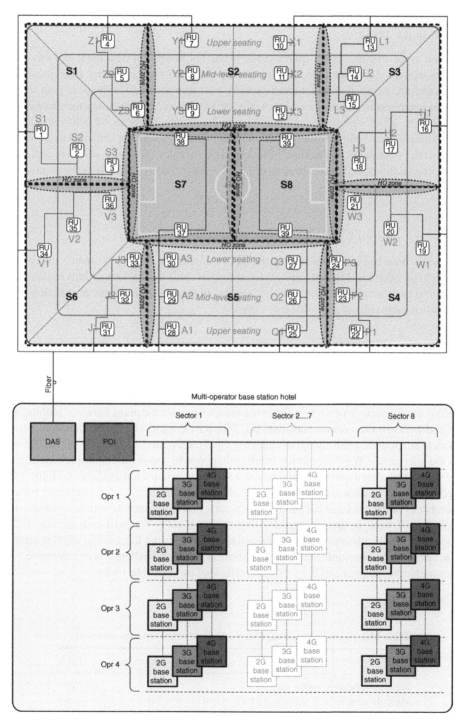

Figure 14.9 Small cells on a sports arena. This is one possible option but the capacity is statically assigned to areas so that cells are permanently assigned to service a fixed area. Utilizing a DAS on a sports arena will give you the flexibility and upgrade path to change your sector (cell) allocation to the different areas. You could even have independent 2G, 3G and 4G plans

Hotspots in the Arena

One could argue that the whole sports venue is one big hotspot of data load, and this would indeed be correct. However, there might be areas with additional challenges and these areas might even have a special need for extra capacity to ensure access to high-speed data services and good voice quality. These would typically be VIP boxes, private suites for sponsors and press lounges, all of which would require special attention in terms of both capacity and quality. Other areas would be office areas and 'back rooms' where the staff will need services to support the operations in the arena. Modern sports venues often also serve as conference and meeting facilities, and these areas would need to be considered as well as the main seating areas.

14.6.4   RF Design Considerations in the Sports Arena

Now that we have had a general look at the main challenge in the sports arena – the capacity and how to address this with the capacity/sector plan – let's have a look at how to design the coverage sections/areas to support the high number of sectors, and to achieve sufficient isolation between these cells to get sufficient capacity.

Small Cells or DAS for the Stadium

In preceding sections, we addressed all the benefits of utilizing a shared base station hotel, and then dividing the arena into different sections via a DAS (which could be connected together in various configurations of sectors, ensuring independent sector plans between mobile operators, and between the various services, 2G, 3G, 4G etc.) and concluded that this gives us freedom to adapt to the ever-changing challenges and requirements of a sports arena. But what about the alternative? Why not use the small cell approach, and deploy multiple remote radio heads (RRHs) or other types of small cells in a similar configuration as show in Figure 14.9?

Well, we could to some extent do this, but often you will have a 1:1 relationship between the number of small cells and the number of sectors, due to the lack of simulcast (where several small cells work in unison, broadcasting the same cell). You will end up with a static solution with small cells, and with multiple RRHs/small cells per location, to support multiple mobile operators and multiple services. The shear installation challenge is considerable.

We summarize the pros and cons of small cells vs. DAS in Table 14.1.

Table 14.1   Comparing small cells vs. DAS for high capacity arenas

	Small cell	DAS
Simulcast, flexible sector plans	No	Yes
Multiple protocols in one remote (2G, 3G, 4G)	No	Yes
Multioperator	No	Yes
Upgradable	(Yes)	Yes
Installation impact	High	Low

So are small cells no good for sport arenas? Well, it all depends – small cells are excellent tools in the toolbox, but for high-capacity sport arenas they are often less than ideal. For small to medium-sized venues, for a single operator, small cells can be a good solution. Just keep in mind that it will be a challenge to upgrade this solution in the future.

Controlling the Noise

In theory, we could just add more and more sectors in order to achieve sufficient capacity in the sports arena, but it is probably not that easy. The main challenge with implementing multiple sectors/cells in a sports arena is the open nature of the venue, and the lack of isolation between multiple sectors. In reality, what will happen is that the more sectors you implement, the more capacity you will get, all the way up to the point where the sectors overlap so much that cannibalization of the capacity due to lack of isolation and overlap will actually decrease the total capacity ('the breakpoint').

Vicious Circle

If we look at Figure 14.9, it is evident that more sectors/cells implemented in the arena will give more capacity; however, at some point the overlap and lack of isolation between the cells will cannibalize more capacity than the added extra cells shown in the HO zones in Figure 14.9, and you actually start to see a fall in the overall arena capacity, as shown in Figure 14.10.

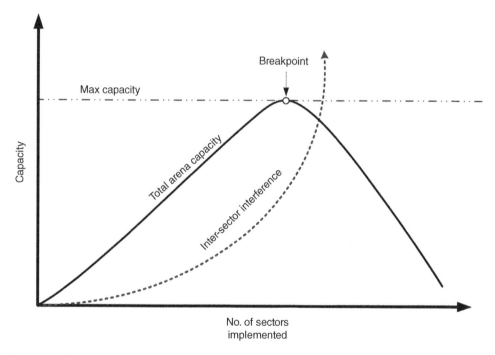

Figure 14.10 The theoretical relationship between the number of implemented sectors/cells in a stadium and the total capacity

If you are able to implement the sectors and antennas in such a way that you have good isolation between the cells, you can stretch the capacity a lot and implement many sectors/cells, typically by implementing many low-power, uniformly spread and optimized antenna locations. If you deploy a few high-power antennas, with limited control of the individual footprint of the sectors/cells, overlaps between the cells will cause handover issues and noise load, and you will only be able to deploy a few sectors before you reach the capacity breakpoint shown in Figure 14.10. We will have a closer look at how to implement many sectors in the arena, while making sure we maintain acceptable isolation between the cells, in Section 14.5.8.

Simulcast Advantage

For a very large, high-capacity venue in an area of the network with a need for general high capacity, it could be a challenge to have a high number of sectors/cells active in the arena. Although we can try our very best to control the footprint of all the sectors deployed in the arena, there might be some leakage to the surrounding area. This might be a compromise you have to accept in view of the extreme capacity needed during a fully loaded stadium. But what about when there are no activities going on in the arena? You might only need a very limited capacity to support the internal staff on-site. With a DAS you could change the simulcast scheme/plan. For example, you could have the stadium divided into the eight sectors/cells shown in Figure 14.9 during the period of full load, whereas during off-peak times you might be able to support the required daily capacity by simulcasting all the areas using one cell only, limiting the interference from the arena to the surrounding network, but maintaining service for the daily users in the arena.

14.6.5   Antenna Locations in the Sports Arena

Let us have a look at a typical stadium, and where antennas are likely to be deployed. In Figure 14.11 we can see a typical example of antenna locations in a cross-section of a typical arena, showing seating areas as wells as 'back rooms' and stairs for access to the various levels. It is important to make sure you get very detailed measurements of the levels of the existing macro signals inside the arena venue, as well as at the HO zones at the entrance to the stadium. This is absolutely essential in order to establish the design target levels to be used in the arena. As we can see in Figure 14.11, a stadium is a combination of indoor and outdoor radio planning. It is ultimately about maximizing the isolation between the cells/sectors, to create a highly focused and well-defined footprint for all the sectors to support the capacity challenge described in Section 14.4.3. It is all about the isolation, and isolation between the different sections (eventually combined into sectors/cells) is the most important parameter to minimize the overlaps and limit the HO zone size (see Figure 14.10), thus maximizing the capacity in the venue. We will address this in more detail in Section 14.5.8.

Let's have a more detailed look at the antennas and their service areas (with reference to Figure 14.11). In general it is highly advisable to use many antennas with relatively low power to ensure a uniform coverage area of each section of the DAS. As described in Section 14.5.3, all the individual sections of areas will be combined into various sectors by simulcasting over multiple sections of the DAS. So please make sure you carefully control each and every

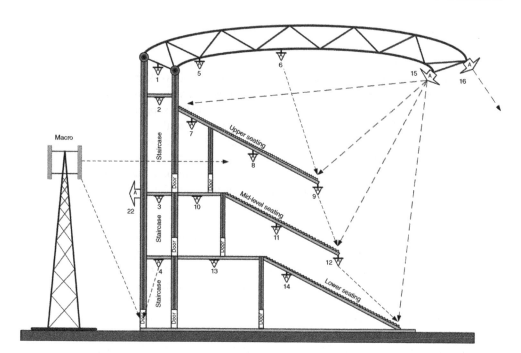

Figure 14.11   A generic example of antenna locations in a cross-section of a typical arena, showing seating areas as wells as 'back rooms' and stairs for access to the various levels

antenna's footprint by carefully selecting the exact antenna locations, types and installation details. Stadium radio planning is a mix of the most challenging indoor and outdoor planning issues, so you will need to use all your experience to plan, test, optimize, and verify.

- **Back rooms**
  These are the areas where you will find restaurants, cafes, toilets, etc. Typically these will be pretty well isolated from the other areas, but watch out for windows and openings that could leak to the other areas of the arena, as well as receive interference from the outside macro network. Use structural elements like concrete beams, walls and so on to limit the footprint of the antennas to cover only the intended areas, and do not assign more transmit power than needed to cover the intended area. In Figure 14.11 these areas will be served by antennas 7, 10, and 13.
- **Staircases, access areas**
  These are the vertical access areas, the staircases. Normally you will prefer to have these as one section (sector/cell) throughout the vertical area, perhaps simulcasted with the upper-most seating area in each vertical 'slice' of the arena, ensuring there is no vertical HO in the staircases when people enter and leave the arena. Pay attention to the footprint of antenna no. 4 and ensure you control the HO zone to and from the macro network without too much leakage. In Figure 14.11 these areas will be served by antennas 1, 2, 3, and 4.
- **Covering just outside the arena**
  Often the areas just outside the arena will be heavily loaded with users before and after the event inside the arena. Sometimes you would include covering these areas as a part of the

DAS used to cover the inside of the arena. You could decide to simulcast the sectors just outside the arena with those already in service inside the venue. However, I do not recommend this. You will probably pick up interference outside the arena that will limit your UL performance and also emit DL interference to the outside network from the high-capacity cell inside the arena. Very often there will be existing macro base stations servicing these areas; the challenge, however, is the sheer capacity needed to support the many users just outside the arena. If you cover these areas from antennas at the stadium walls, beaming away from the venue, such as antenna 22 in Figure 14.11, the benefit is that you radiate away from the high-capacity venue, limiting the impact of interference from the dedicated cell feeding antenna 22, rather than having a nearby macro base station transmitting the high-capacity cell needed towards the arena, with the potential of creating interference inside the venue.

- **Seating areas**

The seating areas are without doubt one of the most challenging parts of the arena. You are trying to achieve sufficient service levels, and want to focus the footprints of the antennas and limit the overlap to adjacent sections (sectors/cells) of the DAS. If you want to have vertical sectors in each section of the stadium, you will be facing a big challenge. If we use the example in Figure 14.11 and you want to design for separate sections (sectors/cells) for upper, mid-level and lower seating areas, you need to watch out for the isolation between antenna 6 (intended to service upper seating) and the mid-level seating. You also need to ensure that antenna 9 (intended for mid-level seating) does not leak upwards to the upper seating or downwards to the lower seating. Also make sure that antenna 12 (intended for lower seating) does not leak upwards to the mid-level seating. You need to ensure that antennas 8, 11, and 14 do not leak into the adjacent seating area above – select locations for the antennas that will maximize the isolation vertically to avoid this issue. Do not rely solely on predictions – make sure to measure the losses to each intended service area (and those unintended) for all typical antenna locations. For example, you want to make sure that antenna 9 services the outermost seats in the mid-level seating area, so it should overpower the signal from the sector above (antenna 6) but without leaking too much signal down to the outermost lower seating areas where antenna 12 is intended to be dominant.

If you run the whole vertical 'slice' of the arena divided into segments of the same sectors (as shown in Figure 14.9) you might want to use directional antennas at location 15, possibly in combination with omni antennas above the seating sections. Make sure to use antennas at location 15 that have a very well-defined footprint, with limited 'side lobes' and 'back lobes'. Special antennas are available for this purpose. In Figure 14.11 these areas will be served by antennas 5, 6, 8, 9, 11, 12, and 15.

- **The open field**

Without doubt, the most challenging part of any arena design will be the sectors/cells you need to support mobile services on the field. These will be sectors S7 and S8 in the example shown in Figure 14.9. In the example in Figure 14.12, antennas 16 and 17 are used to service a common area by simulcasting the same cell. If we try to implement two different sectors/cells to feed antenna 16 and one to feed antenna 17, we will end up with a large HO zone in the middle of the field with a lack of isolation and degraded performance / capacity. The same goes for the next sector, in our case the area between sectors/cells S7 and S8 in the field. It is very hard to design a well-defined HO area this far from the antennas, and it takes very well constructed antennas and optimized installation to optimize these areas. It is also

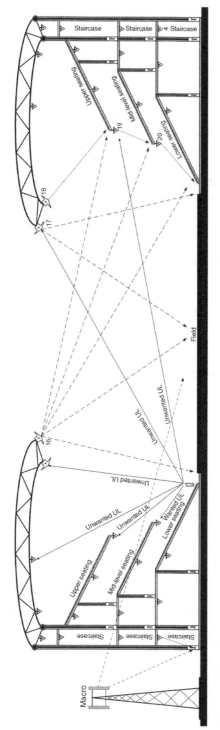

Figure 14.12 Generic example of antenna locations in a typical arena (cross-section), showing opposite sides of the arena and the field, which is often the most challenging area to obtain sufficient isolation

evident that the antennas used to cover the field will be the challenge in terms of isolation to the sectors used in the seating areas. You need to be very careful to select the right antenna type and the right location and spend the necessary time to optimize the exact location and alignment of these antennas (Figure 14.12, antennas 16 and17).

- **Use the structure to improve isolation**
 You need to make sure you utilize the structural elements of the arena to maximize the isolation to areas you do not want to reach in adjacent sections/areas. This means using metal struts, concrete pillars, walls, concrete beams and elements to shade off areas and directions from the antennas. This is a challenge, because it is one thing as a radio planner to come up with the ideal antenna design, and implementation, but it is quite another to obtain permission from the architect and site owner to deploy this ideal solution!

- **Radio planning, verification measurements, and optimization are equally important**
 I cannot state this clearly enough, but you need to 'complete the circle' – radio planning for a sports arena is not a desktop design. Let's look at the typical project flow of a sports arena DAS design:

1. **Draft plan and site visit**
 You need to start with a rough plan and capacity estimate, and then you need to do a very detailed site survey, trying to establish all the most critical antenna locations (the locations that are expected to be the most challenging when it comes to isolation – so in Figure 14.11 these will be locations 6, 9, 12, 15, and 16).

2. **RF survey**
 Then you need to obtain permission to install the antennas at these locations. After this you will perform an RF signal survey, confirming your design calculations in terms of coverage areas and isolation of the actual antennas at the actual locations (expect costs associated with getting the antennas temporarily installed at these often hard-to-get-to locations).

3. **Verify design**
 Once you are happy with the RF survey and have fine-tuned your design, the installation can commence. Often quite a few of the sections of the DAS implementation will be symmetrical, with the same layout of antenna locations repeated in the various sections. Therefore it could be a good idea only to install a few typical sections completely and then repeat the RF survey to confirm coverage and isolations before you commence with the rest of the arena. This can save a lot of cost and grief in case you need to fine-tune the final design.

4. **Pre-activate and optimize**
 Once you have fitted out the complete venue, it is now time to activate the services, bringing all the mobile services live. Do plan to bring the system live weeks prior to any event at the arena. You will need time to verify and optimize antenna directions and fine-tune the system. Given the sheer size and complexity of the DAS, allow a number of weeks for this task.

5. **On-air and fine-tune**
 No matter how you perform the process, you will not know for sure exactly how the implemented system will perform before it is fully loaded up with real users during an actual event. Be on-site during an event and try to verify performance at the most critical locations. Be online with all the statistical KPI tools so you have a 'finger on the pulse' during the first event; you may even be able to optimize some parameter setting during

the event in order to optimize performance. (Have a plan for exactly what you expect to fine-tune during the event – don't just try 'turning the knobs' to see what happens.)

6. **Monitor the performance**

Like any other important solution, you should always monitor the performance closely and take the required actions. The advantage with a sports arena is that you will know when big events are happening before the actual date. This gives you the opportunity, to some extent, to plan for extra capacity, sector layout and so on. Perhaps you will not switch on sectors S7 and S8 in Figure 14.9 unless there is a concert, so as to minimize overlaps and noise. You will carefully monitor the performance after all events, analyze the KPIs and take the actions required.

14.6.6   Interference Across the Sports Arena

The main challenge of implementing many sectors in a dense open area is, as discussed earlier, the isolation between the many sectors. The potential challenge is evident in Figure 14.12. First of all, there is the issue of potential leakage between the vertical sectors if you divide the different seating areas – upper, mid-level and lower – into different sectors vertically. The challenge is that, for example, antenna 9, which is intended to cover the mid-level seating, will also spill signal to the lower and upper seating areas. In addition to this isolation challenge, you will also need to consider the signals from all the other 'unwanted' antennas across the arena. The omni antennas just above the seating areas are rarely a big challenge, if you use many antennas with relatively low power and thus obtain a uniform coverage footprint. The advantage of this approach is that a symmetrical design will be used across the arena, and the distance alone will automatically result in sufficient isolation.

Instead of omni directional antennas above the seating levels, you could consider downward-facing directional antennas, which will offer even better isolation across the arena to other sectors.

The biggest challenge with 'inter-sector interference' in the arena will come from the sectors servicing the field. This is because these antennas will need to reach across the field, at least to the center, and then the simulcast from an antenna on the other side of the stadium will cover the other half. In Figure 14.12 this is achieved by antennas 16 and 17. There are specially designed antennas for this use (antennas with a very clear 'definition' of footprint, and almost no side and back lobes), It is very important to use antennas with a very clear definition of the footprint, so that antenna 16 does not 'hit' the sectors below on the same side of the arena, or hit high signal levels for the antennas across the arena, and vice versa for antenna 17 (Figure 14.12). To deal with the problem of antenna 16 hitting the other side of the arena with unintended signal, you could improve the wanted signal level by deploying some directional antennas in the most challenging part to increase the service level (antenna 18). This strategy could be used on all the seating levels if the interference from unwanted cells across the arena was a problem. Therefore directional antennas facing down and inwards at the outermost installation locations (closest to the field) could be advantageous to (and boost the isolation) the overall capacity and performance.

I highly recommend a solution that will give you the option of only having the sectors intended to cover the field active when you need the mobile services in the field, during concerts etc. Then, during sporting events you would not need the increased capacity in the field and would have much better performance in the seating areas with the 'field sectors' off.

Remember that the final RF performance, hence the data throughput, is dependent on getting the highest margins between the wanted signal and the noise. The noise sources in the sports arena are obviously the active DAS elements, but to a greater extent they are the sum of all the unwanted signals from the multiple cells deployed in and near the stadium. This is especially important for systems operating on the same frequency for each cell, i.e. 3G and 4G where you don't have the luxury of frequency separation, although fractional reuse of the spectrum on 4G will help, but will result in limiting the actual bandwidth, hence defeating the purpose of deploying many cells in the arena to boost the capacity.

Remember the Uplink, It Makes All the Difference

As for all RF systems, we need to consider both the DL and the UL. UL noise increase can be the Achilles heel of stadium design, and many such systems have crashed and burned during real-life traffic loads, frequently due to UL noise load spinning out of control.

Uplink and DL losses are obviously symmetrical, with the same loss on the UL as on the DL. In the previous sections we have focused on the isolation of the cells, and how to overcome the challenges so as to optimize this isolation. Key to controlling the UL performance, and minimizing the noise increase across to other cells (shown in Figure 14.12), when mobiles are on the edge of the serving cell, is to ensure you keep the mobiles transmitting a minimum of power. In Figure 14.12 we can appreciate how easily the mobile UL will hit other sector antennas as well, so keeping the mobile at low transmit power is important for the overall performance. You will have thousands of mobiles active at the same time. Therefore pay attention to your UL link budget, and perhaps limit the allowable UL power level for the mobiles, or even consider intentionally tipping the link balance in favor of the UL. In practice, this often means making sure the system is DL-limited, to ensure that the UL power will be at a minimum. This is one of the key tuning points during the optimization phase with live traffic loading up the arena.

The Macro Mistake

Quite often you will find yourself in a situation where there is already a macro base station nearby or even at the stadium structure itself, as shown in Figure 14.12. This was common practice when rolling out 2G. This ensured very strong signal levels across the venue, and planners would add a lot of TRXs on the cells covering the venue, enabling half rate to gain even more capacity. This approach might have been viable for 2G but would be a big issue if we tried to use these cells to provide 3G and 4G data-centric solutions for high capacity at the arena. The vast overlap of the 3G cells from the macro and any DAS or small cell system deployed later to cater for the data capacity would simply turn the venue into a big soft HO zone with very limited efficiency and capacity. The same would be true for 4G from any powerful macro site; it would make the fractional reuse of the spectrum very inefficient, and degrade the quality, and thus the capacity in the stadium from the small cells or DAS deployed in the venue. Even 2G might struggle due to the limitations of implementing new channels already in-service from the macro's serving cells or adjacent channels from these cells.

The best solution to the problem would be to decommission the existing macro covering the arena with high signals, or alternatively try to realign the antennas, use down tilt etc. to limit the macro signal in the arena itself. I do realize that the macro located on or very close to the venue will also be covering other nearby areas, but you would have to address this separately. Perhaps the solution might be that you turn down the transmit power on the macro during events in the stadium, or even switch one or more sectors off!

14.6.7 Upgrading Old 2G designs, with 3G and 4G Overlay on a Sports Arena

In Figure 14.13 we can see a typical 2G design for a sports arena that we might have done several years back. We have deployed six sectors/cells to support the traffic (typically voice and slow data speed, being a 2G design). The six cells are fed to a few high-power panel antennas, and S3 and S6 are split into two antennas to make the most of the footprint. The advantage with 2G is that the cells are separated in frequency so there not too much to worry about in terms of inter-cell interference. You might even intentionally design for a large overlap of the 2G cells, so you can share the resources of more than one cell. During a concert, the users on the field could benefit from more or less access to capacity from all six cells in this example.

The advantage of 2G is the separation of the cells by different frequencies, so that even in the overlapping area you will be able to control the mobile to stay on a specific cell, and maintain a good signal-to-noise ratio.

For 3G, however, all mobiles in the overlapping area (within a 6 dB delta) would be in soft HO, loading more than one cell. This can be seen in Figure 14.14, where we have used the 2G strategy from Figure 14.13 and applied a 3G overlay as an upgrade using exact same approach.

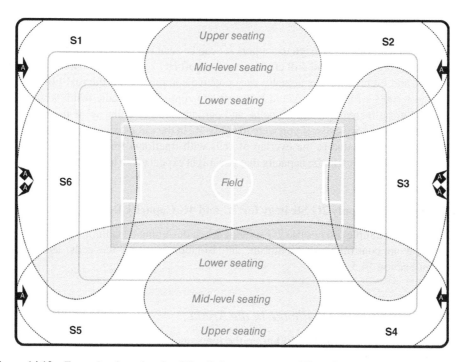

Figure 14.13 Example of overlapping 2G cells in a sports arena. The cells are separated by frequency, and capacity resources can be shared due to large overlaps of the 2G cells

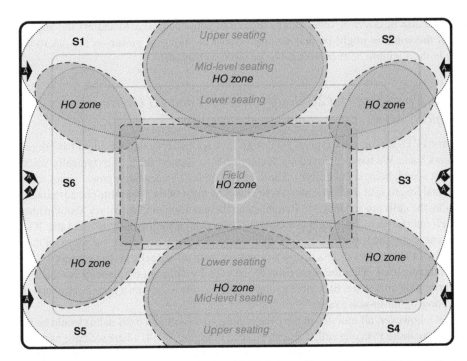

Figure 14.14 Example of overlapping cells. An old 2G design will struggle on 3G/4G due to the large overlaps between cells

The result of using the 2G design to support 3G can be seen in Figure 14.14; the large overlaps of the cells are evident, and will create extended soft HO zones on 3G. This will degrade the capacity and Figure 14.14 is actually very positive compared with the real-life scenario, where almost 100% of the arena would be a soft HO zone. The extreme result would be an uncontrolled noise increase of all the cells, and perhaps even a 'collapse' if the UL on the cells, resulting in dropped calls. If you were to apply 4G in the same scenario, you would basically limit the capacity due to the overlap of cells with fractional frequency reuse and you would not have the factor 6 of the capacity that you might expect when looking at Figure 14.14.

Conclusion: Do Not Apply 2G Strategy for 3G and 4G Capacity Design

The conclusion is simple: do not use 2G strategy when designing a stadium for 3G and 4G.

It is all about controlling the footprint, and isolation between the cells, as discussed throughout Section 14.5.

14.6.8   The HO Zone Challenge in the Arena

If we take a large stadium that is 400 m on the long side, and try to calculate the expected soft HO zones on 2100 MHz and we calculate the 6 dB overlap (+/– 3 dB from the serving cell signal level), it is all about the relative signal level between the wanted (serving) cell

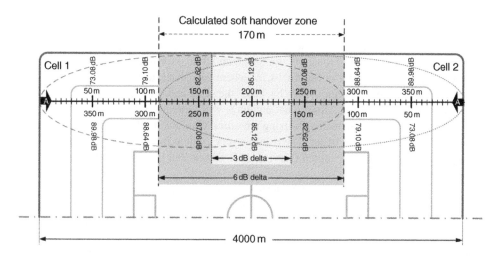

Figure 14.15 Example of two overlapping cells (2100 MHz) fed from two different corner-mounted directional antennas symmetrically installed in the stadium. The figure only shows one side of the arena, two identical antennas, aligned towards each other

and the power of the other cells. We want to have a good margin, as high as possible between the wanted and unwanted signals. The unwanted signals will be the sum of the power of all other cells – in Figure 14.15 we merely consider the neighboring cell for ease of illustration.

The Isolation Issue in the Open Arena

In Figure 14.15 we can see a section of a 400-m-long sports arena, with two symmetrically mounted directional antennas, one at each end, providing excellent signal level in this section of the arena. However, if we were designing for 3G, the challenge would be the overlapping area, the '6 dB window' (+/–3 dB) where, in this example, we will be in soft HO, loading both Cell 1 and Cell 2.

Using a standard free space loss model, with a 50 m breakpoint (see Chapter 9 for more details) we can calculate the link loss (excluding antenna gains, directivity, etc. – as the two antennas are identical, and symmetrically installed, the relative difference will be the same). In Figure 14.15 we can see the link loss from each of the two antennas from 0 to 400 m (shown up to 350 m).

This makes it easy to conclude the relative difference between the link losses, and eventually signal levels – and thus we can plot the expected overlap zone with a given margin between the two cells.

In Figure 14.15 we can see two zones marked, one with a 3 dB delta (+/– 1.5 dB) and one with a 6 dB delta (+/– 3 dB), the expected soft HO zone for 3G. The calculations and the plot in Figure 14.15 clearly show that about 170 m (without adding fading margin) out of the 400 m will be in soft HO on 3G, and will impact capacity on 4G due to fractional reuse; 170 m corresponds to about 42.5% of the area!

Addressing the Isolation Issue in the Open Arena

The challenge with regard to isolation between the cells in Figure 14.15 can be addressed by ensuring a clearer definition of the cells' footprint, and thus improved isolation. One possible solution to the issue is shown in Figure 14.16, where two 'back to back' mounted antennas between the antennas from Figure 14.15 'backfire' the same cell, by simulcasting the cell back into the intended service area. Directional antennas will typically have a very high 'front to back ratio' of at least 30 dB, and by utilizing structural elements in the arena when installing the antennas, this can be even better. The result is very evident in Figure 14.16, where we can see simple calculations of the wanted (serving cell) signals and the unwanted signal from the adjacent cell reaching into our serving cells' coverage area.

We can now calculate the co-channel interference (C/I) margin between the wanted (C) cell and the unwanted (I) cell, and this is plotted in Figure 14.16. It is clear that the expected soft HO zone is considerably smaller than was the case in Figure 14.16. The real size will depend on the actual installation of the antennas: the higher they are above the user, the less well-defined the footprint and the larger the soft HO zone will typically be. Although the principle with the 'backfire' antenna in Figure 14.16 improves the isolation significantly, it is clear that there is still a 'weak point' with isolation of about 9.54 dB in between the two serving antennas.

This can be improved with antenna tilt, and adjustments, but in addition to this we could consider filling in more signal in this area by deploying one or more omni antennas, or down-facing directional antennas in this area. We need to be careful to make sure we don't overdo it, using too much power that could hit other cells in the arena.

Please keep in mind that Figures 14.15 and 14.16 are for a simplified example, and in a real-life design you have the complexity of multiple cells building up the noise load, and multiple mobiles on the UL loading your cells, but the principle should be clear.

Inter-symbol Issues

One important thing to remember when simulcasting from multiple antennas over long distances, as shown in Figure 14.16, is the potential delay differences between the antennas. This is due to long fiber cables that might not be equal for the individual antennas that service the same area.

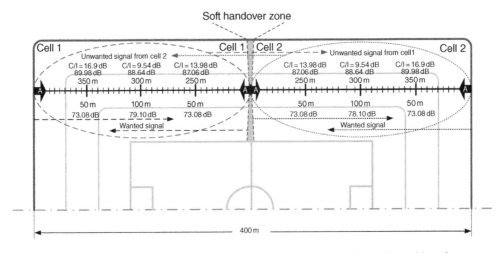

Figure 14.16   Example of two sectors, utilizing 'backfire' isolation. CI, co-channel interference

Depending on the time difference between the two radiating antennas and the mobile, and the type of service (2G, 3G or 4G), this could lead to 'inter-symbol interference'; the two signals will degrade each other. The risk is greater the longer the fiber cables and distances. This is the same issue we have in tunnel installations, so please refer to Section 11.11 for more details.

14.6.9   The Ideal DAS Design for a Stadium

Any DAS deployment in an open sports arena, intended to provide high-quality mobile services and support extreme capacity in an open environment, will be a challenge. The perfect DAS design for this application does not exist; it is always a compromise. At the end of the day, cost also plays a role, so even a theoretically 'perfect DAS' might never be feasible, or might not be possible to deploy in practice. Based on the previous inputs, you could design and implement a DAS in the arena that would be a good compromise and a balance between performance and applicability in practice.

Figure 14.17 shows the principle of a good approach with the following highlights:

- **Backfire antennas**
  As described above we use 'backfire' directional antennas that simulcast the cells into the same area from multiple sides. The high 'front to back' suppression of the directional antennas help to improve the isolation between the cells and minimize the soft HO zone (3G) and limit the throughput on 4G.

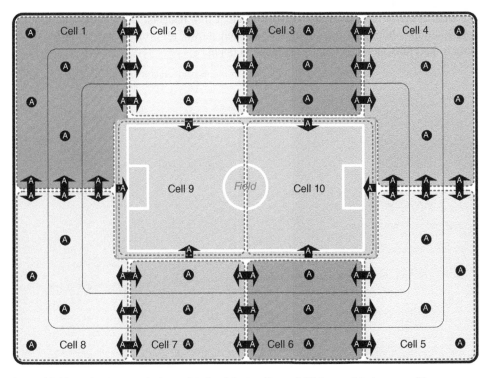

Figure 14.17 'Ideal' arena design, with 'backfire' cells, and fill-in omni in-between. Two sectors to service the field area can be switched on during concerts, but will be non-operational during normal sports events to limit interference with other cells

- **Fill-in antennas**
  Deploying omni or directional antennas beaming down in-between the 'backfire' antennas helps to improve the area in the center of the cell, where the 'backfire' antennas could struggle to dominate.
- **Field sectors**
  The two sectors supporting the users in the field during concerts are also using 'backfire' from multiple antennas. These antennas are installed at the same locations as the other cells' antennas, to improve isolation. The field sectors will only be active during concerts, and will be switched off during sports events to minimize interference from these sectors when they are not needed to support the traffic.
- **Uniform level**
  Deploying multiple antennas in each sector, emitting uniform and controlled footprints of each cell, controls the isolation, the DL quality and improves UL performance, and ensures mobiles will be transmitting low power levels. Controlling the UL noise is key to performance and capacity in the arena.

14.7   Final Remark on Application Examples

I hope the application examples in this chapter have given you some ideas and helped to place a lot of the theory into context. The intention of this chapter and the examples was not to cover all the details, but to highlight some of the main features and provide some practical experience, tips and guidance.

Be sure to verify all your designs with sufficient link budgets, calculations and required documentation. Although a few examples has been shown, all projects will be different and should be carefully and individually designed in detail.

15

Planning Procedure, Installation, Commissioning, and Documentation

Procedure and Documentation Are Very Important

I know that the paperwork and project flow needed to ensure a smooth planning, installation and verification of an indoor radio system are perhaps not as interesting as the details of noise calculations, link budgets and the like. Perhaps this is why the details of the actual procedural definitions, paperwork, and process flow of a successful indoor system are often unappreciated. Nothing could be more incorrect, and a lot of mistakes originate from the lack of a clear process.

The radio planner should be very interested in making sure that the whole process, from start to finish, is flawless, well documented, transparent and traceable. The radio planner might not be responsible for all the details in the complete workflow, but he or she must be very keen to make sure that everything is done to the designed specification, and that everything complies with the design and installation guidelines. In addition, the paperwork and project flow are clearly very important for the purpose of documenting the project, and also in order to ensure that we are constantly improving the design, the processes and the DAS deployments, by allowing us to learn from the experience from past designs and implementations.

I recommend using a well-thought-out, well-documented procedure to control the project flow, as well as traceable documentation from end to end. The process should be defined with improvement and optimization of the workflow in mind. Use only certified installers and personnel who are up to standard on the quality levels and knowledge defined by the supplier. Suppliers will often have certification programs and processes in place that will help to reassure that they have installed and deployed the DAS as per the vendor's guidelines and specifications.

Indoor Radio Planning: A Practical Guide for 2G, 3G and 4G, Third Edition. Morten Tolstrup.
© 2015 John Wiley & Sons, Ltd. Published 2015 by John Wiley & Sons, Ltd.

Also (needless to say, but just to be sure) use *only* calibrated measurement equipment and references. A lot of the RF test and measurement equipment used in the field are very rigid and designed to be dragged around various locations; however, this test type of equipment is highly sensitive and has fragile internal components. A few dB offset in the measurements can have catastrophic consequences for the performance of the designed and implemented DAS, the RF planning levels used in the future based on field measurements and performance, etc.

One of the worst examples I have experienced was a major DAS in one of the world's largest airports. After walk testing the DAS signal levels in this airport, the customer was complaining about the implemented DAS. But after flying halfway across the globe, it turned out that the antenna of their 'test terminal' had broken off and there was literally a hole in the cabinet where the antenna should have been!

Let us now have a look at the procedure from start to finish. We can divide the whole process into three main phases: design, implementation, and verification (see Figure 15.1).

15.1 The Design Phase

This is the start of the project, where the radio planner has more or less full responsibility. The goal is to establish and lock the requirements, and then produce a design to hand over to the installation team.

15.1.1 Design Inputs

The whole design process typically starts with a request from a customer. Often the person making the request is unsure about the exact design inputs, such as signal level in dBm, Erlang per user, data load, etc. So clearly some guidelines and help might be needed, but the basic inputs would be as follows:

- **Floor plans**
  With as much detail as possible in terms of wall types, room use and occupancy, scale of building
- **Coverage requirements**
  What types of services are needed – 2G, 3G, 4G, etc.? What is the required service levels (data speeds)?
  Where do they require service? Do they have any VIP areas? Should there be coverage inside lifts, staircases, parking areas, storage rooms, etc.?
- **Capacity**
  Number and type of users in the various areas. Do they have any high-capacity areas (hotspots)? Do they use Wi-Fi or data offload?
- *Output: Design requirement specification*
  All of these inputs should result in a well-structured document, compiling all the design inputs in a 'design requirement specification' that documents the key performance indicator (KPI) inputs in detail.

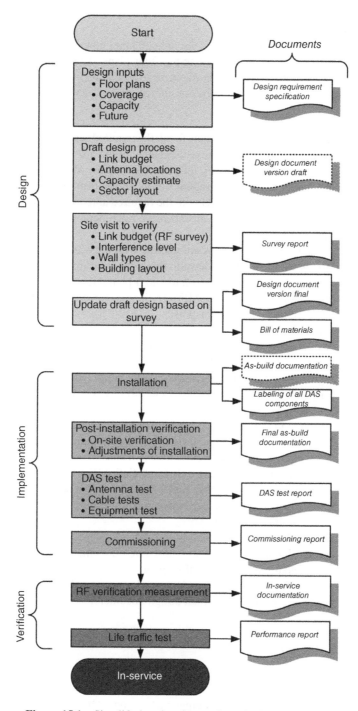

Figure 15.1   Simplified project flow and required documentation

15.1.2   Draft Design Process

Based on the inputs in the previous section, the radio planner now establishes the foundation of the design. The radio planner will be using the floor plans, details of the wall types and material used in the building, information about nearby macro (including predicted signal level) and other base stations that need to be considered, pictures and so on.

- **Link budget**
  The radio planner calculates the link budgets for all the required services, taking into account any nearby base stations. At this stage the radio planner might also decide that an RF survey with a test transmitter is needed to verify the link calculations. At this stage, we also evaluate if there is a possible need for upgrading, for example, from 3G to 4G, within a few years. We might as well plan based on 4G with the required antenna locations to support 4G up-front, rather than having to carry out an expensive full redeployment in the future.
- **Antenna locations**
  Based on the link budget, the floor plans are marked with the estimated antenna locations, bearing in mind that each location will have to be verified during the site survey. Antenna locations that might look logical and easy on the floor plan might not be realistic to deploy in the real-life installation. How about MIMO? Should we provision for this up-front, perhaps only in the expected capacity hotspots?
- **Capacity calculations**
  Both calculations of the required voice capacity and data capacity will be done in detail. Expected future demands of possible higher data and/or voice capacity requirements need to be considered. We might need to leave in some overhead in the active DAS to cope with greater load in the future, making sure we do not exceed the composite power resources of the amplifiers.
- **Sector layout**
  The layout of the sectors in the building is obviously closely related to the capacity calculations above. If, in this early stage of the design process, we plan for more capacity in the future, we can easily provision for a few extra fibers, or runs of coax trunks, thus saving on future expenditure, rather than performing a post-installation. You can easily save a factor of 10 or more on the installation cost of these cables when done as a part of the initial deployment. Obviously there is a healthy balance between being 'future-proof' and overdoing it, making the solutions far too expensive.
- **Output: Draft design document**
  All these calculations should result in a well-structured document, compiling all the coverage and capacity calculations, diagrams, floor plans with antennas, sector outline, and bill of material into a 'draft design document'. This document will eventually be updated throughout the project from a draft to a final design document.

15.1.3   Site Visit – Survey

In most cases it is highly recommended that the radio planner visits the site prior to the final design. This is important in order to verify the assumptions used to calculate the losses in the building, and the selected antenna and equipment locations. Often you will find that there might have been changes to the building itself since the floor plans were produced, such as new walls or new areas added.

It is also often recommended to perform an RF survey of some of the typical antenna locations for the actual design, and also to verify the existing signal levels in and close to the building.

We obviously want to make sure our design works, but it is also important that we do not over-design and add additional cost to the solution. It is a healthy balance, and the site visit/ survey will help to optimize the design. Sometimes the building might not have been built when you are asked to perform the initial design. Obviously the design will then remain a 'draft' until you have the final floor plans. This is important – I know from experience that the internal layout of buildings can change significantly from the initial design and floor plans compared to the final construction.

- **Link budget verification**
  Quite often it is recommended to verify the propagation assumptions used to establish the actual link budget. For obvious reasons, it is important that we fulfill the requirements stated in the design requirement specification. We will verify this by performing a 'CV survey' with a test transmitter (refer to Section 5.2.5 for more details).
- **Interference levels**
  As you probably already know (if you have read the previous chapters in this book) you need to know the noise/interference level in the building/system in order to establish your design levels, and come up with the actual DAS design, the numbers of antennas, the design for the 'isolation' of the system, etc. So please always measure the actual levels from the outdoor network in and close to the building, as described in Section 5.2.4
- **Wall types, building layout and antenna locations**
  You may be surprised how often some buildings will have been changed since the floor plans you have were updated. Some buildings are pretty static and do not change too much over time, while others, such as shopping malls and airports, are more dynamic and are often ongoing construction sites! Therefore it is highly recommended to walk the whole structure before locking the final design, and also make sure that all the antenna locations can actually be implemented, documenting each and every location with photographs. Verify the main wall types. I know it is time-consuming and tedious work, but you will save a lot of time on the project by making sure there are no mistakes as a result of a simple lack of information and documentation.
- **Bring the installer!**
  No site visit should be done without the attendance of the responsible project manager from the installation team. Make sure they are in sync with your vision and ideas regarding the actual deployment, and make sure they verify the required cable ducts etc. to support the deployment of the DAS.

  This is done to ensure that your nice desktop design is actually deployable in real life. There will always be compromises and adjustments to the design during the site survey, so make sure to document this in details – preferably with photographs.

  Working with good installers and project managers means they will easily pick up your concerns and basic understanding of antenna locations, things to watch out for, what to avoid, etc. This knowledge will help to avoid a lot of issues later on, and save on cost by doing it right first time. I recommend sharing some basic RF planning knowledge and guidelines with the installers, as this will help them understand the importance of antenna locations and the like.

- **Output: Survey report**
  All the measurements, photographs and other data are saved and documented in detail in the 'survey report'. This document will help you adjust and verify the design assumptions, and also be an important input to the final design and final documentation needed in the installation process.

15.1.4   Update of Draft Design

The last process in the design phase is to update the draft design based on the inputs from the site survey, and the walk-through during the survey with the installation team.

- **Update documentation and design**
  During the site walk, we verified all antenna and equipment locations, documented these with photos, and updated locations and information on the floor plans. We also measured the propagation loss in the different area types in the building, and we measured the interference and presence of other signals in the building, in order to confirm the final link budget.
- **Output: Final design and bill of materials**
  The outputs we produce after adjusting the link budget and deciding on the final antenna locations are the 'design document, version final' and the 'bill of materials', so we can ensure all the required equipment is ready for the implementation phase.

15.2   The Implementation Phase

This is the part of the process where the responsibility is handed over to the installation team and project manager. The RF planner should still be interested in the process and participate as needed.

15.2.1   Installation

Based on the 'design document, version final' from the design procedure, the installation team will implement the DAS as agreed. The radio planner should, as a minimum, participate in a few installations, and I do mean help perform the actual installation work, in order to get some experience of what the nice lines on your design drawing actually mean in terms of the real-world installation – you might think twice about that 7/8' cable the next time you want to specify it, after having tried to install it yourself.

- **Installation work**
  The installation team will carefully make sure that everything is installed as agreed and specified in the documentation. They will also make sure to comply with all guidelines from the various manufacturers of the DAS components. Attention to issues like passive inter-modulation (PIM) is important (see Section 5.7 for more details). Good installation discipline and attention to the craftsmanship when installing DAS are very important to the performance of the DAS, and will help to avoid many potential issues now and in the future.

- **Output: As-build documentation**

  Based on the actual implementation, the 'as-build documentation' will be produced. These documents contain photos of all installed equipment, and details of the exact locations of every DAS component, cable and piece of equipment marked on the floor plan.

  An often neglected part of the documentation of the installation is the labeling of every single component in the DAS, and I do mean every cable end, every splitter, antenna, and active DAS element with the exact equipment number as per the design document. This labeling will save you many hours of troubleshooting in the future.

15.2.2 Post-installation Verification

After the installation has been completed and the as-build documentation has been reviewed by the radio planner, it is important that the radio planner performs an audit of the installation, on-site, prior to the commissioning and launch of the system.

- **On-site verification**

  The radio planner and the project manager from the installation team should verify all antenna locations. It is also a good idea to do some 'sample verifications' of cable installations, connector and general installation work – perhaps measure the voltage standing wave ratio (VSWR) and PIM, etc.
- **Adjustments to the installation**

  Any deviations in the installation from the design plan will be evaluated, and if need be corrective actions will be taken to bring the installation and documentation to be 100% in sync.
- **Output: Final as-build documentation**

  Based on the above work, the installation and documentation will be a 100% match and the final 'as-build documentation' will be produced. This document contains photos of all installed equipment, with details of the exact locations of every DAS component, cable and piece of equipment marked on the floor plan. It is recommended to have a paper copy on site for any troubleshooting or service work in the future.

15.2.3 DAS Test

Before we carry out the commissioning and set the system on-air, we need to ensure that the DAS performs to specifications on a component/sub-system basis. It would be far too comprehensive to give all the details of the various tests, so please consult the appropriate literature from the equipment and test equipment manufacturers for details.

All the installed equipment and components will already have been subjected to a visual inspection during the post-installation procedure.

- **Antenna tests**

  It is recommended to test all antennas, although I do realize that it is not likely to happen. At least perform sample testing of VSWR sweep on some selected antennas. Do the same with PIM, making sure to perform some sample tests at least of the antennas in the 'hot' part of the DAS where there is higher power.

- **Cable tests**

  (Please consult the documentation and manuals from test equipment and cable manufacturers for detailed descriptions and test procedures.)

  - **50 Ω coax test.** For passive installations, all sections of coax cables should be tested for PIM, sweep over distance (reflection vs. distance), VSWR, etc.
  - **Fiber test** – All fiber connectors should always be carefully cleaned before connecting to any equipment. No vacant ports should ever be left without protection caps.

    Before the fiber plant is taking into use, all fibers should be tested end to end with the optical time domain reflectometer (OTDR) test. This allows you to 'see' the fiber end-to-end by sweeping the complete link; all patches, splices and distances will be shown in detail in terms of reflections and distance losses. (Note that if you use analog active DAS, the fiber will most probably need to be terminated with APC connectors all through the installation to avoid reflections (including optical patch panels)
  - **CAT5/6 test** – Many types of active DAS utilize CAT5/6 ethernet type cables, and these should be tested typically against the TIA/EIA-568A standard. As a minimum you should test the 'wire map' and the loop resistance, to ensure the DC power to the remote unit can be maintained. (I have seen cases where cable we believed to be copper turned out instead to be copper-clad iron, resulting in excessive DC losses, and no power to the remotes!)
  - **CATV test** – Many types of active DAS utilize cable TV-type coax (75 Ω). This cable typically carries an intermediate frequency for the RF signals as well as DC power to the remote antenna units. Again, you will typically perform the same types of tests as for standard coax: PIM, sweep over distance (reflection vs. distance), VSWR, and so forth. DC loop resistance is also important to ensure we can maintain the DC power to the remote antenna units.

- **Equipment test**

  All passive components should be tested, at least on a 'sample basis', the POI (Point Of Interface) must be thoroughly tested from 'port to port' (and test reports from the suppliers should be checked and verified), unused ports must be terminated.

  The Active DAS equipment will often have quite detailed test program internally, these automatic tests would often perform en 'end-to-end' test, even checking signal levels on the various fiber and RF ports of the system. The active equipment will typical also be able to save logs, test reports and measurement files for documentation purposes

- **Output: DAS test report**

  The measurements of all the DAS elements are carefully saved in a 'DAS test report'. This will comprise logs from the active equipment, screenshots and measurement reports from the cable, antenna and fiber tests, etc. This report should have a 'pass/fail' criterion to ensure we will only proceed with the following commissioning once everything is verified and documented.

15.2.4 Commissioning

The commissioning is the point where we finally get the system on-air and are able to verify whether all our design, implementation and verification efforts will result in a system that performs as we want it to. Normally the radio designer will not perform the actual commissioning, but I do recommend taking part in at least a few commissionings in order to get the full experience of the project cycle.

The commissioning would also typically be the time where the project is concluded and 'signed off' as 'delivered as per specification' and when the 'as-build documentation' would be accepted as the actual implemented solution.

- **On-site inspection**
  Normally the commissioning will commence with a quick check of all the installed equipment in the equipment room. We will check the status indicators on all active equipment, such as power supply, battery backup, air-con, base stations, repeaters, and active DAS elements.

  Also the main passive components will swiftly undergo a visual inspection, and the point of interface (POI) and cables connecting the base station will be checked. After verifying that everything is okay, we would normally take a few pictures to document the status and any deviations will obviously be documented in detail and corrected prior to the launch of the system.

- **Preparing the active elements**
  If we are using an active DAS, we will start bringing the system on, perform automatic system tests, ensure the alarm status is okay, and save any log files and screenshots to document the status of the system. The base station or repeater that will feed the DAS is then brought on-air, connected to a dummy load via a power meter analyzer. Once the RF signal is up and stable, we will measure the required power levels and verify the functionality of the serving cell. The RF levels are needed to ensure that we feed the DAS the correct power, and will help us to align gain and attenuation in the active DAS, if used. Normally the signal will pass through a POI, and it is recommended to perform these measurements and tests as close to the DAS as possible to ensure that all components and cables are included. All other active elements such as modems, IP links, and alarm interfaces, must also be checked.

  We will document all the measurements and status logs, screenshots etc. from the test to be used in the commissioning report as a reference

- **On-air**
  Once the RF power leveling and basic test have been completed, we can bring the system on-air. We will level, adjust and verify the active parts of the DAS. We will then verify the functionality of the basic services at the closest DAS antenna, ensuring that calls and data sessions are possible, and that the base station can actually carry traffic (is unlocked in the network). It is very important to do this as soon as possible to ensure that mobile users inside the building are able to use the network.

  After the functionality test, we will perform measurement samples at antenna locations to establish that the coverage and services are as expected. We will typically measure what we expect to be the weakest place in the design, as well sample measurements in all the main sections of the DAS/building.

- **Output: Commissioning report**
  All the measurements conducted during the commissioning are saved and documented in detail in the 'DAS commissioning report'. This will contain logs from the active equipment, screenshots and measurement reports from all the tests, and settings of the active equipment for reference. All sample measurements in the building should be documented on floor plans with an indication of location.

  This report should have a 'pass/fail' criterion to ensure we will only proceed with the following phase (verification) once everything is verified and documented. At the end of the commissioning report, there should be a conclusion to ensure that the design requirements have been verified, as well as concerns and/or future adjustments that might be needed.

15.3   The Verification Phase

This is now the final phase of the system implementation, when the radio planner tests and verifies all the propagation calculations and system performance. It is crucial that you compare measurements and calculations to learn from real-life performance, but also to catch any discrepancies that might arise from mistakes during implementation/tuning of the complete system.

15.3.1   RF Verification

Often the RF verification of the implemented DAS would be completed just after the commissioning, and usually by the same team. The RF verification will also include checking that all the handovers to and from the DAS are functional, and thus all entrances to the building must be verified, as well as inter-cell functionality inside the building

- **RF verification measurements**
  The measurements should preferably be conducted using a tool that can log the RF field measurements whilst tracking the location on a floor plan. This would make it possible to pinpoint the actual location on the floor plan in sync with the measurements. Basic RF signal levels, quality, neighbor information, etc. are logged. It is recommended to activate a 'trace' function in the network at the same time, where specific KPIs on the test mobile are logged, so you get information about uplink performance.
- **Output: In-service documentation**
  All the measurements conducted during the RF verification test are saved and documented in detail in the 'In-service documentation', which will consist of log and trace files, as well as statistical information to support a conclusion on the overall performance such as '98% of measurement samples are above −75 dBm'. A check matrix is recommended to document the handover functionality between all neighboring cells.
- **Do learn from the documentation**
  The in-service measurements are also very important for the radio planner, obviously to conclude that everything is performing at or better than the planned level. But they will also ensure that the system design was not 'overkill', exceeding the actual requirements, and at a much higher cost than was actually needed. So I really recommend using the documentation as an active part of the experience-building in the planning team; it is a great way to share the experience in a structured way.

15.3.2   Live Traffic Test

The key purpose behind designing and implementing the DAS was to improve the wireless services inside the building. And although we will do our best to verify all the KPIs during the RF verification, what really counts is how the users are experiencing and using the DAS. To conclude on this, we will rely on the performance 'dashboards' that are available in the mobile network itself.

- **Live traffic testing**

  The network will have a large database logging various 'counters' of events and measurements related to the specific cells performance. This will include radio performance measurements in terms of level and quality, call completion success rates, handover success, data download statistics, and traffic and voice volume in the cell. All these measurements and events till trigger the statistical counters in the network with a time stamp or in 'bins' of, for example, 15-minute averages and are used as inputs for a complex statistical tool that would be capable of presenting these performance metrics in a graphical dashboards. Often these tools will have user-defined reports, trigger levels for performance alarms, etc.

  The mobile operator will usually have a well-defined 'script' that will run a detailed 'in-service report' after bringing a new cell on-air. This will ensure that a fixed set of monitored parameters over a fixed timescale will be performed and conclude the success of the newly implemented system, before the new DAS is handed over to normal daily operation.

- **Output: Performance report**

  The output of the live traffic testing will be a well-defined 'performance report' and it is important that the radio planner carefully follows the results and concludes on this report prior to handing over the system to 'operations'.

15.4 Conclusion

I hope this chapter was beneficial, and will perhaps inspire you to implement a well-thought-out and planned process for your projects from start to finish. The documentation and traceability are the cornerstone and will help you not only to plan and implement the needed solutions, but also to structure your learning from these planned and deployed systems, ensuring that you constantly improve both your processes and the planning and implementation of these projects.

A lot of the process, from planning through to verification, can be supported by the tool described in Section 9.7.

References

This book is a result of more than 23 years' experience in indoor radio planning, working for a mobile operator as an RF-planner and working for leading suppliers of DAS equipment. The subjects, examples and approaches in this book are based on practical experience and indoor implementations I have been responsible for and directly involved in.

My interest is the RF part of these radio systems and not all the deep background core network or protocols. There are no direct transcripts or copies from other books, but I have to recommend some of the daily reference material I use myself. These are the books that helped me understand the basics, and that I use as my reference.

[1] Michel Mouly and Marie-Bernadette Pautet. *The GSM System for Mobile Communications*. Published by the authors (ISBN 2-9507190-0-7).
[2] P. E. Clint Smith and Curt Gervelis. *Cellular System Design & Optimization*. McGraw-Hill: New York.
[3] Harri Holma and Antti Toskala (eds). *WCDMA for UMTS, Radio Access for Third Generation Mobile Communications*. Wiley: Chichester.
[4] Jaana Laiho, Achim Wacker and Tomáš Novosad (eds). *Radio Network Planning and Optimisation for UMTS*, 2nd edn. Wiley: Chichester.
[5] Harri Holma and Antti Toskala (eds). *HSDPA/HSUPA for UMTS, High Speed Radio Access for Mobile Communications*. Wiley: Chichester.
[6] S. R. Saunders and A. Aragon-Zavala. *Antennas and Propagation for Wireless Communication Systems*, 2nd edn. Wiley: Chichester.
[7] S. Sesia, I. Toufik and M. Baker. *The UMTS Long Term Evolution, From Theory To Practice*. Wiley: Chichester.
[8] E. Dahlman, S. Parkvall, J. Skold and P. Beming. *3G Evolution HSPA and LTE for Mobile Broadband*. Academic Press.
[9] S. R. Saunders, S. Carlaw, A. Giustina, R. R. Bhat, V. S. Rao and R. Siegberg. *Femtocells Opportunities and Challenges for Business and Technology*. Wiley: Chcichester.

Indoor Radio Planning: A Practical Guide for 2G, 3G and 4G, Third Edition. Morten Tolstrup.
© 2015 John Wiley & Sons, Ltd. Published 2015 by John Wiley & Sons, Ltd.

Appendix

Reference Material

Conversions of Watt to dBm, dBm to Watt

$$\text{dBm} = 10\log\left(\frac{power(\text{W})}{1\text{mW}}\right)$$

$$P\text{mW} = 10^{(\text{dBm}/10)}$$

Table A.1

Watt	dBm	Watt	dBm	Watt	dBm	Watt	dBm	Watt	dBm	Watt	dBm
0.001	0.00	2.1	33.22	5.90	37.71	9.70	39.87	13.50	41.30	17.30	42.38
0.002	3.01	2.2	33.42	6.00	37.78	9.80	39.91	13.60	41.34	17.40	42.41
0.003	4.77	2.3	33.62	6.10	37.85	9.90	39.96	13.70	41.37	17.50	42.43
0.004	6.02	2.4	33.80	6.20	37.92	10.00	40.00	13.80	41.40	17.60	42.46
0.005	6.99	2.5	33.98	6.30	37.99	10.10	40.04	13.90	41.43	17.70	42.48
0.006	7.78	2.6	34.15	6.40	38.06	10.20	40.09	14.00	41.46	17.80	42.50
0.007	8.45	2.7	34.31	6.50	38.13	10.30	40.13	14.10	41.49	17.90	42.53
0.008	9.03	2.8	34.47	6.60	38.20	10.40	40.17	14.20	41.52	18.00	42.55
0.009	9.54	2.9	34.62	6.70	38.26	10.50	40.21	14.30	41.55	18.10	42.58
0.010	10.00	3	34.77	6.80	38.33	10.60	40.25	14.40	41.58	18.20	42.60
0.020	13.01	3.1	34.91	6.90	38.39	10.70	40.29	14.50	41.61	18.30	42.62
0.030	14.77	3.2	35.05	7.00	38.45	10.80	40.33	14.60	41.64	18.40	42.65
0.040	16.02	3.3	35.19	7.10	38.51	10.90	40.37	14.70	41.67	18.50	42.67
0.050	16.99	3.4	35.31	7.20	38.57	11.00	40.41	14.80	41.70	18.60	42.70
0.060	17.78	3.5	35.44	7.30	38.63	11.10	40.45	14.90	41.73	18.70	42.72
0.070	18.45	3.6	35.56	7.40	38.69	11.20	40.49	15.00	41.76	18.80	42.74

Indoor Radio Planning: A Practical Guide for 2G, 3G and 4G, Third Edition. Morten Tolstrup.
© 2015 John Wiley & Sons, Ltd. Published 2015 by John Wiley & Sons, Ltd.

Table A.1 *(Continued)*

Watt	dBm	Watt	dBm	Watt	dBm	Watt	dBm	Watt	dBm	Watt	dBm
0.080	19.03	3.7	35.68	7.50	38.75	11.30	40.53	15.10	41.79	18.90	42.76
0.090	19.54	3.80	35.80	7.60	38.81	11.40	40.57	15.20	41.82	19.00	42.79
0.100	20.00	3.90	35.91	7.70	38.86	11.50	40.61	15.30	41.85	19.10	42.81
0.200	23.01	4.00	36.02	7.80	38.92	11.60	40.64	15.40	41.88	19.20	42.83
0.300	24.77	4.10	36.13	7.90	38.98	11.70	40.68	15.50	41.90	19.30	42.86
0.400	26.02	4.20	36.23	8.00	39.03	11.80	40.72	15.60	41.93	19.40	42.88
0.500	26.99	4.30	36.33	8.10	39.08	11.90	40.76	15.70	41.96	19.50	42.90
0.600	27.78	4.40	36.43	8.20	39.14	12.00	40.79	15.80	41.99	19.60	42.92
0.700	28.45	4.50	36.53	8.30	39.19	12.10	40.83	15.90	42.01	19.70	42.94
0.800	29.03	4.60	36.63	8.40	39.24	12.20	40.86	16.00	42.04	19.80	42.97
0.900	29.54	4.70	36.72	8.50	39.29	12.30	40.90	16.10	42.07	19.90	42.99
1.000	30.00	4.80	36.81	8.60	39.34	12.40	40.93	16.20	42.10	20.00	43.01
1.100	30.41	4.90	36.90	8.70	39.40	12.50	40.97	16.30	42.12	25.00	44.77
1.200	30.79	5.00	36.99	8.80	39.44	12.60	41.00	16.40	42.15	30.00	46.99
1.300	31.14	5.10	37.08	8.90	39.49	12.70	41.04	16.50	42.17	35.00	45.44
1.400	31.46	5.20	37.16	9.00	39.54	12.80	41.07	16.60	42.20	40.00	46.02
1.500	31.76	5.30	37.24	9.10	39.59	12.90	41.11	16.70	42.23	50.00	47.00
1.600	32.04	5.40	37.32	9.20	39.64	13.00	41.14	16.80	42.25	60.00	47.78
1.700	32.30	5.50	37.40	9.30	39.68	13.10	41.17	16.90	42.28	70.00	48.45
1.800	32.55	5.60	37.48	9.40	39.73	13.20	41.21	17.00	42.30	80.00	49.03
1.900	32.79	5.70	37.56	9.50	39.78	13.30	41.24	17.10	42.33	90.00	49.54
2.000	33.01	5.80	37.63	9.60	39.82	13.40	41.27	17.20	42.36	100.00	50.00

Noise Power Calculator

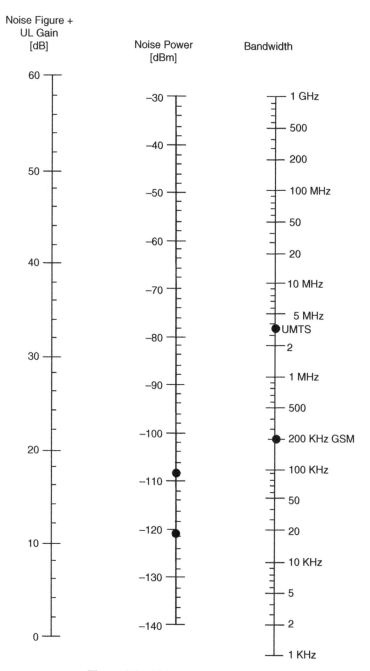

Figure A.1   Noise power normograph

Table A.2   Erlang B table, 1–50 channels, 0.01–5% grade of service

Number of channels	Blocking (GOS)					Number of channels
	0.1%	0.5%	1%	2%	5%	
1	0.0010	0.0050	0.0101	0.0204	0.0526	1
2	0.0458	0.1054	0.1526	0.2235	0.3813	2
3	0.1938	0.3490	0.4555	0.6022	0.8994	3
4	0.4393	0.7012	0.8694	1.0923	1.5246	4
5	0.7621	1.1320	1.3608	1.6571	2.2185	5
6	1.1459	1.6218	1.9090	2.2759	2.9603	6
7	1.5786	2.1575	2.5009	2.9354	3.7378	7
8	2.0513	2.7299	3.1276	3.6271	4.5430	8
9	2.5575	3.3326	3.7825	4.3447	5.3702	9
10	3.0920	3.9607	4.4612	5.0840	6.2157	10
11	3.6511	4.6104	5.1599	5.8415	7.0764	11
12	4.2314	5.2789	5.8760	6.6147	7.9501	12
13	4.8306	5.9638	6.6072	7.4015	8.8349	13
14	5.4464	6.6632	7.3517	8.2003	9.7295	14
15	6.0772	7.3755	8.1080	9.0096	10.633	15
16	6.7215	8.0995	8.8750	9.8284	11.544	16
17	7.3781	8.8340	9.6516	10.656	12.461	17
18	8.0459	9.5780	10.437	11.491	13.385	18
19	8.7239	10.331	11.230	12.333	14.315	19
20	9.4115	11.092	12.031	13.182	15.249	20
21	10.108	11.860	12.838	14.036	16.189	21
22	10.812	12.635	13.651	14.896	17.132	22
23	11.524	13.416	14.470	15.761	18.080	23
24	12.243	14.204	15.295	16.631	19.031	24
25	12.969	14.997	16.125	17.505	19.985	25
26	13.701	15.795	16.959	18.383	20.943	26
27	14.439	16.598	17.797	19.265	21.904	27
28	15.182	17.406	18.640	20.150	22.867	28
29	15.930	18.218	19.487	21.039	23.833	29
30	16.684	19.034	20.337	21.932	24.802	30
31	17.442	19.854	21.191	22.827	25.773	31
32	18.205	20.678	22.048	23.725	26.746	32
33	18.972	21.505	22.909	24.626	27.721	33
34	19.743	22.336	23.772	25.529	28.698	34
35	20.517	23.169	24.638	26.435	29.677	35
36	21.296	24.006	25.507	27.343	30.657	36
37	22.078	24.846	26.378	28.254	31.640	37
38	22.864	25.689	27.252	29.166	32.624	38
39	23.652	26.534	28.129	30.081	33.609	39
40	24.444	27.382	29.007	30.997	34.596	40
41	25.239	28.232	29.888	31.916	35.584	41
42	26.037	29.085	30.771	32.836	36.574	42
43	26.837	29.940	31.656	33.758	37.565	43
44	27.641	30.797	32.543	34.682	38.557	44
45	28.447	31.656	33.432	35.607	39.550	45
46	29.255	32.517	34.322	36.534	40.545	46
47	30.066	33.381	35.215	37.462	41.540	47
48	30.879	34.246	36.109	38.392	42.537	48
49	31.694	35.113	37.004	39.323	43.534	49
50	32.512	35.982	37.901	40.255	44.533	50

Unit Conversions

Decibel Conversion: Power

$$dB = 10\log(P2/P1)$$

Decibels relative to power.

Decibel Conversion: Voltage

$$dB = 20\log[V1/V2]$$

Decibels relative to voltage across the same resistance.

Decibel Conversion: Current

$$dB = 20\log[I1/I2]$$

Decibels relative to current through same resistance.

Decibel Conversion: Milliwatts

$$dBm = 10\log[single(\text{mW})/1\text{mW}]$$

Decibels relative to 1 mW.

Decibel Conversion: Microvolts

$$dB\mu V = 20\log[single(\mu V)/1\mu V]$$

Decibels relative to 1 mV across the same resistance.

Decibel Conversion: Microamps

$$dB\mu A = 20\log[single(\mu A)/1\mu A]$$

Decibels relative to one microamp through the same resistance.

Power Conversion: dBw to dBm

$$dBm = dBw + 30$$

Conversion from dBw to dBm.

Voltage Conversion: dBV to dBµV

$$dB\mu V = dBV + 120$$

Conversion from dBV to dBµV.

Voltage to Power Conversion: dBµV to dBm

$$dBm = dB\mu V - 107$$

where the constant 107 is as follows:
  RF systems are matched to 50 V

$$P = V^2 / R$$
$$10\log_{10}(P) = 20\log_{10}(V) - 10\log_{10}(50\Omega)$$
$$V = (PR)^{0.5} = 0.223\text{V} = 223000\mu\text{V}$$

For a resistance of 50 Ω and a power of 1 mW:

$$20\log_{10}[223000\mu\text{V}] = 107\text{dB}$$

Power Density

$$\text{dBW/m}^2 = 10\log_{10}(\text{V/m} - \text{A/m})$$

Decibel–Watts per square meter.

$$\text{dBm/m}^2 = \text{dBW/m}^2 + 30$$

where the constant 30 is the decibel equivalent of the factor 1000 used to convert between W and mW:

$$10\log_{10}[1000] = 30$$

Electric Field Voltage

$$\text{V/m} = 10^{\{[(\text{dB}\mu\text{V/m}) - 120]/20\}}$$

Electric field voltage in Volts per meter.

Electric Field Current

$$\text{dB}\mu\text{A/m} = \text{dB}\mu\text{V/m} - 51.5$$

where the constant 51.5 is a conversion of the characteristic impedance of free space (120 Ω) into decibels: $20\log_{10}(120\ \Omega) = 51.5$.

$$\text{A/m} = 10^{\{[(\text{dB}\mu\text{A/m}) - 120]/20\}}$$

Electric field current in Amps per meter.

Conversion Units

$dB = decibels\,(\log 10)$

$\mu = micro = 10 \times 10^{-6}$

$m = milli = 10 \times 10^{-3}$

$dBw = decibels\;relative\,to\,1\,W$

$dBm = decibels\;relative\,to\,1\,mW$

$dBV = decibels\;relative\,to\,1\,V$

$dB\mu V = decibels\;relative\,to\,1\,\mu V$

$dB\mu A = decibels\;relative\,to\,1\,\mu A$

V = Volts; A = Amps; I = current; R = resistance V = Ohms (50); W = Watts;
P = Power; m = meters

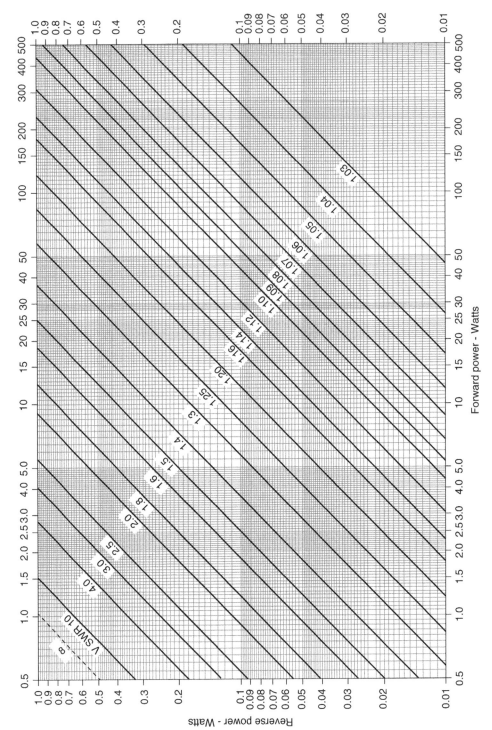

Figure A.2 VSWR chart

Table A.3 GSM 900 channels and frequencies

Channel	UL (MHz)	DL (MHz)	Channel	UL (MHz)	DL (MHz)	Channel	UL (MHz)	DL (MHz)
1	890 200	935 200	41	898 200	943 200	81	906 200	951 200
2	890 400	935 400	42	898 400	943 400	82	906 400	951 400
3	890 600	935 600	43	898 600	943 600	83	906 600	951 600
4	890 800	935 800	44	898 800	943 800	84	906 800	951 800
5	891 000	936 000	45	899 000	944 000	85	907 000	952 000
6	891 200	936 200	46	899 200	944 200	86	907 200	952 200
7	891 400	936 400	47	899 400	944 400	87	907 400	952 400
8	891 600	936 600	48	899 600	944 600	88	907 600	952 600
9	891 800	936 800	49	899 800	944 800	89	907 800	952 800
10	892 000	937 000	50	900 000	945 000	90	908 000	953 000
11	892 200	937 200	51	900 200	945 200	91	908 200	953 200
12	892 400	937 400	52	900 400	945 400	92	908 400	953 400
13	892 600	937 600	53	900 600	945 600	93	908 600	953 600
14	892 800	937 800	54	900 800	945 800	94	908 800	953 800
15	893 000	938 000	55	901 000	946 000	95	909 000	954 000
16	893 200	938 200	56	901 200	946 200	96	909 200	954 200
17	893 400	938 400	57	901 400	946 400	97	909 400	954 400
18	893 600	938 600	58	901 600	946 600	98	909 600	954 600
19	893 800	938 800	59	901 800	946 800	99	909 800	954 800
20	894 000	939 000	60	902 000	947 000	100	910 000	955 000
21	894 200	939 200	61	902 200	947 200	101	910 200	955 200
22	894 400	939 400	62	902 400	947 400	102	910 400	955 400
23	894 600	939 600	63	902 600	947 600	103	910 600	955 600
24	894 800	939 800	64	902 800	947 800	104	910 800	955 800
25	895 000	940 000	65	903 000	948 000	105	911 000	956 000
26	895 200	940 200	66	903 200	948 200	106	911 200	956 200
27	895 400	940 400	67	903 400	948 400	107	911 400	956 400
28	895 600	940 600	68	903 600	948 600	108	911 600	956 600
29	895 800	940 800	69	903 800	948 800	109	911 800	956 800
30	896 000	941 000	70	904 000	949 000	110	912 000	957 000
31	896 200	941 200	71	904 200	949 200	111	912 200	957 200
32	896 400	941 400	72	904 400	949 400	112	912 400	957 400
33	896 600	941 600	73	904 600	949 600	113	912 600	957 600
34	896 800	941 800	74	904 800	949 800	114	912 800	957 800
35	897 000	942 000	75	905 000	950 000	115	913 000	958 000
36	897 200	942 200	76	905 200	950 200	116	913 200	958 200
37	897 400	942 400	77	905 400	950 400	117	913 400	958 400
38	897 600	942 600	78	905 600	950 600	118	913 600	958 600
39	897 800	942 800	79	905 800	950 800	119	913 800	958 800
40	898 000	943 000	80	906 000	951 000	120	914 000	959 000

Table A.4 GSM 1800 channels and frequencies

Channel	UL (MHz)	DL (MHz)	Channel	UL (MHz)	DL (MHz)	Channel	UL (MHz)	DL (MHz)	Channel	UL (MHz)	DL (MHz)
			570	1 721.800	1 816.800	630	1 733.800	1 828.800	690	1 745.800	1 840.800
			571	1 722.000	1 817.000	631	1 734.000	1 829.000	691	1 746.000	1 841.000
512	1 710.200	1 805.200	572	1 722.200	1 817.200	632	1 734.200	1 829.200	692	1 746.200	1 841.200
513	1 710.400	1 805.400	573	1 722.400	1 817.400	633	1 734.400	1 829.400	693	1 746.400	1 841.400
514	1 710.600	1 805.600	574	1 722.600	1 817.600	634	1 734.600	1 829.600	694	1 746.600	1 841.600
515	1 710.800	1 805.800	575	1 722.800	1 817.800	635	1 734.800	1 829.800	695	1 746.800	1 841.800
516	1 711.000	1 806.000	576	1 723.000	1 818.000	636	1 735.000	1 830.000	696	1 747.000	1 842.000
517	1 711.200	1 806.200	577	1 723.200	1 818.200	637	1 735.200	1 830.200	697	1 747.200	1 842.200
518	1 711.400	1 806.400	578	1 723.400	1 818.400	638	1 735.400	1 830.400	698	1 747.400	1 842.400
519	1 711.600	1 806.600	579	1 723.600	1 818.600	639	1 735.600	1 830.600	699	1 747.600	1 842.600
520	1 711.800	1 806.800	580	1 723.800	1 818.800	640	1 735.800	1 830.800	700	1 747.800	1 842.800
521	1 712.000	1 807.000	581	1 724.000	1 819.000	641	1 736.000	1 831.000	701	1 748.000	1 843.000
522	1 712.200	1 807.200	582	1 724.200	1 819.200	642	1 736.200	1 831.200	702	1 748.200	1 843.200
523	1 712.400	1 807.400	583	1 724.400	1 819.400	643	1 736.400	1 831.400	703	1 748.400	1 843.400
524	1 712.600	1 807.600	584	1 724.600	1 819.600	644	1 736.600	1 831.600	704	1 748.600	1 843.600
525	1 712.800	1 807.800	585	1 724.800	1 819.800	645	1 736.800	1 831.800	705	1 748.800	1 843.800
526	1 713.000	1 808.000	586	1 725.000	1 820.000	646	1 737.000	1 832.000	706	1 749.000	1 844.000
527	1 713.200	1 808.200	587	1 725.200	1 820.200	647	1 737.200	1 832.200	707	1 749.200	1 844.200
528	1 713.400	1 808.400	588	1 725.400	1 820.400	648	1 737.400	1 832.400	708	1 749.400	1 844.400
529	1 713.600	1 808.600	589	1 725.600	1 820.600	649	1 737.600	1 832.600	709	1 749.600	1 844.600
530	1 713.800	1 808.800	590	1 725.800	1 820.800	650	1 737.800	1 832.800	710	1 749.800	1 844.800
531	1 714.000	1 809.000	591	1 726.000	1 821.000	651	1 738.000	1 833.000	711	1 750.000	1 845.000
532	1 714.200	1 809.200	592	1 726.200	1 821.200	652	1 738.200	1 833.200	712	1 750.200	1 845.200
533	1 714.400	1 809.400	593	1 726.400	1 821.400	653	1 738.400	1 833.400	713	1 750.400	1 845.400
534	1 714.600	1 809.600	594	1 726.600	1 821.600	654	1 738.600	1 833.600	714	1 750.600	1 845.600
535	1 714.800	1 809.800	595	1 726.800	1 821.800	655	1 738.800	1 833.800	715	1 750.800	1 845.800
536	1 715.000	1 810.000	596	1 727.000	1 822.000	656	1 739.000	1 834.000	716	1 751.000	1 846.000
537	1 715.200	1 810.200	597	1 727.200	1 822.200	657	1 739.200	1 834.200	717	1 751.200	1 846.200
538	1 715.400	1 810.400	598	1 727.400	1 822.400	658	1 739.400	1 834.400	718	1 751.400	1 846.400
539	1 715.600	1 810.600	599	1 727.600	1 822.600	659	1 739.600	1 834.600	719	1 751.600	1 846.600
540	1 715.800	1 810.800	600	1 727.800	1 822.800	660	1 739.800	1 834.800	720	1 751.800	1 846.800
541	1 716.000	1 811.000	601	1 728.000	1 823.000	661	1 740.000	1 835.000	721	1 752.000	1 847.000
542	1 716.200	1 811.200	602	1 728.200	1 823.200	662	1 740.200	1 835.200	722	1 752.200	1 847.200
543	1 716.400	1 811.400	603	1 728.400	1 823.400	663	1 740.400	1 835.400	723	1 752.400	1 847.400
544	1 716.600	1 811.600	604	1 728.600	1 823.600	664	1 740.600	1 835.600	724	1 752.600	1 847.600
545	1 716.800	1 811.800	605	1 728.800	1 823.800	665	1 740.800	1 835.800	725	1 752.800	1 847.800
546	1 717.000	1 812.000	606	1 729.000	1 824.000	666	1 741.000	1 836.000	726	1 753.000	1 848.000
547	1 717.200	1 812.200	607	1 729.200	1 824.200	667	1 741.200	1 836.200	727	1 753.200	1 848.200
548	1 717.400	1 812.400	608	1 729.400	1 824.400	668	1 741.400	1 836.400	728	1 753.400	1 848.400
549	1 717.600	1 812.600	609	1 729.600	1 824.600	669	1 741.600	1 836.600	729	1 753.600	1 848.600
550	1 717.800	1 812.800	610	1 729.800	1 824.800	670	1 741.800	1 836.800	730	1 753.800	1 848.800
551	1 718.000	1 813.000	611	1 730.000	1 825.000	671	1 742.000	1 837.000	731	1 754.000	1 849.000
552	1 718.200	1 813.200	612	1 730.200	1 825.200	672	1 742.200	1 837.200	732	1 754.200	1 849.200
553	1 718.400	1 813.400	613	1 730.400	1 825.400	673	1 742.400	1 837.400	733	1 754.400	1 849.400
554	1 718.600	1 813.600	614	1 730.600	1 825.600	674	1 742.600	1 837.600	734	1 754.600	1 849.600
555	1 718.800	1 813.800	615	1 730.800	1 825.800	675	1 742.800	1 837.800	735	1 754.800	1 849.800
556	1 719.000	1 814.000	616	1 731.000	1 826.000	676	1 743.000	1 838.000	736	1 755.000	1 850.000
557	1 719.200	1 814.200	617	1 731.200	1 826.200	677	1 743.200	1 838.200	737	1 755.200	1 850.200
558	1 719.400	1 814.400	618	1 731.400	1 826.400	678	1 743.400	1 838.400	738	1 755.400	1 850.400
559	1 719.600	1 814.600	619	1 731.600	1 826.600	679	1 743.600	1 838.600	739	1 755.600	1 850.600
560	1 719.800	1 814.800	620	1 731.800	1 826.800	680	1 743.800	1 838.800	740	1 755.800	1 850.800
561	1 720.000	1 815.000	621	1 732.000	1 827.000	681	1 744.000	1 839.000	741	1 756.000	1 851.000
562	1 720.200	1 815.200	622	1 732.200	1 827.200	682	1 744.200	1 839.200	742	1 756.200	1 851.200
563	1 720.400	1 815.400	623	1 732.400	1 827.400	683	1 744.400	1 839.400	743	1 756.400	1 851.400
564	1 720.600	1 815.600	624	1 732.600	1 827.600	684	1 744.600	1 839.600	744	1 756.600	1 851.600
565	1 720.800	1 815.800	625	1 732.800	1 827.800	685	1 744.800	1 839.800	745	1 756.800	1 851.800
566	1 721.000	1 816.000	626	1 733.000	1 828.000	686	1 745.000	1 840.000	746	1 757.000	1 852.000
567	1 721.200	1 816.200	627	1 733.200	1 828.200	687	1 745.200	1 840.200	747	1 757.200	1 852.200
568	1 721.400	1 816.400	628	1 733.400	1 828.400	688	1 745.400	1 840.400	748	1 757.400	1 852.400
569	1 721.600	1 816.600	629	1 733.600	1 828.600	689	1 745.600	1 840.600	749	1 757.600	1 852.600

Table A.5

Channel	UL (MHz)	DL (MHz)	Channel	UL (MHz)	DL (MHz)	Channel	UL (MHz)	DL (MHz)
750	1 757.800	1 852.800	810	1 769.800	1 864.800	870	1 781.800	1 876.800
751	1 758.000	1 853.000	811	1 770.000	1 865.000	871	1 782.000	1 877.000
752	1 758.200	1 853.200	812	1 770.200	1 865.200	872	1 782.200	1 877.200
753	1 758.400	1 853.400	813	1 770.400	1 865.400	873	1 782.400	1 877.400
754	1 758.600	1 853.600	814	1 770.600	1 865.600	874	1 782.600	1 877.600
755	1 758.800	1 853.800	815	1 770.800	1 865.800	875	1 782.800	1 877.800
756	1 759.000	1 854.000	816	1 771.000	1 866.000	876	1 783.000	1 878.000
757	1 759.200	1 854.200	817	1 771.200	1 866.200	877	1 783.200	1 878.200
758	1 759.400	1 854.400	818	1 771.400	1 866.400	878	1 783.400	1 878.400
759	1 759.600	1 854.600	819	1 771.600	1 866.600	879	1 783.600	1 878.600
760	1 759.800	1 854.800	820	1 771.800	1 866.800	880	1 783.800	1 878.800
761	1 760.000	1 855.000	821	1 772.000	1 867.000	881	1 784.000	1 879.000
762	1 760.200	1 855.200	822	1 772.200	1 867.200	882	1 784.200	1 879.200
763	1 760.400	1 855.400	823	1 772.400	1 867.400	883	1 784.400	1 879.400
764	1 760.600	1 855.600	824	1 772.600	1 867.600	884	1 784.600	1 879.600
765	1 760.800	1 855.800	825	1 772.800	1 867.800	885	1 784.800	1 879.800
766	1 761.000	1 856.000	826	1 773.000	1 868.000			
767	1 761.200	1 856.200	827	1 773.200	1 868.200			
768	1 761.400	1 856.400	828	1 773.400	1 868.400			
769	1 761.600	1 856.600	829	1 773.600	1 868.600			
770	1 761.800	1 856.800	830	1 773.800	1 868.800			
771	1 762.000	1 857.000	831	1 774.000	1 869.000			
772	1 762.200	1 857.200	832	1 774.200	1 869.200			
773	1 762.400	1 857.400	833	1 774.400	1 869.400			
774	1 762.600	1 857.600	834	1 774.600	1 869.600			
775	1 762.800	1 857.800	835	1 774.800	1 869.800			
776	1 763.000	1 858.000	836	1 775.000	1 870.000			
777	1 763.200	1 858.200	837	1 775.200	1 870.200			
778	1 763.400	1 858.400	838	1 775.400	1 870.400			
779	1 763.600	1 858.600	839	1 775.600	1 870.600			
780	1 763.800	1 858.800	840	1 775.800	1 870.800			
781	1 764.000	1 859.000	841	1 776.000	1 871.000			
782	1 764.200	1 859.200	842	1 776.200	1 871.200			
783	1 764.400	1 859.400	843	1 776.400	1 871.400			
784	1 764.600	1 859.600	844	1 776.600	1 871.600			
785	1 764.800	1 859.800	845	1 776.800	1 871.800			
786	1 765.000	1 860.000	846	1 777.000	1 872.000			
787	1 765.200	1 860.200	847	1 777.200	1 872.200			
788	1 765.400	1 860.400	848	1 777.400	1 872.400			
789	1 765.600	1 860.600	849	1 777.600	1 872.600			
790	1 765.800	1 860.800	850	1 777.800	1 872.800			
791	1 766.000	1 861.000	851	1 778.000	1 873.000			
792	1 766.200	1 861.200	852	1 778.200	1 873.200			
793	1 766.400	1 861.400	853	1 778.400	1 873.400			
794	1 766.600	1 861.600	854	1 778.600	1 873.600			
795	1 766.800	1 861.800	855	1 778.800	1 873.800			
796	1 767.000	1 862.000	856	1 779.000	1 874.000			
797	1 767.200	1 862.200	857	1 779.200	1 874.200			
798	1 767.400	1 862.400	858	1 779.400	1 874.400			
799	1 767.600	1 862.600	859	1 779.600	1 874.600			
800	1 767.800	1 862.800	860	1 779.800	1 874.800			
801	1 768.000	1 863.000	861	1 780.000	1 875.000			
802	1 768.200	1 863.200	862	1 780.200	1 875.200			
803	1 768.400	1 863.400	863	1 780.400	1 875.400			
804	1 768.600	1 863.600	864	1 780.600	1 875.600			
805	1 768.800	1 863.800	865	1 780.800	1 875.800			
806	1 769.000	1 864.000	866	1 781.000	1 876.000			
807	1 769.200	1 864.200	867	1 781.200	1 876.200			
808	1 769.400	1 864.400	868	1 781.400	1 876.400			
809	1 769.600	1 864.600	869	1 781.600	1 876.600			

Index

Indoor Radio Planning: A Practical Guide for 2G, 3G and 4G, Third Edition. Morten Tolstrup.
© 2015 John Wiley & Sons, Ltd. Published 2015 by John Wiley & Sons, Ltd.